D0138700

NUMERICAL METHODS

NUMERICAL METHODS

DESIGN, ANALYSIS, AND COMPUTER IMPLEMENTATION OF ALGORITHMS

ANNE GREENBAUM
TIMOTHY P. CHARTIER

PRINCETON UNIVERSITY PRESS • PRINCETON AND OXFORD

Published by Princeton University Press, 41 William Street, Princeton, New Jersey 08540

In the United Kingdom: Princeton University Press, 6 Oxford Street, Woodstock, Oxfordshire OX20 1TW

press.princeton.edu

Cover image courtesy of the University of Sheffield and Ansys UK.

Library of Congress Cataloging-in-Publication Data

Greenbaum, Anne.
Numerical methods: design, analysis, and computer implementation of algorithms /
Anne Greenbaum, Timothy P. Chartier.
p. cm.
ISBN 978-0-691-15122-9 (hbk. : alk. paper)
1. Numerical analysis. I. Chartier, Timothy P., 1969–
II. Title.
QA297.G15 2012
518–dc23
2011045732

British Library Cataloging-in-Publication Data is available

This book has been composed in Sabon
Printed on acid-free paper. ∞

Typeset by S R Nova Pvt Ltd, Bangalore, India
Printed in the United States of America

10 9 8 7 6 5 4 3 2 1

To Tanya

for her constant love and support,

whether life is well conditioned or unstable.

—*TC*

CONTENTS

Preface xiii

1 MATHEMATICAL MODELING 1

1.1 Modeling in Computer Animation 2

1.1.1 A Model Robe 2

1.2 Modeling in Physics: Radiation Transport 4

1.3 Modeling in Sports 6

1.4 Ecological Models 8

1.5 Modeling a Web Surfer and Google 11

1.5.1 The Vector Space Model 11

1.5.2 Google's PageRank 13

1.6 Chapter 1 Exercises 14

2 BASIC OPERATIONS WITH MATLAB 19

2.1 Launching MATLAB 19

2.2 Vectors 20

2.3 Getting Help 22

2.4 Matrices 23

2.5 Creating and Running .m Files 24

2.6 Comments 25

2.7 Plotting 25

2.8 Creating Your Own Functions 27

2.9 Printing 28

2.10 More Loops and Conditionals 29

2.11 Clearing Variables 31

2.12 Logging Your Session 31

2.13 More Advanced Commands 31

2.14 Chapter 2 Exercises 32

3 MONTE CARLO METHODS 41

3.1 A Mathematical Game of Cards 41

3.1.1 The Odds in Texas Holdem 42

3.2 Basic Statistics 46

3.2.1 Discrete Random Variables 48

3.2.2 Continuous Random Variables 51

3.2.3 The Central Limit Theorem 53

3.3 Monte Carlo Integration 56

3.3.1 Buffon's Needle 56
3.3.2 Estimating π 58
3.3.3 Another Example of Monte Carlo Integration 60
3.4 Monte Carlo Simulation of Web Surfing 64
3.5 Chapter 3 Exercises 67

4 SOLUTION OF A SINGLE NONLINEAR EQUATION IN ONE UNKNOWN 71
4.1 Bisection 75
4.2 Taylor's Theorem 80
4.3 Newton's Method 83
4.4 Quasi-Newton Methods 89
4.4.1 Avoiding Derivatives 89
4.4.2 Constant Slope Method 89
4.4.3 Secant Method 90
4.5 Analysis of Fixed Point Methods 93
4.6 Fractals, Julia Sets, and Mandelbrot Sets 98
4.7 Chapter 4 Exercises 102

5 FLOATING-POINT ARITHMETIC 107
5.1 Costly Disasters Caused by Rounding Errors 108
5.2 Binary Representation and Base 2 Arithmetic 110
5.3 Floating-Point Representation 112
5.4 IEEE Floating-Point Arithmetic 114
5.5 Rounding 116
5.6 Correctly Rounded Floating-Point Operations 118
5.7 Exceptions 119
5.8 Chapter 5 Exercises 120

6 CONDITIONING OF PROBLEMS; STABILITY OF ALGORITHMS 124
6.1 Conditioning of Problems 125
6.2 Stability of Algorithms 126
6.3 Chapter 6 Exercises 129

7 DIRECT METHODS FOR SOLVING LINEAR SYSTEMS AND LEAST SQUARES PROBLEMS 131
7.1 Review of Matrix Multiplication 132
7.2 Gaussian Elimination 133
7.2.1 Operation Counts 137
7.2.2 LU Factorization 139
7.2.3 Pivoting 141
7.2.4 Banded Matrices and Matrices for Which Pivoting Is Not Required 144

7.2.5 Implementation Considerations for High Performance 148
7.3 Other Methods for Solving $A\mathbf{x} = \mathbf{b}$ 151
7.4 Conditioning of Linear Systems 154
7.4.1 Norms 154
7.4.2 Sensitivity of Solutions of Linear Systems 158
7.5 Stability of Gaussian Elimination with Partial Pivoting 164
7.6 Least Squares Problems 166
7.6.1 The Normal Equations 167
7.6.2 QR Decomposition 168
7.6.3 Fitting Polynomials to Data 171
7.7 Chapter 7 Exercises 175

8 POLYNOMIAL AND PIECEWISE POLYNOMIAL INTERPOLATION 181
8.1 The Vandermonde System 181
8.2 The Lagrange Form of the Interpolation Polynomial 181
8.3 The Newton Form of the Interpolation Polynomial 185
8.3.1 Divided Differences 187
8.4 The Error in Polynomial Interpolation 190
8.5 Interpolation at Chebyshev Points and chebfun 192
8.6 Piecewise Polynomial Interpolation 197
8.6.1 Piecewise Cubic Hermite Interpolation 200
8.6.2 Cubic Spline Interpolation 201
8.7 Some Applications 204
8.8 Chapter 8 Exercises 206

9 NUMERICAL DIFFERENTIATION AND RICHARDSON EXTRAPOLATION 212
9.1 Numerical Differentiation 213
9.2 Richardson Extrapolation 221
9.3 Chapter 9 Exercises 225

10 NUMERICAL INTEGRATION 227
10.1 Newton–Cotes Formulas 227
10.2 Formulas Based on Piecewise Polynomial Interpolation 232
10.3 Gauss Quadrature 234
10.3.1 Orthogonal Polynomials 236
10.4 Clenshaw–Curtis Quadrature 240
10.5 Romberg Integration 242
10.6 Periodic Functions and the Euler–Maclaurin Formula 243
10.7 Singularities 247
10.8 Chapter 10 Exercises 248

11 NUMERICAL SOLUTION OF THE INITIAL VALUE PROBLEM FOR ORDINARY DIFFERENTIAL EQUATIONS 251
11.1 Existence and Uniqueness of Solutions 253
11.2 One-Step Methods 257
11.2.1 Euler's Method 257
11.2.2 Higher-Order Methods Based on Taylor Series 262
11.2.3 Midpoint Method 262
11.2.4 Methods Based on Quadrature Formulas 264
11.2.5 Classical Fourth-Order Runge–Kutta and Runge–Kutta–Fehlberg Methods 265
11.2.6 An Example Using MATLAB's ODE Solver 267
11.2.7 Analysis of One-Step Methods 270
11.2.8 Practical Implementation Considerations 272
11.2.9 Systems of Equations 274
11.3 Multistep Methods 275
11.3.1 Adams–Bashforth and Adams–Moulton Methods 275
11.3.2 General Linear *m*-Step Methods 277
11.3.3 Linear Difference Equations 280
11.3.4 The Dahlquist Equivalence Theorem 283
11.4 Stiff Equations 284
11.4.1 Absolute Stability 285
11.4.2 Backward Differentiation Formulas (BDF Methods) 289
11.4.3 Implicit Runge–Kutta (IRK) Methods 290
11.5 Solving Systems of Nonlinear Equations in Implicit Methods 291
11.5.1 Fixed Point Iteration 292
11.5.2 Newton's Method 293
11.6 Chapter 11 Exercises 295

12 MORE NUMERICAL LINEAR ALGEBRA: EIGENVALUES AND ITERATIVE METHODS FOR SOLVING LINEAR SYSTEMS 300
12.1 Eigenvalue Problems 300
12.1.1 The Power Method for Computing the Largest Eigenpair 310
12.1.2 Inverse Iteration 313
12.1.3 Rayleigh Quotient Iteration 315
12.1.4 The QR Algorithm 316
12.1.5 Google's PageRank 320
12.2 Iterative Methods for Solving Linear Systems 327

12.2.1 Basic Iterative Methods for Solving Linear Systems 327
12.2.2 Simple Iteration 328
12.2.3 Analysis of Convergence 332
12.2.4 The Conjugate Gradient Algorithm 336
12.2.5 Methods for Nonsymmetric Linear Systems 334
12.3 Chapter 12 Exercises 345

13 NUMERICAL SOLUTION OF TWO-POINT BOUNDARY VALUE PROBLEMS 350
13.1 An Application: Steady-State Temperature Distribution 350
13.2 Finite Difference Methods 352
13.2.1 Accuracy 354
13.2.2 More General Equations and Boundary Conditions 360
13.3 Finite Element Methods 365
13.3.1 Accuracy 372
13.4 Spectral Methods 374
13.5 Chapter 13 Exercises 376

14 NUMERICAL SOLUTION OF PARTIAL DIFFERENTIAL EQUATIONS 379
14.1 Elliptic Equations 381
14.1.1 Finite Difference Methods 381
14.1.2 Finite Element Methods 386
14.2 Parabolic Equations 388
14.2.1 Semidiscretization and the Method of Lines 389
14.2.2 Discretization in Time 389
14.3 Separation of Variables 396
14.3.1 Separation of Variables for Difference Equations 400
14.4 Hyperbolic Equations 402
14.4.1 Characteristics 402
14.4.2 Systems of Hyperbolic Equations 403
14.4.3 Boundary Conditions 404
14.4.4 Finite Difference Methods 404
14.5 Fast Methods for Poisson's Equation 409
14.5.1 The Fast Fourier Transform 411
14.6 Multigrid Methods 414
14.7 Chapter 14 Exercises 418

APPENDIX A REVIEW OF LINEAR ALGEBRA 421
A.1 Vectors and Vector Spaces 421
A.2 Linear Independence and Dependence 422
A.3 Span of a Set of Vectors; Bases and Coordinates; Dimension of a Vector Space 423

A.4 The Dot Product; Orthogonal and Orthonormal Sets; the Gram–Schmidt Algorithm 423
A.5 Matrices and Linear Equations 425
A.6 Existence and Uniqueness of Solutions; the Inverse; Conditions for Invertibility 427
A.7 Linear Transformations; the Matrix of a Linear Transformation 431
A.8 Similarity Transformations; Eigenvalues and Eigenvectors 432

APPENDIX B TAYLOR'S THEOREM IN MULTIDIMENSIONS 436

References 439

Index 445

PREFACE

In this book we have attempted to integrate a reasonably rigorous mathematical treatment of elementary numerical analysis with motivating examples and applications as well as some historical background. It is designed for use as an upper division undergraduate textbook for a course in numerical analysis that could be in a mathematics department, a computer science department, or a related area. It is assumed that the students have had a calculus course, and have seen Taylor's theorem, although this is reviewed in the text. It is also assumed that they have had a linear algebra course. Parts of the material require multivariable calculus, although these parts could be omitted. Different aspects of the subject—design, analysis, and computer implementation of algorithms—can be stressed depending on the interests, background, and abilities of the students.

We begin with a chapter on mathematical modeling to make the reader aware of where numerical computing problems arise and the many uses of numerical methods. In a numerical analysis course, one might go through all or some of the applications in this chapter or one might just assign it to students to read. Next is a chapter on the basics of MATLAB [94], which is used throughout the book for sample programs and exercises. Another high-level language such as SAGE [93] could be substituted, as long as it is a language that allows easy implementation of high-level linear algebra procedures such as solving a system of linear equations or computing a QR decomposition. This frees the student to concentrate on the use and behavior of these procedures rather than the details of their programming, although the major aspects of their implementation are covered in the text in order to explain proper interpretation of the results.

The next chapter is a brief introduction to Monte Carlo methods. Monte Carlo methods usually are not covered in numerical analysis courses, but they should be. They are *very* widely used computing techniques and demonstrate the close connection between mathematical modeling and numerical methods. The basic statistics needed to understand the results will be useful to students in almost any field that they enter.

The next chapters contain more standard topics in numerical analysis—solution of a single nonlinear equation in one unknown, floating-point arithmetic, conditioning of problems and stability of algorithms, solution of linear systems and least squares problems, and polynomial and piecewise polynomial interpolation. Most of this material is standard, but we do include some recent results about the efficacy of polynomial interpolation when the interpolation points are *Chebyshev points*. We demonstrate the use of a MATLAB software package called `chebfun` that performs such interpolation, choosing the degree of the interpolating polynomial adaptively to attain a level of accuracy near the machine precision. In the next two chapters, we discuss the application of this approach to numerical differentiation and integration.

We have found that the material through polynomial and piecewise polynomial interpolation can typically be covered in a quarter, while a semester course would include numerical differentiation and integration as well and perhaps some material on the numerical solution of ordinary differential equations (ODEs). Appendix A covers background material on linear algebra that is often needed for review.

The remaining chapters of the book are geared towards the numerical solution of differential equations. There is a chapter on the numerical solution of the initial value problem for ordinary differential equations. This includes a short section on solving systems of nonlinear equations, which should be an easy generalization of the material on solving a single nonlinear equation, assuming that the students have had multivariable calculus. The basic Taylor's theorem in multidimensions is included in Appendix B. At this point in a year long sequence, we usually cover material from the chapter entitled "More Numerical Linear Algebra," including iterative methods for eigenvalue problems and for solving large linear systems. Next come two-point boundary value problems and the numerical solution of partial differential equations (PDEs). Here we include material on the fast Fourier transform (FFT), as it is used in fast solvers for Poisson's equation. The FFT is also an integral part of the `chebfun` package introduced earlier, so we are now able to tell the reader a little more about how the polynomial interpolation procedures used there can be implemented efficiently.

One can arrange a sequence in which each quarter (or semester) depends on the previous one, but it is also fairly easy to arrange independent courses for each topic. This requires a review of MATLAB at the start of each course and usually a review of Taylor's theorem with remainder plus a small amount of additional material from previous chapters, but the amount required from, for example, the linear algebra sections in order to cover, say, the ODE sections is small and can usually be fit into such a course.

We have attempted to draw on the popularity of mathematical modeling in a variety of new applications, such as movie animation and information retrieval, to demonstrate the importance of numerical methods, not just in engineering and scientific computing, but in many other areas as well. Through a variety of examples and exercises, we hope to demonstrate some of the many, many uses of numerical methods, while maintaining the emphasis on analysis and understanding of results. Exercises seldom consist of simply computing an answer; in most cases a computational problem is combined with a question about convergence, order of accuracy, or effects of roundoff. Always an underlying theme is, "How much confidence do you have in your computed result?" We hope that the blend of exciting new applications with old-fashioned analysis will prove a successful one. Software that is needed for some of the exercises can be downloaded from the book's web page, via http://press.princeton.edu/titles/9763.html. Also provided on that site are most of the MATLAB codes used to produce the examples throughout the book.

Acknowledgments. The authors thank Richard Neidinger for his contributions and insights after using drafts of the text in his teaching at Davidson College. We also thank the Davidson College students who contributed ideas for improving the text, with special thanks to Daniel Orr for his contributions to the exercises. Additional exercises were contributed by Peter Blossey and Randall LeVeque of the University of Washington. We also thank Danny Kaplan of Macalester College for using an early version of the text in his classes there, and we thank Dan Goldman for information about the use of numerical methods in special effects.

NUMERICAL METHODS

1

MATHEMATICAL MODELING

Numerical methods play an important role in modern science. Scientific exploration is often conducted on computers rather than laboratory equipment. While it is rarely meant to completely replace work in the scientific laboratory, computer simulation often complements this work.

For example, the aerodynamic simulation of two NASCAR autos pictured in figure 1.1(a) requires the numerical solution of *partial differential equations* (PDEs) that model the flow of air past the car. An auto body must be smooth and sleek, so it is often modeled using cubic (or higher order) *splines*. Similar computations are done in designing aircraft. We will study the numerical issues in using splines and solving PDEs in chapters 8 and 14, respectively.

Other examples occur in the field of mathematical biology, an active area of research in industry, government, and academia. Numerical algorithms play a crucial role in this work. For example, protein-folding models are often solved as large *optimization* problems. Protein arranges itself in such a way as to minimize energy—nature has no trouble finding the right arrangement, but it is not so easy for humans. The field of numerical optimization is an entire subject on its own, so it will not be covered in this book. The numerical methods described here, however, form the core of most optimization procedures.

Before studying issues in the analysis and implementation of efficient and accurate numerical methods, we first look briefly at the topic of mathematical modeling, which turns real-world problems into the sorts of mathematical equations that numerical analysts can tackle. The mathematical formulation usually represents only a *model* of the actual physical situation, and it is often important for the numerical analyst or computational scientist to know something about the origin of the model; in fact, numerical analysts sometimes work directly with scientists and engineers in devising the mathematical model. This interaction is important for a number of reasons. First, many algorithms do not produce the exact solution but only an approximate one. An understanding of the origin of the problem is necessary to determine what constitutes an acceptably good "approximate" solution: an error of a few centimeters might be acceptable in locating an enemy tank, but it would not be acceptable in locating a tumor for laser surgery! Second, even if the algorithm theoretically produces the exact solution, when implemented on a computer using finite-precision arithmetic, the results produced will most

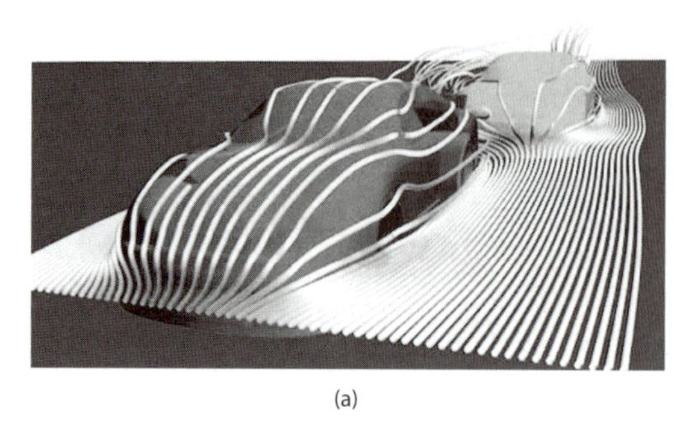

(a)

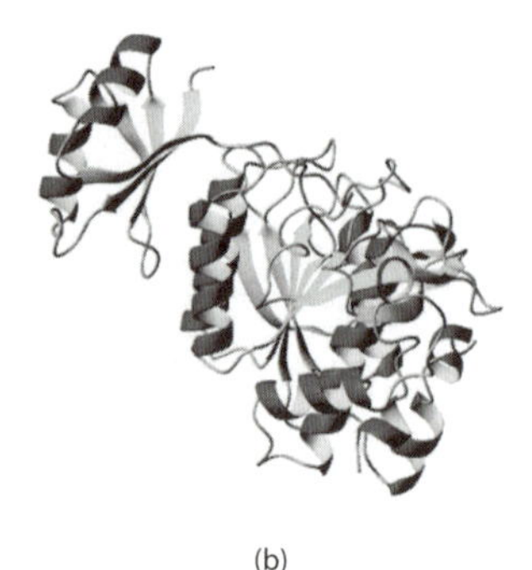

(b)

Figure 1.1. (a) A simulation of two NASCAR autos depicts the streamlines of air produced as a car drafts and is about to pass another. (Simulation performed with STAR-CCM+.) (b) Solving protein-folding models utilizes numerical optimization.

likely be inexact. Part of numerical analysis is the understanding of the impact of finite-precision computations on the accuracy of results. We will look more deeply at the issues in computing in finite precision in chapter 5.

In this chapter we present a variety of applications that involve numerical computation and come from the mathematical modeling of various processes.

1.1 MODELING IN COMPUTER ANIMATION

Many of the computer generated graphics that dominate the silver screen are produced with **dynamic simulation**; that is, a model is created, often using the laws of physics, and numerical methods are then used to compute the results of that model. In this section, we will look at the role of numerics in animation that appeared in the 2002 film *Star Wars: Episode II Attack of the Clones*. In particular, we will take a careful look at some of the special effects used to digitally create the character of Yoda, a Jedi master who first appeared as a puppet in the Star Wars saga in the 1980 film, *The Empire Strikes Back*. In the 2002 film, Yoda was digitally created, which required heavy use of numerical algorithms.

A key aspect of creating a digital Yoda involves producing believable movement of the character. The movement of Yoda's body is described using **key-frame animation**, in which a pose is specified at particular points in time and the computer automatically determines the poses in the intervening frames through interpolation. (We will discuss several interpolation techniques in chapter 8.) Animators have many controls over such movement, with the ability to specify, for instance, velocities and tangents of motion. While animators indicate the movement of Yoda's body, the computer must determine the resulting flow of his robe.

1.1.1 A Model Robe

Referring to figure 1.2, we see that the robe is represented with triangles and the motion of each vertex of the robe must be determined. Each vertex is

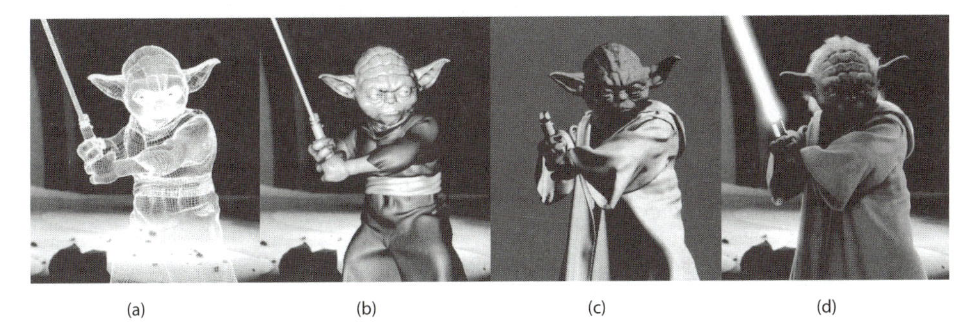

(a) (b) (c) (d)

Figure 1.2. Stages of simulation in the animation of the digitally created Yoda in the fight scene with Count Dooku in *Star Wars Episode II*. Two layers of Yoda's clothing, seen in (b) and (c), were computed separately. A process known as collision detection ensured that the inner layer of clothing did not intersect the outer robe and become visible. A simplified model of cloth illumination created the appearance of a real garment, producing the final rendering of the image in (d) [22]. (Courtesy of Lucasfilm Ltd. *Star Wars: Episode II - Attack of the Clones* ™ & © 2002 Lucasfilm Ltd. All rights reserved. Used under authorization. Unauthorized duplication is a violation of applicable law. Digital Work by Industrial Light & Magic.)

modeled as a particle, which in this context is a pointlike object that has mass, position, and velocity, and responds to forces, but has no size.

The motion of a particle is governed by Newton's second law, which is expressed mathematically by the equation

$$\mathbf{F} = m\mathbf{a} = m(d^2\mathbf{y}/dt^2), \tag{1.1}$$

where $\mathbf{y}$ is a distance function of time t. Note that the equations involve vector-valued functions since our computations are performed in three dimensions. Since a particle has mass ($m \neq 0$), equation (1.1) can be rewritten as the second-order ordinary differential equation (ODE)

$$\frac{d^2\mathbf{y}}{dt^2} = \frac{\mathbf{F}}{m}. \tag{1.2}$$

This ODE is part of an initial value problem since the state of the particle at some initial time is given. In the case of a movie, this is where the scene (which may be several seconds or a fraction of a second in duration) begins.

To keep the shape of the robe, pairs of neighboring particles are attached to each other using a spring force. Hooke's law states that a spring exerts a force F_s that is proportional to its displacement from its rest length x_0. This is expressed mathematically as

$$F_s = -k(x - x_0),$$

where x denotes the current position of the spring and k is the spring constant. For simplicity, we have stated the one-dimensional formulation of Hooke's law, but to model the Jedi's robe a three-dimensional version is used.

Many other forces are computed to animate Yoda's robe, including gravity, wind forces, collision forces, friction, and even completely made-up forces that are invented solely to achieve the motion that the director requests. In the simplest of cases, an *analytic solution* of the model may exist. In computer animation, however, the forces acting on the particle constantly change and finding an analytic solution for each frame—if even possible—would be impractical. Instead, numerical methods are used to find approximate solutions by simulating the motion of the particles over discrete time steps. There is a large body of literature on the numerical solution of initial value problems for ODEs, and some of the methods will be covered in chapter 11.

BENDING THE LAWS OF PHYSICS

When simulating the motion of Jedi robes in *Star Wars Episode II*, animators discovered that if the stunts were performed in reality, the clothing would be ripped apart by such accelerations of the motion. To solve the problem, custom "protection" effects were added to the simulations, which dampened this acceleration. In the end, the clothes on the digital actors were less distorted by their superhuman motion. To the left we see a digital double of Obi-Wan Kenobi (played by Ewan McGregor), performing a stunt too dangerous for a live actor, in *Star Wars Episode II*. Note that the hairs, like the clothing, were simulated as strips of particles connected by springs [22]. (Courtesy of Lucasfilm Ltd. *Star Wars: Episode II - Attack of the Clones* ™ & © 2002 Lucasfilm Ltd. All rights reserved. Used under authorization. Unauthorized duplication is a violation of applicable law. Digital Work by Industrial Light & Magic.)

A goal of movie animation is creating convincing simulations. Accuracy, which is a leading goal of scientific simulation, may conflict with this goal. As such, numerical simulations are used for different goals by scientists and by the entertainment industry, but many of the mathematical tools and strategies are common. Let us now turn our attention to a simulation in science.

1.2 MODELING IN PHYSICS: RADIATION TRANSPORT

The transport of radiation can be described stochastically and modeled using Monte Carlo simulation. Monte Carlo methods will be described in chapter 3. Radiation transport also can be modeled by an integro-differential equation, the Boltzmann transport equation. This is a PDE that also involves an integral

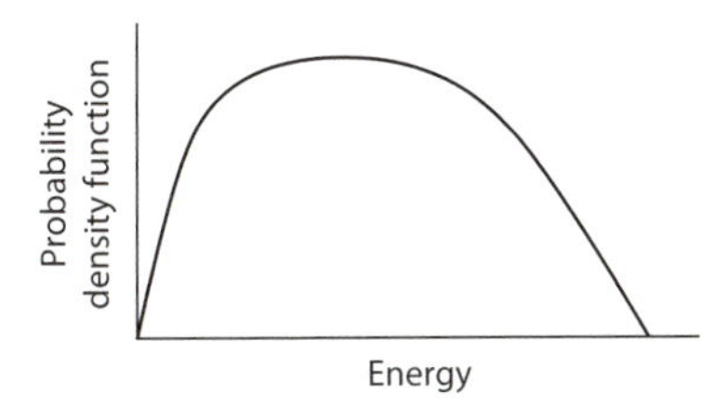

Figure 1.3. An example distribution of the energy of photons.

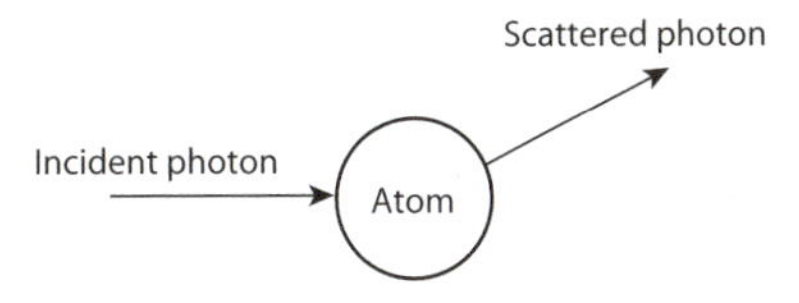

Figure 1.4. A collision between a photon and an atom.

term. In chapter 14 we discuss the numerical solution of PDEs, while chapter 10 describes methods of numerical integration. A combination of these ideas, together with iterative techniques for solving large linear systems (section 12.2), is used to approximate solutions to the Boltzmann transport equation.

What radiation dose does your body receive from a dental X-ray examination? You probably recall that a heavy vest is usually placed over you during the exam and that the dental assistant leaves the room while the X-ray machine is on. The purpose of the covering is to absorb radiation. How can the transport of X-ray photons be modeled mathematically in order to aid in the design of such protective materials and to verify their effectiveness? The photons are produced by an electron beam that is turned on and off as X-rays are needed. The energy and direction of travel of any individual photon cannot be predicted, but the overall distribution of energy and direction of the X-rays can be approximated.

The independent variables in the system are energy, position, direction, and time of production of the photons, and each of these variables can be thought of as random but obeying a certain distribution. The energy of the photons might be distributed as shown in figure 1.3, for example.

It is assumed that particles move in straight lines until they enter matter, where, with a certain probability (depending on the characteristics of the material), they collide with an atom. The collision usually involves an exchange of energy with an electron in the atom, after which the photon emerges with reduced energy and an altered direction of travel. The photon may even give up all of its energy to the electron, in which case it is considered to be absorbed by the atom.

The probabilities of each of these events must be known in order to simulate the situation. These probabilities are deduced from theoretical and measured properties of X-rays and of various materials that might be used as shields. One can then run a computer simulation, often with millions of photons, each following a random path determined by these probability distributions. The history of each photon is followed until it is either absorbed (or loses so much energy that it effectively can be ignored) or travels outside the system to a point from which it will not return. Average results are then taken to determine what dose of radiation is received at a particular location.

Studying Semiconductors

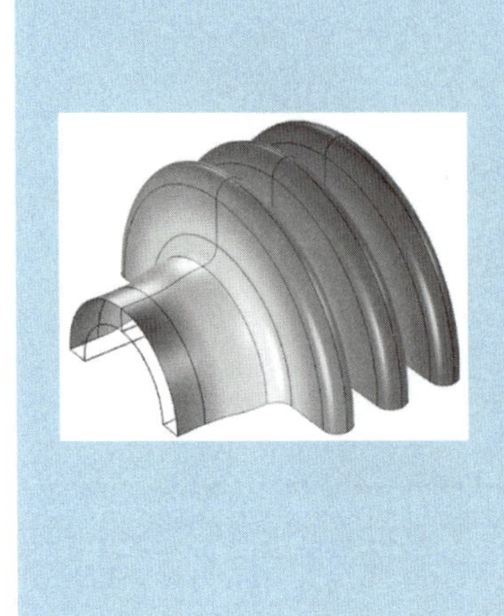

Studying the behavior of electrons in semiconductor materials requires solving the Boltzmann transport equation, which involves complicated integrals. Both deterministic methods and Monte Carlo methods are sometimes used in this case. The picture to the left is a finite element discretization used in a deterministic model for radiation transport problems. (Image reproduced with the kind permission of the Applied Modelling and Computational Group at Imperial College London and EDF Energy. All other rights of the copyright owners are reserved.)

1.3 MODELING IN SPORTS

In FIFA World Cup soccer matches, soccer balls curve and swerve through the air, in the players' attempts to confuse goalkeepers and send the ball sailing to the back of the net. World class soccer players such as Brazil's Roberto Carlos, Germany's Michael Ballack and England's David Beckham have perfected "bending" the ball from a free kick.

According to Computational Fluid Dynamics (CFD) research by the University of Sheffield's Sports Engineering Research Group and Fluent Europe, the shape and surface of the soccer ball, as well as its initial orientation, play a fundamental role in the ball's trajectory through the air. In particular, such CFD research has increased the understanding of the "knuckleball" effect sometimes used to confuse an opposing goalkeeper who stands as the last line of defense. To obtain such results, a soccer ball was digitized down to its stitching as seen in figure 1.5. Note the refinement near the seams, which is required in order to properly model the boundary layer.

Some free kicks in soccer have an initial velocity of almost 70 mph. Wind tunnel experiments demonstrate that a soccer ball moves from laminar to turbulent flow at speeds between 20 and 30 mph, depending on the ball's surface structure and texture.

The techniques developed in Sheffield facilitated detailed analysis of the memorable goal by David Beckham of England versus Greece during the World Cup Qualifiers in 2001. In a sense, Beckham's kick applied sophisticated physics. While the CFD simulations at Sheffield can accurately model turbulent flow only when it is averaged over time, and so cannot yet give realistic trajectories in all cases, such research could affect soccer players from beginner

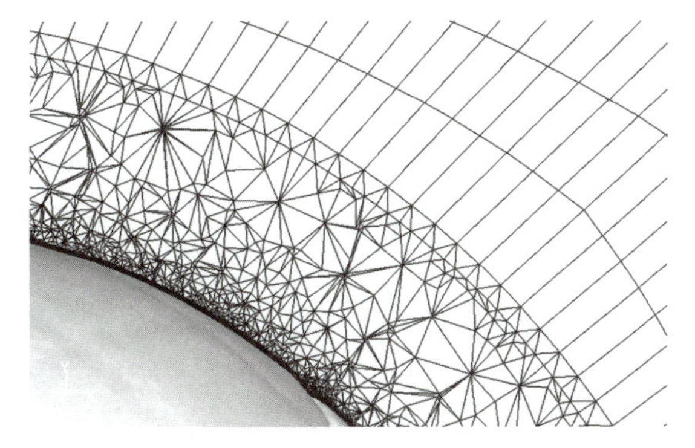

Figure 1.5. An important step in CFD simulations at the University of Sheffield is capturing the geometry of a soccer ball with a three-dimensional noncontact laser scanner. The figure shows part of a soccer ball mesh with approximately 9 million cells. (Courtesy of the University of Sheffield and Ansys UK.)

(a)

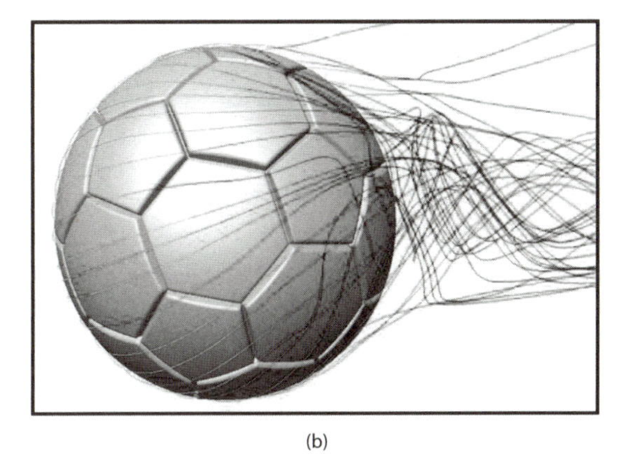

(b)

Figure 1.6. (a) Wind tunnel smoke test of a nonspinning soccer ball. (b) CFD simulation showing wake-flow path lines of a nonspinning soccer ball, air speed of 27 mph. (Courtesy of the University of Sheffield and Ansys UK.)

to professional. For instance, ball manufacturers could exploit such work to produce a more consistent or interesting ball that could be tailored to the needs and levels of players. Such work could also impact the training of players. For more information, see, for instance [6] or [7].

To this end, there is a simulation program called Soccer Sim developed at the University of Sheffield. The program predicts the flight of a ball given input conditions, which can be acquired from CFD and wind tunnel tests, as well as from high-speed videoing of players' kicks. The software can then be used to compare the trajectory of a ball given varying initial orientations of the ball or different spins induced by the kick. Moreover, the trajectory can be compared for different soccer balls.

Note that this application, like that in section 1.1, involves the solution of differential equations. The next application that we discuss involves a discrete phenomenon.

Figure 1.7. High-speed airflow path lines colored by local velocity over the 2006 Teamgeist soccer ball. (Courtesy of the University of Sheffield and Ansys UK.)

1.4 ECOLOGICAL MODELS

Computational biology is a growing field of application for numerical methods. In this section, we explore a simplified example from ecology.

Suppose we wish to study the population of a certain species of bird. These birds are born in the spring and live at most 3 years. We will keep track of the population just before breeding, when there will be three classes of birds, based on age: Age 0 (born the previous spring), Age 1, and Age 2. Let $v_0^{(n)}$, $v_1^{(n)}$ and $v_2^{(n)}$ represent the number of females in each age class in Year n. To model population changes we need to know:

- Survival rates. Suppose 20% of Age 0 birds survive to the next spring, and 50% of Age 1 birds survive to become Age 2.
- Fecundity rates. Suppose females that are 1 year old produce a clutch of a certain size, of which 3 females are expected to survive to the next breeding season. Females that are 2 years old lay more eggs and suppose that of these, 6 females are expected to survive to the next spring.

Then we have the following model of the number of females in each age class:

$$\begin{aligned} v_0^{(n+1)} &= 3v_1^{(n)} + 6v_2^{(n)}, \\ v_1^{(n+1)} &= 0.2v_0^{(n)}, \\ v_2^{(n+1)} &= 0.5v_1^{(n)}. \end{aligned}$$

In matrix–vector form,

$$\begin{aligned} \mathbf{v}^{(n+1)} &= A\mathbf{v}^{(n)} \\ &= \begin{pmatrix} 0 & 3 & 6 \\ 0.2 & 0 & 0 \\ 0 & 0.5 & 0 \end{pmatrix} \begin{pmatrix} v_0^{(n)} \\ v_1^{(n)} \\ v_2^{(n)} \end{pmatrix}. \end{aligned} \tag{1.3}$$

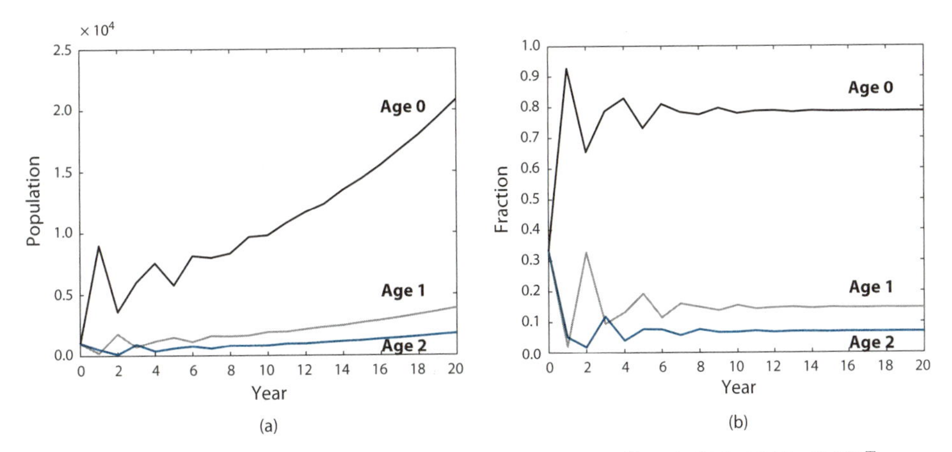

Figure 1.8. Bird population with $a_{32} = 0.5$ and initial data $\mathbf{v}^{(0)} = (1000, 1000, 1000)^T$.

The matrix A in (1.3) can be used to predict future populations from year to year and expected long-term behavior. This type of matrix, reflecting survival rates and fecundities, is called a **Leslie matrix**.

Suppose we start with a population of 3000 females, 1000 of each age. Using (1.3), we find

$$\mathbf{v}^{(0)} = \begin{pmatrix} 1000 \\ 1000 \\ 1000 \end{pmatrix}, \qquad \mathbf{v}^{(1)} = A\mathbf{v}^{(0)} = \begin{pmatrix} 0 & 3 & 6 \\ 0.2 & 0 & 0 \\ 0 & 0.5 & 0 \end{pmatrix} \mathbf{v}^{(0)} = \begin{pmatrix} 9000 \\ 200 \\ 500 \end{pmatrix},$$

$$\mathbf{v}^{(2)} = A\mathbf{v}^{(1)} = \begin{pmatrix} 3600 \\ 1800 \\ 100 \end{pmatrix}, \quad \text{and} \quad \mathbf{v}^{(3)} = A\mathbf{v}^{(2)} = \begin{pmatrix} 6000 \\ 720 \\ 900 \end{pmatrix}.$$

Plotting the population in each age group as a function of the year produces the graph in figure 1.8(a). Clearly the population grows exponentially. Figure 1.8(b) shows the proportion of the population in each age class as a function of the year. Note how the proportion of birds in each age class settles into a steady state.

To be more quantitative, we might look, for instance, at the population vectors $\mathbf{v}^{(20)}$ and $\mathbf{v}^{(19)}$. We find that $\mathbf{v}^{(20)} = (20833, 3873, 1797)^T$, which indicates that in year 20 the fraction of birds in each age class is $(0.7861, 0.1461, 0.068)^T$. Looking at the difference between $\mathbf{v}^{(20)}$ and $\mathbf{v}^{(19)}$, we find that the total population grows by a factor of 1.0760 between year 19 and year 20.

Note that $\mathbf{v}^{(n)} = A\mathbf{v}^{(n-1)} = A^2\mathbf{v}^{(n-2)} = \ldots = A^n\mathbf{v}^{(0)}$. In chapter 12, we will see how this observation indicates that the asymptotic behavior of this system (the behavior after a long time period) can be predicted from the dominant eigenvalue (the one of largest absolute value) and associated eigenvector of A. The eigenvalues of A can be computed to be 1.0759, $-0.5380 + 0.5179i$, and $-0.5380 - 0.5179i$ (where $i = \sqrt{-1}$). The largest magnitude of an eigenvalue is 1.0759, and the associated eigenvector, scaled so that its entries sum to 1, is $\hat{\mathbf{v}} = (0.7860, 0.1461, 0.0679)$. The largest magnitude of an eigenvalue

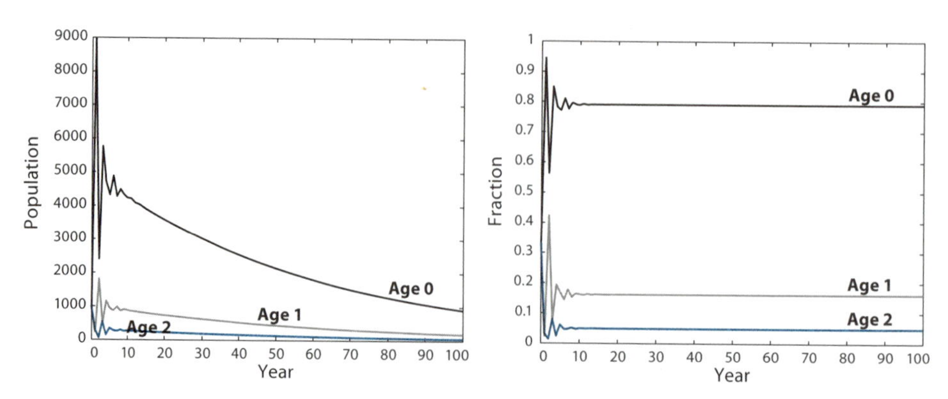

Figure 1.9. Bird population with $a_{32} = 0.3$ and initial data $\mathbf{v}^{(0)} = (1000, 1000, 1000)^T$.

determines the rate of growth of the population after a long period of time, and the entries in $\hat{\mathbf{v}}$ give the fractions of birds in each class, as these fractions approach a steady state. While the eigenvalues and eigenvectors of a 3 by 3 matrix such as A can be determined analytically, we will learn in chapter 12 about methods to numerically approximate eigenvalues and eigenvectors of much larger matrices.

Harvesting Strategies. We can use this model to investigate questions such as the following. Suppose we want to harvest a certain fraction of this population, either to control the exponential growth or to use it as a food source (or both). How much should we harvest in order to control the growth without causing extinction?

Harvesting will decrease the survival rates. We might wish to decrease these rates to a point where the dominant eigenvalue is very close to 1 in absolute value, since this would mean that the total population, after a period of time, would remain fixed. If we decrease the survival rates too much, then all eigenvalues will be less than 1 in magnitude and the population size will decrease exponentially to extinction.

For example, suppose we harvest 20% of the 1-year-olds, so that the survival rate falls from 0.5 to 0.3. In this case, entry a_{32} in matrix A changes from 0.5 to 0.3, and it turns out that the largest magnitude eigenvalue decreases to 0.9830. A simulation with these parameters yields the graphs in figure 1.9.

Note that we now have exponential decay of the population and eventual extinction. The decrease in population is gradual, however, and over the 100-year time span shown here it decreases only from 3000 to 1148. Asymptotically (i.e., after a long period of time), the total population will decrease each year to 0.9830 times that of the previous year. Hence if the population is 1148 after 100 years of this harvesting strategy, the number of additional years before the population drops below 1 (i.e., the species becomes extinct), might be estimated by the value of k that satisfies $1148(0.9830)^k = 1$, which is $k = -\log(1148)/\log(0.9830) \approx 411$ years.

One could try to find the matrix element a_{32} that results in the largest magnitude eigenvalue of A being exactly 1, and while this is an interesting mathematical problem, from the point of view of population control, it would likely be an exercise in futility. Clearly, this model is highly simplified, and even if the assumptions about survival rates and fecundity rates held initially, they might well change over time. Populations would need to be monitored frequently to adjust a particular harvesting strategy to the changing environment.

1.5 MODELING A WEB SURFER AND GOOGLE

Submitting a query to a search engine is a common method of information retrieval. Companies compete to be listed high in the rankings returned by a search engine. In fact, some companies' business is to help raise the rankings of a paying customer's web page. This is done by exploiting knowledge of the algorithms used by search engines. While a certain amount of information about search engine algorithms is publicly available, there is a certain amount that is proprietary. This section discusses how one can rank web pages based on content and is introductory in nature. An interested reader is encouraged to research the literature on search engine analysis, which is an ever-growing field. Here we will consider a simple vector space model for performing a search. This method does not take into account the hyperlink structure of the World Wide Web, and so the rankings from such a model might be aggregated with the results of an algorithm that does consider the Web's hyperlink structure. Google's PageRank algorithm is such a method and will be looked at briefly in this section and in more detail in sections 3.4 and 12.1.5.

1.5.1 The Vector Space Model

The vector space model consists of two main parts: a list of documents and a dictionary. The list of documents consists of those documents on which searches are to be conducted, and the dictionary of terms is a database of keywords. While the dictionary of terms could be all the terms in every document, this may not be desirable or computationally practical. Words not in the dictionary return empty searches.

With a list of n documents and m keywords, the vector space model constructs an m by n *document matrix* A with

$$a_{ij} = \begin{cases} 1 & \text{if document } j \text{ is relevant to term } i, \\ 0 & \text{otherwise.} \end{cases} \qquad (1.4)$$

When a query is issued, a query vector $\mathbf{q} = (q_1, \ldots, q_m)^T$ is then formed, with $q_i = 1$ if the query includes term i and $q_i = 0$ otherwise. The "closeness" of the query to each document is then measured by the cosine of the angle between the query vector $\mathbf{q}$ and the column of A representing that document. If $\mathbf{a}_j$ denotes the jth column of A (and if it is not entirely zero due to document j containing no keywords), then the angle θ_j between $\mathbf{q}$ and $\mathbf{a}_j$ satisfies

$$\cos(\theta_j) = \frac{\mathbf{a}_j^T \mathbf{q}}{\|\mathbf{a}_j\|_2 \|\mathbf{q}\|_2}. \qquad (1.5)$$

TABLE 1.1
The dictionary of terms and documents used in a three-dimensional example.

Dictionary		Documents
electric	I	the art of war
fencing	II	the fencing master
foil	III	fencing techniques of foil, epee, and saber
	IV	hot tips on building electric fencing

Since a larger value of the cosine implies a smaller angle between the vectors, documents are ranked in order of relevance to the query by arranging the cosines in descending order.

A major difficulty in forming the document matrix is determining whether a document is relevant to a term. There are a variety of ways of doing this. For instance, every document can be searched for each keyword. A document is deemed relevant to those keywords that are contained within it. This requires considerable computational expense when one faces a large number of documents and keywords. An alternative that is employed by some search engines is to read and perform analysis on only a portion of a document's text. For example, Google and Yahoo! pull only around 100K and 500K bytes, respectively, of web page text [12]. Other choices that must be made include whether to require exact matching of terms or to allow synonyms, and whether or not to count word order. For example, do we treat queries of `boat show` and `show boat` as the same or different? All of these choices impact which documents are deemed most relevant [36].

To illustrate the method, we give a simple example, in which a document is deemed relevant if its title contains the keyword exactly.

Searching in a Tiny Space

Suppose that our dictionary contains three keywords—"electric", "fencing", and "foil"—and that there are four documents entitled "the art of war", "the fencing master", "fencing techniques of foil, epee, and saber", and "hot tips on building electric fencing". We have written the document titles and dictionary terms in lower case to avoid questions about matching in the presence of such a factor. The information is listed in table 1.1 for easy reference.

To form the document matrix A—where rows 1, 2, and 3 correspond to the terms "electric", "fencing", and "foil", respectively, while columns 1, 2, 3 and 4 correspond to documents I, II, III, and IV, respectively—we use (1.4) to obtain

$$A = \begin{pmatrix} 0 & 0 & 0 & 1 \\ 0 & 1 & 1 & 1 \\ 0 & 0 & 1 & 0 \end{pmatrix}.$$

Figure 1.10 depicts the columns of A (each except the first normalized to have length 1) as vectors in three-space.

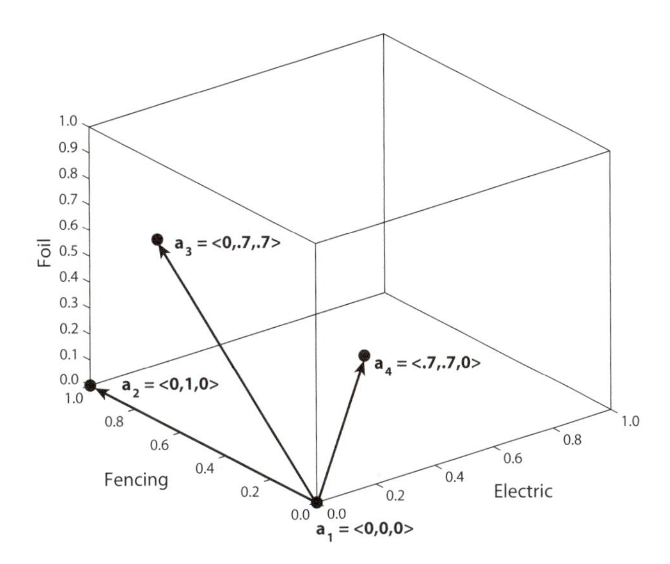

Figure 1.10. Visualization of vector space search in a three-dimensional example.

Suppose our query is `fencing`. Then the query vector is $\mathbf{q} = (0, 1, 0)^T$, and we can see by inspection that it is identical to column 2 of A. Thus, document II would be deemed most relevant to this query. To determine the rankings of the remaining documents, we compute the cosine of the angle between $\mathbf{q}$ and every other nonzero column of A using (1.5) to find

$$\cos(\theta_3) = \frac{\mathbf{a}_3^T\mathbf{q}}{\|\mathbf{a}_3\|_2\|\mathbf{q}\|_2} = \frac{1}{\sqrt{2}},$$

$$\cos(\theta_4) = \frac{\mathbf{a}_4^T\mathbf{q}}{\|\mathbf{a}_4\|_2\|\mathbf{q}\|_2} = \frac{1}{\sqrt{2}}.$$

Thus documents III and IV would be ranked equally, behind document II.

1.5.2 Google's PageRank

Real-world search engines deal with some of the largest mathematical and computer science problems in today's computational world. Interestingly, "Google" is a play on the word "googol", the number 10^{100}, reflecting the company's goal of organizing all information on the World Wide Web.

When you submit a query, such as `numerical analysis`, to Google, how does the search engine distinguish between the web page listed first and the one listed, say, 100th? There are various factors that play into this decision, one of which is the web page's relevance to your query, as discussed previously. Another important component, however, is the "quality" or "importance" of the page; "important" pages, or pages that are pointed to by many other pages, are more likely to be of interest to you. An algorithm called PageRank is used to measure the importance of a web page. This algorithm relies on a model of web-surfing behavior.

Figure 1.11. A representation of the graph of the Internet created by David F. Gleich, based on an image and data from the OPTE project.

As with any model of reality, Google's model of web-surfing behavior is an approximation. An important feature of PageRank is its assumption about the percentage of time that a surfer follows a link on the current web page. The exact assumptions that are made are proprietary, but it is believed that the PageRank algorithm used by Google assumes that a surfer follows a link on a web page about 85% of the time, with any of the links being equally likely. The other 15% of the time the surfer will enter the URL for another web page, possibly the same one that is currently being visited.

These assumptions about surfing behavior affect not only the accuracy of the model but also the efficiency of numerical techniques used to solve the model. Keep in mind that Google indexes billions of web pages, making this one of the largest computational problems ever solved! In chapter 12 we discuss more about the numerical methods used to tackle this problem.

1.6 CHAPTER 1 EXERCISES

1. **The Fibonacci numbers and nature.** The Fibonacci numbers are defined by $F_1 = 1$, $F_2 = 1$, $F_3 = F_2 + F_1 = 2$, $F_4 = F_3 + F_2 = 3$, etc. In general, $F_{j+1} = F_j + F_{j-1}$, $j = 2, 3, \ldots$. Write down F_5, F_6, and F_7.

   It is often observed that the number of petals on a flower or the number of branches on a tree is a Fibonacci number. For example, most daisies have either 34, 55, or 89 petals, and these are the 9th, 10th, and 11th Fibonacci numbers. The reason four-leaf clovers are so rare is that 4 is not a Fibonacci number.

   To see the reason for this, consider the following model of branch growth in trees. We start with the main trunk ($F_1 = 1$), which spends one season growing ($F_2 = 1$), and then after another season, develops two branches ($F_3 = 2$)—a major one and a minor one. After the next season, the major branch develops two branches—a major one and a minor one—while the minor branch grows into a major one, ready to divide in the following season. At this point there are $F_4 = 3$ branches—two major ones and one minor one. At the end of the next season, the two major branches divide, producing two major branches and two minor ones, while the minor one from the previous season grows into a major one. Thus at this point there are $F_5 = 5$ branches, with $F_4 = 3$ of these being major branches ready to divide during the next season. Explain why the Fibonacci numbers arise from this model of branch growth.

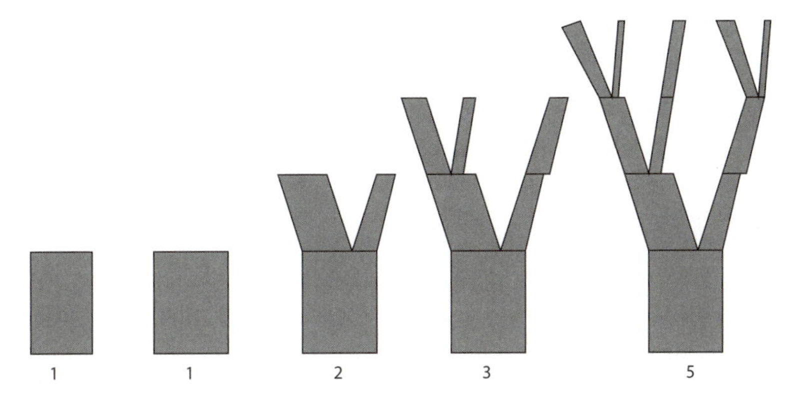

2. **Drunkard's walk.** A drunkard starts at position x in the diagram below and with each step moves right one space with probability .5 and left one space with probability .5. If he reaches the bar, he stays there, drinking himself into oblivion. If he reaches home, he goes to bed and stays there. You wish to know the probability $p(x)$ that he reaches home before reaching the bar.

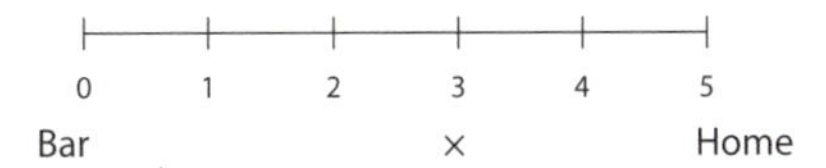

This is a typical Markov chain problem that will be discussed later in the text. Note that $p(0) = 0$ and $p(5) = 1$, since the drunk does not leave either the bar or his home. For $x = 1, 2, 3, 4$, $p(x) = .5\,p(x-1) + .5\,p(x+1)$, since he moves left with probability .5 and right with probability .5.

(a) Let the drunk's starting position be $x = 3$. What are the possible positions that he could be in after one step, and what are the probabilities of each? How about after two steps?

(b) For which initial positions x would you expect him to reach the bar first and for which would you expect him to reach home first, and why?

This is a typical *random walk* problem, and while it is posed as a silly story, it has real physical applications.

Consider an electrical network with equal resistors in series and a unit voltage across the ends.

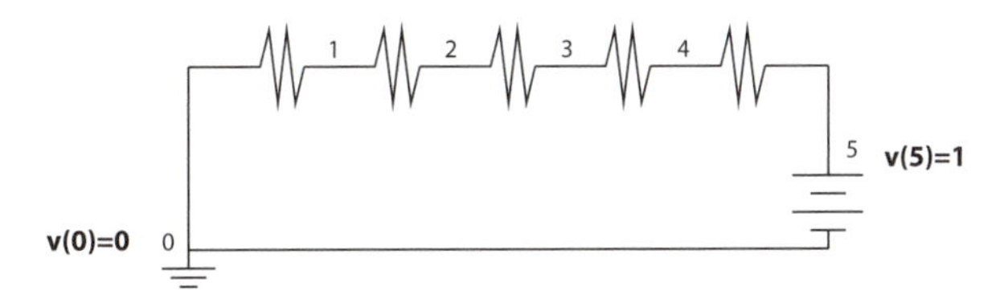

Voltages $v(x)$ will be established at points $x = 0, 1, 2, 3, 4, 5$. We have grounded the point $x = 0$ so that $v(0) = 0$, and there is no resistor between the source and point $x = 5$, so that $v(5) = 1$. By Kirchoff's laws, the current flowing into x must be equal to the current flowing out. By Ohm's law, if points x and y are separated by a resistor of strength R, then the current i_{xy} that flows from x to y is

$$i_{xy} = \frac{v(x) - v(y)}{R}.$$

Thus for $x = 1, 2, 3, 4$, we have

$$\frac{v(x-1) - v(x)}{R} = \frac{v(x) - v(x+1)}{R}.$$

Multiplying by R and combining like terms we see that $v(x) = .5v(x-1) + .5v(x+1)$. This is exactly the same formula (with the same boundary conditions) that we found for $p(x)$ in the drunkard's walk problem.

Can you think of other situations that might be modeled in this same way? The behavior of a stock price perhaps? Suppose you generalize by allowing different probabilities of the drunkard moving right or left: say, the probability of moving right is .6 while that of moving left is .4. What generalization of the resistor problem would this correspond to? [Hint: Consider resistors of different strengths.]

3. **Ehrenfests' urn.** Consider two urns, with N balls distributed between them. At each unit of time, you take a ball from one urn and move it to the other, with the probability of choosing each ball being equal, $1/N$. Thus, the probability of choosing a ball from a given urn is proportional to the number of balls in that urn. Let $X(t)$ denote the number of balls in the left urn at time t, and suppose that $X(0) = 0$. Then $X(1) = 1$, since a ball must be drawn from the right urn and moved to the left one.

   (a) Let $N = 100$. What are the possible values for $X(2)$, $X(3)$, and $X(4)$, and what is the probability of each?
   (a) If this process were carried out for a very long time, what do you think would be the most frequently occurring value of $X(t)$?

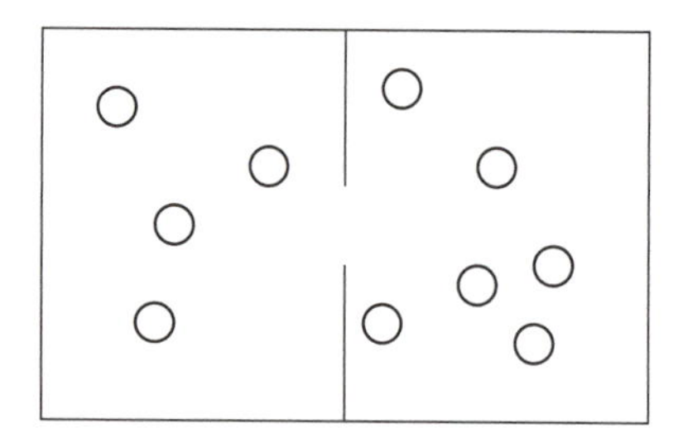

This model was introduced in the early 1900s by Paul Ehrenfest and Tatiana Ehrenfest-Afanassjewa to describe the diffusion of gas molecules through a permeable membrane. See, for example, http://en.wikipedia.org/wiki/Ehrenfest_model for a discussion of its relation to the second law of thermodynamics. In chapter 3 we will discuss Monte Carlo simulation of physical processes using models such as this one. Can you think of other situations that might be modeled by such a process? In addition to physical or chemical processes, you might consider, for example, financial decisions or social interactions.

4. Each year undergraduates participate in the *Mathematical Contest in Modeling (MCM)*. See www.mcm.org. Following is an example of the sort of modeling problems that they tackle:

    > An ornamental fountain in a large open plaza surrounded by buildings squirts water high into the air. On gusty days, the wind blows spray from the fountain onto passersby. The water-flow from the fountain is controlled by a mechanism linked to an anemometer (which measures wind speed and direction) located on top of an adjacent building. The objective of this control is to provide passersby with an acceptable balance between an attractive spectacle and a soaking: The harder the wind blows, the lower the water volume and height to which the water is squirted, hence the less spray falls outside the pool area.
    >
    > Your task is to devise an algorithm which uses data provided by the anemometer to adjust the water-flow from the fountain as the wind conditions change.

    Think about how you might create a mathematical model for this problem and compare your ideas with some of the students' solutions that can be found in: *UMAP: Journal of Undergraduate Mathematics and its Applications*. 2002. 23(3):187–271.

5. Consider the following dictionary of keywords:

    chocolate, ice cream, sprinkles,

    and the following list of documents:

    D1. I eat only the chocolate icing off the cake
    D2. I like chocolate and vanilla ice cream
    D3. Children like chocolate cake with sprinkles
    D4. May I have another scoop of ice cream if you hold both the sprinkles and chocolate sauce

    Form the document matrix A, and, using the vector space model, rank the documents D1, D2, D3, and D4 according to their relevance to the query: `chocolate, ice cream`.

6. When you submit a query to a search engine, an ordered list of web pages is returned. The pages are ranked by their relevance to your query and also by the quality of the page. Consider the small network of web pages in figure 1.12. We will assume this is the set of web pages indexed by our search engine. Each vertex in the graph is a web page. A directed link is drawn from web page i to web page j if web page i has a link to web page j. So, we can see, for example, that web page 1 links to web page 4. The PageRank algorithm, as proposed by Larry Page and Sergey Brin [18], assumes that a surfer follows a link on a web page 85% of the time,

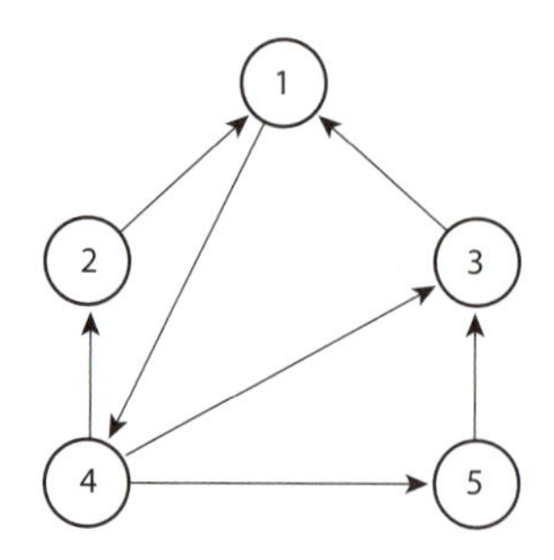

Figure 1.12. A small network of web pages.

with any of the links being equally likely. The other 15% of the time the surfer will enter the URL for another web page, possibly the same one that is currently being visited, again with all pages being equally likely. Let $X_i(t)$ denote the probability of being at web page i after the surfer takes t steps through the network. We will assume the surfer starts at web page 1, so that $X_1(0) = 1$ and $X_i(0) = 0$ for $i = 2, 3, 4, 5$.

(a) Find $X_i(1)$ for $1 \leq i \leq 5$.
(b) Find $X_i(2)$ for $1 \leq i \leq 5$.

As a measure of the quality of a page, PageRank approximates $\lim_{t\to\infty} X_i(t)$ for all i.

2

BASIC OPERATIONS WITH MATLAB

This book is concerned with the understanding of algorithms for problems of continuous mathematics. Part of this understanding is the ability to implement such algorithms. To avoid distracting implementation details, however, we would like to accomplish this implementation in the simplest way possible, even if it is not necessarily the most efficient. One system in which algorithm implementation is especially easy is called MATLAB [94] (short for MATrix LABoratory).

While a helpful academic and instructive tool, MATLAB is used in industry and government, as well. For instance, systems engineers at the NASA Jet Propulsion Laboratory used MATLAB to understand system behavior before launching the Mars Exploration Rover (MER) spacecraft into space.

This chapter contains a short description of basic MATLAB commands, and more commands are described in programs throughout the book. For further information about using MATLAB see, for instance, [52].

2.1 LAUNCHING MATLAB

MATLAB is a high-level programming language that is especially well suited to linear algebra computations, but it can be used for almost any numerical problem. Following are some of the basic features of MATLAB that you will need to carry out the programming exercises in this book. Depending on what system you are using, you will start MATLAB either by double clicking on a MATLAB icon or by typing "matlab" or by some similar means. When MATLAB is ready for input from you it will give a prompt such as >>.

You can use MATLAB like a calculator. For instance, if you type at the prompt

```
>> 1+2*3
```

then MATLAB returns with the answer

```
ans =
     7
```

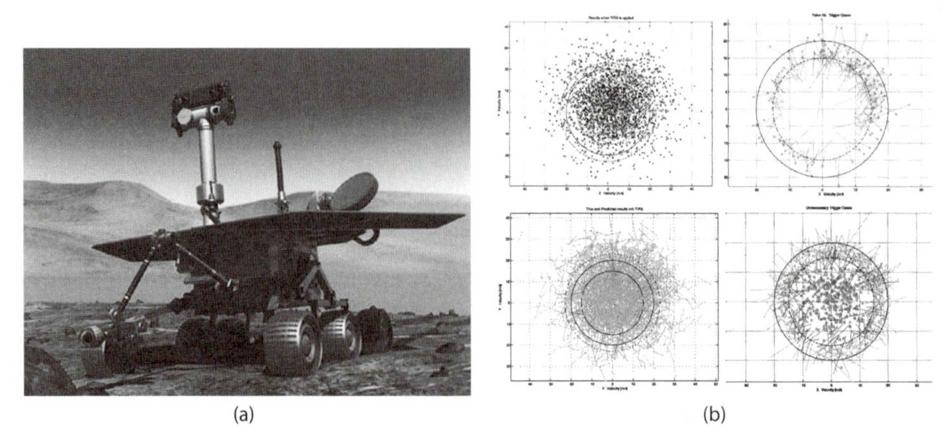

(a) (b)

Figure 2.1. (a) An artist's conception of the Mars rover. (b) Custom statistical MATLAB visualizations that were used to predict how the onboard systems would respond under various atmospheric conditions during descent to the Mars surface. (Images courtesy of NASA/JPL/Cornell University.)

Since you did not give a name to your result, MATLAB stores the result in a variable called ans. You can do further arithmetic using the result in ans:

```
>> ans/4
```

and MATLAB will return with the result

```
ans =
    1.7500
```

2.2 VECTORS

MATLAB can store row or column vectors. The commands

```
>>  v = [1; 2; 3; 4]
v =
     1
     2
     3
     4

>> w = [5, 6, 7, 8]
w =
     5     6     7     8
```

create a column vector v of length 4 and a row vector w of length 4. In general, when defining a matrix or vector, semicolons are used to separate rows, while commas or spaces are used to separate the entries within a row. You can refer to an entry in a vector by giving its index:

```
>> v(2)
ans =
     2

>> w(3)
ans =
     7
```

MATLAB can add two vectors of the same dimension, but it cannot add v and w because v is 4 by 1 and w is 1 by 4. If you try to do this, MATLAB will give an error message:

```
>> v+w
??? Error using ==> +
Matrix dimensions must agree.
```

The transpose of w is denoted w':

```
>> w'
ans =
     5
     6
     7
     8
```

You can add v and w' using ordinary vector addition:

```
>> v + w'
ans =
     6
     8
    10
    12
```

Suppose you wish to compute the sum of the entries in v. One way to do this is as follows:

```
>> v(1) + v(2) + v(3) + v(4)
ans =
    10
```

Another way is to use a for loop:

```
>> sumv = 0;
>> for i=1:4, sumv = sumv + v(i); end;
```

```
>> sumv
sumv =
    10
```

This code initializes the variable sumv to 0. It then loops through each value i = 1, 2, 3, 4 and replaces the current value of sumv with that value plus v(i). The line with the for statement actually contains three separate MATLAB commands. It could have been written in the form

```
for i=1:4
  sumv = sumv + v(i);
end
```

MATLAB allows one line to contain multiple commands, provided they are separated by commas or semicolons. Hence in the one-line version of the for loop, we had to put a comma (or a semicolon) after the statement for i=1:4. This could have been included in the three-line version as well, but it is not necessary. Note also that in the three-line version, we have *indented* the statement(s) inside the for loop. This is not necessary, but it is good programming practice. It makes it easy to see which statements are inside and which are outside the for loop. Note that the statement sumv = 0 is followed by a semicolon, as is the statement sumv = sumv + v(i) inside the for loop. Following a statement by a semicolon suppresses printing of the result. Had we not put the semicolon at the end of the first statement, MATLAB would have printed out the result sumv = 0. Had we not put a semicolon after the statement sumv = sumv + v(i), then each time through the for loop, MATLAB would have printed out the current value of sumv. In a loop of length 4, this might be acceptable; in a loop of length 4 million, it probably would not be! To see the value of sumv at the end, we simply type sumv without a semicolon and MATLAB prints out its value. Of course, if the answer is not what we were expecting, then we might go back and omit the semicolon after the statement sumv = sumv + v(i), since then we could see the result after each step. Extra output is often useful as a program debugging tool.

2.3 GETTING HELP

Actually, the entries in a vector are most easily summed using a built-in MATLAB function called sum. If you are unsure of how to use a MATLAB function or command, you can always type help followed by the command name, and MATLAB will provide an explanation of how the command works:

```
>> help sum
 SUM Sum of elements.
     For vectors, SUM(X) is the sum of the elements of X. For
     matrices, SUM(X) is a row vector with the sum over each
```

```
    column. For N-D arrays, SUM(X) operates along the first
    non-singleton dimension.

   SUM(X,DIM) sums along the dimension DIM.

   Example: If X = [0 1 2
                    3 4 5]

   then sum(X,1) is [3 5 7] and sum(X,2) is [ 3
                                              12];

   See also PROD, CUMSUM, DIFF.
```

In general, you can type help in MATLAB and receive a summary of the classes of commands for which help is available. If you are not sure of the command name for which you are looking, there are two other helpful commands. First, typing helpdesk displays the help browser, which is a very *helpful* tool. Additionally, you can type doc for the same help browser, or doc sum for the hypertext documentation on the MATLAB sum command. Second, if you are interested in commands related to summing, you can type lookfor sum. With this command, MATLAB searches for the specified keyword "sum" in all help entries. This command results in the following response:

```
>> lookfor sum
TRACE  Sum of diagonal elements.
CUMSUM Cumulative sum of elements.
SUM Sum of elements.
SUMMER Shades of green and yellow colormap.
UIRESUME Resume execution of blocked M-file.
UIWAIT Block execution and wait for resume.
RESUME Resumes paused playback.
RESUME Resumes paused recording.
```

It may take some searching to find precisely the topic and/or command that you are looking for.

2.4 MATRICES

MATLAB also works with matrices:

```
>> A = [1, 2, 3; 4, 5, 6; 7, 8, 0]
A =
     1     2     3
     4     5     6
     7     8     0

>> b = [0; 1; 2]
```

```
b =
     0
     1
     2
```

There are many built-in functions for solving matrix problems. For example, to solve the linear system $Ax = b$, type `A\b`:

```
>> x = A\b
x =
    0.6667
   -0.3333
    0.0000
```

Note that the solution is printed out to only four decimal places. It is actually stored to about sixteen decimal places (see chapter 5). To see more decimal places, you can type

```
>> format long
>> x
x =
    0.66666666666667
   -0.33333333333333
    0.00000000000000
```

Other options include `format short e` and `format long e` to display numbers using scientific notation.

You can check this answer by typing `b - A*x`. The notation `A*x` denotes standard matrix–vector multiplication, and standard vector subtraction is used when you subtract the result from `b`. This should give a vector of 0s, if `x` solves the system exactly. Since the machine carries only about 16 decimal digits, we do not expect it to be exactly zero, but, as we will see later, it should be just a moderate size multiple of 10^{-16}:

```
>> format short
>> b - A*x
ans =
   1.0e-15 *

   -0.0740
   -0.2220
         0
```

This is a good result!

2.5 CREATING AND RUNNING .M FILES

Typing MATLAB commands at the keyboard is fine if you are doing a computation once and will never need to make modifications and run it again.

Once you `exit` MATLAB, however, all of the commands that you typed may be lost. To save the MATLAB commands that you type so that they can be executed again, you must enter them into a file called *filename*`.m`. Then, in MATLAB, if you type *filename*, it will run the commands from that file. We will refer to such files as M-files. The M-file can be produced using any text editor, such as the one that comes up as part of the MATLAB window. Once you save this file, it will be available for future use. Before attempting to execute an M-file from MATLAB, you must remember to change the working directory of MATLAB to the directory in which that file resides.

2.6 COMMENTS

Adding documentation to your MATLAB code allows you and others to maintain your code for future use. Many a programmer has coded what appears to be a crystal clear implementation of an algorithm and later returned to be lost in the listing of commands. Comments can help to alleviate this problem. Adding comments to MATLAB code is easy. Simply adding a % makes the remaining portion of that line a comment. In this way, you can make an entire line into a comment:

```
% Solve Ax=b
```

or you can append a comment after a MATLAB command, as in

```
x = A\b;     % This solves Ax=b and stores the result in x.
```

The text following the % is simply a comment for the programmer and is ignored by MATLAB.

2.7 PLOTTING

Tables and figures are usually more helpful than long strings of numerical results. Suppose you are interested in viewing a plot of $\cos(50x)$ for $0 \leq x \leq 1$. You can create two vectors, one consisting of x values and the other consisting of the corresponding $y = \cos(50x)$ values, and use the MATLAB `plot` command. To plot the values of $\cos(50x)$ at $x = 0, 0.1, 0.2, \ldots, 1$, type

```
>> x = 0:0.1:1;     % Form the (row) vector of x values.
>> y = cos(50*x); % Evaluate cos(50*x) at each of the x values.
>> plot(x,y)        % Plot the result.
```

Note that the statement `x = 0:0.1:1;` behaves just like the `for` loop

```
>> for i=1:11, x(i) = 0.1*(i-1); end;
```

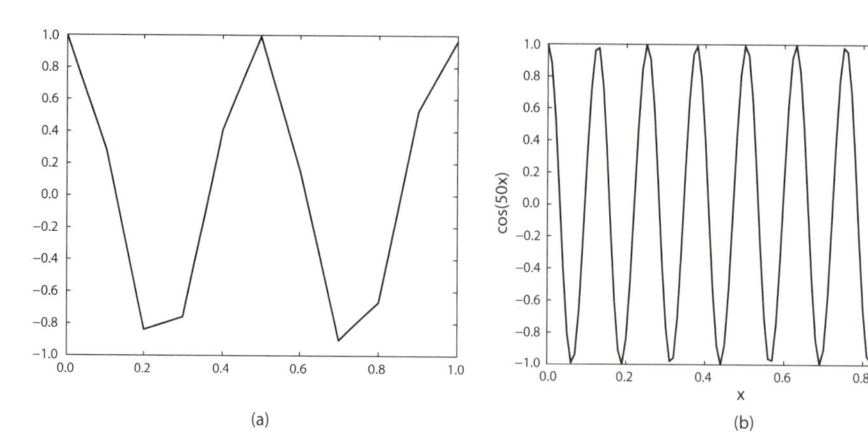

Figure 2.2. Basic MATLAB plots of cosine.

Note also that, unless otherwise specified, each of these statements produces a *row* vector. In order to produce a column vector, one could replace `x(i)` in the above `for` loop by `x(i,1)`, or one could replace the statement in the original code by `x = [0:0.1:1]';`. Clearly, this small number of evaluation points will not produce very high resolution, as is seen in figure 2.2(a).

In the next piece of code we change `x` to include more points and we also include a title and labels for the axes.

```
>> x = 0:0.01:1;        % Create a vector of 101 x values.
>> plot(x,cos(50*x)) % Plot x versus cos(50*x).
>> title('Plot of x versus cos(50x)')
>> ylabel('cos(50x)')
>> xlabel('x')
```

Figure 2.2(b) contains the plot resulting from these commands.

It is also possible to plot more than one function on the same graph. To plot the two functions $f(x) = \cos(50x)$ and $g(x) = x$ on the same graph, type

```
>> plot(x,cos(50*x),x,x)
```

The result is shown in figure 2.3. Note the small boxed legend on the plot; this was added with the command `legend('cos(50x)','x')`.

Another way to plot two functions on the same graph is to first plot one, then type `hold on`, and then plot the other. If you do not type `hold on`, then the second plot will replace the first, but this command tells MATLAB to keep plots on the screen after they are created. To go back to the default of removing old plots before new ones are added, type `hold off`. Again, type `help plot` for more information on plotting or type `doc plot` which posts the Helpdesk documentation for the `plot` command. Other useful commands are `axis` and `plot3`. For a bit of fun, type the commands

```
>> x = 0:0.001:10;
>> comet(x,cos(3*x))
```

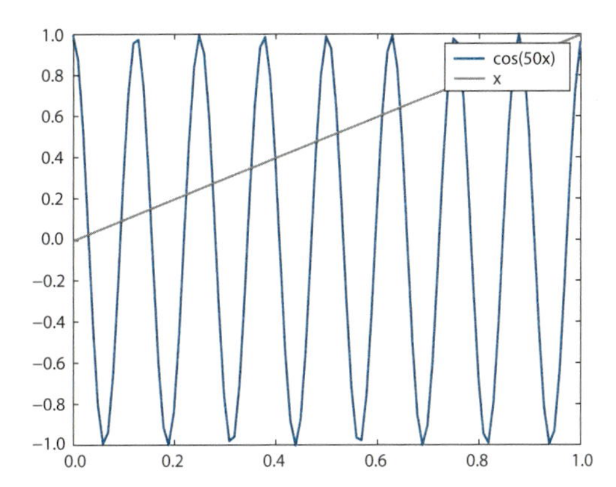

Figure 2.3. A basic MATLAB plot of cosine and a line along with a legend.

For more information on options available for plots, type the commands `hndlgraf`, `hndlaxis` and `ardemo`.

2.8 CREATING YOUR OWN FUNCTIONS

You can create your own functions to use in MATLAB. If your function is simple (e.g., $f(x) = x^2+2x$), then you may enter it using the command `inline`:

```
>> f = inline('x.^2 + 2*x')
f =
     Inline function:
     f(x) = x.^2 + 2*x
```

Note the `.^2` notation. The expression `x^2` produces the square of `x` if `x` is a scalar, but it gives an error message if `x` is a vector, since standard vector multiplication is defined only if the inner dimensions of the vectors are the same (i.e., the first vector is 1 by n and the second is n by 1 or the first is n by 1 and the second is 1 by m). The operation `.^` applied to a vector, however, squares each entry individually. Since we may wish to evaluate the function at each entry in a vector of `x` values, we must use the `.^` operation. To evaluate `f` at the integers between 0 and 5, type

```
>> f([0:5])
ans =
     0     3     8    15    24    35
```

Similarly, you can create an anonymous function by typing

```
>> f = @(x)(x.^2 + 2*x)
f =
    @(x)(x.^2+2*x)
```

If the function is more complicated, you may create a file whose name ends in `.m` which tells MATLAB how to compute the function. Type `help function` to see the format of the function file. Our function here could be computed using the following file (called `f.m`):

```
function output = f(x)
output = x.^2 + 2*x;
```

This function is called from MATLAB in the same way as above; that is, `f(x)`, where `x` can be a scalar or vector.

2.9 PRINTING

While graphical output can be an important visualization tool, numerical results are often presented in tables. There are several ways to print output to the screen in MATLAB. First, to simply display a variable's contents, use the command `display`.

```
>> x = 0:.5:2;
>> display(x)
x =
         0    0.5000    1.0000    1.5000    2.0000
```

In many cases, the extra carriage return imposed by the `display` command clutters printed results. Therefore, another helpful command is `disp`, which is similar to the `display` command.

```
>> disp(x)
         0    0.5000    1.0000    1.5000    2.0000
```

You may still wish to have the variable name printed. Concatenating an array of text for output accomplishes this purpose.

```
>> disp(['x = ',num2str(x)])
x = 0         0.5           1           1.5           2
```

For more information, type `help num2str`.

Tables such as the following can be created using `disp`.

```
>> disp('    Score 1   Score 2   Score 3'), disp(rand(5,3))
    Score 1   Score 2   Score 3
    0.4514    0.3840    0.6085
    0.0439    0.6831    0.0158
    0.0272    0.0928    0.0164
    0.3127    0.0353    0.1901
    0.0129    0.6124    0.5869
```

You may find the fprintf command easier to use for tables of results. For instance, consider the simple loop

```
>> fprintf('    x              sqrt(x)\n=====================\n')
for i=1:5, fprintf('%f       %f\n',i,sqrt(i)), end
   x          sqrt(x)
=====================
1.000000      1.000000
2.000000      1.414214
3.000000      1.732051
4.000000      2.000000
5.000000      2.236068
```

The fprintf command takes format specifiers and variables to be printed in those formats. The %f format indicates that a number will be printed in fixed point format in that location of the line, and the \n forces a carriage return after the two quantities i and sqrt(i) are printed. You can specify the total field width and the number of places to be printed after the decimal point by replacing %f by, say, %8.4f to indicate that the entire number is to be printed in 8 spaces, with 4 places printed after the decimal point. You can send your output to a file instead of the screen by typing

```
fid = fopen('sqrt.txt','w');
fprintf(fid,'    x          sqrt(x)\n=================\n');
for i=1:5, fprintf(fid,'%4.0f        %8.4f\n',i,sqrt(i)); end
```

which prints the following table in a file called sqrt.txt.

```
   x        sqrt(x)
=================
   1        1.0000
   2        1.4142
   3        1.7321
   4        2.0000
   5        2.2361
```

Again, for more information, refer to MATLAB documentation.

2.10 MORE LOOPS AND CONDITIONALS

We have already seen how for loops can be used in MATLAB to execute a set of commands a given number of times. Suppose, instead, that one wishes to execute the commands until some condition is satisfied. For example, one might approximate a root of a given function f(x) by first plotting the function on a coarse scale where one can see the approximate root, then plotting appropriate sections on finer and finer scales until one can identify the root to the precision needed. This can be accomplished with the following MATLAB code.

```
xmin = input(' Enter initial xmin: ');
xmax = input(' Enter initial xmax: ');
tol = input(' Enter tolerance: ');
while xmax-xmin > tol,
  x = [xmin:(xmax-xmin)/100:xmax];
  y = f(x);
  plot(x,y)
  xmin = input(' Enter new value for xmin: ');
  xmax = input(' Enter new value for xmax: ');
end;
```

The user looks at each plot to determine a value xmin that is just left of the root and a value xmax that is just right of the root. The next plot then contains only this section, so that closer values xmin and xmax can be determined. In chapter 4 we discuss more efficient ways of finding a root of f(x).

Another important statement is the conditional if statement. In the above code segment, one might wish to let the user know if the code happens to find a point at which the absolute value of f is less than some other tolerance, say, delta. This could be accomplished by inserting the following lines after the statement y = f(x);:

```
[ymin,index] = min(abs(y));
% This finds the minimum absolute value of y and its index.
if ymin < delta,
 fprintf(' f( %f ) = %f\n', x(index), y(index))
 % This prints the x and y values at this index.
end;
```

The if statement may also contain an else clause. For example, to additionally write a message when ymin is greater than or equal to delta, one could modify the above if statement to say:

```
if ymin < delta,
  fprintf(' f( %f ) = %f\n', x(index), y(index))
  % This prints the x and y values at this index.
else
  fprintf(' No points found where |f(x)| < %f\n', delta)
end;
```

2.11 CLEARING VARIABLES

You may clear a particular variable by typing

```
>> clear x
```

or all variables with

```
>> clear all
```

This is important when you want to be sure that all variable names have been erased from memory.

2.12 LOGGING YOUR SESSION

You can keep a record of your MATLAB session by typing

```
>> diary('hw1.txt')
... some other commands ...
>> diary off
```

This command records all subsequent commands that you type and all responses that MATLAB returns in a file named `hw1.txt`. You will want to name the file by replacing `hw1.txt` with a more descriptive name related to your work. Note, however, that you *cannot* then run the file from MATLAB; this is simply a device for recording what happened during your keyboard session.

Note also that if you execute an M-file in a session logged with the `diary` command, you may want to type `echo on` before executing the M-file. In this way, the commands in the M-file are echoed along with MATLAB's response. Otherwise, the diary file will contain only the responses, not the commands.

2.13 MORE ADVANCED COMMANDS

It is perhaps apparent from the reference at the beginning of this chapter to the work on the Mars Exploration Rover, that MATLAB has a large number of commands. To close this chapter, we demonstrate some of the graphical capabilities through an example.

The commands

```
[X,Y] = meshgrid(-3:.125:3);
Z = peaks(X,Y);
meshc(X,Y,Z);
axis([-3 3 -3 3 -10 5])
```

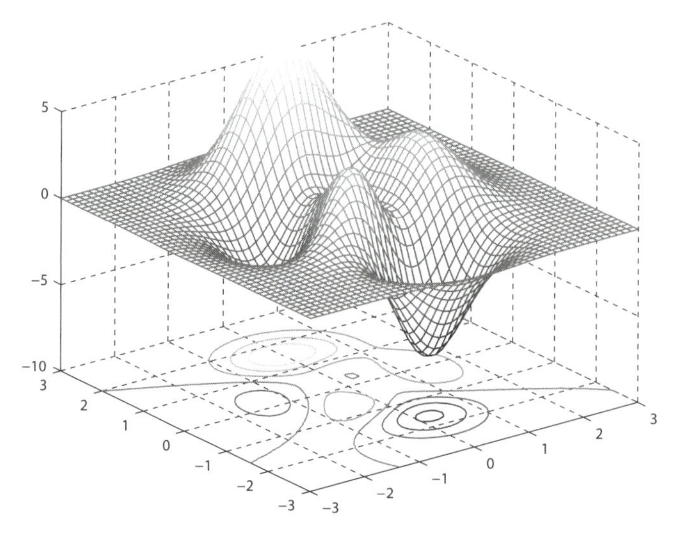

Figure 2.4. A three-dimensional MATLAB plot.

produce the plot in figure 2.4. In order to understand this plot, search MATLAB documentation for the commands `meshgrid`, `peaks`, `meschc`, and `axis`. While this chapter will get you started using MATLAB, effective use of the MATLAB documentation will be the key to proceeding to more complicated programs.

2.14 CHAPTER 2 EXERCISES

1. Run the examples in this chapter using MATLAB to be sure that you see the same results.
2. With the matrices and vectors

$$A = \begin{pmatrix} 10 & -3 \\ 4 & 2 \end{pmatrix}, \quad B = \begin{pmatrix} 1 & 0 \\ -1 & 2 \end{pmatrix}, \quad \mathbf{v} = \begin{pmatrix} 1 \\ 2 \end{pmatrix}, \quad \mathbf{w} = \begin{pmatrix} 1 \\ 1 \end{pmatrix},$$

compute the following *both* by hand and in MATLAB. For the MATLAB computations, use the `diary` command to record your session.

(a) $\mathbf{v}^T\mathbf{w}$
(b) $\mathbf{v}\mathbf{w}^T$
(c) $A\mathbf{v}$
(d) $A^T\mathbf{v}$
(e) AB
(f) BA
(g) A^2 $(= AA)$
(h) the vector $\mathbf{y}$ for which $B\mathbf{y} = \mathbf{w}$
(i) the vector $\mathbf{x}$ for which $A\mathbf{x} = \mathbf{v}$

3. Use MATLAB to produce a single plot displaying the graphs of the functions $\sin(kx)$ across $[0, 2\pi]$, for $k = 1, \ldots, 5$.
4. Use MATLAB to print a table of values x, $\sin x$, and $\cos x$, for $x = 0, \frac{\pi}{6}, \frac{2\pi}{6}, \ldots, 2\pi$. Label the columns of your table.

MATLAB's Origins

Cleve Moler is chairman and chief scientist at MathWorks. In the mid to late 1970s, he was one of the authors of LINPACK and EISPACK, Fortran libraries for numerical computing. To give easy access to these libraries to his students at the University of New Mexico, Dr. Moler invented MATLAB. In 1984, he cofounded The MathWorks with Jack Little to commercialize the MATLAB program. Cleve Moler received his bachelor's degree from California Institute of Technology, and a Ph.D. from Stanford University. Moler was a professor of math and computer science for almost 20 years at the University of Michigan, Stanford University and the University of New Mexico. Before joining The MathWorks full-time in 1989, he also worked for the Intel Hypercube Company and Ardent Computer. Dr. Moler was elected to the National Academy of Engineering in 1997. He also served as president of the Society for Industrial and Applied Mathematics (SIAM) during 2007–2008 [68]. (Photo courtesy of Cleve Moler.)

5. Download the file `plotfunction1.m` from the book's web page and execute it. This should produce the two plots on the next page. The top plot shows the function $f(x) = 2\cos(x) - e^x$ for $-6 \leq x \leq 3$, and from this plot it appears that $f(x)$ has three roots in this interval. The bottom plot is a zoomed view near one of these roots, showing that $f(x)$ has a root near $x = -1.454$. Note the different vertical scale as well as the different horizontal scale of this plot. Note also that when we zoom in on this function it looks nearly *linear* over this short interval. This will be important when we study numerical methods for approximating roots.

   (a) Modify this script so that the bottom plot shows a zoomed view near the leftmost root. Write an estimate of the value of this root to at least 3 decimal places. You may find it useful to first use the zoom feature in MATLAB to see approximately where the root is and then to choose your axis command for the second plot appropriately.

   (b) Edit the script from part (a) to plot the function

   $$f(x) = \frac{4x\sin x - 3}{2 + x^2}$$

   over the range $0 \leq x \leq 4$ and also plot a zoomed view near the leftmost root. Write an estimate of the value of the root from the plots that is accurate to 3 decimal places. Note that once you have defined the vector x properly, you will need to use appropriate componentwise

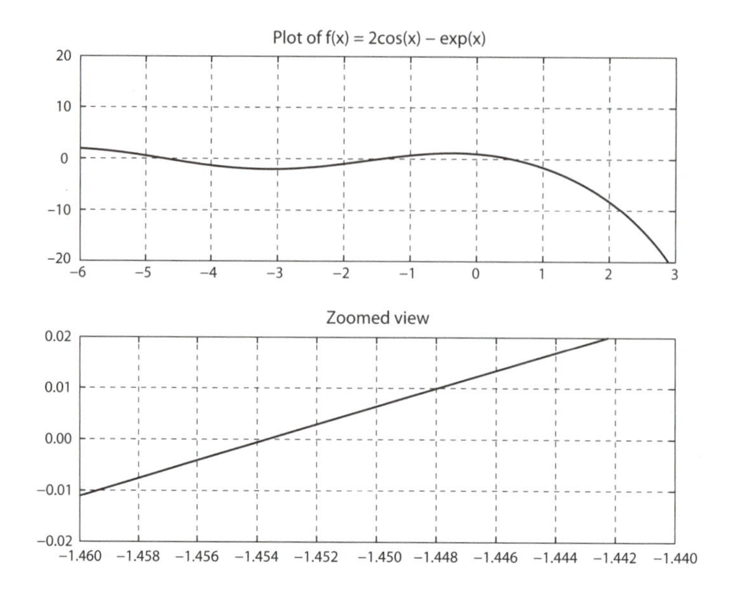

multiplication and division to evaluate this expression:

```
y = (4*x.*sin(x) - 3) ./ (2 + x.^2);
```

6. Plot each of the functions below over the range specified. Produce four plots on the same page using the `subplot` command.

   (a) $f(x) = |x-1|$ for $-3 \le x \le 3$. (Use abs in MATLAB.)
   (b) $f(x) = \sqrt{|x|}$ for $-4 \le x \le 4$. (Use sqrt in MATLAB.)
   (c) $f(x) = e^{-x^2} = \exp(-x^2)$ for $-4 \le x \le 4$. (Use exp in MATLAB.)
   (d) $f(x) = \dfrac{1}{10x^2+1}$ for $-2 \le x \le 2$.

7. Use MATLAB to plot the circles

$$(x-2)^2 + (y-1)^2 = 2,$$
$$(x-2.5)^2 + y^2 = 3.5$$

   and zoom in on the plot to determine approximately where the circles intersect.
   [Hint: One way to plot the first circle is:

```
theta = linspace(0, 2*pi, 1000);
r = sqrt(2);
x = 2 + r*cos(theta);
y = 1 + r*sin(theta);
plot(x,y)
axis equal     % so the circles look circular!
```

   Use the command `hold on` after this to keep this circle on the screen while you plot the second circle in a similar manner.]

8. In this exercise, you will plot the initial stages of a process that creates a fractal known as *Koch's snowflake*, which is depicted below.

This exercise uses the MATLAB M-file `koch.m`, which you will find on the web page. The M-file contains all the necessary commands to create the fractal, except for the necessary plotting commands. Edit this M-file so that each stage of the fractal is plotted. [Hint: This can be accomplished by adding a plot command just before the completion of the outer `for` loop.] Add the following commands to keep consistency between plots in the animation.

```
axis([-0.75 0.75 -sqrt(3)/6 1]);
axis equal
```

Note that the `cla` command clears the axes. Finally, add the command `pause(0.5)` in appropriate places to slow the animation. (The `fill` command, as opposed to `plot`, produced the filled fractals depicted above.) We will create fractals using Newton's method in chapter 4.

9. A magic square is an arrangement of the numbers from 1 to n^2 in an n by n matrix, where each number occurs exactly once, and the sum of the entries in any row, any column, or any main diagonal is the same. The MATLAB command `magic(n)` creates an n by n (where $n > 2$) magic square. Create a 5 by 5 magic square and verify using the `sum` command in MATLAB that the sums of the columns, rows and diagonals are equal. Create a log of your session that records your work. [Hint: To find the sums of the diagonals, read the documentation for the `diag` and the `flipud` commands.]
10. More advanced plotting commands can be useful in MATLAB programming.

    (a) In the MATLAB command window type

    ```
    [X,Y,Z] = peaks(30);
    surf(X,Y,Z);
    ```

    (b) Give this plot the title "3-D shaded surface plot".
    (c) Type `colormap hot` and observe the change in the plot.
    (d) Print the resulting plot with the given title.

11. Computer graphics make extensive use of matrix operations. For example, rotating an object is a simple matrix–vector operation. In two dimensions, a curve can be rotated counterclockwise through an angle θ about the

origin by multiplying every point that lies on the curve by the rotation matrix

$$R = \begin{pmatrix} \cos\theta & -\sin\theta \\ \sin\theta & \cos\theta \end{pmatrix}.$$

As an example, let us rotate the rectangle with vertex coordinates $[1, 0]$, $[0, 1]$, $[-1, 0]$, and $[0, -1]$ through an angle $\theta = \pi/4$. In MATLAB, type the following code to generate the original and rotated squares plotted one on top of the other.

```
% create matrix whose columns contain the coordinates of
% each vertex.
U = [1, 0, -1, 0; 0, 1, 0, -1];

theta = pi/4;

% Create a red unit square
% Note U(1,:) denotes the first row of U
fill(U(1,:),U(2,:),'r')

% Retain current plot and axis properties so that
% subsequent graphing commands add to the existing graph
hold on

% Set the axis
axis([-2 2 -2 2]);

% Perform rotation.
R = [cos(theta) -sin(theta); sin(theta) cos(theta)];
V = R*U;

fill(V(1,:), V(2,:),'b');

axis equal tight, grid on
```

Note that the `fill` command in MATLAB plots a filled polygon determined by two vectors containing the x- and y-coordinates of the vertices.

(a) Adapt this code to plot a triangle with vertices $(5, 0)$, $(6, 2)$ and $(4, 1)$. Plot the triangles resulting from rotating the original triangle by $\pi/2$, π, and $3\pi/2$ radians about the origin. Plot all four triangles on the same set of axes.
(b) Rotating by θ radians and then rotating by $-\theta$ radians leaves the figure unchanged. This corresponds to multiplying first by $R(\theta)$ and then by $R(-\theta)$. Using MATLAB, verify that these matrices are inverses of each other (i.e., their product is the identity) for $\theta = \pi/3$ and for $\theta = \pi/4$.
(c) Using the trigonometric identities $\cos\theta = \cos(-\theta)$ and $-\sin\theta = \sin(-\theta)$, prove that $R(\theta)$ and $R(-\theta)$ are inverses of each other for any θ.

(d) Let R be the matrix that rotates counterclockwise through the angle $\pi/8$, and let $\hat{R} = 0.9*R$. Then the matrix $\hat{R}$ simultaneously rotates and shrinks an object. Edit the code above to repeatedly rotate and shrink (by the same amounts on each step) the square (again originally with coordinates $[1, 0]$, $[0, 1]$, $[-1, 0]$, and $[0, -1]$) for 50 iterations. The plot should show all of the 51 squares on the same set of axes.

(e) Apply $\hat{R}$ but now after each rotation translate the square by 1 unit in the x direction and 2 units in the y direction. Note, you must apply the rotation and translation to all the vertices of the square. The resulting squares should visually suggest the existence of a fixed point. Such a point satisfies the equation

$$\begin{pmatrix} x \\ y \end{pmatrix} = \hat{R}\begin{pmatrix} x \\ y \end{pmatrix} + \begin{pmatrix} 1 \\ 2 \end{pmatrix}.$$

Solve this equation to find the numerical value of this fixed point.

12. The previous exercise used rotation matrices in two dimensions. Now we explore the speed of matrix operations in a computer graphics model in three dimensions. In figure 2.5, we see a model of Yoda. The tessellation contains 33,862 vertices. Let V be a matrix with 3 columns and 33,862 rows, where row i contains the x-, y-, and z-coordinates of the ith vertex in the model. The image can be translated by t units in the y direction by using a translation matrix T where

$$T = \begin{pmatrix} 0 & t & 0 \\ 0 & t & 0 \\ \vdots & \vdots & \vdots \\ 0 & t & 0 \end{pmatrix}.$$

If $V_t = V + T$, then V_t contains the vertex information for the model after a translation of t units in the y direction.

Download the files `yoda.m` and `yodapose_low.mat` from the web page. Run the file `yoda.m` in MATLAB. You will see an animation of the model being translated in space using matrix addition.

(a) The image can be rotated by θ radians about the y-axis by multiplying V on the right by R_y where

$$R_y = \begin{pmatrix} \cos\theta & 0 & -\sin\theta \\ 0 & 1 & 0 \\ \sin\theta & 0 & \cos\theta \end{pmatrix}.$$

Edit the code to continuously rotate the image by $\pi/24$ radians until the image has made one full rotation about the y-axis.

(b) How many multiplications are performed when you use matrix multiplication (with R_y) to rotate the image once by $\pi/24$ radians? (Remember that V is a 33,862 by 3 matrix.) Keep this in mind as you watch how fast MATLAB performs the calculations and displays the results.

Figure 2.5. A model of Yoda created with 33,862 vertices. (Model created by Kecskemeti B. Zoltan. Courtesy of Lucasfilm Ltd. *Star Wars: Episode II - Attack of the Clones* TM & © 2002 Lucasfilm Ltd. All rights reserved. Used under authorization. Unauthorized duplication is a violation of applicable law. Digital Work by Industrial Light & Magic.)

13. You can see a picture of a mandrill by typing the MATLAB commands

```
load mandrill, image(X), colormap(map), axis off equal
```

Each row of the 220 by 3 matrix `map` corresponds to a color with the first, second, and third elements specifying the intensity of red, green, and blue, respectively.

(a) Write down a 3 by 3 matrix T which, when applied to `map` on the right, will reverse the order of the columns. Use the MATLAB commands `map2 = map*T; colormap(map2)` to see the effect. Explain what happened and why.

(b) Write down a 3 by 3 matrix S which, when applied to `map` on the right, leaves columns one and two unchanged but replaces column three by a column of 0s. Use the MATLAB commands `map3=map*S; colormap(map3)` to see the effect of this change. Explain what happened and why. You may need to type `help colormap` to find out exactly what this new color map does.

14. In this exercise, we will create a fractal coastline. Fractals have many uses, and here we see how they can be used to create qualitatively realistic-looking pictures.

We will use the following iterative algorithm to create two-dimensional fractal landscapes.

0. Begin with one straight line segment.
1. For each line segment in the current figure, find the midpoint, denoted by a solid diamond in the picture below.

2. Create a new point by moving a random amount in the x and y directions from that midpoint as seen below. The size of the random displacement will be adjusted at each iteration.

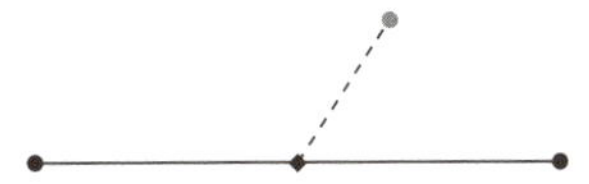

3. Connect the endpoints of the original line with the new point.

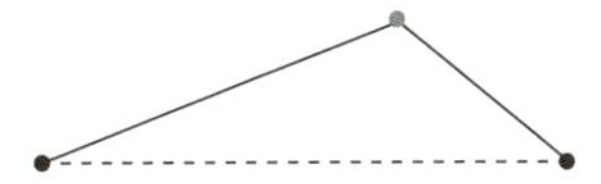

4. If the picture looks good then stop, else adjust the random displacement size and go to step 1.

You will need to determine a suitable range for the random displacement at each iteration to obtain a realistic-looking picture. One such choice resulted in the figures below.

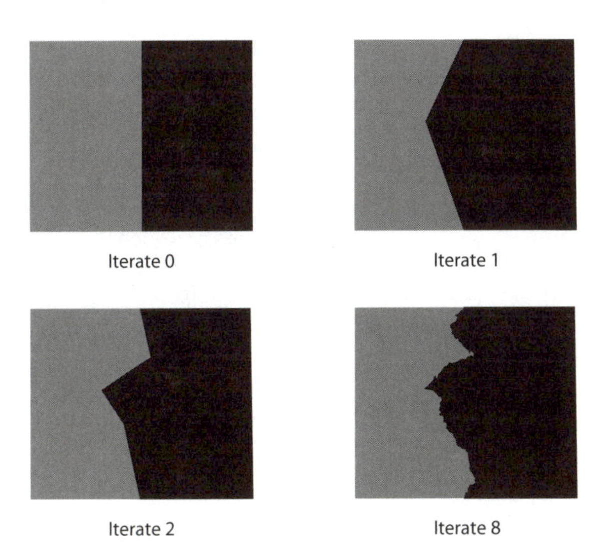

For the exercise, you need not color the land and water masses (although you may do so by using the `fill` command), but simply generate a realistic-looking coastline. If your implementation stores the x values and y values for points on the fractal coastline in the vectors `xValues` and yValues, respectively, then the MATLAB commands:

```
plot(xValues,yValues)
axis equal
```

will plot the fractal coastline with a 1:1 aspect ratio for the axes.

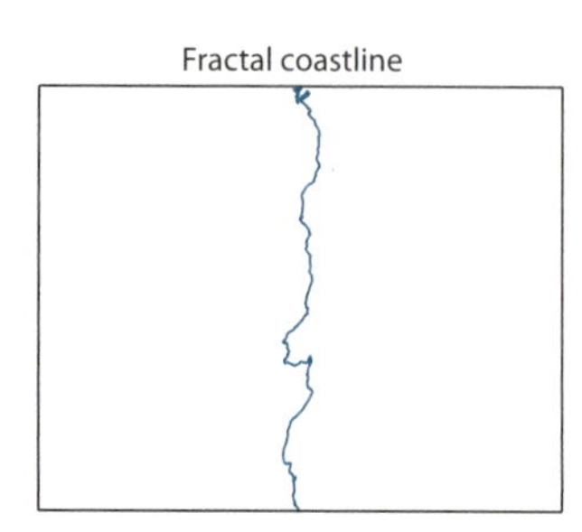

(a) Write a program to create fractal coastlines using the algorithm above.
(b) Describe how your implementation adjusts the range of the random numbers used for displacements in step 2 at each iteration.
(c) Create at least two fractal coastlines with your code.

15. Find documentation on the `movie` command by typing `helpdesk` or `doc movie`. At the bottom of the documentation on this command, you will find the code:

```
Z = peaks;
surf(Z);
axis tight
set(gca,'nextplot','replacechildren');
for j = 1:20
    surf(sin(2*pi*j/20)*Z,Z)
    F(j) = getframe;
end movie(F,5)
```

Cut and paste this code into the MATLAB command window and describe the results. Proficiency in MATLAB programming can increase dramatically through effective use of the available documentation.

3

MONTE CARLO METHODS

Having completed a chapter on mathematical modeling, we now explore one of the simplest types of models—simulating natural phenomena based on probabilities. Methods that do this make extensive use of random numbers and are called **Monte Carlo methods**.

Monte Carlo methods are essentially carefully designed games of chance that enable predictions of interesting phenomena. While no essential link exists between Monte Carlo methods and electronic computers, computer implementation enormously enhances the effectiveness of such methods. While one might be able to perform an experiment tens or hundreds or maybe even a few thousand times by hand, a computer code will likely be able to simulate the same experiment hundreds of thousands or even millions of times in a matter of seconds. Because these computations use random numbers, it may seem surprising that meaningful results about deterministic phenomena can be obtained. Yet Monte Carlo methods are used effectively in a wide range of problems from physics, mechanics, economics, and many other areas.

3.1 A MATHEMATICAL GAME OF CARDS

While Monte Carlo methods have been used for centuries, important advances in their use as a research tool occurred during World War II, with mathematicians including John von Neumann and Stanislaw Ulam and the beginnings of the modern computer.

Ulam came up with a Monte Carlo method in 1946 while pondering the probabilities of winning a card game of solitaire. Assumptions must be made about the way the cards are shuffled and about the decisions that a player makes during the course of the game, but even after such assumptions are made, an analytic computation of the probability of winning would be extremely complex. Instead, Ulam explored another route. He would play the game a large number of times and see the percentage of times that he won. However, Ulam would play the game via an early mainframe computer that he programmed to simulate solitaire. He ran hundreds of trials and computed the proportion of times that he won [68].

Rather than explore the game of solitaire, we will turn our attention to a more modern card game called Texas Holdem.

3.1.1 The Odds in Texas Holdem

"Texas Holdem—It takes a minute to learn and a lifetime to master."
— Mike Sexton, World Poker Tour

Or does it? Since a computer can simulate hundreds of thousands of games in a matter of seconds, can we decrease the time to mastery? If mastering the game means learning to read your opponents effectively while preventing them from reading you, then perhaps not. But if we are interested simply in knowing the odds of winning, given the knowledge that we have, then a computer can be used to determine those odds just by simulating thousands or millions of random games and counting the fraction of wins, losses, and ties for each player. Such a simulation is called a Monte Carlo simulation.

The rules of the game are as follows. Two cards are dealt face down to each player. Then five community cards are revealed, face up. Each player takes the best five-card poker hand from the two down cards and the five community cards, and the player with the best hand wins. During the process of dealing, there are several rounds of betting, and much of the strategy in Texas Holdem comes from betting. Even if one has a poor hand, one may bet a large amount in hopes of convincing other players to fold (to give up and lose all of the money that they have bet so far, fearing that they will lose more if they continue). This is called **bluffing**. While some attempts have been made to develop computer programs that bet and bluff wisely, this is an extremely difficult task and far beyond the scope of this book. Here we simply ask about the odds of winning from a given two-card starting hand, assuming that no players fold. When the game is shown on television, these odds are quickly posted on the screen. Are they correct? How are these odds determined?

Table 3.1 shows the odds of a win, loss, or tie in a two-person game, given any two starting cards and assuming that your opponent's two cards and the five community cards are chosen randomly from the remaining cards. Again, this assumes that both players stay in the hand until the end. This data is plotted in figure 3.1. The probability of winning or losing is plotted as a continuous line, but points on the curve corresponding to the top 8 hands and the bottom 3 hands are marked with o's.

Figure 3.2 shows most of a MATLAB code used to generate the data for this table and figure. The code uses `function whowins` (not shown) to compare the players' hands and decide who wins.

The dealing of cards is simulated using the MATLAB random number generator `rand`. This generates a random number from a uniform distribution between 0 and 1. But the cards are numbered from 1 (2 of clubs) to 52 (ace of spades). To determine which card comes out next, we use the expression `fix(52*rand) + 1`. If the output from `rand` is in the interval $[k/52, (k+1)/52)$ for some nonnegative integer k, then `52*rand` lies in the interval $[k, k+1)$. The MATLAB function `fix` takes the integer part of its argument, and so

TABLE 3.1
Odds for a two-player game assuming neither player folds. W=win, L=lose, T=tie; (s)=suited, (u)=unsuited.

First two	W	L	T	First two	W	L	T	First two	W	L	T	First two	W	L	T
A,A	.85	.15	.01	J,10(s)	.56	.41	.03	K,2(u)	.48	.47	.04	9,5(u)	.40	.55	.05
K,K	.82	.17	.01	A,5(u)	.55	.41	.04	Q,2(s)	.48	.48	.04	10,3(u)	.40	.55	.05
Q,Q	.79	.20	.01	A,2(s)	.56	.41	.04	J,5(s)	.48	.48	.04	9,2(s)	.40	.55	.05
J,J	.77	.23	.01	Q,10(u)	.56	.41	.03	Q,5(u)	.48	.48	.04	7,6(u)	.39	.55	.05
10,10	.75	.24	.01	4,4	.56	.42	.02	J,7(u)	.48	.48	.04	7,4(s)	.39	.55	.05
9,9	.72	.28	.01	A,4(u)	.55	.41	.04	10,8(u)	.48	.48	.04	8,5(u)	.39	.56	.05
8,8	.69	.30	.01	K,6(s)	.55	.42	.04	Q,4(u)	.47	.49	.04	10,2(u)	.39	.56	.05
A,K(s)	.66	.32	.02	Q,8(s)	.54	.42	.03	9,7(s)	.47	.49	.04	6,4(s)	.38	.56	.06
A,Q(s)	.65	.33	.02	K,8(u)	.54	.42	.03	10,6(s)	.47	.49	.04	5,4(s)	.38	.56	.06
7,7	.65	.33	.01	A,3(u)	.54	.42	.04	J,4(s)	.47	.49	.04	8,3(s)	.39	.56	.05
A,J(s)	.65	.33	.02	K,5(s)	.54	.42	.04	Q,3(u)	.46	.50	.04	9,4(u)	.38	.57	.05
A,K(u)	.64	.34	.02	J,9(s)	.54	.43	.03	8,7(s)	.46	.50	.04	7,5(u)	.38	.57	.06
A,10(s)	.64	.34	.02	Q,9(u)	.54	.43	.03	9,8(u)	.46	.50	.04	8,2(s)	.38	.57	.05
A,Q(u)	.63	.35	.02	J,10(u)	.54	.43	.03	J,3(s)	.46	.50	.04	9,3(u)	.37	.57	.05
A,J(u)	.63	.35	.02	K,7(u)	.53	.43	.04	10,7(u)	.46	.50	.04	7,3(s)	.37	.57	.05
K,Q(s)	.62	.36	.02	K,4(s)	.53	.43	.04	J,6(u)	.46	.50	.04	6,5(u)	.37	.57	.06
6,6	.63	.36	.01	A,2(u)	.53	.43	.04	J,2(s)	.45	.50	.04	5,3(s)	.37	.57	.06
A,9(s)	.61	.36	.03	K,6(u)	.52	.44	.04	Q,2(u)	.45	.50	.04	8,4(u)	.37	.58	.06
A,10(u)	.62	.36	.02	J,8(s)	.52	.44	.03	J,5(u)	.45	.50	.05	6,3(s)	.37	.58	.06
K,J(s)	.61	.36	.02	K,3(s)	.52	.44	.04	9,6(s)	.45	.50	.05	9,2(u)	.36	.58	.05
A,8(s)	.60	.37	.03	Q,7(s)	.52	.44	.04	10,5(s)	.45	.51	.05	4,3(s)	.36	.59	.06
K,10(s)	.60	.37	.02	10,9(s)	.52	.44	.03	10,6(u)	.44	.51	.04	7,4(u)	.36	.59	.06
K,Q(u)	.60	.38	.02	3,3	.53	.45	.02	10,4(s)	.44	.51	.05	7,2(s)	.35	.59	.05
A,7(s)	.59	.37	.03	Q,8(u)	.52	.45	.03	J,4(u)	.44	.51	.05	5,4(u)	.35	.59	.06
K,J(u)	.60	.38	.02	Q,6(s)	.51	.45	.04	9,7(u)	.44	.52	.04	6,4(u)	.35	.59	.06
A,9(u)	.59	.38	.03	K,5(u)	.51	.45	.04	8,6(s)	.44	.52	.05	6,2(s)	.35	.59	.06
5,5	.60	.39	.01	K,2(s)	.51	.45	.04	9,5(s)	.43	.52	.05	5,2(s)	.35	.59	.06
Q,J(s)	.59	.39	.02	J,9(u)	.52	.45	.03	J,3(u)	.43	.52	.05	8,3(u)	.35	.60	.05
K,9(s)	.59	.39	.03	Q,5(s)	.51	.45	.04	10,3(s)	.43	.52	.05	4,2(s)	.34	.60	.06
A,5(s)	.58	.38	.04	K,4(u)	.50	.45	.04	7,6(s)	.43	.52	.05	8,2(u)	.34	.61	.06
K,10(u)	.59	.39	.03	J,7(s)	.51	.46	.04	10,2(s)	.43	.53	.05	7,3(u)	.33	.61	.06
A,6(s)	.58	.39	.04	10,8(s)	.50	.46	.04	8,7(u)	.43	.53	.05	5,3(u)	.33	.61	.06
A,8(u)	.58	.39	.03	Q,4(s)	.50	.46	.04	8,5(s)	.42	.53	.05	6,3(u)	.33	.61	.06
Q,10(s)	.58	.39	.03	Q,7(u)	.50	.46	.04	9,6(u)	.42	.53	.05	3,2(s)	.33	.61	.06
A,4(s)	.57	.39	.04	10,9(u)	.50	.47	.03	J,2(u)	.42	.53	.05	4,3(u)	.32	.62	.06
A,7(u)	.57	.40	.03	J,8(u)	.50	.47	.04	9,4(s)	.42	.54	.05	7,2(u)	.32	.63	.06
A,3(s)	.56	.40	.04	K,3(u)	.49	.46	.04	10,5(u)	.41	.54	.05	5,2(u)	.31	.63	.06
K,8(s)	.56	.40	.03	Q,3(s)	.49	.47	.04	7,5(s)	.41	.54	.05	6,2(u)	.31	.63	.06
Q,J(u)	.57	.41	.03	Q,6(u)	.49	.47	.04	10,4(u)	.41	.54	.05	4,2(u)	.30	.64	.06
K,9(u)	.56	.41	.03	J,6(s)	.49	.47	.04	9,3(s)	.41	.54	.05	3,2(u)	.29	.65	.06
Q,9(s)	.56	.41	.03	9,8(s)	.49	.47	.04	8,6(u)	.41	.54	.05				
K,7(s)	.56	.41	.03	10,7(s)	.49	.47	.04	6,5(s)	.40	.54	.06				
A,6(u)	.56	.41	.04	2,2	.50	.49	.02	8,4(s)	.40	.55	.05				

`fix(52*rand)` generates the integer k. Since k will be between 0 and 51, we add 1 to this to obtain the number of the card. Of course, cards that have already been dealt cannot appear again, so we keep track of which cards have already appeared using the vector `taken`. If a random card is generated but it is discovered that this card has already been marked as taken, then we simply retry until we generate a random card that has not yet appeared. Another

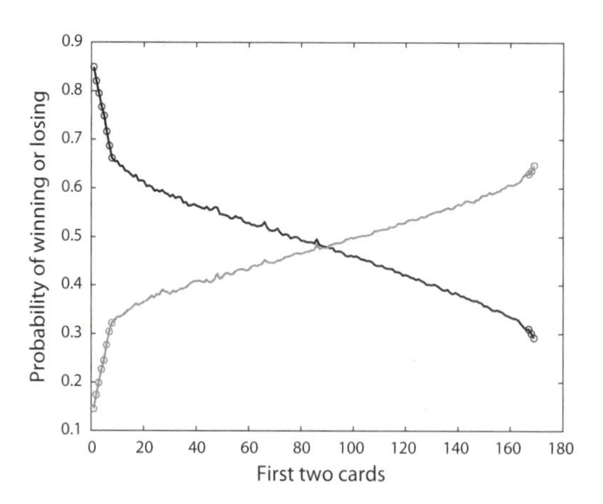

Figure 3.1. Probability of winning (black) and losing (gray) for each of the two-card starting hands in table 3.1. The horizontal axis is the number of the hand from the top (1=A,A) to the bottom (169=3,2(u)). The top 8 hands and the bottom 3 are marked on the curve with o's.

approach would be to use the MATLAB command `randperm`, which uses `rand` to generate a random permutation of the entire deck of 52 cards. With this approach one would not have to worry about duplicate cards, but it might take longer if only a moderate number of cards from the deck are actually used. On the other hand, it would certainly be more efficient if there were a large number of players and many duplicate cards were generated using the code in figure 3.2.

If you study table 3.1 carefully, you will find a few apparent anomalies. The hands are sorted from best to worst based on `wins - losses`, where each hand was played 100,000 times. Although the probabilities are printed to only two decimal places, the numbers of wins, losses, and ties were counted exactly and were used as the basis of this ordering.

It is sometimes claimed that 2,7(u) is the worst possible starting hand; the reason given is that although 2,3(u), . . . ,2,6(u) are lower cards, with each of these hands there is a possibility of a straight if the right community cards are dealt, but there is no such possibility with 2,7(u). Yet table 3.1 indicates that 2,3(u) is the worst starting hand; its probability of winning is just .29, compared to .32 for 2,7(u). Is this difference real? Or are 100,000 simulations insufficient to determine which starting hand is better when the difference is this small?

Note that A,5(s) is rated higher than A,6(s); perhaps this is because one might get a straight (A,2,3,4,5) when one starts with A,5; but then why is A,6(u) rated higher than A,5(u)? Perhaps here the difference is so small that 100,000 simulations are not sufficient to determine which starting hand is better. Similar possible anomalies occur with 5,3(s or u) being rated higher than 6,3(s or u, correspondingly), with 5,4(u) being rated higher than 6,4(u), and with 5,2(u) being rated higher than 6,2(u). Perhaps there is a subtle bug in the code. Monte Carlo codes can be notoriously difficult to debug, since there is a certain amount of variation in the results from one run to the next. Assuming that there is no bug in the code, how does one know if such results are real or if they are within the expected statistical error?

```
% User specifies his first two cards, the number of players, and the number of
% simulations to run. This routine runs the desired number of simulations of
% Texas Holdem and reports the number of wins, losses, and ties.
%
% Cards:
%   rank: 2=2, ..., 10=10, J=11, Q=12, K=13, A=14.
%   suit: C=1, D=2, H=3, S=4.
%   cardno = (rank-2)*4 + suit    (1--52)
%
taken = zeros(52,1);    % Keep track of cards that have alreadybeen dealt.

% Get input from user.
no_players = input(' Enter number of players: ');
first_two = input(' Enter first two cards: [cardno1; cardno2]');
  taken(first_two(1,1)) = 1;  taken(first_two(2,1)) = 1;   % Mark user's cards
                                                            % as taken.
no_plays = input(' Enter number of times to play this hand: ');

% Loop over simulations, counting wins, losses, and ties.
wins = 0;  ties = 0;  losses = 0; takeninit = taken;
for play=1:no_plays,

%   Deal first two cards to other players.
  for m=2:no_players,
    for i=1:2,
      first_two(i,m) = fix(52*rand) + 1;     % Use uniform random distribution.
      while taken(first_two(i,m))==1,        % If card is already taken, try again.
        first_two(i,m) = fix(52*rand) + 1;
      end;
      taken(first_two(i,m)) = 1;               % Mark this card as taken.
    end;
  end;
%   Deal the flop, the turn, and the river card.
  show_cards = zeros(5,1);
  for i=1:5,
    show_cards(i) = fix(52*rand) + 1;     % Use uniform random distribution.
    while taken(show_cards(i))==1,        % If card is already taken, try again.
      show_cards(i) = fix(52*rand) + 1;
    end;
    taken(show_cards(i)) = 1;             % Mark this card as taken.
  end;

%   See if hand 1 wins, loses, or ties.
  [winner, tiedhands] = whowins(no_players, first_two, show_cards);
  if winner==1, wins = wins+1;  elseif tiedhands(1)==1,
  ties = ties+1;
  else losses = losses+1; end;
  taken = takeninit;                      % Prepare for new hand.
end;
wins, losses, ties                        % Print results.
```

Figure 3.2. MATLAB code to simulate Texas Holdem.

We will not be able to answer this question definitively, but we will state some theorems about the expected distribution of results from a Monte Carlo code that give us an indication of how much confidence to put in a computed result.

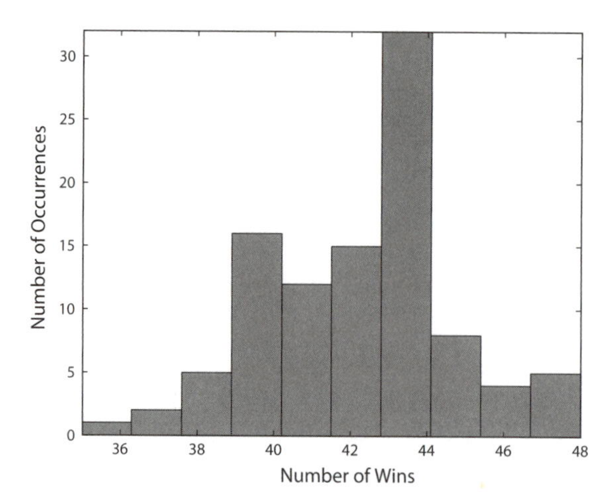

Figure 3.3. Histogram of number of wins from 50 simulations of Texas Holdem with an initial hand of two aces.

Before turning our attention to the basic statistics needed to understand the Monte Carlo results, however, let us run the code and see what sorts of patterns emerge. Using the MATLAB files, holdem.m, whowins.m, and findhand.m, we simulated 50 games, each time starting with the ace of spades and the ace of hearts, while the opponent was dealt two random cards. One such experiment produced 44 wins, 6 losses, and 0 ties. This suggests an 88% chance of winning with a starting hand of two aces. We then repeated the 50-game experiment and this time had 36 wins and 14 losses, which suggests only a 72% chance of winning. Figure 3.3 shows a histogram of the number of wins in each of 100 experiments, where each experiment simulated 50 games with the starting hand of two aces versus a random starting hand.

From the histogram, we see that the computed probabilities of a win varied from 72% to 96%. This is a rather large variation, and one might guess that this variation could be reduced by simulating more games in each experiment.

Suppose we increase the number of games in each experiment to $10,000$. Figure 3.4 shows a histogram of the number of wins in each of 100 experiments, where one experiment consisted of $10,000$ games. Note that the estimated probability of a win now varies only between about 84% and 86%. In the next section we will state a quantitative result about the reduction in variation due to increasing the number of games per experiment. Note also how most of the results fall towards the center of the histogram, with the number of experiments yielding results significantly different from this center value trailing off as we move away from the center. This observation also will be quantified by the Central Limit Theorem.

3.2 BASIC STATISTICS

We begin with a discussion of the basic statistics needed to understand the results of a Monte Carlo simulation.

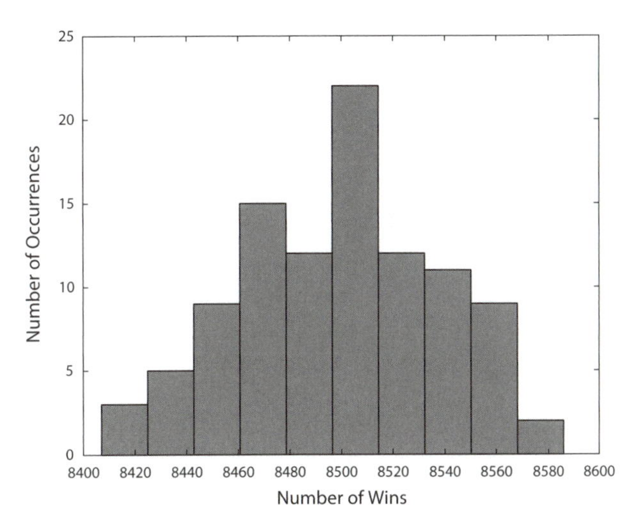

Figure 3.4. Histogram of number of wins from 10,000 simulations of Texas Holdem with an initial hand of two aces.

MONTY HALL PROBLEM

The Monty Hall problem originated in a letter by Steve Selvin to the *American Statistician* in 1975 [89]. However, it was the problem's appearance in Marilyn vos Savant's column "Ask Marilyn" in *Parade* magazine in 1990 that caught wide attention. The problem was stated as having goats and a car behind the doors. In her column, vos Savant asserted that switching is advantageous, which led to thousands of letters with 92% of them insisting she was wrong. Vos Savant called upon "math classes all across the country" to estimate the probabilities using pennies and paper cups, which were later reported to match her results [86, 100–102]. In the exercises at the end of the chapter, you will get a chance to confirm vos Savant's claim using Monte Carlo simulation. (Door image courtesy of the Real Carriage Door Company.)

3.2.1 Discrete Random Variables

A discrete random variable X takes values x_i with probability p_i, $i = 1, \ldots, m$, where $\sum_{i=1}^{m} p_i = 1$.

Example 3.2.1. Roll a fair die and let X be the value that appears. Then X takes on the values 1 through 6, each with probability 1/6.

Example 3.2.2. (Monty Hall problem). You are told that there is a hundred dollar bill behind one of three doors and that there is nothing behind the other two. Choose one of the doors and let X be the amount of money that you find behind your door. Then X takes on the value 100 with probability 1/3 and 0 with probability 2/3.

Now suppose that after choosing a door, but before opening it, you are told one of the other doors that does *not* contain the money; that is, suppose the hundred dollars is actually behind door number one. If you guessed one, then you are told either that it is not behind door number two or that it is not behind door number three. If you guessed two, you are told that it is not behind door number three, and if you guessed three then you are told that it is not behind door number two. You may now change your guess to the remaining door—the one that you did not choose the first time and that you were not told did not contain the hundred dollars. Let Y be the amount of money that you find if you change your guess. Then Y takes on the value 100 with probability 2/3 and 0 with probability 1/3. Do you see why?

If your initial guess was correct and you change it then you will get nothing. This happens with probability 1/3. But if your initial guess was *wrong*, which happens with probability 2/3, you have now been told which other door does *not* contain the money, so that if you switch to the one remaining door you will get the hundred dollars. If you still do not believe it, try playing the game 10 or 20 times with a friend. (But don't play for money!) First try staying with your original guess and then try changing doors when the additional information is revealed, and see which strategy works better.

The **expected value** of a discrete random variable X is defined as

$$E(X) \equiv \langle X \rangle = \sum_{i=1}^{m} p_i x_i .$$

This is also sometimes called the **mean** of the random variable X and denoted as μ.

In Example 3.2.1,

$$E(X) = \frac{1}{6} \cdot 1 + \frac{1}{6} \cdot 2 + \frac{1}{6} \cdot 3 + \frac{1}{6} \cdot 4 + \frac{1}{6} \cdot 5 + \frac{1}{6} \cdot 6 = \frac{7}{2}.$$

In Example 3.2.2,

$$E(X) = \frac{1}{3} \cdot 100 + \frac{2}{3} \cdot 0 = 33\frac{1}{3},$$

$$E(Y) = \frac{2}{3} \cdot 100 + \frac{1}{3} \cdot 0 = 66\frac{2}{3}.$$

If X is a discrete random variable and g is any function, then $g(X)$ is a discrete random variable and

$$E(g(X)) = \sum_{i=1}^{m} p_i g(x_i).$$

Example 3.2.3. $g(X) = aX + b$, a and b constants.

$$\begin{aligned} E(g(X)) &= \sum_{i=1}^{m} p_i(ax_i + b) \\ &= a \sum_{i=1}^{m} p_i x_i \; + \; b \quad \left(\text{since } \sum_{i=1}^{m} p_i = 1 \right) \\ &= a \cdot E(X) + b. \end{aligned}$$

Example 3.2.4. $g(X) = X^2$. Then $E(g(X)) = \sum_{i=1}^{m} p_i x_i^2$.
In Example 3.2.1,

$$E(X^2) = \frac{1}{6} \cdot 1^2 + \frac{1}{6} \cdot 2^2 + \frac{1}{6} \cdot 3^2 + \frac{1}{6} \cdot 4^2 + \frac{1}{6} \cdot 5^2 + \frac{1}{6} \cdot 6^2 = \frac{91}{6}.$$

Recall $\mu = E(X)$ is the expected value of X. The expected value of the *square of the difference* between X and μ is

$$\begin{aligned} E((X - \mu)^2) &= \sum_{i=1}^{m} p_i(x_i - \mu)^2 \\ &= \sum_{i=1}^{m} p_i(x_i^2 - 2\mu x_i + \mu^2) \\ &= \sum_{i=1}^{m} p_i x_i^2 - 2\mu \sum_{i=1}^{m} p_i x_i + \mu^2 \\ &= E(X^2) - \mu^2 \\ &= E(X^2) - (E(X))^2. \end{aligned}$$

The quantity $E(X^2) - (E(X))^2$ is called the **variance** of the random variable X and is denoted $\text{var}(X)$. Note that the variance of a scalar c times a random variable X is $c^2\text{var}(X)$, since $E((cX)^2) - (E(cX))^2 = c^2[E(X^2) - (E(X))^2] = c^2\text{var}(X)$. The square root of the variance, $\sigma \equiv \sqrt{\text{var}(X)}$, is called the **standard deviation**. In Example 3.2.1,

$$\text{var}(X) = \frac{91}{6} - \left(\frac{7}{2}\right)^2 = \frac{35}{12}, \quad \sigma(X) = \sqrt{\frac{35}{12}} \approx 1.708.$$

Let X and Y be two random variables. Then the expected value of the sum $X + Y$ is the sum of the expected values: $E(X + Y) = E(X) + E(Y)$. To see this, suppose X takes on values x_i with probability p_i, $i = 1, \ldots, m$, while Y takes on values y_j with probability q_j, $j = 1, \ldots, n$. Then

$$\begin{aligned}
E(X+Y) &= \sum_{i=1}^{m}\sum_{j=1}^{n}(x_i + y_j)\cdot \text{Prob}(X = x_i \text{ and } Y = y_j) \\
&= \sum_{i=1}^{m} x_i \sum_{j=1}^{n} \text{Prob}(X = x_i \text{ and } Y = y_j) \\
&\quad + \sum_{j=1}^{n} y_j \sum_{i=1}^{m} \text{Prob}(X = x_i \text{ and } Y = y_j) \\
&= \sum_{i=1}^{m} x_i p_i + \sum_{j=1}^{n} y_j q_j = E(X) + E(Y).
\end{aligned}$$

The last line follows because $\sum_{j=1}^{n} \text{Prob}(X = x_i \text{ and } Y = y_j)$ is the probability that $X = x_i$ and Y takes on one of the values $y_1, \ldots, y_n$, which it does with probability 1. Hence this sum is just $\text{Prob}(X = x_i) = p_i$, and similarly, $\sum_{i=1}^{m} \text{Prob}(X = x_i \text{ and } Y = y_j)$ is the probability that $Y = y_j$, which is q_j.

The variance of the sum of two random variables is a bit more complicated. Again let X and Y be two random variables and let c_1 and c_2 be constants. Then the variance of $c_1 X + c_2 Y$ can be expressed as follows, using the rules for expected values:

$$\begin{aligned}
\text{var}(c_1 X + c_2 Y) &= E((c_1 X + c_2 Y)^2) - (E(c_1 X + c_2 Y))^2 \\
&= E(c_1^2 X^2 + 2c_1 c_2 XY + c_2^2 Y^2) - (c_1 E(X) + c_2 E(Y))^2 \\
&= c_1^2 E(X^2) + 2c_1 c_2 E(XY) + c_2^2 E(Y^2) - \\
&\quad [c_1^2 (E(X))^2 + 2c_1 c_2 E(X)E(Y) + c_2^2 (E(Y))^2] \\
&= c_1^2 \text{var}(X) + c_2^2 \text{var}(Y) + 2c_1 c_2 (E(XY) - E(X)E(Y)).
\end{aligned}$$

The **covariance** of X and Y, denoted $\text{cov}(X, Y)$, is the quantity $E(XY) - E(X)E(Y)$.

Two random variables X and Y are said to be **independent** if the value of one does not depend on that of the other; that is, if the probability that $X = x_i$ is the same regardless of the value of Y and the probability that $Y = y_j$ is the same regardless of the value of X. Equivalently, the probability that $X = x_i$ and $Y = y_j$ is the *product* of the probability that $X = x_i$ and the probability that $Y = y_j$.

Example 3.2.5. Toss two fair coins. There are four equally probable outcomes: HH, HT, TH, TT. Let X equal 1 if the first coin is heads, 0 if the first coin is tails. Let Y equal 1 if the second coin is heads, 0 if the second coin is tails. Then X and Y are independent because, for example,

$$\text{Prob}(X = 1 \text{ and } Y = 0) = \frac{1}{4} = \frac{1}{2}\cdot\frac{1}{2} = \text{Prob}(X = 1)\cdot\text{Prob}(Y = 0),$$

and similarly for all other possible values, $\text{Prob}(X = x_i \text{ and } Y = y_j) = \text{Prob}(X = x_i) \cdot \text{Prob}(Y = y_j)$. In contrast, if we define Y to be 0 if the outcome is TT and 1 otherwise, then X and Y are *not* independent because, for example, $\text{Prob}(X = 1 \text{ and } Y = 0) = 0$, yet $\text{Prob}(X = 1) = 1/2$ and $\text{Prob}(Y = 0) = 1/4$.

If X and Y are independent random variables, then $\text{cov}(X, Y) = 0$, and

$$\text{var}(c_1 X + c_2 Y) = c_1^2 \text{var}(X) + c_2^2 \text{var}(Y).$$

3.2.2 Continuous Random Variables

If a random variable X can take on any of a continuum of values, say, any value between 0 and 1, then we cannot define it by listing values x_i and giving the probability p_i that $X = x_i$; for any single value x_i, $\text{Prob}(X = x_i)$ is zero! Instead we can define the **cumulative distribution function,**

$$F(x) \equiv \text{Prob}(X < x),$$

or the **probability density function (pdf),**

$$\rho(x)\, dx \equiv \text{Prob}(X \in [x, x + dx]) = F(x + dx) - F(x).$$

Dividing each side by dx and letting $dx \to 0$, we find

$$\rho(x) = F'(x), \quad F(x) = \int_{-\infty}^{x} \rho(t)\, dt.$$

The expected value of a continuous random variable X is then defined by

$$E(X) = \int_{-\infty}^{\infty} x\rho(x)\, dx.$$

Note that, by definition, $\int_{-\infty}^{\infty} \rho(x)\, dx = 1$. The expected value of X^2 is

$$E(X^2) = \int_{-\infty}^{\infty} x^2 \rho(x)\, dx,$$

and the variance is again defined as $E(X^2) - (E(X))^2$.

Example 3.2.6. Uniform distribution in $[0, 1]$:

$$F(x) = \begin{cases} 0 \text{ if } x < 0, \\ x \text{ if } 0 \le x \le 1, \\ 1 \text{ if } x > 1, \end{cases} \quad \rho(x) = \begin{cases} 0 \text{ if } x < 0, \\ 1 \text{ if } 0 \le x \le 1, \\ 0 \text{ if } x > 1, \end{cases}$$

$$E(X) = \int_{-\infty}^{\infty} x\rho(x)\, dx = \int_0^1 x\, dx = \frac{1}{2},$$

$$\text{var}(X) = \int_0^1 x^2\, dx - \left(\frac{1}{2}\right)^2 = \frac{1}{3} - \frac{1}{4} = \frac{1}{12}.$$

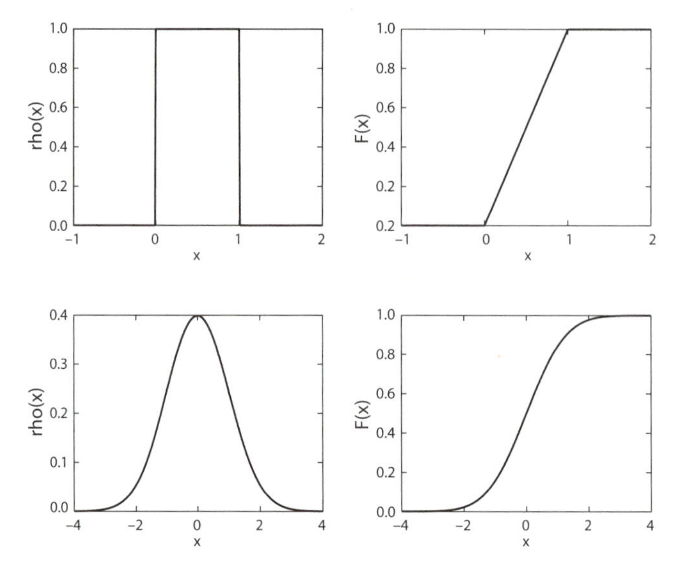

Figure 3.5. Probability density function, $\rho(x)$, and cumulative distribution function, $F(x)$, for a uniform distribution in $[0, 1]$ and a normal distribution with mean 0, variance 1.

Example 3.2.7. Normal (Gaussian) distribution, mean μ, variance σ^2:

$$\rho(x) = \frac{1}{\sigma\sqrt{2\pi}} \exp\left(-\frac{(x-\mu)^2}{2\sigma^2}\right),$$

$$F(x) = \frac{1}{\sigma\sqrt{2\pi}} \int_{-\infty}^{x} \exp\left(-\frac{(t-\mu)^2}{2\sigma^2}\right) dt.$$

In MATLAB:

`rand` generates random numbers from a uniform distribution between 0 and 1.

Suppose you need random numbers uniformly distributed between -1 and 3. How can you use `rand` to obtain such a distribution?
Answer: `4*rand` is a random number from a uniform distribution between 0 and 4, so `4*rand-1` gives a random number from a uniform distribution between -1 and 3.

`randn` generates random numbers from a normal distribution with mean 0, variance 1.

Suppose you need random numbers from a normal distribution with mean 6 and variance 4. How can you use `randn` to obtain such a distribution?
Answer: Since the variance of a scalar times a random variable is the *square* of that scalar times the variance of the random variable, in order to get a normal distribution with variance 4, we must take

2*randn. This generates a random number from a normal distribution with mean 0 and variance 4. To shift the mean to 6, we simply add 6: 2*randn+6 gives the desired result.

3.2.3 The Central Limit Theorem

Let $X_1, \ldots, X_N$ be independent and identically distributed (iid) random variables, with mean μ and variance σ^2. Consider the average value, $A_N = \frac{1}{N}\sum_{i=1}^{N} X_i$. According to the Law of Large Numbers, this average approaches the mean μ as $N \to \infty$, with probability 1.

Example 3.2.8. If you toss a fair coin many, many times, the fraction of heads will approach $\frac{1}{2}$.

The Central Limit Theorem states that, for N sufficiently large, values of the random variable A_N are *normally distributed* about μ, with variance σ^2/N. The expression for the variance follows from the rules we derived for the variance of sums and products:

$$\text{var}(A_N) = \frac{1}{N^2}\sum_{i=1}^{N} \text{var}(X_i) = \frac{\sigma^2}{N}.$$

Theorem 3.2.1 (Central Limit Theorem). Let $X_1, X_2, \ldots, X_N$ be iid random variables. Then the sample average A_N tends to a normal distribution as the sample size tends to infinity. The mean of this distribution will be the mean μ of each X_i, and the variance of this distribution will be the variance of each X_i divided by the sample size N.

Take a moment and consider the power of this theorem. The theorem allows *any* iid random variables X_i. Two random variables X_i and Y_i may have widely differing distributions, such as a uniform distribution and a highly skewed distribution. Yet, if multiple, independent experiments occur for each, the resulting distribution of the sample average always tends towards a normal (Gaussian) distribution as the sample size increases.

As an example, suppose we toss a fair coin 1000 times and record the fraction of times that it comes up heads. We can think of this fraction A_{1000} as the average value of 1000 random variables X_i defined by

$$X_i = \begin{cases} 1 & \text{if } i\text{th toss is heads,} \\ 0 & \text{if } i\text{th toss is tails.} \end{cases}$$

Although the individual random variables X_i take on just two discrete values 0 and 1, each with probability .5, the Central Limit Theorem tells us that their average $A_{1000} = \frac{1}{1000}\sum_{i=1}^{1000} X_i$ will be approximately normally distributed; that is, if we performed this 1000 coin toss experiment, say, 100 times, each time recording the fraction of heads, and made a bar graph of the results, it would look much like a normal distribution, with the majority of results landing close to the mean 0.5 and the number of results landing in other regions trailing off as we move away from the mean. Of course, all values of A_{1000} will be between 0 and 1, so this is not exactly a normal distribution, but the Central Limit Theorem guarantees that if the number of experiments is large enough then the sample average will be close to normally distributed.

We can be even more specific about the normal distribution that the values A_{1000} will follow. It has mean 0.5 (the mean of the individual X_i's) and variance equal to $\mathrm{var}(X_i)/1000$. Since the random variables X_i take on only the values 0 and 1, it is easy to compute their variance. The random variable X_i^2 also takes on only the values 0 and 1, and is 1 if X_i is 1 and 0 if X_i is 0, so its expected value is the same as that of X_i, namely, 0.5. Hence $\mathrm{var}(X_i) = E(X_i^2) - (E(X_i))^2 = 0.5 - 0.5^2 = 0.25$. It follows that the variance of A_{1000} is 0.00025 and its standard deviation is the square root of this number, or, about 0.0158. A normally distributed random variable lands within one standard deviation of its mean about 68% of the time, within two standard deviations about 95% of the time, and within three standard deviations about 99.7% of the time. This means that the value of A_{1000} will be between 0.4842 and 0.5158 about 68% of the time, between 0.4684 and 0.5316 about 95% of the time, and between 0.4526 and 0.5474 about 99.7% of the time.

How can the Central Limit Theorem be used in practice? Suppose one runs an experiment a large number of times N, generating values of iid random variables X_i, $i = 1, \ldots, N$. Let x_i be the value obtained in the ith experiment and let $a_N = \frac{1}{N}\sum_{i=1}^{N} x_i$ be the average value obtained. The value of a_N can be taken as an approximation to the mean μ of each individual random variable X_i. But how good an approximation is it? According to the Central Limit Theorem it is a sample of a normally distributed random variable A_N with variance $\mathrm{var}(X_i)/N$.

If one knows $\mathrm{var}(X_i)$, then this gives an indication of how much confidence can be placed in the result a_N. In most cases, however, one does not know the variance of the individual random variables X_i. Still, this variance can be estimated using experimental data:

$$\mathrm{var}(X_i) = E((X_i - \mu)^2) \approx \frac{1}{N-1}\sum_{i=1}^{N}(x_i - a_N)^2. \tag{3.1}$$

The factor $1/(N-1)$ instead of $1/N$ may seem surprising. This is known as **Bessel's correction**, after the mathematician Friedrich Bessel, and it is designed to correct the bias in the estimate of variance, due to the use of an estimated mean μ obtained from the same data. For further details, consult any good statistics book or look up "Bessel's correction" on Wikipedia. For large N it makes little difference anyway. By keeping track of $\sum_{i=1}^{N} x_i^2$ as well as $\sum_{i=1}^{N} x_i$, one can estimate the variance and hence the standard deviation of A_N and thus determine confidence intervals for the result a_N. Note that since the standard deviation is the square root of the variance,

$$\sigma(A_N) = \frac{\sigma(X_i)}{\sqrt{N}},$$

and so the width of these confidence intervals decreases only like $1/\sqrt{N}$ as N increases. Monte Carlo methods are slow to converge (i.e., to get to a point where the results do not change appreciably with increasing N).

Let us consider again some of the questions raised in section 3.1.1. Is 3,2(u) really a worse starting hand than 7,2(u), as indicated in table 3.1, or are the probabilities of winning or losing with each hand so close that they cannot be distinguished with the 100, 000 experiments run? Suppose we start with 3,2(u). Let X_i be a random variable that takes the value 1 if we win the ith game, −1 if we lose the ith game, and 0 if we tie the ith game. Let Y_i be defined analogously when we start with 7,2(u). The precise numbers obtained in the experiments were:

3,2(u) 29039 wins 64854 losses 6107 ties wins−losses = −35815,

7,2(u) 31618 wins 62662 losses 5720 ties wins−losses = −31044.

We conclude that the expected value of X_i is the sample average $a_{X100,000} = -0.35815$, while that of Y_i is $a_{Y100,000} = -0.31044$. We do not know the variance of X_i and Y_i, but they can be estimated using the experimental data. Using formula (3.1) with computed sample averages,

$$\mathrm{var}(X_i) \approx \frac{1}{99999}\big[29039(1+0.35815)^2 + 64854(-1+0.35815)^2 \\ +6107(0+0.35815)^2\big] \approx 0.82,$$

$$\mathrm{var}(Y_i) \approx \frac{1}{99999}\big[31618(1+0.31044)^2 + 62662(-1+0.31044)^2 \\ +5720(0+0.31044)^2\big] \approx 0.85.$$

From this we conclude that the variances in the sample averages $a_{X100,000}$ and $a_{Y100,000}$ are about 0.82×10^{-5} and 0.85×10^{-5}, respectively, giving standard deviations of about 0.0029 in each case. The sample averages differ by about 0.048, or about 16 standard deviations. This is a significant amount! It is highly unlikely that the expected value of X_i is actually larger than that of Y_i; that is, that 3,2(u) is really a better starting hand than 7,2(u). Put another way, if we ran many, many sets of experiments, each time taking the average of the x_i's and y_i's over 100,000 games, the vast majority of those sets of experiments would yield a sample average for X_i that is smaller than that for Y_i.

Some of the other possible anomalies in table 3.1 are more difficult to verify. Is A,5(s) really a better starting hand that A,6(s)? Here the precise numbers obtained in the experiment were

A,5(s) 58226 wins 38147 losses 3627 ties wins−losses = 20079,

A,6(s) 57928 wins 38550 losses 3522 ties wins−losses = 19378.

Again, we can let X_i be a random variable that takes the value 1 if we win the ith game, −1 if we lose the ith game, and 0 if we tie the ith game, with starting hand A,5(s), and we can let Y_i be defined analogously for the starting hand A,6(s). In this case, however, the sample average for X_i is 0.20079 while that for Y_i is 0.19378. These differ by about 0.007, which is only about twice their estimated standard deviations of 0.003. Probably the ordering is correct, but to gain more confidence we would need to run a larger number of experiments.

Increasing the number of games by a factor of 10 to one million would lower the standard deviation by a factor of $\sqrt{10}$ to about 0.001, and this might be enough to make the determination.

Unfortunately, the results of one such experiment—in fact an experiment with 10 million games—were inconclusive. With 10 million games the variance of sample averages should be only about 0.0003. However, in this case the results were

A,5(s) 5806365 wins 3823200 losses 370435 ties wins−losses = 1983165,

A,6(s) 5814945 wins 3839292 losses 345763 ties wins−losses = 1975653.

The difference between sample averages is $0.1983165 - 0.1975653 = 0.0007512$, a difference that is still just slightly greater than two standard deviations. One can try averaging over still larger numbers of games, but this may require large amounts of computer time. Another approach is to combine computation with some analysis and simulate large numbers of games only in regimes where it is likely to reveal a significant difference between the two starting hands. This is the basic idea of **importance sampling**, which we will discuss further in connection with Monte Carlo integration.

3.3 MONTE CARLO INTEGRATION

While the previous discussion of Monte Carlo methods centered around probabilities, the probability of an event occurring can really be thought of as a ratio of volumes or integrals: the volume of the set of outcomes that constitute a success divided by the volume of the set of all possible outcomes. Thus the Monte Carlo method is really a technique for numerical integration. In chapter 10, we will discuss other techniques that are *far* more efficient for approximating integrals in a few dimensions—say, one or two or three—but for integrals over, say, ten or more dimensions, Monte Carlo methods may be the only practical approach. The error in the approximation still decreases at the rate $1/\sqrt{N}$, where N is the number of points sampled, but this may be quite good compared to deterministic methods that typically require at least a moderate number of points in each dimension: 10 integration points in each of 10 dimensions means a total of 10^{10} points!

We start with a very early example of using the Monte Carlo method to estimate π.

3.3.1 Buffon's Needle

In 1773, Georges-Louis Leclerc, Comte de Buffon, described an experiment that relates the real number π to a random event. The original formulation of the problem describes a game in which two gamblers drop a loaf of French bread on a wide-board floor. They bet on whether or not the loaf falls across a crack in the floor. Buffon posed the question as to what length of the loaf of bread, relative to the width of the floorboards, would lead to a fair game, and he answered this question in 1777. Buffon also considered the case of a

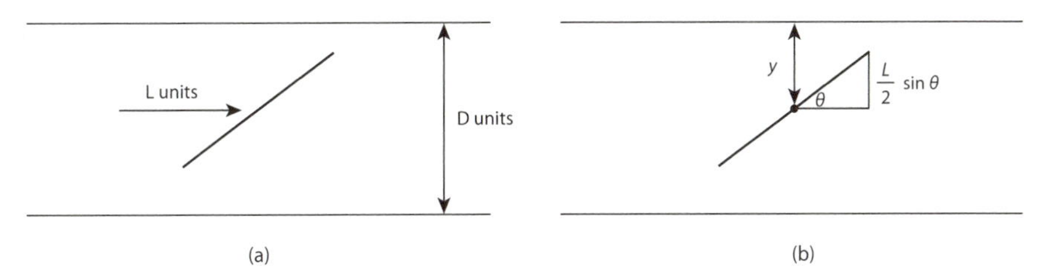

Figure 3.6. Buffon's needle experiment: (a) depicts the experiment where a needle of length L is randomly dropped between two lines a distance D apart. In (b), y denotes the distance between the needle's midpoint and the closest line; θ is the angle of the needle to the horizontal.

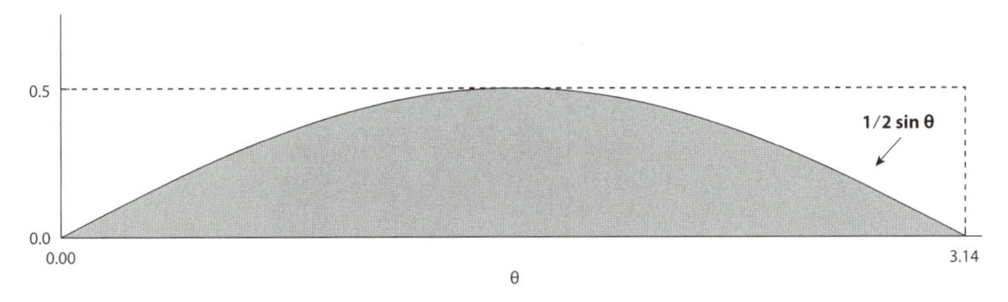

Figure 3.7. Plot of $\frac{1}{2}\sin\theta$ aids in the analysis of Buffon's needle.

checkerboard floor. While Buffon proposed the wrong answer for this second game, the correct answer was developed by Laplace [103, 104].

A simplified version of Buffon's game involves a straight needle of length L being dropped at random onto a plane ruled with straight lines a distance D apart as depicted in figure 3.6 (a). What is the probability p that the needle will intersect one of the lines? Buffon manually repeated the experiment to determine p. Such hand computation limited the feasibility of many mathematical experiments before the advent of the digital computer.

Buffon also derived the value of p mathematically. Let y denote the distance between the needle's midpoint and the closest line, and let θ be the angle of the needle to the horizontal. (Refer to figure 3.6 (b).) Here both y and θ are random variables, assumed to be uniformly distributed between 0 and $D/2$ and between 0 and π, respectively. The needle crosses a line if

$$y \leq \frac{L}{2}\sin\theta.$$

For simplicity, take $L = D = 1$. Consider the plot of $\frac{1}{2}\sin\theta$ shown in figure 3.7. The probability of the needle intersecting a line is the ratio of the area of the shaded region to the area of the rectangle. The area of the rectangle is clearly $\pi/2$. The area of the shaded region is

$$\int_0^\pi \frac{1}{2}\sin\theta \, d\theta = 1,$$

which implies that the probability of a hit is

$$p = \frac{1}{\left(\frac{\pi}{2}\right)} = \frac{2}{\pi}.$$

TABLE 3.2
Estimation of π in one simulation of Buffon's needle experiment from successively dropping a needle of length $L = 1$ between lines a distance $D = 1$ apart.

Total drops	Estimate	Error
10^2	3.12500	1.7×10^{-2}
10^3	3.28407	-1.4×10^{-1}
10^4	3.16857	-2.7×10^{-2}
10^5	3.14936	-7.8×10^{-3}
10^6	3.14112	4.7×10^{-4}
10^7	3.14028	1.3×10^{-3}

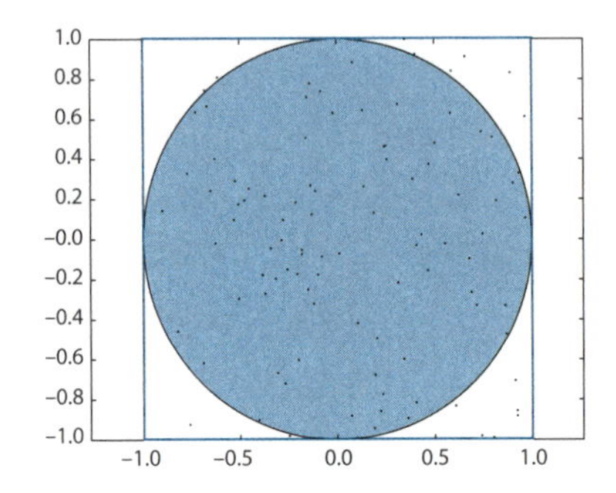

Figure 3.8. Random points dropped in a square of area 4. The fraction that land inside the circle of radius 1 is an estimate of $\pi/4$.

Hence, an estimate of π is

$$\pi \approx 2\left(\frac{\text{total number of drops}}{\text{total number of hits}}\right).$$

How many times must this experiment be performed to obtain a reasonable estimate of π? Ignoring the difficulties of distinguishing short line crossings from near misses, the answer is that the experiment must be performed *many, many* times before an accurate estimate of π emerges. Table 3.2 shows that after 10 million drops of the needle (simulated on a computer now), the estimate is 3.14028. The method displays the expected $O(1/\sqrt{N})$ convergence, although many jumps and dips in the error occur due to the random nature of the experiments.

3.3.2 Estimating π

Another simple way to estimate π using Monte Carlo methods on a computer—and one that easily extends to give estimates of areas or volumes of complicated regions—is as follows.

Generate points randomly in the square $[-1, 1] \times [-1, 1]$, which has area 4 (see figure 3.8). The area inside the circle of radius 1, which is enclosed by the square, is π. Therefore the fraction of points that land inside the circle gives an estimate of $\pi/4$, or, 4 times this fraction is an estimate of π. Following is a simple MATLAB code to do this.

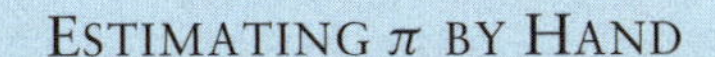

ESTIMATING π BY HAND

In 1864, American astronomer Asaph Hall (*left*), who discovered the two moons of Mars and calculated their orbits, encouraged his friend Captain O. C. Fox to conduct Buffon's needle experiment while he recovered in a military hospital from a severe wound in the American Civil War. Fox constructed a wooden surface ruled with equidistant parallel lines on which he threw a fine steel wire. His first experiments reportedly led to the estimate 3.1780 for π. Fox wanted to decrease any bias in his throws that might result from his position or how he might hold the rod over the surface. So, he added a slight rotatory motion before dropping the rod which reportedly led to improved estimates of 3.1423 and 3.1416. Each experiment involved approximately 500 throws [49].

(Given the small number of throws and the limitations in determining near crossings, it is unlikely that he actually would find such an accurate estimate of π. In fact, in his last, most accurate experiment, the following can be deduced: If he had observed one fewer crossing, his estimate of π would have been 3.1450 (a considerably less accurate approximation), and if he had observed one more crossing his estimate of π would have been the even less accurate number 3.1383. Perhaps there was bias, not in his throwing, but in the conduct or reporting of his experiment.) (Photo: Professor Asaph Hall, Sr. taken at Equatorial Building, August 1899. Courtesy of the Navel Meteorology and Oceanography Command.)

```
N = input('Enter number of points: ');

numberin = 0;          % Count number inside circle.
for i=1:N,             % Loop over points.
  x = 2*rand - 1;      % Random point in the square.
  y = 2*rand - 1;
  if x^2 + y^2 < 1,    % See if point is inside the circle.
    numberin = numberin + 1;    % If so, increment counter.
  end;
end;
pio4 = numberin/N;     % Approximation to pi/4.
piapprox = 4*pio4      % Approximation to pi.
```

As always, while this code returns an estimate of π, we would like to know how good an estimate it is. Of course, since we actually know the value of π, we could compare it to this value (which can be generated in MATLAB by simply typing `pi`), but this is cheating since the idea is to develop a method for approximating areas that we do not already know!

The above code can be thought of as taking an average of the iid random variables

$$X_i = \begin{cases} 1 & \text{if point } i \text{ is inside the circle,} \\ 0 & \text{otherwise.} \end{cases}$$

The mean of X_i is $\pi/4$. Since X_i takes on only the values 0 and 1, the random variable X_i^2 is equal to X_i. Hence the variance of X_i is $\text{var}(X_i) = E(X_i^2) - (E(X_i))^2 = (\pi/4) - (\pi/4)^2$. But, once again, we are assuming that we do not know π, so how does this help? The answer is that we can use the approximated value `pio4` to estimate the variance of X_i. Using formula (3.1), or, what is almost the same (except with a factor $1/N$ in place of $1/(N-1)$ in (3.1)), the expression `pio4 - pio4^2`, as an estimate of the variance of each X_i, we can then use the Central Limit Theorem to determine confidence intervals for our approximation `pio4` to $\pi/4$, and from this we can determine confidence intervals for our approximation `piapprox` to π. Remember, however, that the variance of $4X_i$ is $4^2 = 16$ times the variance of X_i. Thus, if we wish to add some MATLAB code to estimate the standard deviation in our approximation to π, we might add the lines

```
varpio4 = (pio4 - pio4^2)/N;
% Variance in approximation to pi/4.
varpi = 16*varpio4;            % Variance in approximation to pi.
                               %  Note factor of 16.
stdpi = sqrt(varpi)            % Estimated standard deviation in
                               %  approximation to pi.
```

3.3.3 Another Example of Monte Carlo Integration

In figure 3.9, we see a three-dimensional region that is described by the inequalities

$$xyz \leq 1,$$

and

$$-5 \leq x \leq 5, \quad -5 \leq y \leq 5, \quad -5 \leq z \leq 5.$$

Suppose we wish to find the mass of this object,

$$\iiint_{\text{volume}} \gamma(x, y, z)\, dx\, dy\, dz,$$

where $\gamma(x, y, z)$ is a given density. Such an integral would be difficult or impossible to compute analytically, but it is easily approximated using Monte Carlo integration.

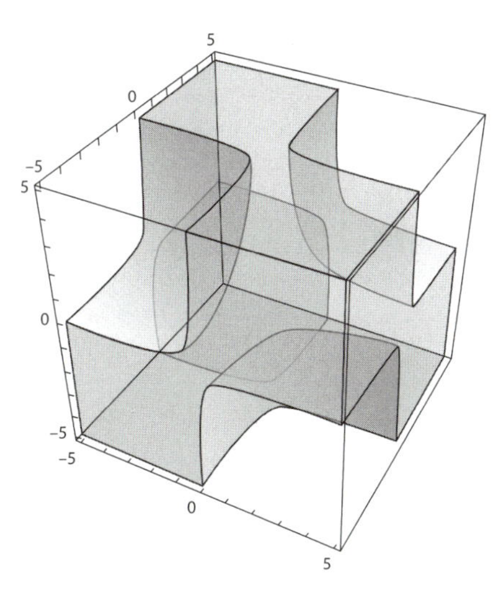

Figure 3.9. A region whose volume and mass will be computed using Monte Carlo integration. The limits of integration are not easily written in closed form.

Suppose the density function is $\gamma(x, y, z) = e^{0.5z}$. The MATLAB code in figure 3.10 estimates the volume and mass of the object, along with the standard deviation of the random variables that approximate these quantities.

If you run this code, you will find that the standard deviation in the estimate of mass is significantly higher than that in the estimate of volume—using 100,000 sample points, we obtained an estimate of about 7.8 for the standard deviation in mass compared to 1.6 for volume. To see why, look at figure 3.11, which is colored according to the density function $\gamma(x, y, z) = e^{0.5z}$. The density increases exponentially in the direction of positive z, so that most of the mass of the object comes from points with high z values; those with low z values contribute little to the calculated mass and so might be sampled less frequently without significantly affecting the computed result. Put another way, one might make a change of variable, say, $u = e^{0.5z}$, so that u ranges between $e^{-2.5} \approx 0.08$ and $e^{+2.5} \approx 12.2$ as z ranges between -5 and 5, and a uniform distribution in u gives points that mostly correspond to positive z values. With this change of variable, since $du = 0.5e^{0.5z}\,dz$, the mass integral becomes

$$\int_{-5}^{5}\int_{-5}^{5}\int_{-5}^{5} \begin{cases} 0 & \text{if } xyz > 1 \\ e^{0.5z} & \text{if } xyz \le 1 \end{cases} dx\,dy\,dz =$$

$$2\int_{e^{-2.5}}^{e^{+2.5}}\int_{-5}^{5}\int_{-5}^{5} \begin{cases} 0 & \text{if } 2xy\ln(u) > 1 \\ 1 & \text{if } 2xy\ln(u) \le 1 \end{cases} dx\,dy\,du.$$

It is an exercise to modify the code in figure 3.10 to use this change of variable and again estimate the mass of the object and the standard deviation of this estimate. Using 100,000 sample points, we computed a standard deviation of about 3.8 using the new formulation, compared to 7.8 with the previous method. Recall that halving the standard deviation in a Monte Carlo calculation normally requires *quadrupling* the number of sample points (and

```
N = input(' Enter number of sample points: ');
gamma = inline('exp(0.5*z)');      % Here gamma is a function of z only.

volumeOfBox = 10*10*10;           % Volume of surrounding box.

vol = 0;  mass = 0;               % Initialize volume and mass of object.
volsq = 0;  masssq = 0;           % Initialize their squares.
                                  % (to be used in computing variance and
                                  %  standard deviation).

for i=1:N,                        % Loop over sample points.
  x = -5 + 10*rand;               % Generate a point from surrounding box.
  y = -5 + 10*rand;
  z = -5 + 10*rand;
  if x*y*z <= 1,                  % Check if point is inside object.
    vol = vol + 1;                %    If so, add to vol and mass.
    mass = mass + gamma(z);
    volsq = volsq + 1;            %    Also add to square of vol and mass.
    masssq = masssq + gamma(z)^2;
  end;
end;

volumeOfObject = (vol/N)*volumeOfBox   % Fraction of pts inside times vol of box
volvar = (1/N)*((volsq/N) - (vol/N)^2)*volumeOfBox^2;  % Variance in vol of object:
                                       % (Expected value of volsq - square of
                                       % expected value of vol) divided by N and
                                       % multiplied by square of factor volumeOfBox.
volstd = sqrt(volvar)                  % Standard deviation in volume of object.

massOfObject = (mass/N)*volumeOfBox    % Average mass times volume of box
massvar = (1/N)*((masssq/N) - (mass/N)^2)*volumeOfBox^2; % Variance in mass:
                                       % (Expected value of masssq - square of
                                       % expected value of mass) divided by N and
                                       % multiplied by square of factor volumeOfBox.
massstd = sqrt(massvar)                % Standard deviation in mass of object.
```

Figure 3.10. MATLAB code to calculate the volume and mass of the object in figure 3.9 using Monte Carlo integration.

hence quadrupling the computation time); here a clever change of variable led to the same error reduction with no increase in computation time.

While the Monte Carlo method is very well suited to computing the volume or mass of a complicated object such as that in figure 3.9, it also can be used for more general multidimensional integrals. For example, to estimate

$$\int_{a_1}^{b_1} \int_{a_2}^{b_2} \cdots \int_{a_m}^{b_m} f(x_1, x_2, \ldots, x_m)\, dx_m \ldots dx_2\, dx_1,$$

where the endpoints of integration of the inner integrals may depend on the values of the outer variables (e.g., $a_2 \equiv a_2(x_1)$ or $b_m \equiv b_m(x_1, \ldots, x_{m-1})$), one can generate random m-tuples $(x_1, x_2, \ldots, x_m)$ uniformly distributed throughout a volume containing the region of integration. For example, if $A_j \le \min_{x_1,\ldots,x_{j-1}} a_j(x_1, \ldots, x_{j-1})$ and $B_j \ge \max_{x_1,\ldots,x_{j-1}} b_j(x_1, \ldots, x_{j-1})$, then the cross-product region $[A_1, B_1] \times \ldots \times [A_m, B_m]$ contains the region of integration, so one might generate points uniformly distributed throughout this

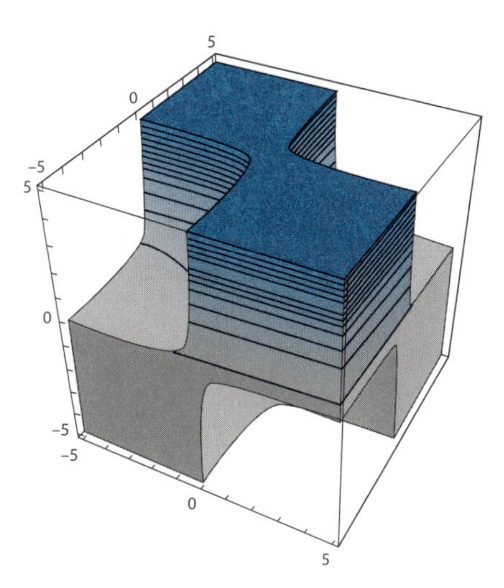

Figure 3.11. The same region as in figure 3.9 but colored by its density $\gamma = e^{0.5z}$.

volume. If a point lands inside the region of integration then it is assigned the value of f at that point, while if it lands outside then it is assigned the value 0. The average of these values times the volume $(B_1 - A_1) \cdots (B_m - A_m)$ is then an approximation to the desired integral, since the process can be thought of as approximating the integral over the entire volume of a function g that is equal to f inside the region of integration and 0 outside. To approximate this integral, one sums sample values of g and multiplies by the volume increment $(B_1 - A_1) \cdots (B_m - A_m)/(\#\text{ samples})$.

Example 3.3.1. Use the Monte Carlo method to estimate

$$\int_0^1 \left[\int_x^1 \left(\int_{xy}^2 \cos(xy\, e^z)\, dz \right) dy \right] dx.$$

The following MATLAB code generates random points in the region $[0, 1] \times [0, 1] \times [0, 2]$, checks to see if the point is within the region of integration and, if so, adds $\cos(xy\, e^z)$ to the variable `int`. It then divides by the number of points and multiplies by the volume of the region from which the points were drawn ($(1 - 0)(1 - 0)(2 - 0) = 2$) to obtain an approximation to the integral.

```
f = inline('cos(x*y*exp(z))');      % Define integrand.

N = input(' Enter number of sample points: ');
int = 0;                            % Initialize integral.
for i=1:N,                          % Loop over sample points.
% Generate a point from surrounding box.
  x = rand; y = rand; z = 2*rand;
  if x <= y & x*y <= z,   % Check if point is inside region.
```

```
    int = int + f(x,y,z);          % If so, add to int.
  end;
end;
int = (int/N)*2                    % Approximation to integral.
```

Three runs with $N = 10,000$ points yielded the estimates 0.5201, 0.5275, and 0.5212, while a run with $N = 1,000,000$ points (for which the standard deviation is one tenth of that with $N = 10,000$ points) yielded an estimate of 0.5206.

3.4 MONTE CARLO SIMULATION OF WEB SURFING

When you submit a query to a search engine, it returns an ordered list of web pages. There are various factors that determine the ordering, one of which is the web page's relevance to your query. A technique to measure this was presented in section 1.5. Another important component is the "quality" or "importance" of the page. An algorithm called PageRank, introduced by Google's founders Larry Page and Sergey Brin, supplies a measure of such quality through a model of web-surfing behavior.

The algorithm presented by Brin and Page, in their original formulation of PageRank, assumes that 85% of the time a surfer follows a link available on the web page that is currently being visited. The other 15% of the time the surfer randomly visits any web page available in the network. If a surfer visits

CATCHING RAYS WITH MONTE CARLO

Raytracing, a method of producing photorealistic digital images, can be done with Monte Carlo integration. For example, the Monte Carlo method is used to compute the lighting at a point from an area light source. As the number of samples at a particular pixel becomes large, the shadows increase in smoothness. This image was rendered using the POV-Ray freeware available at http://www.povray.org or http://mac.povray.org. (Image courtesy of Taylor McLemore)

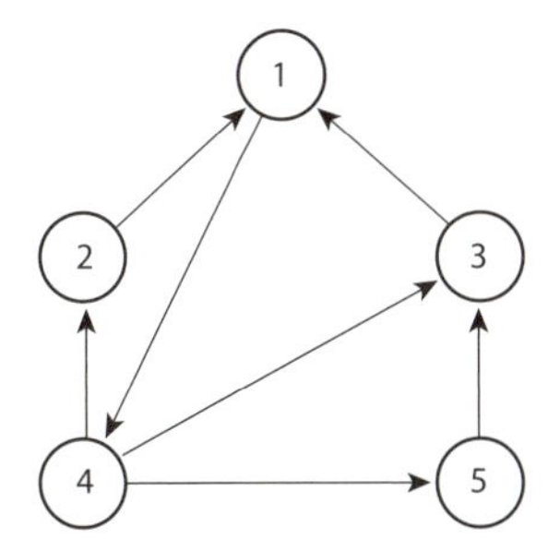

Figure 3.12. A small network of web pages.

TABLE 3.3
Computed probabilities from up to 5000 simulations of a surfer's movements on the network in figure 3.12.

Simulations	Page 1	Page 2	Page 3	Page 4	Page 5
1000	.297	.106	.191	.295	.111
2000	.294	.106	.197	.288	.115
3000	.294	.104	.201	.284	.117
4000	.294	.107	.201	.284	.113
5000	.294	.106	.203	.281	.116

a web page with no links on it, then at the next step the surfer is assumed to randomly visit any web page available in the network (with the likelihood of each being equal).

Let's look at the small network depicted in figure 3.12. If the surfer is at web page 1, then 85% of the time, the surfer will follow the link to web page 4. The other 15% of the time, the surfer will input with equal likelihood the web address for a random web page in the network. Thus, there is a $0.85 \cdot 1 + 0.15 \cdot (1/5) = 0.88$ chance that a surfer will go from web page 1 to web page 4, since the only outgoing link on page 1 is to page 4, while there is a $0.15 \cdot (1/5) = 0.03$ chance that the surfer will go to each of the other web pages, 1, 2, 3, and 5.

Using this model one might ask, over a long period of time, what proportion of the time will a surfer spend visiting web page 1? Answering this question for every web page enables us to determine the most visited or "popular" web page. This proportion of time or the probability of visiting a web page is the PageRank of the web page. Google uses such a concept in its rankings of popularity or importance of web pages. In section 12.1.5, we will discuss this model more fully.

For now, we will determine these probabilities using Monte Carlo simulation. The MATLAB code in figure 3.13 simulates such a surfing session for the network in figure 3.12.

A run of this program produced the results in table 3.3. The simulation appears to have achieved sufficient accuracy for us to rank the pages, but one might want to check this by estimating the variance or running more simulations. We know by other means to be described later in the text that

```
numsims = 5000;  % Parameter controlling the number of simulations

% Connectivity matrix of the network
G = [0 0 0 1 0; 1 0 0 0 0; 1 0 0 0 0; 0 1 1 0 1; 0 0 1 0 0];
n = length(G);  % size of network

state = 1;       % initial state of the system

p = 0.85;        % probability of following link on web page
M = zeros(n,n); % create the matrix of probabilities
for i = 1:n;
    for j = 1:n;
        M(i,j) = p*G(i,j)/sum(G(i,:)) + (1-p)/n;
    end
end

pages = zeros(1,n);   % vector to hold number of times page visited

fprintf('\n\n');
fprintf('=====================================================\n');
fprintf('Computed Probabilities from %d simulations\n\n',numsims);
fprintf('Simulations  Page 1  Page 2  Page 3  Page 4  Page 5 \n');
fprintf('=====================================================\n');

% Simulate a surfer's session
for i=1:numsims
    prob = rand;
    if prob < M(state,1)
        state = 1;
    elseif prob < M(state,1) + M(state,2)
        state = 2;
    elseif prob < M(state,1) + M(state,2) + M(state,3)
        state = 3;
    elseif prob < M(state,1) + M(state,2) + M(state,3) + M(state,4)
        state = 4;
    else state = 5;
    end;

    pages(state) = pages(state) + 1;

    if ~mod(i,numsims/5)
       fprintf('%6d  %5.3f  %5.3f  %5.3f  %5.3f  %5.3f \n',
                 i,pages/sum(pages));
    end
end
```

Figure 3.13. MATLAB code to simulate surfing over the network in figure 3.12.

the relative ranking would not change. (Actually, pages 2 and 5 would be tied as the least popular web pages in the network since they are each pointed to only by web page 4.)

From these results, we see that a surfer is expected to visit web page 1 approximately 30% of the time, and this makes web page 1 the most popular web page. Imagine a larger network with, say, 1 billion web pages or even more, as is indexed by Google for its search engine. In this case, one would expect *far* more simulations to be needed to attain suitable accuracy. In section 12.1.5 we will see an alternative approach to finding these probabilities that involves Markov chains (as opposed to Monte Carlo simulation).

3.5 CHAPTER 3 EXERCISES

1. Suppose you have a pseudorandom number generator (called `rand`) that generates random numbers uniformly distributed between 0 and 1. How could you use it to generate a discrete random variable that takes on the values 1 through 6, each with equal probability?
2. Write a MATLAB code to simulate rolling of a fair die; that is, it should generate a discrete random variable that takes on the values 1 through 6, each with probability 1/6. Run the code 1000 times and plot a bar graph showing the number of 1s, 2s, 3s, etc. Your code might look something like the following:

```
bins = zeros(6,1);
for count=1:1000,
%   Generate a random no. k with value 1,2,3,4,5, or 6,
%   each with prob. 1/6.
%   (You fill in this section.)

% Count the times each number is generated.
  bins(k) = bins(k) + 1;
end;
bar(bins)                           % Plot the bins.
```

Turn in your plot and a listing of your code. Determine the expected value and the variance of the random variable `k`.

Now let A_{1000} be the average value generated. Run the code 100 times, each time recording the average value A_{1000}. For example, you might modify the above code as follows:

```
A = zeros(100,1);                  % Initialize average values.
for n_tries=1:100,                 % Try the experiment 100 times.

  for count=1:1000,
%   Generate a random no. k with value 1,2,3,4,5, or 6,
%   each with prob. 1/6.
%   (You fill in this section.)
```

```
    A(n_tries) = A(n_tries) + k; % Add each value k.
  end;
  A(n_tries) = A(n_tries)/1000;
% Divide by 1000 to get the average of k.

end;
hist(A)
% This plots a histogram of average values.
```

Turn in your histogram and a listing of your code. How does the random variable A_{1000} appear to be distributed (uniform, normal, or some other distribution)? Approximately what is its mean, variance, and standard deviation? [Hint: Use the Central Limit Theorem.]

3. Recall the Monty Hall problem described in section 3.2.1. Marilyn vos Savant called upon "math classes all across the country" to estimate the probabilities of winning with and without switching doors, using pennies and paper cups. Write a Monte Carlo simulation to verify these probabilities.
4. In his online article "Devlin's Angle", Keith Devlin poses the following variation on the Monty Hall problem [34].

   > Suppose you are playing a seven door version of the game. You choose three doors. Monty now opens three of the remaining doors to show you that there is no prize behind any of them. He then says, "Would you like to stick with the three doors you have chosen, or would you prefer to swap them for the one other door I have not opened?" What do you do? Do you stick with your three doors or do you make the 3-for-1 swap he is offering?

   Write a Monte Carlo simulation to find the probability of winning if you stick with your original 3 doors and the probability if you switch to the 1 remaining door. Also see if you can derive these probabilities analytically.
5. Monty Hall problems have also been posed in other settings. For instance, suppose either Alice or Betty is equally likely to be in the shower. Then you hear the showerer singing. You know that Alice always sings in the shower, while Betty only sings 1/4 of the time. What is the probability that Alice is in the shower? Write a Monte Carlo simulation to answer this question.
6. Consider the simplest form of the game of craps. In this game, we roll a pair of dice. If the sum is 7 or 11 on the first throw, then we win right away, while if the sum is 2, 3, or 12, then we lose right away. With any other total (i.e., 4, 5, 6, 8, 9, or 10), we continue rolling the dice until we get either a 7 (in which case we lose), or the same total rolled on the first throw (in which case we win). Assuming that the dice are fair, write a Monte Carlo simulation to determine the odds of winning.
7. How many people do we need in a room so that the probability that two people share the same birthday is at least 50%? While this probability can be computed analytically, write a Monte Carlo simulation to answer

the question. Ignore people with birthdays on February 29, and assume that birthdays are uniformly distributed over the 365 days in a year. Let the number of people m be an input to your code, and for each value m that you try, run, say, 10,000 experiments in which you assign random birthdays to the people. Then determine the fraction of these experiments in which two people have the same birthday. Vary m until this fraction is just over 0.5.

8. Suppose you have a pseudorandom number generator, such as `rand` in MATLAB, that generates random numbers uniformly distributed between 0 and 1. You wish to generate random numbers from a distribution X, where

$$\text{Prob}(X \leq a) = \sqrt{a}, \quad 0 \leq a \leq 1.$$

How could you use the uniform random number generator to obtain random numbers with this distribution?

9. In Buffon's needle experiment, derive a formula that estimates π when you use a needle of length L and parallel lines a distance D apart, where $L \leq D$. [Note: The approximation for $L > D$ is more complicated and involves arcsin's.]

10. Estimate the area inside the ellipse

$$\frac{x^2}{4} + y^2 = 1$$

by generating random points uniformly distributed in the rectangle $[-2, 2] \times [-1, 1]$, and seeing what fraction lie inside the ellipse. The fraction that fall inside the ellipse times the area of the rectangle gives an estimate of the area inside the ellipse.

(a) First use MATLAB to plot the ellipse. Type `hold on` afterwards to keep the ellipse on the screen.

(b) Next generate 1000 random points (x_i, y_i), where x_i comes from a uniform distribution between -2 and 2 and y_i comes from a uniform distribution between -1 and 1. Count the number of points that land inside the ellipse. Plot each of these points on the same graph with the ellipse.

(c) Determine the fraction of points that landed inside the ellipse and use this to estimate the area inside the ellipse.

(d) Let X_i be a random variable that is 1 if the ith point lies inside the ellipse and 0 otherwise. Based on your results, estimate the mean and variance of X_i.

(e) Let $A_{1000} = \frac{1}{1000} \sum_{i=1}^{1000} X_i$. Using the Central Limit Theorem, estimate the mean, variance, and standard deviation of A_{1000}.

(f) Using the results of the previous part, estimate the mean, variance, and standard deviation of $8A_{1000}$, which is the approximation to the area of the ellipse. Based on this result, how much confidence do you have in your answer? (Is it likely to be off by, say, 0.001, 0.01, 0.1, 1, 10, or some other amount?)

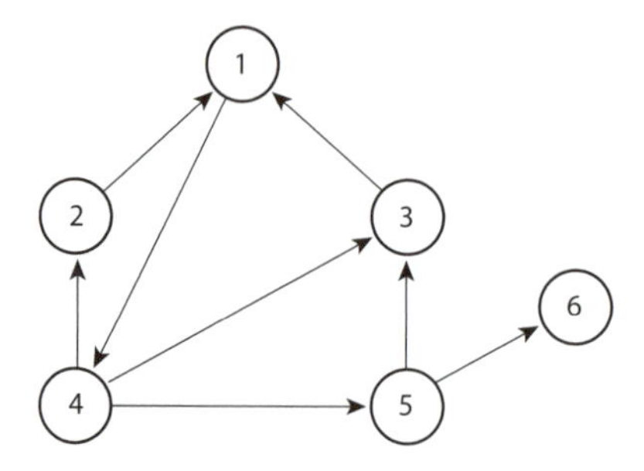

Figure 3.14. A network of web pages.

11. Modify the code in figure 3.10 to use the change of variable $u = e^{0.5z}$ and estimate the mass of the object and its standard deviation.
12. The following 10-dimensional integral can be determined analytically:

$$\int_0^2 \int_0^2 \cdots \int_0^2 x_1 x_2 \cdots x_{10}\, dx_1\, dx_2 \cdots dx_{10}.$$

   (a) Determine the value of this integral analytically.
   (b) Write a Monte Carlo code to estimate the value of the integral and the standard deviation in your approximation.
   (c) Run the code with $N = 10,000$, $N = 40,000$, and $N = 160,000$ points and find the number of standard deviations by which your computed result differs from the true answer found in part (a).

13. Edit the MATLAB code in figure 3.13 to rank the web pages in figure 3.14.

4

SOLUTION OF A SINGLE NONLINEAR EQUATION IN ONE UNKNOWN

The previous chapter dealt with Monte Carlo methods, a widely used class of methods for simulating physical phenomena. Monte Carlo methods involve probabilistic modeling and the main tool for their analysis is *statistics*. The remainder of this text deals with deterministic methods and more traditionally falls under the heading of *numerical analysis*, although such a division is somewhat artificial; both types of approach are important in computational science. We start with a very basic problem of solving a single nonlinear equation in one unknown. Even this most basic problem requires consideration of a number of issues that have not yet been addressed: Does a solution exist? If so, is it unique? If there is more than one solution, is one particular solution to be preferred to others because it is physically correct or more desirable? The same questions of accuracy that arise in Monte Carlo methods will arise in deterministic computations, although the element of randomness is gone. What constitutes an acceptably good approximate solution to a nonlinear equation? And given an algorithm for creating such an approximation, how does the computation time grow with the level of accuracy achieved?

Following are a few sample applications that require the solution of a single nonlinear equation in one unknown:

1. **Projectile motion.** An artillery officer wishes to fire his cannon on an enemy brigade camped d meters away. At what angle θ to the horizontal should he aim the cannon in order to strike the target?

   He knows that the cannonball will leave the muzzle of the gun at an initial speed of v_0 m/s. Neglecting air resistance, the only force acting on the cannonball is then gravity, which imparts a downward acceleration of $g \approx 9.8$ m/s^2. Hence the height $y(t)$ of the cannonball at any time $t > 0$ satisfies the differential equation

$$y''(t) = -g.$$

Assuming that the initial height is $y(0) = 0$ (an approximation, since the cannon barrel is not exactly at ground level) and the initial velocity in the vertical direction is $v_0 \sin\theta$, this equation uniquely determines the height at any time $t > 0$.

One can solve this differential equation by integrating

$$y'(t) = -gt + c_1,$$

where $c_1 = v_0 \sin\theta$, since $y'(0) = v_0 \sin\theta$. Integrating again gives

$$y(t) = -\frac{1}{2}gt^2 + v_0 \sin\theta \, t + c_2,$$

where $c_2 = 0$ since $y(0) = 0$.

Now that we have a formula for the height at time $t > 0$, we can determine the amount of time before the cannonball hits the ground:

$$0 = -\frac{1}{2}gt^2 + v_0 \sin\theta \, t \quad \Rightarrow \quad t = 0 \ \text{ or } \ t = \frac{2v_0 \sin\theta}{g}.$$

Finally, neglecting air resistance, the cannonball travels at a constant horizontal speed $v_0 \cos\theta$, so that it hits the ground at a distance

$$\frac{2v_0^2 \sin\theta \cos\theta}{g}$$

meters from where it was fired. The problem thus becomes one of solving the nonlinear equation

$$\frac{2v_0^2 \sin\theta \cos\theta}{g} = d \tag{4.1}$$

for the angle θ at which to aim the cannon.

While one could solve this problem numerically, perhaps using an available software package for nonlinear equations, there are several things that one should take into account before returning to the gunner with a definitive value for θ.

(a) The model developed above is only an approximation. Consequently, even if one returns with the exact solution θ to (4.1), the fired cannon may end up missing the target. A strong tailwind might cause the shot to sail over the encamped enemy. Or the failure to account for air resistance might make the shot fall short. Such factors may make adjustments to the model necessary.

(b) Equation (4.1) may not have a solution, and this is important to know. The maximum value of $(\sin\theta \cos\theta)$ occurs at $\theta = \pi/4$, where $(\sin\theta \cos\theta) = \frac{1}{2}$. If $d > \frac{v_0^2}{g}$, then the problem has no solution, and the target is simply out of range.

(c) If a solution does exist, it may not be unique. If θ solves problem (4.1), then so does $\theta \pm 2k\pi$, $k = 1, 2, \ldots$. Also $\frac{\pi}{2} - \theta$ solves the problem. Either θ or $\frac{\pi}{2} - \theta$ might be preferable as a solution, depending on what lies in front of the cannon!

(d) Actually this problem can be solved analytically. Using the trigonometric identity $2\sin\theta\cos\theta = \sin(2\theta)$, one finds

$$\theta = \frac{1}{2}\arcsin\frac{dg}{v_0^2}.$$

In light of item (a) above, however, it may be prudent to go ahead and write a code to handle the problem numerically, since more realistic models may require approximate solution by numerical methods.

For example, one might model the effects of air resistance by replacing the above differential equation for $y(t)$ by the equation

$$y''(t) = -g - ky'(t), \tag{4.2}$$

where we now assume that air resistance is proportional to the velocity of the object and that the constant of proportionality is $k > 0$. Similarly, $x(t)$ would satisfy the equation $x''(t) = -kx'(t)$. The equation for $y(t)$ can again be solved analytically:

$$y'(t) = e^{-kt}v_0\sin\theta - \frac{g}{k}(1 - e^{-kt}),$$

$$y(t) = -\frac{1}{k}e^{-kt}v_0\sin\theta - \frac{g}{k}\left(t + \frac{1}{k}e^{-kt}\right) + c_3, \tag{4.3}$$

where $c_3 = (1/k)v_0\sin\theta + g/k^2$, so that $y(0) = 0$. Now, however, one cannot write down an analytical formula for the positive value of t that satisfies $y(t) = 0$. Hence one cannot analytically determine the value of x at this time and use this to find the appropriate angle θ. With even more realistic models it may be impossible to even write down a formula for $y(t)$, let alone find t such that $y(t) = 0$. Still, for a given angle θ, one should be able to approximate $y(t)$ and $x(t)$ numerically, and vary θ until the time at which $y = 0$ coincides with that at which $x = d$.

(e) To solve a nonlinear equation numerically, we usually start by writing it in the form $f(\theta) = 0$. Equation (4.1) can be written as

$$f(\theta) \equiv \frac{2v_0^2\sin\theta\cos\theta}{g} - d = 0.$$

This function is simple enough that it can be differentiated analytically. We will see in the following sections that the root(s) of such functions can be found using Newton's method. For the more complicated models where $f(\theta)$ can only be approximated numerically, more appropriate root finders include the bisection method, the secant method, and quasi-Newton methods, to be described in this chapter.

2. **Inverse problems.** A frequent numerical problem is to solve a differential equation that models some physical phenomenon, using a given set of parameters. For example, in the model (4.2), one might solve for $y(t)$ given k, as was done above, and look for the time $T > 0$ at which

$y(T) = 0$. Alternatively, if the parameter k is unknown, then one might run an experiment to see how long it takes the cannonball to fall to the ground, when aimed at a given angle θ. Using this value in (4.3) leads to a nonlinear equation for k:

$$0 \equiv y(T) = -\frac{1}{k} e^{-kT} v_0 \sin\theta - \frac{g}{k}\left(T + \frac{1}{k} e^{-kT}\right) + c_3.$$

Problems such as this are sometimes called **inverse problems**. They arise frequently in medical applications and imaging science, where doctors and researchers try to reconstruct an image of something inside the human body from minimally invasive and nondestructive measurements. Using X-ray tomography or magnetic resonance imaging, for example, a known source is applied and measurements are made of what comes out in order to deduce properties of the material inside.

3. **Financial mathematics.** Nonlinear equations frequently arise in problems involving loans, interest rates, and savings. Often these can be solved analytically, but sometimes this is not the case.

   Suppose you borrow \$100,000 at interest rate $R\%$ per year to buy a house. Each month you make a payment of p dollars, and the loan is to be paid off in 30 years. Given the interest rate R, one can determine the monthly payment p. Alternatively, you might have decided that you are willing to pay a certain amount p each month and you might ask at what interest rate you can afford the loan. Let $r = R/1200$ denote the fraction of interest accrued to the loan monthly (so that if $R = 6\%$, then $r = 0.06/12 = 0.005$, for example). To derive the equations for this problem, note that after one month the amount you will owe is

$$A_1 = 100{,}000(1 + r) - p,$$

   since interest has brought the amount owed up to $100{,}000(1 + r)$, but you have now paid p. After two months, this becomes

$$A_2 = A_1(1 + r) - p = 100{,}000(1 + r)^2 - p[1 + (1 + r)].$$

   Continuing in this way, we find that after 360 months (30 years), the amount owed is

$$\begin{aligned} A_{360} &= 100{,}000(1 + r)^{360} - p\left[1 + (1 + r) + (1 + r)^2 + \cdots + (1 + r)^{359}\right] \\ &= 100{,}000(1 + r)^{360} - p\frac{(1 + r)^{360} - 1}{r}. \end{aligned}$$

   Setting A_{360} to 0, one can easily solve for p in terms of r:

$$p = 100{,}000\, r \frac{(1 + r)^{360}}{(1 + r)^{360} - 1},$$

   but given p, one can only approximate the solution r to this equation.

4. **Computer graphics.** The problem of hidden lines and surfaces is important in rendering computer graphics. These problems involve finding intersections of objects in two or three dimensions and usually require solving systems of nonlinear equations. As a simple example, suppose we

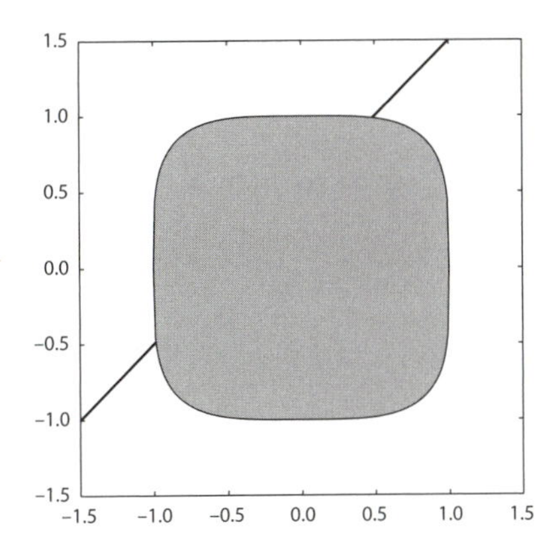

Figure 4.1. Rendering in computer graphics can involve detecting when a line disappears behind an object.

wish to plot an object defined by the inequality

$$x^4 + y^4 \leq 1$$

in the plane, and we wish to show a straight line, say, $y = x + 0.5$, going behind the object, as pictured in figure 4.1.

To determine what piece of the line to show, one must determine where it intersects the object. The line hits the boundary of the object at points where x satisfies

$$x^4 + (x + 0.5)^4 = 1.$$

This quartic equation has four solutions, but only two of them are real and correspond to actual intersection points. Once the x-coordinates of the intersection points are known, the y-coordinates are easily computed as $x + 0.5$, but to determine the x values requires solving the above polynomial equation. Actually the roots of a quartic (but *not* necessarily a quintic) polynomial can be written down using radicals, but the formula is complicated. It is probably easier to use a more general polynomial equation solver to find the relevant roots.

4.1 BISECTION

How would one solve some of the nonlinear equations of the previous section? Long before Galileo explained projectile motion through the law of inertia, artillery men had learned through trial and error how to aim their cannons. If a shot fell too short, they adjusted the angle; if the next shot was too long, they moved to an angle somewhere in between the two previous ones, perhaps the average of the two previous angles, to try and hone in on the target. Today, these adjustments are done by computer, but the basic idea is the same. This simple procedure is the method of **bisection**, and it can be used to solve any equation $f(x) = 0$, provided f is a continuous function of x and we have starting values x_1 and x_2 where f has different signs.

Bisecting a Phone Book

Have a friend choose a name at random from your local phone book. Bet your friend that in less than thirty yes-no questions, you will be able to guess the name that was chosen. This task is simple. Place about half of the pages in your left hand (or right if you prefer) and ask your friend, "Is the name in this half of the phone book?" In one question, you have eliminated approximately half of the names. Holding half of the noneliminated pages in your hand, ask the question again. Repeat this process until you have discovered the page which contains the name. Continue to ask questions which halve the number of possible names until you have discovered the name. Like the bisection algorithm, you use the speed of exponential decay (halving the number of remaining names with each question) to your advantage. Since $(1/2)^{30}$ is about 9.3×10^{-10} and the population of the United States is approximately 280 million, you can see that any city phone book will suffice (and should leave you a comfortable margin for error).

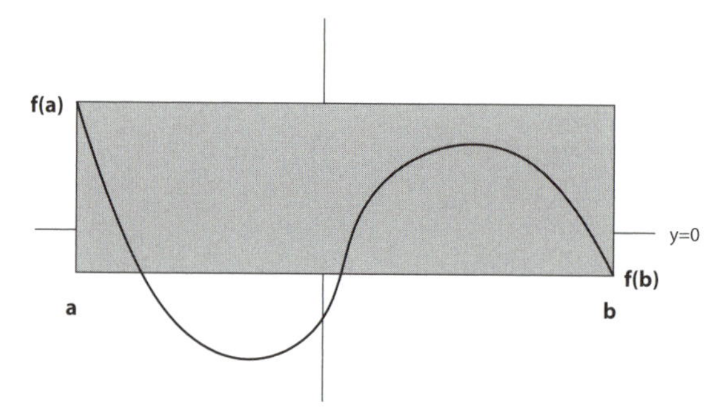

Figure 4.2. The bisection method forms a rectangular region which must contain a root of the continuous function.

The basis for the bisection method is the Intermediate Value Theorem:

Theorem 4.1.1 (Intermediate Value Theorem). If f is continuous on $[a, b]$ and y lies between $f(a)$ and $f(b)$, then there is a point $x \in [a, b]$ where $f(x) = y$.

If one of $f(a)$ and $f(b)$ is positive and the other negative, then there is a solution to $f(x) = 0$ in $[a, b]$. In the bisection algorithm, we start with an interval $[a, b]$ where $\text{sign}(f(a)) \neq \text{sign}(f(b))$. We then compute the midpoint $\frac{a+b}{2}$ and evaluate $f\left(\frac{a+b}{2}\right)$. We then replace the endpoint (a or b) where f has the same sign as $f\left(\frac{a+b}{2}\right)$ by $\frac{a+b}{2}$, and repeat.

Following is a simple MATLAB code that implements the bisection algorithm.

```
% This routine uses bisection to find a zero of a user-supplied
% continuous function f. The user must supply two points a and
% b such that f(a) and f(b) have different signs. The user also
```

```
% supplies a convergence tolerance delta.

fa = f(a);
if fa==0, root = a; break; end;  % Check to see if a is a root.
fb = f(b);
if fb==0, root = b; break; end;  % Check to see if b is a root.

if sign(fa)==sign(fb),           % Check for an error in input.
  display('Error: f(a) and f(b) have same sign.')
  break
end;

while abs(b-a) > 2*delta,
  % Iterate until interval width <= 2*delta.
  c = (a+b)/2;                   % Bisect interval.
  fc = f(c);                     % Evaluate f at midpoint.
  if fc==0, root = c; break; end; % Check to see if c is a root.

  if sign(fc)==sign(fa),         % Replace a by c, fa by fc.
    a = c;
    fa = fc;
  else                           % Replace b by c, fb by fc.
    b = c;
    fb = fc;
  end;
end;

root = (a+b)/2;    % Take midpoint as approximation to root.
```

There are several things to note about this code. First, it checks to see if the user has made an error in the input and supplied two points where the sign of `f` is the same. When writing code for others to use, it is always good to check for errors in input. Of course, typing in such things is time-consuming for the programmer, so when we write codes for ourselves, we sometimes omit these things. (And sometimes live to regret it!) Next, it checks to see if it has happened upon an "exact" root; that is, a point where f evaluates to exactly zero. If this happens, it returns with that point as the computed root. This is not necessary; it could continue to bisect the interval until the interval width reaches `2*delta`. The MATLAB `sign` function returns 1 if the argument is positive, −1 if it is negative, and 0 if it is 0, so if `fc` were 0, the code would end up replacing `b` by `c`.

Example 4.1.1. Use bisection to find a root of $f(x) = x^3 - x^2 - 1$ in the interval $[a, b]$, where $a = 0$ and $b = 2$.

Table 4.1 shows a list of results from the bisection algorithm. A graphical depiction of these results is given in figure 4.3.

TABLE 4.1
Results of 8 steps of the bisection algorithm for $f(x) = x^3 - x^2 - 1$. Left endpoint of the interval a_n, right endpoint b_n, midpoint r_n, and value of f at r_n.

n	a_n	b_n	r_n	$f(r_n)$
1	0	2	1	−1
2	1	2	1.5	0.125
3	1	1.5	1.25	−0.609
4	1.25	1.5	1.375	−0.291
5	1.375	1.5	1.4375	−0.096
6	1.4375	1.5	1.46875	0.011
7	1.4375	1.46875	1.453125	−0.043
8	1.453125	1.46875	1.4609375	−0.016

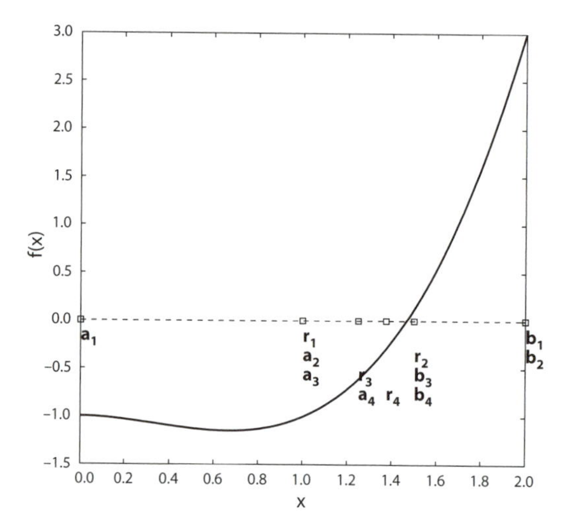

Figure 4.3. A graphical depiction of the bisection algorithm for $f(x) = x^3 - x^2 - 1$.

Rate of Convergence. What can be said about the *rate* of convergence of the bisection algorithm? Since the interval size is reduced by a factor of 2 at each step, the interval size after k steps is $|b - a|/2^k$, which converges to zero as $k \to \infty$. To obtain an interval of size 2δ we need

$$\frac{|b-a|}{2^k} \leq 2\delta \Leftrightarrow 2^{k+1} \geq \frac{|b-a|}{\delta} \Leftrightarrow k \geq \log_2\left(\frac{|b-a|}{\delta}\right) - 1.$$

If the approximate root is taken to be the midpoint of this interval, then it differs from an actual root by at most δ.

Example 4.1.2. If $a = 1$ and $b = 2$ and we want to guarantee an error less than or equal to 10^{-4}, how many iterations do we need to take?

$$\frac{|b-a|}{2^k} \leq 2 \cdot 10^{-4} \quad \Longleftrightarrow \quad 2^{k+1} \geq \frac{|b-a|}{10^{-4}} \quad \Longleftrightarrow \quad k \geq \log_2\left(\frac{|b-a|}{10^{-4}}\right) - 1.$$

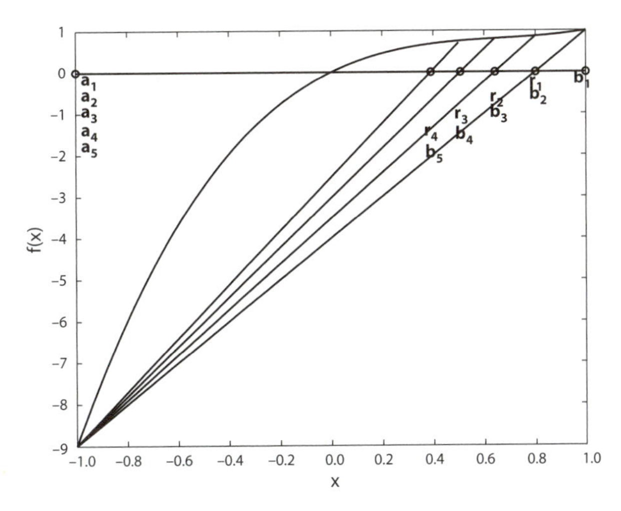

Figure 4.4. A graphical depiction of regula falsi for $f(x) = 2x^3 - 4x^2 + 3x$.

Since $b - a = 1$, then $k \geq 12.29$ iterations are required to guarantee an error less than or equal to 10^{-4}. After 13 iterations, your answer could have an error much less than 10^{-4} but will not have an error greater than 10^{-4}.

Since each step reduces the error (i.e., half the bracket size) by a constant factor (i.e., a factor of 2), the bisection algorithm is said to converge **linearly**.

One might try more sophisticated strategies than simply bisecting the interval to choose a point between the two current points a and b at which one expects f to have a root. For example, if $|f(a)| >> |f(b)|$, then it is reasonable to expect the root to lie closer to b than to a. In fact, one might draw a line between the points $(a, f(a))$ and $(b, f(b))$ and choose the next point c to be the point where this line intersects the x-axis. This can be thought of as a combination of the secant method (to be described in section 4.4.3) and bisection, and it is known as the *regula falsi* algorithm (method of false position). Note, however, that this method does not reduce the interval size by a factor of 2 at each step and can, in fact, be slower than bisection. Figure 4.4 illustrates the performance of regula falsi in finding a root of $f(x) = 2x^3 - 4x^2 + 3x$ in the interval $[-1, 1]$.

This one-sided convergence behavior is typical of regula falsi, with one endpoint of the interval always being replaced but the other remaining fixed. To avoid this problem, one can modify the formula for choosing the next point (using, say, the midpoint of the interval instead of the intersection of the secant line with the x-axis) if one endpoint remains the same for several consecutive steps.

The most difficult part of using the bisection algorithm or regula falsi or some combination of the two may be in finding an initial interval where the sign of f is different at the endpoints. Once this is found the algorithms are guaranteed to converge. Note that these algorithms cannot be used for a problem like $x^2 = 0$, pictured in figure 4.5, since the sign of the function does not change.

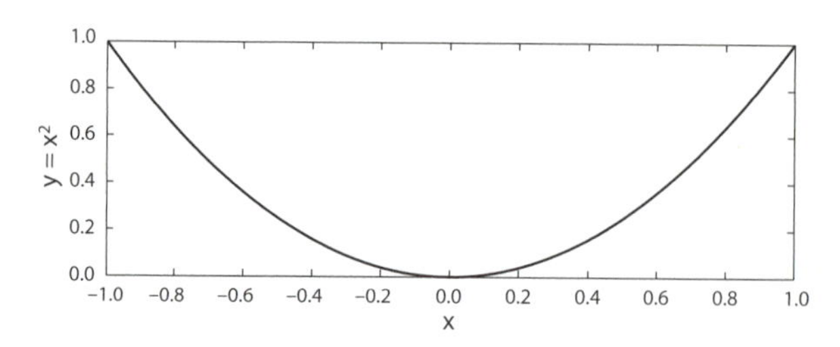

Figure 4.5. A graph such as $x^2 = 0$ would pose a problem for the bisection algorithm or regula falsi.

Figure 4.6. Brook Taylor (1685–1731). (Illustration from *The Universal Historical Dictionary* by George Crabb, published in 1825 (engraving). English School (19th century) (after) Private Collection/© Look and Learn/The Bridgeman Art Library International.)

Methods described in the following sections, such as Newton's method, may be used for such problems, but before describing Newton's method we review *Taylor's theorem*.

4.2 TAYLOR'S THEOREM

Probably the *most-used* theorem in numerical analysis is Taylor's theorem, especially Taylor's theorem with remainder. It is worth reviewing this important result.

Theorem 4.2.1 (Taylor's theorem with remainder). Let f, f', $\ldots$, $f^{(n)}$ be continuous on $[a, b]$ and let $f^{(n+1)}(x)$ exist for all $x \in (a, b)$. Then there is a number $\xi \in (a, b)$ such that

$$f(b) = f(a) + (b-a) f'(a) + \frac{(b-a)^2}{2!} f''(a) + \cdots + \frac{(b-a)^n}{n!} f^{(n)}(a) + \frac{(b-a)^{n+1}}{(n+1)!} f^{(n+1)}(\xi). \tag{4.4}$$

When the two points a and b in (4.4) are close to each other, one often writes b in the form $b = a + h$ for some small number h, and since the interval over which f is smooth may be much larger than the interval between the two points involved in (4.4), one often writes (4.4) for arbitrary points x and $x + h$

in this interval:

$$f(x+h) = f(x)+hf'(x)+\frac{h^2}{2!}f''(x)+\cdots+\frac{h^n}{n!}f^{(n)}(x)+\frac{h^{n+1}}{(n+1)!}f^{(n+1)}(\xi), \quad (4.5)$$

where $\xi \in (x, x+h)$. To emphasize the fact that the expansion is about a fixed point a and that it holds for any x in the interval where f is sufficiently smooth, one can also write

$$f(x) = f(a) + (x-a)f'(a) + \frac{(x-a)^2}{2!}f''(a) + \cdots + \frac{(x-a)^n}{n!}f^{(n)}(a) + \frac{(x-a)^{n+1}}{(n+1)!}f^{(n+1)}(\xi), \quad (4.6)$$

where ξ is between x and a. Each of the forms (4.5) and (4.6) is equivalent to (4.4), just using slightly different notation. Formula (4.4) (or (4.5) or (4.6)) can also be written using summation notation:

$$f(b) = \sum_{j=0}^{n} \frac{(b-a)^j}{j!}f^{(j)}(a) + \frac{(b-a)^{n+1}}{(n+1)!}f^{(n+1)}(\xi).$$

The remainder term in the Taylor series expansion (4.6) of $f(x)$ about the point a,

$$R_n(x) \equiv f(x) - \sum_{j=0}^{n} \frac{(x-a)^j}{j!}f^{(j)}(a) = \frac{(x-a)^{n+1}}{(n+1)!}f^{(n+1)}(\xi),$$

is sometimes written as $O((x-a)^{n+1})$, since there is a finite constant C (namely, $f^{(n+1)}(a)/(n+1)!$) such that

$$\lim_{x\to a} \frac{R_n(x)}{(x-a)^{n+1}} = C.$$

Sometimes the $O(\cdot)$ notation is used to mean, in addition, that there is no higher power $j > n+1$ such that $\lim_{x\to a} R_n(x)/(x-a)^j$ is finite. This will be the case if the constant C above is nonzero.

If f is infinitely differentiable, then the Taylor series expansion of $f(x)$ about the point a is

$$\sum_{j=0}^{\infty} \frac{(x-a)^j}{j!}f^{(j)}(a).$$

Depending on the properties of f, the Taylor series may converge to f everywhere or throughout some interval about a or only at the point a itself. For very well-behaved functions like e^x (which we will also denote as $\exp x$), $\sin x$ and $\cos x$, the Taylor series converges everywhere. The Taylor series about $x = 0$ occurs often and is sometimes called a Maclaurin series.

The Taylor series contains an infinite number of terms. If we are interested only in the behavior of the function in a neighborhood around $x = a$, then only the first few terms of the Taylor series may be needed:

$$f(x) \approx P_n(x) = \sum_{j=0}^{n} \frac{(x-a)^j}{j!} f^{(j)}(a).$$

The **Taylor polynomial** $P_n(x)$ may serve as an approximation to the function. This can be useful in situations where one can deal with polynomials but not with arbitrary functions f; for example, it might not be possible to integrate f analytically, but one can always integrate a polynomial approximation to f.

Example 4.2.1. Find the Taylor series of $e^x = \exp(x)$ about 1. First, find the value of the function and its derivatives at $x = 1$:

$$\begin{aligned} f(1) &= \exp(1) \\ f'(1) &= \exp(1) \\ f''(1) &= \exp(1) \\ &\vdots \end{aligned}$$

Next, construct the series using these values and $a = 1$:

$$\begin{aligned} \exp(x) &= \underbrace{f(1)}_{=\exp(1)} + (x-1)\underbrace{f'(1)}_{=\exp(1)} + \frac{(x-1)^2}{2!}\underbrace{f''(1)}_{=\exp(1)} + \frac{(x-1)^3}{3!}\underbrace{f^{(3)}(1)}_{=\exp(1)} + \cdots \\ &= \exp(1)\left[1 + (x-1) + \frac{(x-1)^2}{2!} + \frac{(x-1)^3}{3!} + \cdots\right] \\ &= \exp(1)\sum_{k=0}^{\infty} \frac{(x-1)^k}{k!} \end{aligned}$$

Example 4.2.2. Find the Maclaurin series for $\cos x$. First, find the value of the function $f(x) = \cos x$ and its derivatives at $x = 0$:

$$\begin{aligned} f(0) &= \cos(0) = 1 & f^{(3)}(0) &= \sin(0) = 0 \\ f'(0) &= -\sin(0) = 0 & f^{(4)}(0) &= \cos(0) = 1 \\ f''(0) &= -\cos(0) = -1 & &\vdots \end{aligned}$$

Next, construct the series using these values and $a = 0$:

$$\begin{aligned} \cos x &= \underbrace{f(0)}_{=1} + x\underbrace{f'(0)}_{=0} + \frac{x^2}{2!}\underbrace{f''(0)}_{=-1} + \frac{x^3}{3!}\underbrace{f^{(3)}(0)}_{=0} + \frac{x^4}{4!}\underbrace{f^{(4)}(0)}_{=1} + \cdots \\ &= 1 - \frac{x^2}{2} + \frac{x^4}{4!} + \cdots = \sum_{k=0}^{\infty} \frac{(-1)^k x^{2k}}{(2k)!} \end{aligned}$$

Example 4.2.3. Find the fourth-order *Taylor polynomial* $P_4(x)$ of $\cos x$ about $x = 0$. Using the work from the previous example, we find

$$P_4(x) = 1 - \frac{1}{2}x^2 + \frac{1}{24}x^4$$

Figure 4.7. Isaac Newton (1643–1727).

In the following section, Taylor's theorem will be used to derive a numerical method. At the end of the chapter, Exercise 19 discusses how John Machin, a contemporary of Brook Taylor, used Taylor's theorem to find 100 decimal places of π.

4.3 NEWTON'S METHOD

Isaac Newton, a founder of calculus, described a method for solving equations in his *Method of Fluxions and Infinite Series* [20, 77]. What is currently called *Newton's method*, however, may more properly be credited to Thomas Simpson who first described a root-finding algorithm that was both iterative and expressed in terms of derivatives [60].

We start with a geometric derivation of Newton's method. Suppose we wish to solve $f(x) = 0$ and we are given an initial guess x_0 for the solution. We evaluate $f(x_0)$ and then construct the tangent line to f at x_0 and determine where it hits the x-axis. If f were linear, this would be the root of f. In general, it might give a better approximation to a root than x_0 does, as illustrated in figures 4.8(a) and (b). The slope of this tangent line is $f'(x_0)$, and it goes through the point $(x_0, f(x_0))$, so its equation is $y - f(x_0) = f'(x_0)(x - x_0)$. It hits the x-axis ($y = 0$) at $x = x_0 - f(x_0)/f'(x_0)$, assuming $f'(x_0) \neq 0$. Letting x_1 be the point where this tangent line hits the x-axis, this process can be repeated by constructing the tangent line to f at x_1, determining where it hits the x-axis and calling that point x_2, etc. This process is known as *Newton's method*.

Newton's method for solving $f(x) = 0$.
Given an initial guess x_0, for $k = 0, 1, 2, \ldots,$

$$\text{Set } x_{k+1} = x_k - \frac{f(x_k)}{f'(x_k)}.$$

As a historical note, Newton explained his method in 1669 using numerical examples and did not use the geometric motivation given here involving the

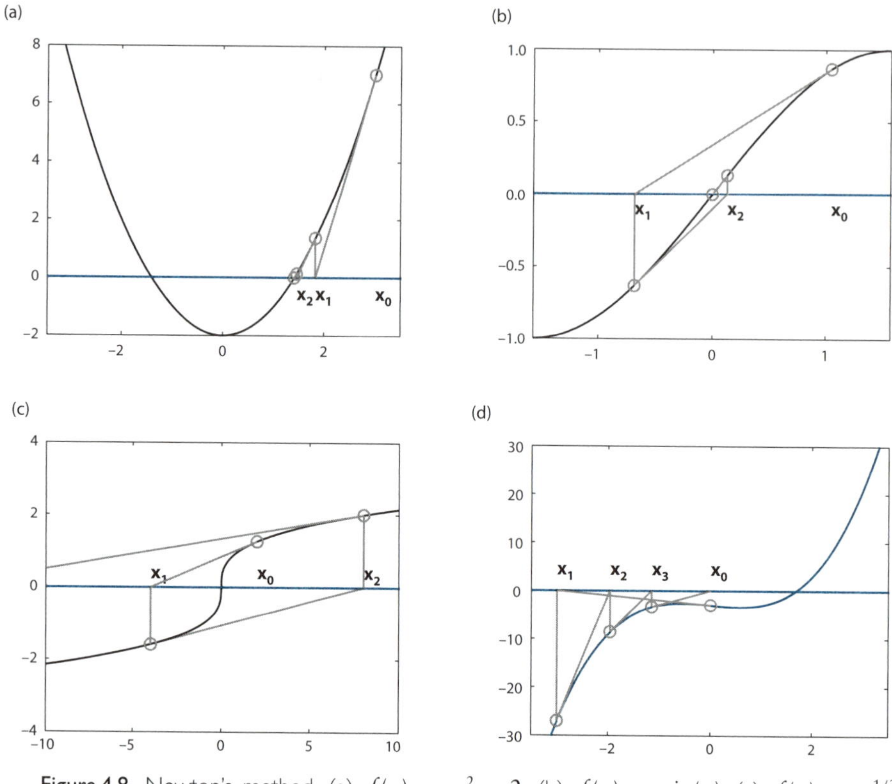

Figure 4.8. Newton's method. (a) $f(x) = x^2 - 2$, (b) $f(x) = \sin(x)$, (c) $f(x) = x^{1/3}$, (d) $f(x) = x^3 - x - 3$.

approximation of a curve with its tangent line. His work also did not develop the recurrence relation given above, which was developed by the English mathematician Joseph Raphson in 1690. It is for this reason that the method is often called the *Newton–Raphson method*. But, as noted earlier, the method should perhaps be called Simpson's method, since in 1740 he appears to have given the first description of the method that includes both iteration and derivatives [60].

Unfortunately, Newton's method does not always converge. Can you think of some situations that might cause problems for Newton's method? Figures 4.8(c) and (d) illustrate cases in which Newton's method fails. In case (c), Newton's method diverges, while in (d) it cycles between points, never approaching the actual zero of the function.

One might combine Newton's method with bisection, to obtain a more reliable algorithm. Once an interval is determined where the function changes sign, we know by the Intermediate Value Theorem that the interval contains a root. Hence if the next Newton step would land outside that interval, then do not accept it but do a bisection step instead. With this modification, the method is guaranteed to converge *provided* it finds an interval where the function has different signs at the endpoints. This combination would solve case (c) in figure 4.8 because once x_0 and x_1 were determined, the next Newton iterate x_2 would be rejected and replaced by $(x_0 + x_1)/2$. Case (d) in figure 4.8 would still fail because the algorithm does not locate any pair of points where the function has opposite signs.

Newton's method can also be derived from Taylor's theorem. According to that theorem,

$$f(x) = f(x_0) + (x - x_0) f'(x_0) + \frac{(x - x_0)^2}{2} f''(\xi),$$

for some ξ between x_0 and x. If $f(x_*) = 0$, then

$$0 = f(x_0) + (x_* - x_0) f'(x_0) + \frac{(x_* - x_0)^2}{2} f''(\xi), \tag{4.7}$$

and if x_0 is close to the root x_*, then the last term involving $(x_* - x_0)^2$ can be expected to be small in comparison to the others. Neglecting this last term, and defining x_1, instead of x_*, to be the point that satisfies the equation when this term is omitted, we have

$$0 = f(x_0) + (x_1 - x_0) f'(x_0),$$

or,

$$x_1 = x_0 - \frac{f(x_0)}{f'(x_0)}.$$

Repeating this process for $k = 0, 1, 2, \ldots$ gives

$$x_{k+1} = x_k - \frac{f(x_k)}{f'(x_k)}. \tag{4.8}$$

This leads to the following theorem about the convergence of Newton's method.

Theorem 4.3.1. If $f \in C^2$, if x_0 is *sufficiently close* to a root x_* of f, and if $f'(x_*) \neq 0$, then Newton's method converges to x_*, and *ultimately* the convergence rate is *quadratic*; that is, there exists a constant $C_* = |f''(x_*)/(2f'(x_*))|$ such that

$$\lim_{k\to\infty} \frac{|x_{k+1} - x_*|}{|x_k - x_*|^2} = C_*. \tag{4.9}$$

Before proving this theorem, we first make a few observations. The quadratic convergence rate described in (4.9) means that for k large enough, convergence will be very rapid. If C is any constant greater than C_*, it follows from (4.9) that there exists K such that for all $k \geq K$,

$$|x_{k+1} - x_*| \leq C|x_k - x_*|^2.$$

If, say, $C = 1$ and $|x_K - x_*| = 0.1$, then we will have $|x_{K+1} - x_*| \leq 10^{-2}$, $|x_{K+2} - x_*| \leq 10^{-4}$, $|x_{K+3} - x_*| \leq 10^{-8}$, etc. There are some important limitations, however. First, the theorem requires that the initial guess x_0 be *sufficiently close* to the desired root x_*, but it does not spell out exactly what *sufficiently close* means nor does it provide a way to check whether a given initial guess is sufficiently close to a desired root. The proof will make clearer why it is difficult to say just how close is close enough. Second, it shows that the *ultimate* rate of convergence is quadratic, but it does not say how many steps of the iteration might be required before this quadratic convergence rate is achieved. This too will depend on how close the initial guess is to the solution.

Proof of Theorem. It follows from equation (4.7), with x_0 replaced by x_k, that

$$x_* = x_k - \frac{f(x_k)}{f'(x_k)} - \frac{(x_* - x_k)^2}{2} \frac{f''(\xi_k)}{f'(x_k)},$$

for some ξ_k between x_k and x_*. Subtracting this from equation (4.8) for x_{k+1} gives

$$x_{k+1} - x_* = \frac{f''(\xi_k)}{2f'(x_k)} (x_k - x_*)^2. \tag{4.10}$$

Now since f'' is continuous and $f'(x_*) \neq 0$, if $C_* = |f''(x_*)/(2f'(x_*))|$, then for any $C > C_*$, there is an interval about x_* in which $|f''(\xi)/(2f'(x))| \leq C$, for x and ξ in this interval. If, for some K, x_K lies in this interval, and if also $|x_K - x_*| < 1/C$ (which is sure to hold, even for $K = 0$, if x_0 is sufficiently close to x_*), then (4.10) implies that $|x_{K+1} - x_*| \leq C|x_K - x_*|^2 < |x_K - x_*|$, so that x_{K+1} also lies in this interval and satisfies $|x_{K+1} - x_*| < 1/C$. Proceeding by induction, we find that all iterates x_k, $k \geq K$, lie in this interval and so, by (4.10), satisfy

$$\begin{aligned}
|x_{k+1} - x_*| &\leq C|x_k - x_*|^2 \\
&\leq (C|x_k - x_*|) \cdot |x_k - x_*| \\
&\leq (C|x_k - x_*|)(C|x_{k-1} - x_*|) \cdot |x_{k-1} - x_*| \\
&\;\;\vdots \\
&\leq (C|x_k - x_*|) \ldots (C|x_K - x_*|) \cdot |x_K - x_*| \\
&\leq (C|x_K - x_*|)^{k+1-K} |x_K - x_*|.
\end{aligned}$$

Since $C|x_K - x_*| < 1$, it follows that $(C|x_K - x_*|)^{k+1-K} \to 0$ as $k \to \infty$, and this implies that $x_k \to x_*$ as $k \to \infty$. Finally, since x_k and hence ξ_k in (4.10) converge to x_*, it follows from (4.10) that

$$\frac{|x_{k+1} - x_*|}{|x_k - x_*|^2} = \left| \frac{f''(\xi_k)}{2f'(x_k)} \right| \to C_*.$$

□

As noted previously, the hypothesis of x_0 being *sufficiently close* to a root x_* (where "sufficiently close" can be taken to mean that x_0 lies in a neighborhood of x_* where $|f''/(2f')|$ is less than or equal to some constant C and $|x_0 - x_*| < 1/C$) is usually impossible to check since one does not know the root x_*. There are a number of other theorems giving different sufficient conditions on the initial guess to guarantee convergence of Newton's method, but as with this theorem, the conditions are usually difficult or impossible to check, or they may be far more restrictive than necessary, so we do not include these results here. Early mathematicians looking at conditions under which Newton's method converges include Mourraille in 1768 and later Lagrange. In 1818, Fourier looked at the question of the rate of convergence and in 1821 and again in 1829 Cauchy addressed this question. For more information about the historical development of the method, see [20].

Example 4.3.1. Compute $\sqrt{2}$ using Newton's method; that is, solve $x^2 - 2 = 0$.
Since $f(x) = x^2 - 2$, we have $f'(x) = 2x$, and Newton's method becomes

$$x_{k+1} = x_k - \frac{x_k^2 - 2}{2x_k}, \quad k = 0, 1, \ldots.$$

Starting with $x_0 = 2$, and letting e_k denote the error $x_k - \sqrt{2}$, we find:

$$\begin{aligned} x_0 &= 2, & e_0 &= 0.59, \\ x_1 &= 1.5, & e_1 &= 0.086, \\ x_2 &= 1.4167, & e_2 &= 0.0025, \\ x_3 &= 1.4142157, & e_3 &= 2.1 \times 10^{-6} \approx 0.35 e_2^2. \end{aligned}$$

The constant $|f''(x_*)/(2f'(x_*))|$ for this problem is $1/(2\sqrt{2}) \approx 0.3536$. Note that for this problem Newton's method fails if $x_0 = 0$. For $x_0 > 0$ it converges to $\sqrt{2}$ and for $x_0 < 0$ it converges to $-\sqrt{2}$.

Example 4.3.2. Suppose $f'(x_*) = 0$. Newton's method may still converge, but only linearly. Consider the problem $f(x) \equiv x^2 = 0$. Since $f'(x) = 2x$, Newton's method becomes

$$x_{k+1} = x_k - \frac{x_k^2}{2x_k} = \frac{1}{2} x_k.$$

Starting, for example, with $x_0 = 1$, we find that $e_0 = 1$, $x_1 = e_1 = \frac{1}{2}$, $x_2 = e_2 = \frac{1}{4}$, ..., $x_k = e_k = \frac{1}{2^k}$. Instead of squaring the error in the previous step, each step reduces the error by a factor of 2.

Example 4.3.3. Consider the problem $f(x) \equiv x^3 = 0$. Since $f'(x) = 3x^2$, Newton's method becomes

$$x_{k+1} = x_k - \frac{x_k^3}{3x_k^2} = \frac{2}{3} x_k.$$

Clearly the error (the difference between x_k and the true root $x_* = 0$) is multiplied by the factor $\frac{2}{3}$ at each step.

Example 4.3.4. Suppose $f(x) \equiv x^j$. What is the error reduction factor for Newton's method?

To understand Example 4.3.2 recall formula (4.10) from the proof of theorem 4.3.1:

$$x_{k+1} - x_* = \frac{f''(\xi_k)}{2f'(x_k)}(x_k - x_*)^2.$$

Since $f'(x_*) = 0$ in Example 4.3.2, the factor multiplying $(x_k - x_*)^2$ becomes larger and larger as x_k approaches x_*, so we cannot expect quadratic convergence. However, if we expand $f'(x_k)$ about x_* using a Taylor series, we find

$$f'(x_k) = f'(x_*) + (x_k - x_*) f''(\eta_k) = (x_k - x_*) f''(\eta_k),$$

for some η_k between x_* and x_k. Substituting this expression for $f'(x_k)$ above gives

$$x_{k+1} - x_* = \frac{f''(\xi_k)}{2 f''(\eta_k)}(x_k - x_*).$$

If $f''(x_*) \neq 0$, then this establishes linear convergence (under the assumption that x_0 is sufficiently close to x_*) with the convergence factor approaching $\frac{1}{2}$. By retaining more terms in the Taylor series expansion of $f'(x_k)$ about x_*, this same approach can be used to explain Examples 4.3.3 and 4.3.4.

Theorem 4.3.2. If $f \in C^{p+1}$ for $p \geq 1$, if x_0 is sufficiently close to a root x_* of f, and if $f'(x_*) = \ldots = f^{(p)}(x_*) = 0$ but $f^{(p+1)}(x_*) \neq 0$, then Newton's method converges linearly to x_*, with the error ultimately being reduced by about the factor $p/(p+1)$ at each step; that is,

$$\lim_{k\to\infty} \frac{|x_{k+1} - x_*|}{|x_k - x_*|} = \frac{p}{p+1}.$$

Proof. Taking formula (4.8) and subtracting x_* from each side gives

$$x_{k+1} - x_* = x_k - x_* - \frac{f(x_k)}{f'(x_k)}. \tag{4.11}$$

Now expanding $f(x_k)$ and $f'(x_k)$ about x_* gives

$$f(x_k) = \frac{(x_k - x_*)^{p+1}}{(p+1)!} f^{(p+1)}(\xi_k),$$

$$f'(x_k) = \frac{(x_k - x_*)^{p}}{p!} f^{(p+1)}(\eta_k),$$

for some points ξ_k and η_k between x_k and x_*, since the first $p+1$ terms in the first Taylor series expansion and the first p terms in the second are zero. Making these substitutions in (4.11) gives

$$x_{k+1} - x_* = (x_k - x_*)\left(1 - \frac{1}{p+1}\frac{f^{(p+1)}(\xi_k)}{f^{(p+1)}(\eta_k)}\right). \tag{4.12}$$

Since $f^{(p+1)}(x_*) \neq 0$ and $f^{(p+1)}$ is continuous, for any $\epsilon > 0$, there is an interval about x_* in which the factor $1 - (1/(p+1)) f^{(p+1)}(\xi)/f^{(p+1)}(\eta)$ is within ϵ of its value at $\xi = \eta = x_*$ (namely, $1 - 1/(p+1) = p/(p+1)$), whenever ξ and η lie in this interval. In particular, for ϵ sufficiently small, this factor will be bounded above by a constant $C < 1$. It follows from (4.12) that if x_k lies in this interval, then $|x_{k+1} - x_*| \leq C|x_k - x_*|$, so x_{k+1} also lies in this interval, and, if x_0 lies in this interval then, by induction, $|x_{k+1} - x_*| \leq C^{k+1}|x_0 - x_*|$, so $x_k \to x_*$ as $k \to \infty$. The limiting value of the error reduction factor is $p/(p+1)$. □

As another example, consider the functions $f_1(x) = \sin x$ and $f_2(x) = \sin^2 x$, both of which have $x_* = \pi$ as a root. Note that $f_1'(x) = \cos x \neq 0$ at $x_* = \pi$, while $f_2'(x) = 2 \sin x \cos x = 0$ at $x_* = \pi$. Note, however, that $f_2''(\pi) \neq 0$. Therefore, we expect quadratic and linear convergence (with a convergence factor of about $1/2$) for Newton's method applied to $f_1(x)$ and

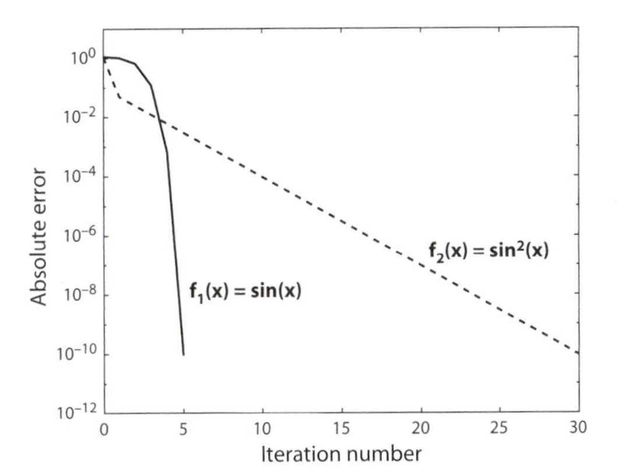

Figure 4.9. Convergence of Newton's method for a function with a root of multiplicity 1 versus a function with a root of multiplicity 2.

$f_2(x)$, respectively. This is seen graphically in figure 4.9. For f_1, the root $x_* = \pi$ is said to have multiplicity 1 or to be a simple root, while for f_2 it is a root of multiplicity 2. Newton's method converges only linearly for a root of multiplicity greater than 1.

4.4 QUASI-NEWTON METHODS

4.4.1 Avoiding Derivatives

Newton's method has the *very* nice property of quadratic convergence, when sufficiently close to a root. This means that one can hone in on a root from a nearby point very quickly, much faster than with the bisection method, for example. The price is that one must evaluate both f and its derivative at each step. For the simple examples presented here, this is not difficult. In many problems, however, the function f is quite complicated. It might not be one that can be written down analytically. Instead, one might have to run a program to evaluate $f(x)$. In such cases, differentiating f analytically is difficult or impossible, and, even if a formula can be found, it may be very expensive to evaluate. For such problems, one would like to avoid computing derivatives or, at least, compute as few of them as possible, while maintaining something close to quadratic convergence. Iterations of the form

$$x_{k+1} = x_k - \frac{f(x_k)}{g_k}, \quad \text{where } g_k \approx f'(x_k), \tag{4.13}$$

are sometimes called **quasi-Newton** methods.

4.4.2 Constant Slope Method

One idea is to evaluate f' once at x_0 and then to set g_k in (4.13) equal to $f'(x_0)$ for all k. If the slope of f does not change much as one iterates, then one might expect this method to mimic the behavior of Newton's method. This method is called the **constant slope method:**

$$x_{k+1} = x_k - \frac{f(x_k)}{g}, \quad \text{where } g = f'(x_0). \tag{4.14}$$

To analyze this iteration, we once again use Taylor's theorem. Expanding $f(x_k)$ about the root x_*, we find

$$f(x_k) = (x_k - x_*) f'(x_*) + O((x_k - x_*)^2),$$

using the $O(\cdot)$ notation, instead of writing out the remainder term explicitly. Subtracting x_* from each side of (4.14) and again letting $e_k \equiv x_k - x_*$ denote the error in x_k, we have

$$e_{k+1} = e_k - \frac{f(x_k)}{g} = e_k \left(1 - \frac{f'(x_*)}{g}\right) + O(e_k^2).$$

If $|1 - f'(x_*)/f'(x_0)| < 1$, then for x_0 sufficiently close to x_* (close enough so that the $O(e_k^2)$ term above is negligible), the method converges and convergence is *linear*. In general, we cannot expect better than linear convergence from the constant slope method.

A variation on this idea is to update the derivative occasionally. Instead of taking g_k in (4.13) to be $f'(x_0)$ for all k, one might monitor the convergence of the iteration and, when it seems to be slowing down, compute a new derivative $f'(x_k)$. This requires more derivative evaluations than the constant slope method, but it may converge in fewer iterations. Choosing between methods for solving nonlinear equations usually involves a trade-off between the cost of an iteration and the number of iterations required to reach a desired level of accuracy.

4.4.3 Secant Method

The **secant method** is defined by taking g_k in (4.13) to be

$$g_k = \frac{f(x_k) - f(x_{k-1})}{x_k - x_{k-1}}. \tag{4.15}$$

Note that in the limit as x_{k-1} approaches x_k, g_k approaches $f'(x_k)$, so that g_k can be expected to be a reasonable approximation to $f'(x_k)$ when x_{k-1} is close to x_k. It is actually the slope of the line through the points $(x_{k-1}, f(x_{k-1}))$ and $(x_k, f(x_k))$, called a **secant line** to the curve f since it intersects the curve at two points. The secant method is then defined by

$$x_{k+1} = x_k - \frac{f(x_k)(x_k - x_{k-1})}{f(x_k) - f(x_{k-1})}, \quad k = 1, 2, \ldots. \tag{4.16}$$

To begin the secant method, one needs *two* starting points x_0 and x_1. An illustration of the secant method for finding a root of $f(x) = x^3 - x^2 - 1$ with $x_0 = 0.5$ and $x_1 = 2$ is given in figure 4.10.

Example 4.4.1. Consider again the problem $f(x) \equiv x^2 - 2 = 0$. Taking $x_0 = 1$ and $x_1 = 2$, the secant method generates the following approximations and

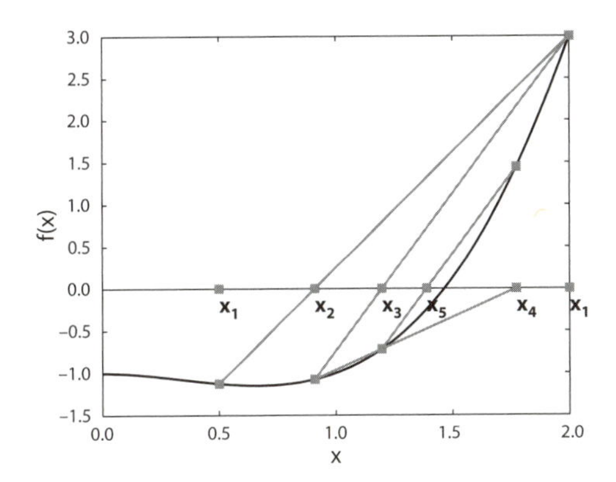

Figure 4.10. Finding a root of $f(x) = x^3 - x^2 - 1$ using $x_0 = 0.5$ and $x_1 = 2$ with the secant method.

errors:

$$\begin{aligned} x_2 &= 1.3333, & e_2 &= -0.0809, \\ x_3 &= 1.4, & e_3 &= -0.0142, \\ x_4 &= 1.4146, & e_4 &= 4.2 \times 10^{-4}, \\ x_5 &= 1.4142114, & e_5 &= -2.1 \times 10^{-6}. \end{aligned}$$

From the results in Example 4.4.1, it is difficult to determine exactly what the convergence rate of the secant method might be. It seems faster than linear: the ratio $|e_{k+1}/e_k|$ is getting smaller with k. But it seems slower than quadratic: $|e_{k+1}/e_k^2|$ seems to be growing larger with k. Can you guess what the order of convergence is?

It turns out that the order of convergence of the secant method is $\frac{1+\sqrt{5}}{2} \approx 1.62$. You probably didn't guess that! This result is obtained from the following lemma.

Lemma 4.4.1. If $f \in C^2$, if x_0 and x_1 are sufficiently close to a root x_* of f, and if $f'(x_*) \neq 0$, then the error e_k in the secant method satisfies

$$\lim_{k\to\infty} \frac{e_{k+1}}{e_k e_{k-1}} = C_*, \tag{4.17}$$

where $C_* = f''(x_*)/(2f'(x_*))$.

Statement (4.17) means that for k large enough

$$e_{k+1} \approx C_* e_k e_{k-1}, \tag{4.18}$$

where the approximate equality can be made as close as we like by choosing k large enough.

Before proving the lemma, let us demonstrate why the stated convergence rate follows from this result. Assume that there exist constants a and α such that, for all k sufficiently large,

$$|e_{k+1}| \approx a|e_k|^\alpha,$$

where again we can make this as close as we like to an actual equality by choosing k large enough. (In a formal proof, this assumption would need justification, but we will just make the assumption in our argument.) Since also $|e_k| \approx a|e_{k-1}|^\alpha$, we can write

$$|e_{k-1}| \approx (|e_k|/a)^{1/\alpha}.$$

Now, assuming that (4.18) holds, substituting these expressions for e_{k-1} and e_{k+1} into (4.18) gives

$$a|e_k|^\alpha \approx C|e_k|(|e_k|/a)^{1/\alpha} = C|e_k|^{1+1/\alpha}a^{-1/\alpha}.$$

Moving the constant terms to one side, this becomes

$$a^{1+1/\alpha}C^{-1} \approx |e_k|^{1+1/\alpha-\alpha}.$$

Since the left-hand side is a constant independent of k, the right-hand side must be as well, which means that the exponent of $|e_k|$ must be zero:

$$1 + 1/\alpha - \alpha = 0.$$

Solving for α, we find

$$\alpha = \frac{1 \pm \sqrt{5}}{2}.$$

In order for the method to converge, α must be positive, so the order of convergence must be the positive value $\alpha = (1 + \sqrt{5})/2$.

Proof of Lemma. Subtracting x_* from each side of (4.16) gives

$$e_{k+1} = e_k - \frac{f(x_k)(e_k - e_{k-1})}{f(x_k) - f(x_{k-1})},$$

where we have substituted $e_k - e_{k-1} = (x_k - x_*) - (x_{k-1} - x_*)$ for $x_k - x_{k-1}$. Combining terms, this becomes

$$\begin{aligned} e_{k+1} &= \frac{f(x_k)e_{k-1} - f(x_{k-1})e_k}{f(x_k) - f(x_{k-1})} \\ &= e_k e_{k-1}\left[\frac{f(x_k)/e_k - f(x_{k-1})/e_{k-1}}{x_k - x_{k-1}} \cdot \frac{x_k - x_{k-1}}{f(x_k) - f(x_{k-1})}\right], \end{aligned} \tag{4.19}$$

where the last factor inside the brackets in (4.19) converges to $1/f'(x_*)$ as x_k and x_{k-1} approach x_*. The first factor can be written in the form

$$\frac{\frac{f(x_k)-f(x_*)}{x_k - x_*} - \frac{f(x_{k-1})-f(x_*)}{x_{k-1}-x_*}}{x_k - x_{k-1}}.$$

If we define

$$g(x) \equiv \frac{f(x) - f(x_*)}{x - x_*}, \tag{4.20}$$

then the first factor inside the brackets in (4.19) is $(g(x_k) - g(x_{k-1}))/(x_k - x_{k-1})$, which converges to $g'(x_*)$ as x_k and x_{k-1} approach x_*. Differentiating expression (4.20), we find that

$$g'(x) = \frac{(x - x_*)f'(x) - (f(x) - f(x_*))}{(x - x_*)^2},$$

and taking the limit as $x \to x_*$, using L'Hôpital's rule, gives

$$\lim_{x \to x_*} g'(x) = \lim_{x \to x_*} \frac{(x - x_*) f''(x)}{2(x - x_*)} = \frac{f''(x_*)}{2}.$$

Thus, *assuming* that e_k and e_{k-1} converge to 0 as $k \to \infty$, it follows from (4.19) that

$$\lim_{k \to \infty} \frac{e_{k+1}}{e_k e_{k-1}} = \frac{1}{2} \frac{f''(x_*)}{f'(x_*)}.$$

Finally, the assumption that x_0 and x_1 are sufficiently close to x_* means that for any $C > \frac{1}{2}|f''(x_*)/f'(x_*)|$, we can assume that x_0 and x_1 lie in an interval about x_* in which $|e_2| \leq C|e_1| \cdot |e_0|$ and that, say, $|e_1| \leq 1/(2C)$ and $|e_0| \leq 1/(2C)$. It follows that $|e_2| \leq \frac{1}{2} \min\{|e_1|, |e_0|\} \leq 1/(4C)$, so x_2 also lies in this interval. By induction it follows that all iterates x_k lie in this interval and that $|e_k| \leq \frac{1}{2}|e_{k-1}| \leq \cdots \leq \frac{1}{2^k}|e_0|$. Hence e_k and e_{k-1} converge to 0 as $k \to \infty$. □

As noted earlier, the secant method, like Newton's method, can be combined with bisection *if* one finds a pair of points x_k and x_{k-1} at which f has opposite signs. In the secant method, this means that instead of automatically working with the new pair of points x_{k+1} and x_k, one works with x_{k+1} and either x_k or x_{k-1}, depending on whether $f(x_k)$ or $f(x_{k-1})$ has the opposite sign from $f(x_{k+1})$. In this way, one always maintains an interval that is known to contain a root of f. This is called the regula falsi algorithm and while it is guaranteed to converge, as we saw earlier, it may have only a slow linear rate of convergence. It is somewhat surprising that moving *outside* an interval that is known to contain a root may actually lead to faster convergence to that root!

4.5 ANALYSIS OF FIXED POINT METHODS

Sometimes, instead of writing a nonlinear equation in the form $f(x) = 0$, one writes it in the form of a **fixed point problem**: $x = \varphi(x)$; that is, the problem is to find a point x that remains fixed under the mapping φ. A natural approach to solving such a problem is to start with an initial guess x_0, and then iterate according to

$$x_{k+1} = \varphi(x_k). \tag{4.21}$$

Since many problems are most naturally expressed as fixed point problems, it is desirable to understand general iterations of the form (4.21).

There are many ways that one can translate from an equation of the form $f(x) = 0$ to one of the form $x = \varphi(x)$ and vice versa. The equation $f(x) = 0$, for example, can be written as $\varphi(x) \equiv x + f(x) = x$, or it can be written as $\varphi(x) \equiv x - f(x) = x$. In the opposite direction, the equation $x = \varphi(x)$ can be written as $f(x) \equiv x - \varphi(x) = 0$, or it can be written as $f(x) \equiv \exp(x - \varphi(x)) - 1 = 0$. In any of these cases, $f(x) = 0$ if and only if $\varphi(x) = x$.

Both Newton's method and the constant slope method can be thought of as fixed point iterations. Although we have already analyzed these methods,

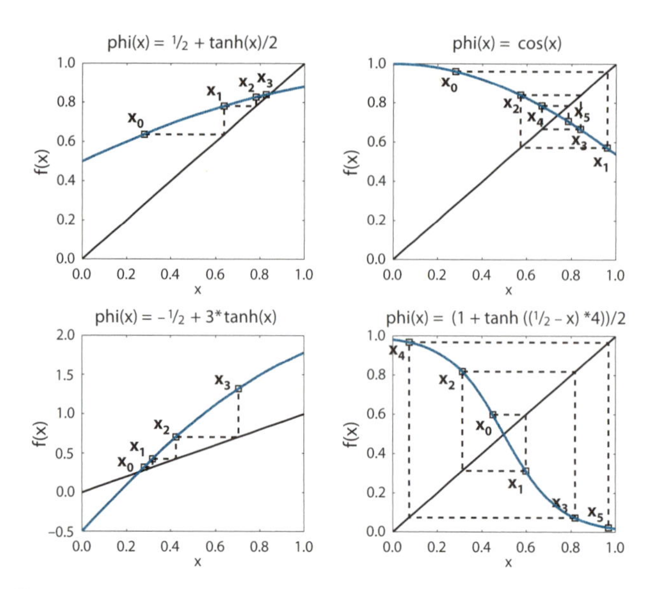

Figure 4.11. Fixed point iteration. The iteration may display monotonic convergence (*upper left*), oscillatory convergence (*upper right*), monotonic divergence (*lower left*), or oscillatory divergence (*lower right*).

the analysis of this section will apply as well, provided the function $\varphi(x)$ is defined appropriately. For Newton's method, since $x_{k+1} = x_k - f(x_k)/f'(x_k)$, the function φ whose fixed point is being sought is $\varphi(x) = x - f(x)/f'(x)$. Note that $\varphi(x) = x$ if and only if $f(x) = 0$ (assuming that $f'(x) \neq 0$). For the constant slope method, since $x_{k+1} = x_k - f(x_k)/g$, we have $\varphi(x) = x - f(x)/g$, where $g = f'(x_0)$. Note that for this function as well, $\varphi(x) = x$ if and only if $f(x) = 0$. The secant method, $x_{k+1} = x_k - f(x_k)(x_k - x_{k-1})/(f(x_k) - f(x_{k-1}))$ does not fit this general pattern since the right-hand side is a function of *both* x_k and x_{k-1}.

Figure 4.11 illustrates cases where fixed point iteration converges and diverges. In particular, the iteration can produce monotone convergence, oscillatory convergence, periodic behavior, or chaotic/quasi-periodic behavior.

Geometrically, fixed points lie at the intersection of the line $y = x$ and the curve $y = \varphi(x)$, as seen in figure 4.11. Mathematically, such points occur when the output of a function is equal to its input: $\varphi(x) = x$. Fixed points correspond to equilibria of dynamical systems; that is, to points where the solution neither grows nor decays.

Example 4.5.1. Find the fixed points of $\varphi(x) = x^2 - 6$.

First, plot $y = x^2 - 6$ and $y = x$ on the same axes to get an idea of where the intersection points are; see figure 4.12. To determine the fixed points mathematically, solve the quadratic equation $x^2 - 6 = x$, or, $x^2 - x - 6 = (x - 3)(x + 2) = 0$. The solutions are $x = 3$ and $x = -2$.

What determines whether a fixed point iteration will converge? The following theorem gives a *sufficient* condition for convergence.

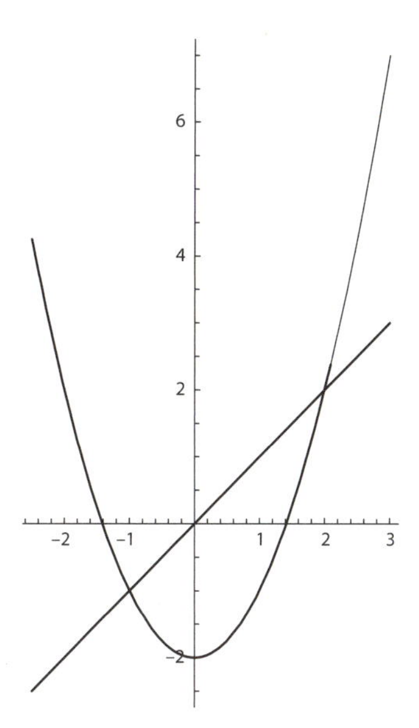

Figure 4.12. A plot of $y = x^2 - 6$ and $y = x$.

Theorem 4.5.1. Assume that $\varphi \in C^1$ and $|\varphi'(x)| < 1$ in some interval $[x_* - \delta, x_* + \delta]$ centered about a fixed point x_* of φ. If x_0 is in this interval then the fixed point iteration (4.21) converges to x_*.

Proof. Expanding $\varphi(x_k)$ in a Taylor series about x_* gives

$$x_{k+1} = \varphi(x_k) = \varphi(x_*) + (x_k - x_*)\varphi'(\xi_k) = x_* + (x_k - x_*)\varphi'(\xi_k),$$

for some ξ_k between x_k and x_*. Subtracting x_* from each side, and denoting the error as $e_k \equiv x_k - x_*$, this becomes

$$e_{k+1} = e_k\varphi'(\xi_k),$$

and taking absolute values on each side gives

$$|e_{k+1}| = |\varphi'(\xi_k)|\,|e_k|.$$

Thus if $|\varphi'(x)| < 1$ for all x in an interval centered about x_*, and if x_0 is in this interval, then future iterates remain in this interval and $|e_k|$ decreases at each step by at least the factor $\max_{x \in [x_* - \delta, x_* + \delta]} |\varphi'(x)|$. The error reduction factor approaches $|\varphi'(x_*)|$:

$$\lim_{k \to \infty} \frac{|e_{k+1}|}{|e_k|} = |\varphi'(x_*)|. \qquad \square$$

Note that the interval in theorem 4.5.1 must be *centered* about the fixed point x_* in order to guarantee that new iterates remain in this interval. Given a more general interval $[a, b]$ containing the fixed point x_* and on which $|\varphi'(x)| < 1$, one could make the assumption that φ maps the interval $[a, b]$ into $[a, b]$; that is, if $x \in [a, b]$, then $\varphi(x) \in [a, b]$. This also would ensure that if one starts with an initial guess x_0 in $[a, b]$, then future iterates remain in this

TABLE 4.2
Fixed point iteration applied to three different functions derived from the equation $x^3 + 6x^2 - 8 = 0$.

n	$\varphi_1(x) = x$	$\varphi_2(x) = x$	$\varphi_3(x) = x$
0	1.5	1.5	1.5
1	10.375	1.032795559	0.8779711461
2	1764.990234	1.066549452	1.104779810
3	5516973759	1.063999177	1.052898680
4	1.679×10^{29}	1.064191225	1.067142690
5	4.734×10^{87}	1.064176759	1.063386479
6	1.061×10^{263}	1.064177849	1.064388114
7		1.064177767	1.064121800
8		1.064177773	1.064192663
9		1.064177772	1.064173811
10			1.064178826
11			1.064177492
12			1.064177847
13			1.064177753
14			1.064177778
15			1.064177771
16			1.064177773
17			1.064177772

interval. The arguments in the theorem then show that the error $|e_k|$ is reduced at each step by at least the factor $\max_{x \in [a,b]} |\varphi'(x)|$.

Example 4.5.2. As an example, let us consider three fixed point iteration schemes that can be derived from the equation $f(x) \equiv x^3 + 6x^2 - 8 = 0$, which has a solution in the interval $[1, 2]$, since $f(1) = -1 < 0$ and $f(2) = 24 > 0$. The functions whose fixed points we seek are

1. $\varphi_1(x) = x^3 + 6x^2 + x - 8$,

2. $\varphi_2(x) = \sqrt{\dfrac{8}{x+6}}$, and

3. $\varphi_3(x) = \sqrt{\dfrac{8 - x^3}{6}}$.

Simple algebra shows that each of these functions has a fixed point in the interval $[1, 2]$ at the point(s) where $f(x) = x^3 + 6x^2 - 8$ has a root; for example, $\varphi_2(x) = x$ if $x \geq 0$ and $x^2 = 8/(x+6)$, or, $x^3 + 6x^2 - 8 = 0$. [Exercise: Show that $\varphi_3(x) = x$ if $x \in [0, 2]$ and $f(x) = 0$.]

Table 4.2 shows the behavior of fixed point iteration applied to each of these functions with initial guess $x_0 = 1.5$.

These results can be explained using theorem 4.5.1:

(a) First consider $\varphi_1(x) = x^3 + 6x^2 + x - 8$, a function for which fixed point iteration quickly diverged. Note that $\varphi_1'(x) = 3x^2 + 12x + 1 > 1$ for all

$x > 0$. Starting at $x_0 = 1.5$, one must expect the iterates to grow at each step and so the procedure diverges.

(b) Next consider $\varphi_2(x) = \sqrt{\frac{8}{x+6}}$. In this case, $|\varphi_2'(x)| = \sqrt{2}/(x+6)^{3/2}$, which is less than 1 for $x > 2^{1/3} - 6 \approx -4.74$. Since we know that there is a fixed point x_* somewhere in the interval $[1, 2]$, the distance from $x_0 = 1.5$ to this fixed point is at most 0.5; hence the interval $[x_* - 0.5, x_* + 0.5] \subset [0.5, 2.5]$ contains x_0 and is an interval on which $|\varphi_2'(x)| < 1$. Therefore convergence is guaranteed. One also could establish convergence by noting that φ_2 maps the interval $[1, 2]$ that contains a fixed point to the interval $[1, \sqrt{8/7}] \subset [1, 2]$. Therefore, by the discussion after theorem 4.5.1, the iteration must converge.

(c) Finally consider $\varphi_3(x) = \sqrt{\frac{8-x^3}{6}}$. In this case $|\varphi_3'(x)| = (x^2/4) \times \sqrt{6/(8-x^3)}$. It can be checked that this is less than 1 for x less than about 1.6, but if we know only that there is a fixed point in the interval $[1, 2]$, we cannot conclude that fixed point iteration applied to φ_3 will converge. Since we know from running fixed point iteration with φ_2, however, that the fixed point x_* is at about 1.06, we can again conclude that in the interval $[x_* - 0.5, x_* + 0.5]$, which contains the initial guess $x_0 = 1.5$, we have $|\varphi_3'(x)| < 1$; hence the fixed point iteration will converge to x_*.

Actually, $\varphi'(x_*)$ need not exist in order to have convergence. The iteration (4.21) will converge to a fixed point x_* if φ is a **contraction**; that is, if there exists a constant $L < 1$ such that for all x and y

$$|\varphi(x) - \varphi(y)| \leq L|x - y|. \tag{4.22}$$

Theorem 4.5.2. If φ is a contraction (on all of $\mathbf{R}$), then it has a unique fixed point x_* and the iteration $x_{k+1} = \varphi(x_k)$ converges to x_* from any x_0.

Proof. We will show that the sequence $\{x_k\}_{k=0}^{\infty}$ is a Cauchy sequence and hence converges. To see this, let j and k be positive integers with $k > j$ and use the triangle inequality to write

$$|x_k - x_j| \leq |x_k - x_{k-1}| + |x_{k-1} - x_{k-2}| + \cdots + |x_{j+1} - x_j|. \tag{4.23}$$

Each difference $|x_m - x_{m-1}|$ can be written as $|\varphi(x_{m-1}) - \varphi(x_{m-2})|$, which is bounded by $L|x_{m-1} - x_{m-2}|$, where $L < 1$ is the constant in (4.22). Repeating this argument for $|x_{m-1} - x_{m-2}|$, we find that $|x_m - x_{m-1}| \leq L^2|x_{m-2} - x_{m-3}|$, and continuing in this way gives $|x_m - x_{m-1}| \leq L^{m-1}|x_1 - x_0|$. Making these substitutions in (4.23) we find

$$|x_k - x_j| \leq (L^{k-1} + L^{k-2} + \cdots + L^j)|x_1 - x_0| = L^j \frac{1 - L^{k-j}}{1 - L}|x_1 - x_0|.$$

If k and j are both greater than or equal to some positive integer N, we will have

$$|x_k - x_j| \leq L^N \frac{1}{1-L} |x_1 - x_0|,$$

and this quantity goes to 0 as $N \to \infty$. This shows that the sequence x_k, $k = 0, 1, \ldots$ is a Cauchy sequence and hence converges.

To show that x_k converges to a fixed point of φ, we first note that the fact that φ is a contraction implies that it is continuous. Hence $\varphi(\lim_{k\to\infty} x_k) = \lim_{k\to\infty} \varphi(x_k)$. Letting x_* denote $\lim_{k\to\infty} x_k$, we have

$$\varphi(x_*) \equiv \varphi\left(\lim_{k\to\infty} x_k\right) = \lim_{k\to\infty} \varphi(x_k) = \lim_{k\to\infty} x_{k+1} = x_*,$$

so that the limit of the sequence x_k is indeed a fixed point. Finally, if y_* is also a fixed point, then since φ is a contraction, we must have

$$|x_* - y_*| = |\varphi(x_*) - \varphi(y_*)| \leq L|x_* - y_*|,$$

where $L < 1$. This can hold only if $y_* = x_*$, so the fixed point is unique. □

Theorem 4.5.2 could be modified to assume only that φ is a contraction on some interval $[a, b]$, where φ maps $[a, b]$ into itself. The conclusion would then be that there is a unique fixed point x_* in $[a, b]$ and that if fixed point iteration is started with an initial guess $x_0 \in [a, b]$, then it converges to x_*.

4.6 FRACTALS, JULIA SETS, AND MANDELBROT SETS

Fixed point iteration and Newton's method can be used for problems defined in the complex plane as well. Consider, for example, the problem of finding a fixed point of $\varphi(z) \equiv z^2$. It is easy to predict the behavior of the fixed point iteration $z_{k+1} = z_k^2$. If $|z_0| < 1$, then the sequence z_k converges to the fixed point 0. If $|z_0| > 1$, then the iterates grow in modulus and the method diverges. If $|z_0| = 1$, then $|z_k| = 1$ for all k. If $z_0 = 1$ or $z_0 = -1$, the sequence quickly settles down to the fixed point 1. On the other hand, if, say, $z_0 = e^{2\pi i/3}$, then $z_1 = e^{4\pi i/3}$ and $z_2 = e^{8\pi i/3} = e^{2\pi i/3} = z_0$, so the cycle repeats. It can be shown that if $z_0 = e^{2\pi i\alpha}$ where α is irrational then the sequence of points z_k never repeats but becomes dense on the unit circle. The sequence of points z_0, $z_1 = \varphi(z_0)$, $z_2 = \varphi(\varphi(z_0))$, $\ldots$ is called the **orbit** of z_0 under φ.

If $\varphi(z)$ is a polynomial function, then the set of points z_0 for which the orbit remains bounded is called the **filled Julia set** for φ, and its boundary is called the **Julia set**. Thus the filled Julia set for z^2 is the closed unit disk, while the Julia set is the unit circle. These sets are named after the French mathematician Gaston Julia. In 1918, he and Pierre Fatou independently investigated the behavior of these sets, which can be *far* more interesting than the simple example presented above. Their work received renewed attention in the 1980s when the advent of computer graphics made numerical experimentation and visualization of the results easy and fun [33].

JULIA OF THE JULIA SET

Gaston Julia (1893–1978) was only 25 years of age when he published his 199 page, *Mémoire sur l'itération des fonctions rationelles* [56]. The publication led to fame in the 1920s but was virtually forgotten until Benoit Mandelbrot's study of fractals. The leather strap seen covering a portion of Julia's face was due to a severe injury in World War I that led to the loss of his nose.

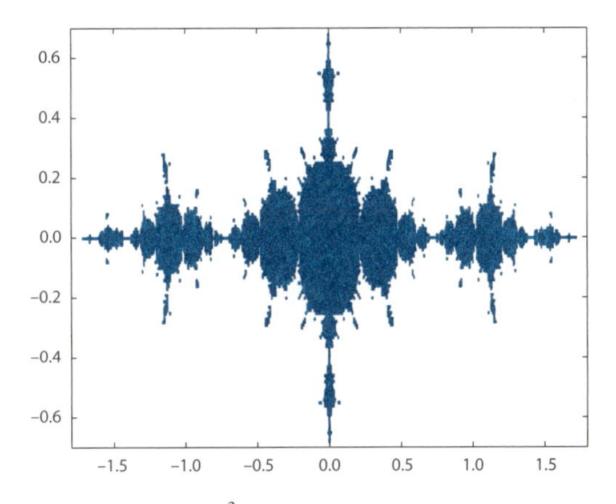

Figure 4.13. Filled Julia set for $\varphi(z) = z^2 - 1.25$.

Figure 4.13 displays the filled Julia set for the function $\varphi(z) = z^2 - 1.25$. The boundary of this set is a **fractal**, meaning that it has dimension neither one nor two but some fraction in between. The figure is also **self-similar**, meaning that if one repeatedly zooms in on a small subset of the picture, the pattern still looks like the original.

The plot in figure 4.13 was obtained by starting with many different points z_0 throughout a rectangular region containing the two fixed points of φ, and running the fixed point iteration until one of three things occurred: Either $|z_k| \geq 2$, in which case we conclude that the orbit is unbounded and z_0 is not in the set; or z_k comes within 10^{-6} of a fixed point and stays within that distance for 5 iterations, in which case we conclude that the iterates are converging to the fixed point and z_0 is in the set; or the number of iterations k reaches 100 before either of the previous two conditions occur, in which case we conclude that the sequence z_k is not converging to a fixed point but does remain bounded and hence z_0 is again in the set. Note that since $|z_{k+1}| \geq |z_k|^2 - 1.25$, if $|z_k| \geq 2$, then $|z_{k+1}| \geq 4 - 1.25 = 2.75$, $|z_{k+2}| \geq 2.75^2 - 1.25 \approx 6.3$, etc., so that the sequence really is unbounded. The code to produce this plot is given in figure 4.14.

```
phi = inline('z^2 - 1.25');    % Define the function whose fixed points we seek.
fixpt1 = (1 + sqrt(6))/2;      % These are the fixed points.
fixpt2 = (1 - sqrt(6))/2;

colormap([1 0 0; 1 1 1]);      % Points numbered 1 (inside) will be colored red;
                               %  those numbered 2 (outside) will be white.
M = 2*ones(141,361);           % Initialize array of point colors to 2 (white).

for j=1:141,                   % Try initial values with imaginary parts between
  y = -.7 + (j-1)*.01;         %   -0.7 and 0.7
  for i=1:361,                 % and with real parts between
    x = -1.8 + (i-1)*.01;      %   -1.8 and 1.8.
    z = x + 1i*y;              % 1i is the MATLAB symbol for sqrt(-1).
    zk = z;
    iflag1 = 0;                % iflag1 and iflag2 count the number of iterations
    iflag2 = 0;                %   when a root is within 1.e-6 of a fixed point;
    kount = 0;                 % kount is the total number of iterations.

    while kount < 100 & abs(zk) < 2 & iflag1 < 5 & iflag2 < 5,
      kount = kount+1;
      zk = phi(zk);            % This is the fixed point iteration.

      err1 = abs(zk-fixpt1);   % Test for convergence to fixpt1.
      if err1 < 1.e-6, iflag1 = iflag1 + 1; else, iflag1 = 0; end;

      err2 = abs(zk-fixpt2);   % Test for convergence to fixpt2.
      if err2 < 1.e-6, iflag2 = iflag2 + 1; else, iflag2 = 0; end;

    end;
    if iflag1 >= 5 | iflag2 >= 5 | kount >= 100,   % If orbit is bounded, set
      M(j,i) = 1;                                  %   point color to 1 (red).
    end;
  end;
end;

image([-1.8 1.8],[-.7 .7],M),  % This plots the results.
axis xy                        % If you don't do this, vertical axis is inverted.
```

Figure 4.14. MATLAB code to compute the filled Julia set for $\varphi(z) = z^2 - 1.25$.

While this small bit of analysis enables us to determine a sufficient condition for the orbit to be unbounded, we really have only an educated guess as to the points for which the orbit is bounded. It is possible that the iterates remain less than 2 in modulus for the first 100 steps but then grow unboundedly, or that they appear to be converging to one of the fixed points but then start to diverge. Without going through further analysis, one can gain confidence in the computation by varying some of the parameters. Instead of assuming that the sequence is bounded if the iterates remain less than 2 in absolute value for 100 iterations, one could try 200 iterations. To improve the efficiency of the code, one might also try a lower value, say, 50 iterations. If reasonable choices of parameters yield the same results, then that suggests that the computed sets may be correct, while if different parameter choices yield different sets then one must conclude that the numerical results are suspect. Other ways to test the computation include comparing with other results in the literature and

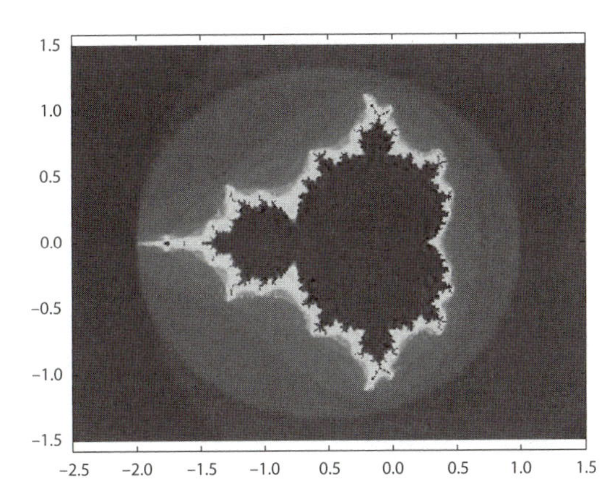

Figure 4.15. Mandelbrot set.

comparing the computed results with what is known analytically about these sets. This comparison of numerical results and theory is often important in both pure and applied mathematics. One uses the theory to check numerical results, and in cases where theory is lacking, one uses numerical results to suggest what theory might be true.

If one investigates Julia sets for all functions of the form $\varphi(z) = z^2 + c$, where c is a complex constant, one finds that some of these sets are **connected** (one can move between any two points in the set without leaving the set), while others are not. Julia and Fatou simultaneously found a simple criterion for the Julia set associated with $\varphi(z)$ to be connected: It is connected if and only if 0 is in the filled Julia set. In 1982, Benoit Mandelbrot used computer graphics to study the question of which values of c give rise to Julia sets that are connected, by testing different values of c throughout the complex plane and running fixed point iteration as above, with initial value $z_0 = 0$. The astonishing answer was the **Mandelbrot set,** depicted in figure 4.15. The black points are the values of c for which the Julia set is connected, while the shading of the other points indicates the rate at which the fixed point iteration applied to $z^2 + c$ with $z_0 = 0$ diverges.

Newton's method can also be applied to problems in the complex plane, and plots of points from which it is or is not convergent are just as beautiful and interesting as the ones obtained from fixed point iteration. Figure 4.16 shows the behavior of Newton's method when applied to the problem $f(z) \equiv z^3 + 1 = 0$. This equation has three roots: $z = -1$, $z = e^{\pi i/3}$, and $z = e^{-\pi i/3}$. Initial points from which Newton's method converges to the first root are lightest, those from which it converges to the second root are darkest, and those from which it converges to the third root are in between. The boundary of this image is also a fractal and one can zoom in on different pieces of the figure to see a pattern like that of the original.

Once again, we should stress that further analysis is needed to verify figure 4.16. The figure was obtained by trying many different initial guesses z_0, iterating until 10 consecutive Newton iterates were within 10^{-6} of one of the roots and then coloring that initial guess according to which root was

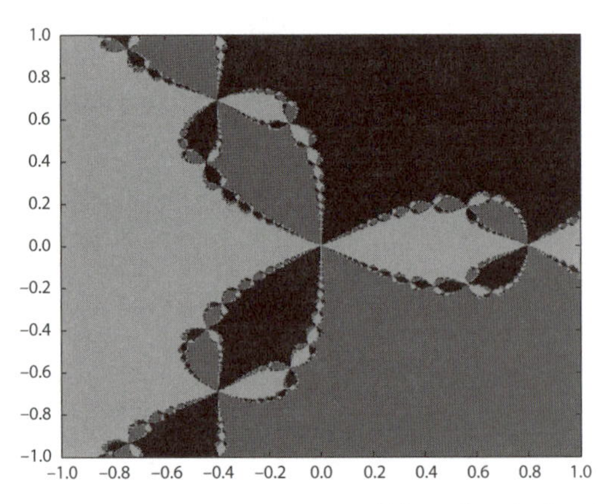

Figure 4.16. Convergence of Newton's method for $f(z) = z^3 + 1$. The lightest colored initial points converge to -1, the darkest colored initial points converge to $e^{\pi i/3}$, and the ones in between converge to $e^{-\pi i/3}$.

approximated. It is possible that 10 Newton iterates could be within 10^{-6} of one root while later iterates would converge to a different root or diverge. Further analysis can verify the figure, but that is beyond the scope of this book. For an excellent elementary discussion of Julia sets and Newton's method in the complex plane, see [69].

4.7 CHAPTER 4 EXERCISES

1. (a) Write the equation for the tangent line to $y = f(x)$ at $x = p$.
   (b) Solve for the x-intercept of the line in equation (a). What formula have you derived, with what roles for p and x?
   (c) Write the equation of the line that intersects the curve $y = f(x)$ at $x = p$ and $x = q$.
   (d) Solve for the x-intercept of the line in equation (c). What formula have you derived, with what roles for p, q, and x?
2. Use MATLAB to plot the function $f(x) = (5 - x)\exp(x) - 5$, for x between 0 and 5. (This function is associated with the *Wien radiation law*, which gives a method to estimate the surface temperature of a star.)

   (a) Write a bisection routine or use routine `bisect` available from the book's web page to find a root of $f(x)$ in the interval $[4, 5]$, accurate to six decimal places (i.e., find an interval of width at most 10^{-6} that contains the root, so that the midpoint of this interval is within 5×10^{-7} of the root). At each step, print out the endpoints of the smallest interval known to contain a root. Without running the code further, answer the following: How many steps would be required to reduce the size of this interval to 10^{-12}? Explain your answer.
   (b) Write a routine to use Newton's method or use routine `newton` available from the book's web page to find a root of $f(x)$, using initial guess $x_0 = 5$. Print out your approximate solution x_k and the value of $f(x_k)$ at each step and run until $|f(x_k)| \leq 10^{-8}$. Without running the code further, but perhaps using information from your code about the

rate at which $|f(x_k)|$ is reduced, can you estimate how many more steps would be required to make $|f(x_k)| \leq 10^{-16}$ (assuming that your machine carried enough decimal places to do this)? Explain your answer.

(c) Take your routine for doing Newton's method and modify it to run the secant method. Repeat the run of part (b), using, say, $x_0 = 4$ and $x_1 = 5$, and again predict (without running the code further) how many steps would be required to reduce $|f(x_k)|$ below 10^{-16} (assuming that your machine carried enough decimal places to do this) using the secant method. Explain your answer.

3. Newton's method can be used to compute reciprocals, without division. To compute $1/R$, let $f(x) = x^{-1} - R$ so that $f(x) = 0$ when $x = 1/R$. Write down the Newton iteration for this problem and compute (by hand or with a calculator) the first few Newton iterates for approximating $1/3$, starting with $x_0 = 0.5$, and not using any division. What happens if you start with $x_0 = 1$? For positive R, use the theory of fixed point iteration to determine an interval about $1/R$ from which Newton's method will converge to $1/R$.
4. Use Newton's method to approximate $\sqrt{2}$ to 6 decimal places.
5. Write down the first few iterates of the secant method for solving $x^2 - 3 = 0$, starting with $x_0 = 0$ and $x_1 = 1$.
6. Consider the function $h(x) = x^4/4 - 3x$. In this problem, we will see how to use Newton's method to find the *minimum* of the function $h(x)$.

   (a) Derive a function f that has a root at the point where h achieves its minimum. Write down the formula for Newton's method applied to f.
   (b) Take one step (by hand) with Newton's method starting with a guess of $x_0 = 1$.
   (c) Take two steps (by hand) towards the root of f (i.e., the minimum of h) using bisection with $a = 0$, $b = 4$.

7. In finding a root with Newton's method, an initial guess of $x_0 = 4$ with $f(x_0) = 1$ leads to $x_1 = 3$. What is the derivative of f at x_0?
8. In using the secant method to find a root, $x_0 = 2$, $x_1 = -1$ and $x_2 = -2$ with $f(x_1) = 4$ and $f(x_2) = 3$. What is $f(x_0)$?
9. Can the bisection method be used to find the roots of the function $f(x) = \sin x + 1$? Why or why not? Can Newton's method be used to find the roots (or a root) of this function? If so, what will be its order of convergence and why?
10. The function

$$f(x) = \frac{x^2 - 2x + 1}{x^2 - x - 2}$$

has exactly one zero in the interval $[0, 3]$, at $x = 1$. Using $a = 0$ and $b = 3$, run the bisection method on $f(x)$ with a stopping tolerance of `delta = 1e-3`. Explain why it does not appear to converge to the root.

Why does the Intermediate Value Theorem (theorem 4.1.1) *not* guarantee that there is a root in $[0, 1.5]$ or in $[1.5, 3]$?

Use MATLAB to plot this function over the interval in a way that makes it clear what is going on. [Hint: You may want to use the plot command over two different x intervals separately in order to show the behavior properly.]

11. Write out the Newton iteration formula for each of the functions below. For each function and given starting value, write a MATLAB script that takes 5 iterations of Newton's method and prints out the iterates with format `%25.15e` so you see all digits.

    (a) $f(x) = \sin x, \quad x_0 = 3$;
    (b) $f(x) = x^3 - x^2 - 2x, \quad x_0 = 3$;
    (c) $f(x) = 1+0.99x-x, \quad x_0 = 1$. In exact arithmetic, how many iterations does Newton's method need to find a root of this function?

12. Let function $\varphi(x) = (x^2 + 4)/5$.

    (a) Find the fixed point(s) of $\varphi(x)$.
    (b) Would the fixed point iteration, $x_{k+1} = \varphi(x_k)$, converge to a fixed point in the interval $[0, 2]$ for all initial guesses $x_0 \in [0, 2]$?

13. Consider the equation $a = y - \epsilon \sin y$, where $0 < \epsilon < 1$ is given and $a \in [0, \pi]$ is given. Write this in the form of a fixed point problem for the unknown solution y and show that it has a unique solution. [Hint: You can show that $|\sin y - \sin x| \leq |y - x|$ for all x and y by using Taylor's theorem with remainder to express $\sin y$ as $\sin x$ plus a remainder term.]
14. If you enter a number into a handheld calculator and repeatedly press the cosine button, what number (approximately) will eventually appear? Provide a proof. [Note: Set your calculator to interpret numbers as radians rather than degrees; otherwise you will get a different answer.]
15. The function $\varphi(x) = \frac{1}{2}\left(-x^2 + x + 2\right)$ has a fixed point at $x = 1$. Starting with $x_0 = 0.5$, we use $x_{n+1} = \varphi(x_n)$ to obtain the sequence $\{x_n\} = \{0.5, 1.1250, 0.9297, 1.0327, 0.9831, \ldots\}$. Describe the behavior of the sequence. (If it converges, how does it converge? If it diverges, how does it diverge?)
16. Steffensen's method for solving $f(x) = 0$ is defined by

$$x_{k+1} = x_k - \frac{f(x_k)}{g_k},$$

where

$$g_k = \frac{f(x_k + f(x_k)) - f(x_k)}{f(x_k)}.$$

Show that this is quadratically convergent, under suitable hypotheses.

17. Returning to the problem facing an artillery officer during battle, he needs to determine the angle θ at which to aim the cannon. Again, the desired θ

is a solution to the nonlinear equation

$$\frac{2v_0^2 \sin\theta \cos\theta}{g} = d. \tag{4.24}$$

(a) Show that

$$\theta = \frac{1}{2} \arcsin \frac{dg}{v_0^2}$$

solves equation (4.24), assuming that v_0 and d are fixed and that $dg/v_0^2 \leq 1$.

(b) Use Newton's method to find θ (to within two decimal places) when $v_0 = 126$ m/s, $d = 1200$ m and $g = 9.8$ m/s^2. As an initial guess use $\theta = \pi/6$.

18. Recall that the Taylor series expansion of a function $f(x)$ about a point x_0 is

$$f(x) = f(x_0) + f'(x_0)(x - x_0) + \frac{1}{2!} f''(x_0)(x - x_0)^2 + \frac{1}{3!} f'''(x_0)(x - x_0)^3 + \cdots$$

From this we can define a sequence of polynomials of increasing degree that approximate the function near the point x_0. The Taylor polynomial of degree n is

$$P_n(x) = f(x_0) + f'(x_0)(x - x_0) + \frac{1}{2} f''(x_0)(x - x_0)^2 + \frac{1}{6} f'''(x_0)(x - x_0)^3 + \cdots + \frac{1}{n!} f^{(n)}(x_0)(x - x_0)^n,$$

where $f^{(n)}(x_0)$ means the nth derivative of f evaluated at x_0.

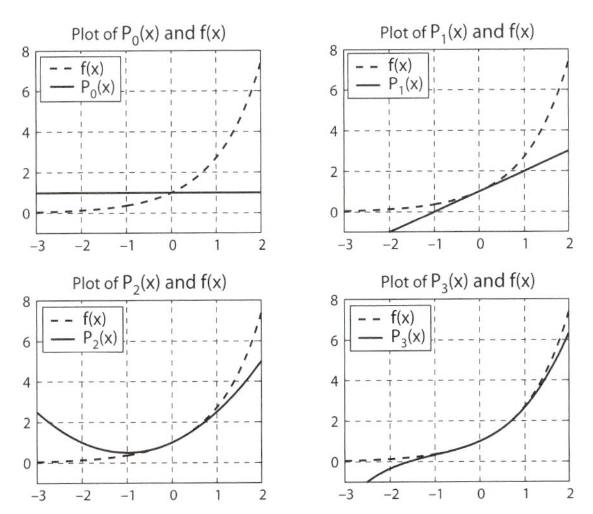

The figure shows the function $f(x) = e^x$ and the first four Taylor polynomials from the expansion about the point $x_0 = 0$. Since all derivatives of $f(x) = e^x$ are again just e^x, the nth-order Taylor polynomial from a series expansion about a general point x_0 is

$$P_n(x) = e^{x_0} + e^{x_0}(x - x_0) + \frac{1}{2}e^{x_0}(x - x_0)^2 + \frac{1}{6}e^{x_0}(x - x_0)^3 + \cdots + \frac{1}{n!}e^{x_0}(x - x_0)^n.$$

For $x_0 = 0$, we find

$$P_0(x) = 1,$$
$$P_1(x) = 1 + x,$$
$$P_2(x) = 1 + x + \frac{1}{2}x^2,$$
$$P_3(x) = 1 + x + \frac{1}{2}x^2 + \frac{1}{6}x^3,$$

and these are the functions plotted in the figure, along with $f(x)$.

Produce a similar set of four plots for the Taylor polynomials arising from the Taylor series expansion of $f(x) = e^{1-x^2}$ about the point $x_0 = 1$. You can start with the M-file (`plotTaylor.m`) used to create the figure above, which can be found on the web page for this book. Choose the `x` values and `axis` parameters so that the plots are over a reasonable range of values to exhibit the functions well. The polynomials should give good approximations to the function near the point $x_0 = 1$.

19. Finding the area of 96-sided polygons that inscribe and circumscribe a circle of diameter 1, Greek mathematician Archimedes (287–212 BC) determined that $\frac{223}{71} < \pi < \frac{22}{7}$. Almost two thousand years later, John Machin (1680–1752) exploited a recent discovery of his contemporary Brook Taylor, the Taylor series. He also employed the following trigonometric identity that he discovered:

$$\pi = 16 \arctan\left(\frac{1}{5}\right) - 4 \arctan\left(\frac{1}{239}\right). \tag{4.25}$$

(a) Find the Maclaurin series for $\arctan x$.

(b) Write down $P_n(x)$, the nth-degree Taylor polynomial for $\arctan x$ centered at $x = 0$.

(c) Approximate π by using $P_n(x)$ and (4.25). More specifically, use the approximation

$$\pi \approx T_n = 16P_n\left(\frac{1}{5}\right) - 4P_n\left(\frac{1}{239}\right).$$

Using MATLAB, find T_n and the absolute error $|T_n - \pi|$ for $n = 1, 3, 5, 7$, and 9. How many decimal places of accuracy did you find for $n = 9$?

For centuries, Machin's formula served as a primary tool for π-hunters. As late as 1973, one million digits of π were evaluated on a computer by Guilloud and Bouyer using a version of Machin's formula [5, 10].

5

FLOATING-POINT ARITHMETIC

Having studied and implemented a few computational algorithms, we now look at the effects of computer arithmetic on these algorithms. In most cases, this effect is minimal and numerical analysts are happy to program their procedures, run the programs on a computer, and trust the answers that it returns, at least to the extent that they trust their models to be correct and their codes to be bug-free. If a code returns clearly incorrect results, one immediately suspects an error in the program or input data, or a misunderstanding of the algorithm or physical model, and in the vast majority of cases, this is indeed the source of the problem.

However, there is another possible source of error when algorithms are implemented on a computer. Unless one uses a symbolic system, which quickly becomes impractical for large computations, the arithmetic that is performed is not exact. It is typically rounded to about 16 decimal places. This seems negligible if one is interested in only a few decimal places of the answer, but sometimes rounding errors can accumulate in a disastrous manner. While such occurrences are rare, they can be extremely difficult to diagnose and understand. In order to do so, one must first understand something about the way in which computers perform arithmetic operations.

Most computers store numbers in *binary* (base 2) format and since there is limited space in a computer word, not all numbers can be represented exactly; they must be *rounded* to fit the word size. This means that arithmetic operations are not performed exactly. Although the error made in any one operation is usually negligible (of relative size about 10^{-16} using double precision), a poorly designed algorithm may magnify this error to the point that it destroys all accuracy in the computed solution. For this reason it is important to understand the effects of rounding errors on computed results.

To see the effects that roundoff can have, consider the following iteration:

$$x_{k+1} = \begin{cases} 2x_k, & x_k \in [0, \frac{1}{2}], \\ 2x_k - 1, & x_k \in (\frac{1}{2}, 1]. \end{cases} \tag{5.1}$$

Let $x_0 = 1/10$. Then $x_1 = 2/10$, $x_2 = 4/10$, $x_3 = 8/10$, $x_4 = 6/10$, and $x_5 = x_1$. The iteration cycles periodically between these values. When implemented on a computer, however, using floating-point arithmetic, this is not the case. As seen

TABLE 5.1
Computed results from iteration (5.1). After 55 iterations, the computed value is 1 and it remains there for all subsequent iterations.

k	True x_k	Computed x_k
0	0.10000	0.10000
1	0.20000	0.20000
2	0.40000	0.40000
3	0.80000	0.80000
4	0.60000	0.60000
5	0.20000	0.20000
10	0.40000	0.40000
20	0.60000	0.60000
40	0.60000	0.60001
42	0.40000	0.40002
44	0.60000	0.60010
50	0.40000	0.40625
54	0.40000	0.50000
55	0.80000	1.00000

in table 5.1, computed results agree with exact results to at least five decimal places until iteration 40, where accumulated error becomes visible in the fifth decimal digit. As the iterations continue, the error grows until, at iteration 55, the computed result takes on the value 1 and remains there for all subsequent iterations. Later in the chapter, we will see why such errors occur.

We begin this chapter with two stories that demonstrate the potential cost of overlooking the effects of rounding errors.

5.1 COSTLY DISASTERS CAUSED BY ROUNDING ERRORS

The Intel Pentium Flaw. [37, 55, 76, 78, 82] In the summer of 1994, Intel anticipated the commercial success of its new Pentium chip. The new chip was twice as fast at division as previous Intel chips running at the same clock rate. Concurrently, Professor Thomas R. Nicely, a mathematician at Lynchburg College in Virginia, was computing the sum of the reciprocals of prime numbers using a computer with the new Pentium chip. The computational and theoretical results differed significantly. However, results run on a computer using an older 486 CPU calculated correct results. In time, Nicely tracked the error to the Intel chip. Having contacted Intel and received little response to his initial queries, Nicely posted a general notice on the Internet asking for others to confirm his findings. The posting (dated October 30, 1994) with subject line *Bug in the Pentium FPU* [78] began:

> It appears that there is a bug in the floating point unit (numeric coprocessor) of many, and perhaps all, Pentium processors.

This email began a furor of activity, so much so that only weeks later on December 13, IBM halted shipment of their Pentium machines, and in late

Figure 5.1. A bug in the Pentium chip cost Intel millions of dollars. (Courtesy of CPU Collection Konstantin Lanzet.)

December, Intel agreed to replace all flawed Pentium chips upon request. The company put aside a reserve of \$420 million to cover costs, a major investment for a flaw. With a flurry of Internet activity between November 29 and December 11, Intel had become a laughingstock on the Internet joke circuit, but it wasn't funny to Intel. On Friday, December 16, Intel stock closed at \$59.50, down \$3.25 for the week.

What type of error could the chip make in its arithmetic? The *New York Times* printed the following example of the Pentium bug: Let $A = 4{,}195{,}835.0$ and $B = 3{,}145{,}727.0$, and consider the quantity

$$A - (A/B) * B.$$

In exact arithmetic, of course, this would be 0, but the Pentium computed 256, because the quotient A/B was accurate to only about 5 decimal places. Is this *close enough*? For many applications it probably is, but we will see in later sections that we need to be able to count on computers to do better than this.

While such an example can make one wonder how Intel missed such an error, it should be noted that subsequent analysis confirmed the subtlety of the mistake. Alan Edelman, professor of mathematics at Massachusetts Institute of Technology, writes in his article of 1997 published in *SIAM Review* [37]:

> We also wish to emphasize that, despite the jokes, the bug is far more subtle than many people realize....The bug in the Pentium was an easy mistake to make, and a difficult one to catch.

Ariane 5 Disaster. [4,45] The Ariane 5, a giant rocket capable of sending a pair of three-ton satellites into orbit with each launch, took 10 years and 7 billion dollars for the European Space Agency to build. Its maiden launch was met with eager anticipation as the rocket was intended to propel Europe far into the lead in the commercial space business.

On June 4, 1996, the unmanned rocket took off cleanly but veered off course and exploded in just under 40 seconds after liftoff. Why? To answer this question, we must step back into the programming of the onboard computers.

During the design of an earlier rocket, programmers decided to implement an additional "feature" that would leave the horizontal positioning function (designed for positioning the rocket on the ground) running after the count-down had started, anticipating the possibility of a delayed takeoff. Since the

(a)

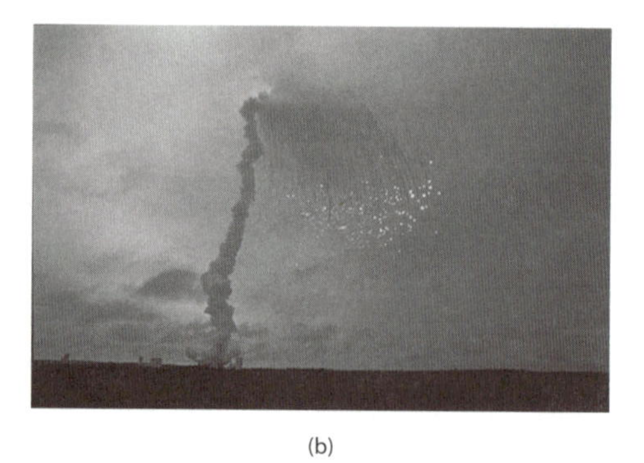

(b)

Figure 5.2. The Ariane 5 rocket (a) lifting off and (b) flight 501 self-destructing after a numerical error on June 4, 1996. (Courtesy of ESA/CNES.)

expected deviation while on the ground was minimal, only a small amount of memory (16 bits) was allocated to the storage of this information. After the launch, however, the horizontal deviation was large enough that the number could not be stored correctly with 16 bits, resulting in an exception error. This error instructed the primary unit to shut down. Then, all functions were transferred to a backup unit, created for redundancy in case the primary unit shut down. Unfortunately, the backup system contained the same bug and shut itself down. Suddenly, the rocket was veering off course causing damage between the solid rocket boosters and the main body of the rocket. Detecting the mechanical failure, the master control systems triggered a self-destruct cycle, as had been programmed in the event of serious mechanical failure in flight. Suddenly, the rocket and its expensive payloads were scattered over about 12 square kilometers east of the launch pad. Millions of dollars would have been saved if the data had simply been saved in a larger variable rather than the 16-bit memory location allocated in the program.

As seen from these examples, it is important for numerical analysts to understand the impact of rounding errors on their calculations. This chapter covers the basics of computer arithmetic and the IEEE standard. Later we will see more about how to apply this knowledge to the analysis of algorithms. For an excellent and very readable book on computer arithmetic and the IEEE standard, see [79].

5.2 BINARY REPRESENTATION AND BASE 2 ARITHMETIC

Most computers today use *binary* or *base 2* arithmetic. This is natural since on/off gates can represent a 1 (on) or a 0 (off), and these are the only two digits in base 2. In base 10, a natural number is represented by a sequence of digits from 0 to 9, with the rightmost digit representing 1s (or 10^0s), the next representing 10s (or 10^1s), the next representing 100s (or 10^2s), etc. In base 2, the digits are 0 and 1, and the rightmost digit represents 1s (or 2^0s), the next represents 2s (or 2^1s), the next 4s (or 2^2s), etc. Thus, for example, the decimal number 10 is written in base 2 as 1010_2: one $2^3 = 8$, zero 2^2, one $2^1 = 2$, and zero 2^0. The decimal number 27 is 11011_2: one $2^4 = 16$, one $2^3 = 8$, zero 2^2, one $2^1 = 2$, and one $2^0 = 1$. The binary representation of a positive integer is determined by first finding the highest power of 2 that is less than or equal to the number; in the case of 27, this is 2^4, so a 1 goes in the fifth position from the right of the number: 1____. One then subtracts 2^4 from 27 to find that the remainder is 11. Since 2^3 is less than 11, a 1 goes in the next position to the right: 11___. Subtracting 2^3 from 11 leaves 3, which is less than 2^2, so a 0 goes in the next position: 110__. Since 2^1 is less than 3, a 1 goes in the next position, and since $3 - 2^1 = 1$, another 1 goes in the rightmost position to give $27 = 11011_2$.

Binary arithmetic is carried out in a similar way to decimal arithmetic, except that when adding binary numbers one must remember that $1 + 1$ is 10_2. To add the two numbers 10 and 27, we align their binary digits and do the addition as below. The top row shows the digits that are *carried* from one column to the next.

```
11 1
  1010
+11011
 -----
100101
```

You can check that 100101_2 is equal to 37. Subtraction is similar, with *borrowing* from the next column being necessary when subtracting 1 from 0. Multiplication and division follow similar patterns.

Just as we represent rational numbers using decimal expansions, we can also represent them using *binary* expansions. The digits to the right of the decimal point in base 10 represent 10^{-1}s (tenths), 10^{-2}s (hundredths), etc., while those to the right of the binary point in base 2 represent 2^{-1}s (halves) 2^{-2}s (fourths), etc. For example, the fraction 11/2 is 5.5 in base 10, while it is 101.1_2 in base 2: one 2^2, one 2^0, and one 2^{-1}. Not all rational numbers can be represented with *finite* decimal expansions. The number 1/3, for example, is $0.33\overline{3}$, with the bar over the 3 meaning that this digit is repeated infinitely many times. The same is true for binary expansions, although the numbers that require an infinite binary expansion may be different from the ones that require an infinite decimal expansion. For example, the number $1/10 = 0.1$ in base 10 has the repeating binary expansion: $0.0\overline{0011}00_2$. To see this, one can do binary long division in a similar way to base 10 long division:

```
           .0001100
          ----------
1010 /1.00000000
          1010
          ----
           1100
           1010
           ----
             10000
```

Irrational numbers such as $\pi \approx 3.141592654$ can only be approximated by decimal expansions, and the same holds for binary expansions.

5.3 FLOATING-POINT REPRESENTATION

A computer word consists of a certain number of bits, which can be either on (to represent 1) or off (to represent 0). Some early computers used **fixed-point representation**, where one bit is used to denote the sign of a number, a certain number of the remaining bits are used to store the part of the binary number to the left of the binary point, and the remaining bits are used to store the part to the right of the binary point. The difficulty with this system is that it can store numbers only in a very limited range. If, say, 16 bits are used to store the part of the number to the left of the binary point, then the leftmost bit represents 2^{15}, and numbers greater than or equal to 2^{16} cannot be stored. Similarly, if, say, 15 bits are used to store the part of the number to the right of the binary point, then the rightmost bit represents 2^{-15} and no positive number smaller than 2^{-15} can be stored.

A more flexible system is **floating-point representation**, which is based on scientific notation. Here a number is written in the form $\pm m \times 2^E$, where $1 \leq m < 2$. Thus, the number $10 = 1010_2$ would be written as $1.010_2 \times 2^3$, while $\frac{1}{10} = 0.0\overline{0011}00_2$ would be written as $1.1\overline{0011}00_2 \times 2^{-4}$. The computer word consists of three fields: one for the sign, one for the exponent E, and one for the significand m. A **single-precision** word consists of 32 bits: 1 bit for the sign (0 for $+$, 1 for $-$), 8 bits for the exponent, and 23 bits for the significand. Thus the number $10 = 1.010_2 \times 2^3$ would be stored in the form

0	E=3	1.010...0

while $5.5 = 1.011_2 \times 2^2$ would be stored in the form

0	E=2	1.0110...0

(We will explain later how the exponent is actually stored.) A number that can be stored exactly using this scheme is called a **floating-point number**. The number $1/10 = 1.1\overline{0011}00_2 \times 2^{-4}$ is not a floating-point number since it must be **rounded** in order to be stored.

Before describing rounding, let us consider one slight improvement on this basic storage scheme. Recall that when we write a binary number in the form $m \times 2^E$, where $1 \leq m < 2$, the first bit to the left of the binary point is always 1. Hence there is no need to store it. If the significand is of the form $b_0.b_1 \ldots b_{23}$, then instead of storing $b_0, b_1, \ldots, b_{22}$, we can keep one extra place by storing $b_1, \ldots, b_{23}$, knowing that b_0 is 1. This is called **hidden-bit representation.** Using this scheme, the number $10 = 1.010_2 \times 2^3$ is stored as

0	E=3	010...0

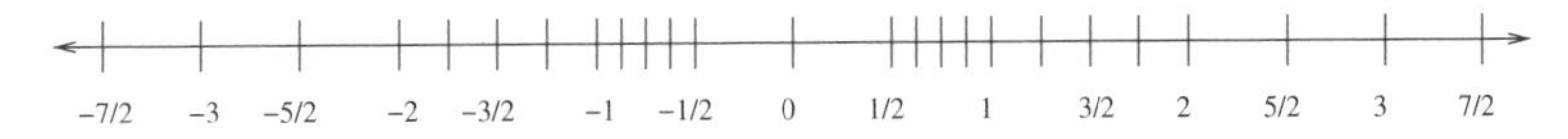

Figure 5.3. Number line for toy floating-point system.

The number $1/10 = 1.\overbrace{10011001100110011001100}^{23\text{ bits}}\overline{1100}_2 \times 2^{-4}$ is approximated by either

0	E=-4	10011001100110011001100

or

0	E=-4	10011001100110011001101

depending on the rounding mode being used (see section 5.5). There is one difficulty with this scheme, and that is how to represent the number 0. Since 0 cannot be written in the form $1.b_1 b_2 \ldots \times 2^E$, we will need a special way to represent 0.

The gap between 1 and the next larger floating-point number is called the **machine precision** and is often denoted by ϵ. (In MATLAB, this number is called eps.) [Note: Some sources define the machine precision to be $\frac{1}{2}$ times this number.] In single precision, the next floating-point number after 1 is $1+2^{-23}$, which is stored as

0	E=0	00000000000000000000001

Thus, for single precision, we have $\epsilon = 2^{-23} \approx 1.2 \times 10^{-7}$.

The default precision in MATLAB is not single but **double precision**, where a word consists of 64 bits: 1 for the sign, 11 for the exponent, and 52 for the significand. Hence in double precision the next larger floating-point number after 1 is $1 + 2^{-52}$, so that the machine precision is $2^{-52} \approx 2.2 \times 10^{-16}$.

Since single- and double-precision words contain a large number of bits, it is difficult to get a feel for what numbers can and cannot be represented. It is instructive instead to consider a toy system in which only significands of the form $1.b_1 b_2$ can be represented and only exponents 0, 1, and -1 can be stored; this system is described in [79]. What are the numbers that can be represented in this system? Since $1.00_2 = 1$, we can represent $1 \times 2^0 = 1$, $1 \times 2^1 = 2$, and $1 \times 2^{-1} = \frac{1}{2}$. Since $1.01_2 = \frac{5}{4}$, we can store the numbers $\frac{5}{4}$, $\frac{5}{2}$, and $\frac{5}{8}$. Since $1.10_2 = \frac{3}{2}$, we can store the numbers $\frac{3}{2}$, 3, and $\frac{3}{4}$. Finally, $1.11_2 = \frac{7}{4}$ gives us the numbers $\frac{7}{4}$, $\frac{7}{2}$, and $\frac{7}{8}$. These numbers are plotted in figure 5.3.

There are several things to note about this toy system. First, the machine precision ϵ is 0.25, since the number just right of 1 is $\frac{5}{4}$. Second, in general, the gaps between representable numbers become larger as we move away from the origin. This is acceptable since the *relative gaps*, the difference between two consecutive numbers divided by, say, their average, remains of reasonable size. Note, however, that the gap between 0 and the smallest positive number is *much* larger than the gap between the smallest and next smallest positive number. This is the case with single- and double-precision floating-point numbers as well. The smallest positive (normalized) floating-point number in any such system is 1.0×2^{-E}, where $-E$ is the smallest representable exponent; the next smallest number is $(1+\epsilon) \times 2^{-E}$, where ϵ is the machine precision, so the gap between these two numbers is ϵ times the gap between 0 and the first positive number. This gap can be filled in using **subnormal** numbers. Subnormal numbers have less precision than normalized floating-point numbers and will be described in the next section.

5.4 IEEE FLOATING-POINT ARITHMETIC

In the 1960s and 1970s, each computer manufacturer developed its own floating-point system, resulting in inconsistent program behavior across machines. Most computers used binary arithmetic, but the IBM 360/70 series used hexadecimal (base 16), and Hewlett-Packard calculators used decimal arithmetic. In the early 1980s, due largely to the efforts of W. Kahan, computer manufacturers adopted a standard: the IEEE (Institute of Electrical and Electronics Engineers) standard. This standard required:

- Consistent representation of floating-point numbers across machines.
- Correctly rounded arithmetic.
- Consistent and sensible treatment of exceptional situations such as divide by 0.

In order to comply with the IEEE standard, computers represent numbers in the way described in the previous section. A special representation is needed for 0 (since it cannot be represented with the standard hidden bit format), and also for $\pm\infty$ (the result of dividing a nonzero number by 0), and also for NaN (Not a Number; e.g., $0/0$). This is done with special bits in the exponent field, which slightly reduces the range of possible exponents. Special bits in the exponent field are also used to signal subnormal numbers.

There are 3 standard precisions. As noted previously, a **single-precision** word consists of 32 bits, with 1 bit for the sign, 8 for the exponent, and 23 for the significand. A **double-precision** word consists of 64 bits, with 1 bit for the sign, 11 for the exponent, and 52 for the significand. An **extended-precision** word consists of 80 bits, with 1 bit for the sign, 15 for the exponent, and 64 for the significand. [Note, however, that numbers stored in extended precision do not use hidden-bit storage.]

Table 5.2 shows the floating-point numbers, subnormal numbers, and exceptional situations that can be represented using IEEE double precision.

It can be seen from the table that the smallest positive normalized floating-point number that can be stored is $1.0_2 \times 2^{-1022} \approx 2.2 \times 10^{-308}$, while the largest is $1.1\ldots1_2 \times 2^{1023} \approx 1.8 \times 10^{308}$. The exponent field for normalized floating-point numbers represents the actual exponent plus 1023, so, using the 11 available exponent bits (and setting aside two possible bit configurations for special situations) we can represent exponents between -1022 and $+1023$.

TABLE 5.2
IEEE double precision.

If exponent field is	Then number is	Type of number:
00000000000	$\pm(0.b_1 \ldots b_{52})_2 \times 2^{-1022}$	0 or subnormal
$00000000001 = 1_{10}$	$\pm(1.b_1 \ldots b_{52})_2 \times 2^{-1022}$	Normalized number
$00000000010 = 2_{10}$	$\pm(1.b_1 \ldots b_{52})_2 \times 2^{-1021}$	
⋮	⋮	Exponent field is
$01111111111 = 1023_{10}$	$\pm(1.b_1 \ldots b_{52})_2 \times 2^{0}$	(actual exponent) + 1023
⋮	⋮	
$11111111110 = 2046_{10}$	$\pm(1.b_1 \ldots b_{52})_2 \times 2^{1023}$	
11111111111	$\pm\infty$ if $b_1 = \ldots = b_{52} = 0$, NaN otherwise	Exception

The two special exponent field bit patterns are all 0s and all 1s. An exponent field consisting of all 0s signals either 0 or a subnormal number. Note that subnormal numbers have a 0 in front of the binary point instead of a 1 and are always multiplied by 2^{-1022}. Thus the number $1.1_2 \times 2^{-1024} = 0.011_2 \times 2^{-1022}$ would be represented with an exponent field string of all 0s and a significand field $0110\ldots0$. Subnormal numbers have less precision than normalized floating-point numbers since the significand is shifted right, causing fewer of its bits to be stored. The smallest positive subnormal number that can be represented has fifty-one 0s followed by a 1 in its significand field, and its value is $2^{-52} \times 2^{-1022} = 2^{-1074}$. The number 0 is represented by an exponent field consisting of all 0s and a significand field of all 0s. An exponent field consisting of all 1s signals an exception. If all bits in the significand are 0, then it is $\pm\infty$. Otherwise it represents NaN.

WILLIAM KAHAN

William Kahan is an eminent mathematician, numerical analyst and computer scientist who has made important contributions to the study of accurate and efficient methods of solving numerical problems on a computer with finite precision. Among his many contributions, Kahan was the primary architect behind the IEEE 754 standard for floating-point computation. Kahan has received many recognitions for his work, including the Turing Award in 1989 and being named an ACM Fellow in 1994. Kahan was a professor of mathematics and computer science and electrical engineering at the University of California, Berkeley, and he continues his contributions to the ongoing revision of IEEE 754. (Photo courtesy of Peg Skorpinski.)

5.5 ROUNDING

There are four rounding modes in the IEEE standard. If x is a real number that cannot be stored exactly, then it is replaced by a nearby floating-point number according to one of the following rules:

- Round down—round(x) is the largest floating-point number that is less than or equal to x.
- Round up—round(x) is the smallest floating-point number that is greater than or equal to x.
- Round towards 0—round(x) is either round-down(x) or round-up(x), whichever lies between 0 and x. Thus if x is positive then round(x) = round-down(x), while if x is negative then round(x) = round-up(x).
- Round to nearest—round(x) is either round-down(x) or round-up(x), whichever is closer. In case of a tie, it is the one whose least significant (rightmost) bit is 0.

The default is *round to nearest.*

Using double precision, the number $\frac{1}{10} = 1.0\overline{0011}00_2 \times 2^{-4}$ is replaced by

0	01111111011	1001100110011001100110011001100110011001100110011001

using round down or round towards 0, while it becomes

0	01111111011	1001100110011001100110011001100110011001100110011010

using round up or round to nearest. [Note also the exponent field which is the binary representation of 1019, or, 1023 plus the exponent −4.]

The **absolute rounding error** associated with a number x is defined as $|\text{round}(x) - x|$. In double precision, if $x = \pm(1.b_1 \ldots b_{52}b_{53} \ldots)_2 \times 2^E$, where E is within the range of representable exponents (−1022 to 1023), then the absolute rounding error associated with x is less than $2^{-52} \times 2^E$ for any rounding mode; the worst rounding errors occur if, for example, $b_{53} = b_{54} = \ldots = b_n = 1$ for some large number n, and round towards 0 is used. For round to nearest, the absolute rounding error is less than or equal to $2^{-53} \times 2^E$, with the worst case being attained if, say, $b_{53} = 1$ and $b_{54} = \ldots = 0$; in this case, if $b_{52} = 0$, then x would be replaced by $1.b_1 \ldots b_{52} \times 2^E$, while if $b_{52} = 1$, then x would be replaced by this number plus $2^{-52} \times 2^E$. Note that 2^{-52} is machine ϵ for double precision. In single and extended precision we have the analogous result that the absolute rounding error is less than $\epsilon \times 2^E$ for any rounding mode, and less than or equal to $\frac{\epsilon}{2} \times 2^E$ for round to nearest.

Usually one is interested not in the absolute rounding error but in the **relative rounding error**, defined as $|\text{round}(x) - x|/|x|$. Since we have seen that $|\text{round}(x) - x| < \epsilon \times 2^E$ when x is of the form $\pm m \times 2^E$, $1 \leq m < 2$, it follows that the relative rounding error is less than $\epsilon \times 2^E/(m \times 2^E) \le \epsilon$. For round to nearest, the relative rounding error is less than or equal to $\frac{\epsilon}{2}$. This means that for any real number x (in the range of numbers that can be represented by normalized floating-point numbers), we can write

$$\text{round}(x) = x(1+\delta), \qquad \text{where } |\delta| < \epsilon \quad (\text{or} \le \tfrac{\epsilon}{2} \text{ for round to nearest}).$$

The IEEE standard requires that *the result of an operation (addition, subtraction, multiplication, or division) on two floating-point numbers must be the correctly rounded value of the exact result.* For numerical analysts, this is the most important statement in this chapter. It means that if a and b are floating-point numbers and $\oplus$, $\ominus$, $\otimes$, and $\oslash$ represent floating-point addition, subtraction, multiplication, and division, then we will have

$$\begin{aligned}
a \oplus b &= \text{round}(a+b) = (a+b)(1+\delta_1),\\
a \ominus b &= \text{round}(a-b) = (a-b)(1+\delta_2),\\
a \otimes b &= \text{round}(ab) = (ab)(1+\delta_3),\\
a \oslash b &= \text{round}(a/b) = (a/b)(1+\delta_4),
\end{aligned}$$

where $|\delta_i| < \epsilon$ (or $\le \epsilon/2$ for round to nearest), $i = 1, \ldots, 4$. This is important in the analysis of many algorithms.

Taking Stock in Vancouver [65, 70, 84]

On Friday, November 25, 1983, investors and brokers watched as an apparent bear market continued to slowly deplete the Vancouver Stock Exchange (VSE). The following Monday, almost magically, the VSE opened over 550 points higher than it had closed on Friday, even though stock prices were unchanged from Friday's closing, as no trading had occurred over the weekend. What *had* transpired between Friday evening and Monday morning was a correction of the index resulting from three weeks of work by consultants from Toronto and California. From the inception of the VSE in January 1982 at a level of 1,000, an error had resulted in a loss of about a point a day or 20 points a month.

Representing 1,500 stocks, the index was recalculated to four decimal places after each recorded transaction and then summarily truncated to only three decimal places. Thus if the index was calculated as 560.9349, it would be truncated to 560.934 instead of rounded to the nearer number 560.935.

With an activity level often at the rate of 3,000 index changes per day, the errors accumulated, until the VSE fell to 520 on November 25, 1983. That next Monday, it was recalculated properly to be 1098.892, thus correcting truncation errors that had been compounding for 22 months.

5.6 CORRECTLY ROUNDED FLOATING-POINT OPERATIONS

While the idea of correctly rounded floating-point operations sounds quite natural and reasonable (Why would anyone implement *incorrectly* rounded floating-point operations?!), it turns out that it is not so easy to accomplish. Here we will give just a flavor of some of the implementation details.

First consider floating-point addition and subtraction. Let $a = m \times 2^E$, $1 \le m < 2$, and $b = p \times 2^F$, $1 \le p < 2$, be two positive floating-point numbers. If $E = F$, then $a + b = (m + p) \times 2^E$. This result may need further normalization if $m + p \ge 2$. Following are two examples, where we retain three digits after the binary point:

$$1.100_2 \times 2^1 + 1.000_2 \times 2^1 = 10.100_2 \times 2^1 \longrightarrow 1.010_2 \times 2^2,$$

$$1.101_2 \times 2^0 + 1.000_2 \times 2^0 = 10.101_2 \times 2^0 \longrightarrow 1.010_2 \times 2^1.$$

The second example required rounding, and we used round to nearest.

Next suppose that $E \neq F$. In order to add the numbers we must first shift one of the numbers to align the significands, as in the following example:

$$1.100_2 \times 2^1 + 1.100_2 \times 2^{-1} = 1.100_2 \times 2^1 + 0.011_2 \times 2^1 = 1.111_2 \times 2^1.$$

As another example, consider adding the two single-precision numbers 1 and $1.11_2 \times 2^{-23}$. This is illustrated below, with the part of the number after the 23rd bit shown on the right of the vertical line:

1.00000000000000000000000		$\times 2^0$
+.00000000000000000000001	11	$\times 2^0$
1.00000000000000000000001	11	$\times 2^0$

Using round to nearest, the result becomes $1.0 \ldots 010_2$.

Now let us consider subtracting the single-precision number $1.1 \ldots 1_2 \times 2^{-1}$ from 1:

1.00000000000000000000000		$\times 2^0$
−.11111111111111111111111	1	$\times 2^0$
0.00000000000000000000000	1	$\times 2^0$

The result is $1.0_2 \times 2^{-24}$, a perfectly good floating-point number, so the IEEE standard requires that we compute this number exactly. In order to do this a **guard bit** is needed to keep track of the 1 to the right of the register after the second number is shifted. Cray computers used to get this wrong because they had no guard bit.

It turns out that correctly rounded arithmetic can be achieved using 2 guard bits and a sticky bit to flag some tricky cases. Following is an example of a tricky case. Suppose we wish to subtract $1.0\ldots01_2 \times 2^{-25}$ from 1:

$$
\begin{array}{r|rr}
1.00000000000000000000000 & & \times 2^0 \\
-.00000000000000000000000 & 0100000000000000000000001 & \times 2^0 \\
\hline
0.11111111111111111111111 & 1011111111111111111111111 & \times 2^0
\end{array}
$$

Renormalizing and using round to nearest, the result is $1.1\ldots1_2 \times 2^{-1}$. This is the correctly rounded value of the true difference, $1 - 2^{-25} - 2^{-48}$. With only 2 guard bits, however, or with any number of guard bits less than 25, the computed result is

$$
\begin{array}{r|rr}
1.00000000000000000000000 & & \times 2^0 \\
-.00000000000000000000000 & 01 & \times 2^0 \\
\hline
0.11111111111111111111111 & 11 & \times 2^0
\end{array}
$$

which, using round to nearest, gives the incorrect answer $1.0_2 \times 2^0$. A sticky bit is needed to flag problems of this sort. In practice, floating-point operations are often carried out using the 80-bit extended-precision registers, in order to minimize the number of special cases that must be flagged.

Floating-point multiplication is relatively easy compared to addition and subtraction. The product $(m \times 2^E) \times (p \times 2^F)$ is $(m \times p) \times 2^{E+F}$. This result may need to be renormalized, but no shifting of the factors is required.

5.7 EXCEPTIONS

Usually when one divides by 0 in a code, it is due to a programming error, but there are occasions when one would actually like to do this and to work with the result as if it were the appropriate mathematical quantity, namely either $\pm\infty$ if the numerator is nonzero or NaN if the numerator is 0. In the past, when a division by 0 occurred, computers would either stop with an error message or continue by setting the result to the largest floating-point number. The latter had the unfortunate consequence that two mistakes could cancel each other out: $1/0 - 2/0 = 0$. Now $1/0$ would be set to ∞, $2/0$ would be set to ∞, and the difference, $\infty - \infty$, would be NaN. The quantities $\pm\infty$ and NaN obey the standard mathematical rules; for example, $\infty \times 0 = \text{NaN}$, $0/0 = \text{NaN}$, $\infty + a = \infty$ and $a - \infty = -\infty$ for a a floating-point number.

Overflow is another type of exceptional situation. This occurs when the true result of an operation is greater than the largest floating-point number ($1.1\ldots1_2 \times 2^{1023} \approx 1.8 \times 10^{308}$ for double precision). How this is handled depends on the rounding mode. Using round up or round to nearest, the result is set to ∞; using round down or round towards 0, it is set to the largest floating-point number. **Underflow** occurs when the true result is less than the smallest floating-point number. The result is stored as a subnormal number if it is in the range of the subnormal numbers, and otherwise it is set to 0.

5.8 CHAPTER 5 EXERCISES

1. Write down the IEEE double-precision representation for the following decimal numbers:

   (a) 1.5, using round up.
   (b) 5.1, using round to nearest.
   (c) -5.1, using round towards 0.
   (d) -5.1, using round down.

2. Write down the IEEE double-precision representation for the decimal number 50.2, using round to nearest.
3. What is the gap between 2 and the next larger double-precision number?
4. What is the gap between 201 and the next larger double-precision number?
5. How many different normalized double-precision numbers are there? Express your answer using powers of 2.
6. Describe an algorithm to compare two double-precision floating-point numbers a and b to determine whether $a < b$, $a = b$, or $a > b$, by comparing each of their bits from left to right, stopping as soon as a differing bit is encountered.
7. What is the largest decimal number x that has the IEEE double-precision representation

0	10000000011	111000000000000000000000000000000000000 0000000001000

   using round to nearest. Explain your answer.
8. Consider a very limited system in which significands are only of the form $1.b_1b_2b_3$ and the only exponents are 0, 1, and -1. What is the machine precision ϵ for this system? Assuming that subnormal numbers are not used, what is the smallest positive number that can be represented in this system, and what is the largest number that can be represented? Express your answers in decimal form.
9. Consider IEEE double-precision floating-point arithmetic, using round to nearest. Let a, b, and c be normalized double-precision floating-point numbers, and let $\oplus$, $\ominus$, $\otimes$, and $\oslash$ denote correctly rounded floating-point addition, subtraction, multiplication, and division.

   (a) Is it necessarily true that $a \oplus b = b \oplus a$? Explain why or give an example where this does not hold.
   (b) Is it necessarily true that $(a \oplus b) \oplus c = a \oplus (b \oplus c)$? Explain why or give an example where this does not hold.
   (c) Determine the maximum possible relative error in the computation $(a \otimes b) \oslash c$, assuming that $c \neq 0$. (You may omit terms of order $O(\epsilon^2)$

and higher.) Suppose $c = 0$; what are the possible values that $(a \otimes b) \oslash c$ could be assigned?

10. Let a and b be two positive floating-point numbers with the same exponent. Explain why the computed difference $a \ominus b$ is always exact using IEEE arithmetic.
11. Explain the behavior seen in (5.1) at the beginning of this chapter. First note that as long as the exponent in the binary representation of x_k is less than -1 (so that $x_k < \frac{1}{2}$), the new iterate x_{k+1} is formed just by multiplying x_k by 2. How is this done using IEEE double-precision arithmetic? Are there any rounding errors? Once the exponent of x_k reaches -1 (assuming that its significand has at least one nonzero bit), the new iterate x_{k+1} is formed by multiplying x_k by 2 (i.e., increasing the exponent to 0) and then subtracting 1. What does this do to the binary representation of the number when the number is renormalized to have the form $1.b_1 \ldots b_{52} \times 2^E$? Based on these observations, can you explain why, starting with any $x_0 \in (0, 1]$, the computed iterates eventually reach 1 and remain there?
12. The total resistance, T, of an electrical circuit with two resistors connected in parallel, with resistances R_1 and R_2 respectively, is given by the formula

$$T = \frac{1}{\frac{1}{R_1} + \frac{1}{R_2}}.$$

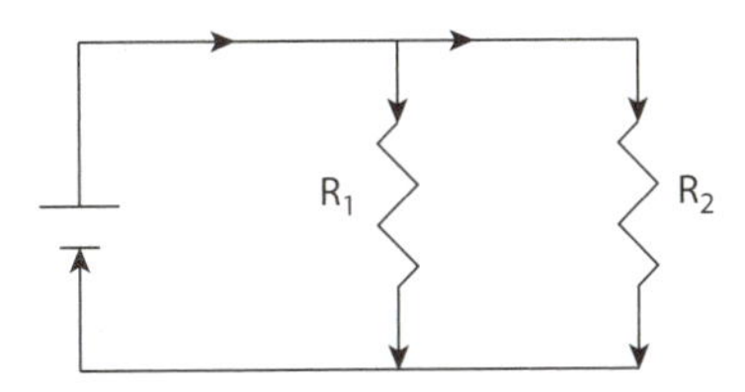

If $R_1 = R_2 = R$, then the total resistance is $R/2$ since half of the current flows through each resistor. On the other hand, if R_2 is much less than R_1, then most of the current will flow through R_2 though a small amount will still flow through R_1, so the total resistance will be slightly less than R_2. What if $R_2 = 0$? Then T should be 0 since if $R_1 \neq 0$, then all of the current will flow through R_2, and if $R_1 = 0$ then there is no resistance anywhere in the circuit. If the above formula is implemented using IEEE arithmetic, will the correct answer be obtained when $R_2 = 0$? Explain your answer.
13. In the 7th season episode *Treehouse of Horror VI* of *The Simpsons*, Homer has a nightmare in which the following equation flies past him:

$$1782^{12} + 1841^{12} = 1922^{12}. \qquad (5.2)$$

Note that this equation, if true, would contradict Fermat's last theorem, which states: For $n \geq 3$, there do not exist any natural numbers x, y and z that satisfy the equation $x^n + y^n = z^n$. Did Homer dream up a counterexample to Fermat's last theorem?

(a) Compute $\sqrt[12]{1782^{12} + 1841^{12}}$ by typing the following into MATLAB:

```
format short
(1782^12 + 1841^12)^(1/12)
```

What result does MATLAB report? Now look at the answer using `format long`.

(b) Determine the *absolute* and *relative* error in the approximation $1782^{12} + 1841^{12} \approx 1922^{12}$. (Such an example is called a Fermat near miss because of the small relative error. This example was created by *The Simpsons* writer David S. Cohen with the intent of catching the eye of audience members with a mathematical interest.)

(c) Note that the right-hand side of equation (5.2) is even. Use this to *prove* that the equation cannot be true.

(d) In a later episode entitled *The Wizard of Evergreen Terrace*, Homer writes the equation $3987^{12} + 4365^{12} = 4472^{12}$. Can you debunk this equation?

14. In the 1999 movie *Office Space*, a character creates a program that takes fractions of cents that are truncated in a bank's transactions and deposits them to his own account. This is not a new idea, and hackers who have actually attempted it have been arrested. In this exercise we will simulate the program to determine how long it would take to become a millionaire this way.
Assume that we have access to 50,000 bank accounts. Initially we can take the account balances to be uniformly distributed between, say, \$100 and \$100,000. The annual interest rate on the accounts is 5%, and interest is compounded daily and added to the accounts, except that fractions of a cent are truncated. These will be deposited to an illegal account that initially has balance \$0.
Write a MATLAB program that simulates the *Office Space* scenario. You can set up the initial accounts with the commands

```
accounts = 100 + (100000-100)*rand(50000,1);
% Sets up 50,000 accounts
% with balances
% between $100 and $100000.
accounts = floor(100*accounts)/100;
% Deletes fractions of a cent
% from initial balances.
```

(a) Write a MATLAB program that increases the accounts by (5/365)% interest each day, truncating each account to the nearest penny and placing the truncated amount into an account, which we will call the illegal account. Assume that the illegal account can hold fractional amounts (i.e., do not truncate this account's values) and let the illegal account also accrue daily interest. Run your code to determine how many days it would take to become a millionaire assuming the illegal account begins with a balance of zero.

(b) Without running your MATLAB code, answer the following questions: On average, about how much money would you expect to be added to the illegal account each day due to the embezzlement? Suppose you had access to 100,000 accounts, each initially with a balance of, say, $5000. About how much money would be added to the illegal account each day in this case? Explain your answers.

Note that this type of rounding corresponds to *fixed-point* truncation rather than *floating-point*, since only two places are allowed to the right of the decimal point, regardless of how many or few decimal digits appear to the left of the decimal point.

15. In the 1991 Gulf War, the Patriot missile defense system failed due to roundoff error. The troubles stemmed from a computer that performed the tracking calculations with an internal clock whose integer values in tenths of a second were converted to seconds by multiplying by a 24-bit binary approximation to one tenth:

$$0.1_{10} \approx 0.00011001100110011001100_2. \qquad (5.3)$$

(a) Convert the binary number in (5.3) to a fraction. Call it x.
(b) What is the absolute error in this number; that is, what is the absolute value of the difference between x and $\frac{1}{10}$?
(c) What is the time error in seconds after 100 hours of operation (i.e., the value of $|\, 360{,}000 - 3{,}600{,}000\, x|$)?
(d) During the 1991 war, a Scud missile traveled at approximately Mach 5 (3750 mph). Find the distance that a Scud missile would travel during the time error computed in (c).

On February 25, 1991, a Patriot battery system, which was to protect the Dhahran Air Base, had been operating for over 100 consecutive hours. The roundoff error caused the system not to track an incoming Scud missile, which slipped through the defense system and detonated on US Army barracks, killing 28 American soldiers.

6

CONDITIONING OF PROBLEMS; STABILITY OF ALGORITHMS

Errors of many sorts are almost always present in scientific computations:

1. Replacing the physical problem by a mathematical model involves approximations.
2. Replacing the mathematical model by a problem that is suitable for numerical solution may involve approximations; for example, truncating an infinite Taylor series after a finite number of terms.
3. The numerical problem often requires some input data, which may come from measurements or from other sources that are not exact.
4. Once an algorithm is devised for the numerical problem and implemented on a computer, rounding errors will affect the computed result.

It is important to know the effects of such errors on the final result. In this book, we deal mainly with items 2–4, leaving item 1 to the specific application area. In this chapter, we will assume that the mathematical problem has been posed in a way that is amenable to numerical solution and study items 3–4: the effects of errors in the input data on the (exact) solution to the numerical problem, and the effects of errors in computation (e.g., rounding all quantities to 16 decimal places) on the output of an algorithm designed to solve the numerical problem.

There are different ways of measuring error. One can talk about the *absolute error* in a computed value $\hat{y}$, which is the absolute value of the difference between $\hat{y}$ and the true value, y: $|\hat{y} - y|$. One can also measure error in a *relative* sense, where the quantity of interest is $|\hat{y} - y|/|y|$. It is usually the relative error that is of interest in applications. In chapter 7, we will deal with numerical problems and algorithms in which the answer is a vector of values rather than a single number. In this case, we will introduce different *norms* in which the error can be measured, again in either an absolute or a relative sense.

6.1 CONDITIONING OF PROBLEMS

The **conditioning** of a problem measures how sensitive the answer is to small changes in the input. [Note that this is *independent* of the algorithm used to compute the answer; it is an intrinsic property of the problem.]

Let f be a scalar-valued function of a scalar argument x, and suppose that $\hat{x}$ is close to x (e.g., $\hat{x}$ might be equal to round(x)). How close is $y = f(x)$ to $\hat{y} = f(\hat{x})$? We can ask this question in an *absolute* sense: If

$$|\hat{y} - y| \approx C(x)|\hat{x} - x|,$$

then $C(x)$ might be called the **absolute condition number** of the function f at the point x. We also can ask the question in a *relative* sense: If

$$\left|\frac{\hat{y} - y}{y}\right| \approx \kappa(x)\left|\frac{\hat{x} - x}{x}\right|,$$

then $\kappa(x)$ might be called the **relative condition number** of f at x.

To determine a possible expression for $C(x)$, note that

$$\hat{y} - y = f(\hat{x}) - f(x) = \frac{f(\hat{x}) - f(x)}{\hat{x} - x} \cdot (\hat{x} - x),$$

and for $\hat{x}$ very close to x, $(f(\hat{x}) - f(x))/(\hat{x} - x) \approx f'(x)$. Therefore $C(x) = |f'(x)|$ is defined to be the absolute condition number. To define the relative condition number $\kappa(x)$, note that

$$\frac{\hat{y} - y}{y} = \frac{f(\hat{x}) - f(x)}{\hat{x} - x} \cdot \frac{\hat{x} - x}{x} \cdot \frac{x}{f(x)}.$$

Again we use the approximation $(f(\hat{x}) - f(x))/(\hat{x} - x) \approx f'(x)$ to determine

$$\kappa(x) = \left|\frac{xf'(x)}{f(x)}\right|.$$

Example 6.1.1. Let $f(x) = 2x$. Then $f'(x) = 2$, so that $C(x) = 2$ and $\kappa(x) = (2x)/(2x) = 1$. This problem is *well conditioned*, in the sense that $C(x)$ and $\kappa(x)$ are of moderate size. Exactly what counts as being of "moderate size" depends on the context, but usually one hopes that if x is changed by an anticipated amount (e.g., 10^{-16} if the change is due to roundoff in double precision, or perhaps a much larger relative amount if the change is due to measurement error), then the change in f will be negligible for the application being considered.

Example 6.1.2. Let $f(x) = \sqrt{x}$. Then $f'(x) = \frac{1}{2}x^{-1/2}$, so that $C(x) = \frac{1}{2}x^{-1/2}$ and $\kappa(x) = 1/2$. This problem is well conditioned in a relative sense. In an absolute sense, it is well conditioned if x is not too close to 0, but if, say, $x = 10^{-16}$, then $C(x) = 0.5 \times 10^8$, so a small absolute change in x (say, changing x from 10^{-16} to 0) results in an absolute change in $\sqrt{x}$ of about 10^8 times the change in x (i.e., from 10^{-8} to 0).

Example 6.1.3. Let $f(x) = \sin x$. Then $f'(x) = \cos x$, so that $C(x) = |\cos x| \leq 1$ and $\kappa(x) = |x \cot x|$. The relative condition number for this function is large if x is near $\pm\pi$, $\pm 2\pi$, etc., and it also is large if $|x|$ is very large and $|\cot x|$ is not extremely small. As $x \to 0$, we find $\kappa(x) \to \lim_{x\to 0} |\frac{x\cos x}{\sin x}| = \lim_{x\to 0} |\frac{\cos x - x\sin x}{\cos x}| = 1$.

In chapter 7, when we deal with vector-valued functions of a vector of arguments—for example, solving a linear system $A\mathbf{y} = \mathbf{b}$—we will define *norms* in order to measure differences in the input vector $\|\mathbf{b} - \hat{\mathbf{b}}\|/\|\mathbf{b}\|$ and in the output vector $\|\mathbf{y} - \hat{\mathbf{y}}\|/\|\mathbf{y}\|$.

6.2 STABILITY OF ALGORITHMS

Suppose we have a well-conditioned problem and an algorithm for solving this problem. Will our algorithm give the answer to the expected number of places when implemented in floating-point arithmetic? An algorithm that achieves the level of accuracy defined by the conditioning of the problem is called **stable,** while one that gets *unnecessarily inaccurate* results due to roundoff is sometimes called **unstable.** To determine the stability of an algorithm one may do a *rounding error analysis.*

Example 6.2.1 (Computing sums). If x and y are two real numbers and they are rounded to floating-point numbers and their sum is computed on a machine with unit roundoff ϵ, then

$$\text{fl}(x+y) \equiv \text{round}(x) \oplus \text{round}(y) = (x(1+\delta_1) + y(1+\delta_2))(1+\delta_3), \quad |\delta_i| \le \epsilon \quad (\epsilon/2 \text{ for round to nearest}), \tag{6.1}$$

where $\text{fl}(\cdot)$ denotes the floating-point result.

Forward Error Analysis. Here one asks the question: How much does the computed value differ from the exact solution? Multiplying the terms in (6.1), we find

$$\text{fl}(x+y) = x + y + x(\delta_1 + \delta_3 + \delta_1\delta_3) + y(\delta_2 + \delta_3 + \delta_2\delta_3).$$

Again one can ask about the *absolute error*

$$|\text{fl}(x+y) - (x+y)| \le (|x| + |y|)(2\epsilon + \epsilon^2),$$

or the *relative error*

$$\left|\frac{\text{fl}(x+y) - (x+y)}{x+y}\right| \le \frac{(|x|+|y|)(2\epsilon + \epsilon^2)}{|x+y|}.$$

Note that if $y \approx -x$, then the relative error can be large! The difficulty is due to the initial rounding of the real numbers x and y, not the rounding of the sum of the two floating-point numbers. (In fact, this sum may be computed exactly; see Exercise 10 in chapter 5.) In the case where $y \approx -x$, however, this would be considered an ill-conditioned problem since small changes in x and y can make a large relative change in their sum; the algorithm for adding the two numbers does as well as one could hope. On the other hand, the subtraction of two nearly equal numbers might be only one step in an algorithm for solving a well-conditioned problem; in this case, the algorithm is likely to be unstable, and one should look for an alternative algorithm that avoids this difficulty.

Backward Error Analysis. Here one tries to show that the computed value is the exact solution to a nearby problem. If the given problem is ill conditioned (i.e., if a small change in the input data makes a large change in the solution), then probably the best one can hope for is to compute the exact solution of a problem with slightly different input data. For the problem of summing the two numbers x and y, we have from (6.1):

$$fl(x+y) = x(1+\delta_1)(1+\delta_3) + y(1+\delta_2)(1+\delta_3),$$

so the computed value is the exact sum of two numbers that differ from x and y by relative amounts no greater than $2\epsilon + \epsilon^2$. The algorithm for adding two numbers is *backward stable*.

Example 6.2.2. Compute $\exp(x)$ using the Taylor series expansion

$$e^x = 1 + x + \frac{x^2}{2!} + \frac{x^3}{3!} + \cdots .$$

The following MATLAB code can be used to compute $\exp(x)$:

```
oldsum = 0;
newsum = 1;
term = 1;
n = 0;
while newsum ~= oldsum,   % Iterate until next term is negligible
   n = n + 1;
   term = term * x/n;     % x^n/n! = (x^{n-1}/(n-1)!) * x/n
   oldsum = newsum;
   newsum = newsum + term;
end;
```

This code adds terms in the Taylor series until the next term is so small that adding it to the current sum makes no change in the floating-point number that is stored. The code works fine for $x > 0$. The computed result for $x = -20$, however, is 5.6219×10^{-9}; the correct result is 2.0612×10^{-9}. The size of the terms in the series increases to 4.3×10^7 (for $n = 20$) before starting to decrease, as seen in table 6.1. This results in double-precision rounding errors of size about $4.3 \times 10^7 \times 10^{-16} = 4.3 \times 10^{-9}$, which lead to completely wrong results when the true answer is on the order of 10^{-9}.

Note that the *problem* of computing $\exp(x)$ for $x = -20$ is well conditioned in both the absolute and relative sense: $C(x)|_{x=-20} = \frac{d}{dx}\exp(x)|_{x=-20} = \exp(-20) << 1$ and $\kappa(x)|_{x=-20} = \left|x\exp(x)/\exp(x)|_{x=-20}\right| = 20$. Hence the difficulty here is with the *algorithm*; it is *unstable*.

Exercise: How can you modify the above code to produce accurate results when x is negative? [Hint: Note that $\exp(-x) = 1/\exp(x)$.]

TABLE 6.1
Terms added in an unstable algorithm for computing e^x.

n	nth term of series	n	nth term of series
1	-20	25	$-2.16 \times 10^{+07}$
2	200	30	$4.05 \times 10^{+06}$
3	$-1.33 \times 10^{+03}$	40	$1.35 \times 10^{+04}$
4	$6.67 \times 10^{+03}$	50	$3.70 \times 10^{+00}$
5	$-2.67 \times 10^{+04}$	60	1.39×10^{-04}
10	$2.82 \times 10^{+06}$	70	9.86×10^{-10}
15	$-2.51 \times 10^{+07}$	80	1.69×10^{-15}
20	$4.31 \times 10^{+07}$	90	8.33×10^{-22}

Example 6.2.3 (Numerical differentiation). Using Taylor's theorem with remainder, one can approximate the derivative of a function $f(x)$:

$$f(x+h) = f(x) + hf'(x) + \frac{h^2}{2} f''(\xi), \quad \xi \in [x, x+h],$$

$$f'(x) = \frac{f(x+h) - f(x)}{h} - \frac{h}{2} f''(\xi).$$

The term $-\frac{h}{2} f''(\xi)$ is referred to as the **truncation error** or **discretization error** when the approximation $(f(x+h) - f(x))/h$ is used for $f'(x)$. The truncation error in this case is of order h, denoted $O(h)$, and the approximation is said to be **first-order accurate.**

Suppose one computes this finite difference quotient numerically. In the best possible case, one may be able to store x and $x+h$ exactly, and suppose that the only errors made in evaluating $f(x)$ and $f(x+h)$ come from rounding the results at the end. Then, ignoring any other rounding errors in subtraction or division, one computes

$$\frac{f(x+h)(1+\delta_1) - f(x)(1+\delta_2)}{h} = \frac{f(x+h) - f(x)}{h} + \frac{\delta_1 f(x+h) - \delta_2 f(x)}{h}.$$

Since $|\delta_1|$ and $|\delta_2|$ are less than ϵ, the absolute error due to roundoff is less than or equal to about $2\epsilon|f(x)|/h$, for small h. Note that the truncation error is proportional to h, while the rounding error is proportional to $1/h$. Decreasing h lowers the truncation error but increases the error due to roundoff.

Suppose $f(x) = \sin x$ and $x = \pi/4$. Then $f'(x) = \cos x$ and $f''(x) = -\sin x$, so the truncation error is about $\sqrt{2}h/4$, while the rounding error is about $\sqrt{2}\epsilon/h$. The most accurate approximation is obtained when the two errors are approximately equal:

$$\frac{\sqrt{2}h}{4} = \frac{\sqrt{2}\epsilon}{h} \Rightarrow h = 2\sqrt{\epsilon}.$$

In this case, the error is on the order of the square root of the machine precision, far less than what one might have hoped for!

Once again, the fault is with the *algorithm*, not the problem of determining $\frac{d}{dx}\sin x|_{x=\pi/4} = \cos x|_{x=\pi/4} = \sqrt{2}/2$. This problem is well conditioned since if the input argument $x = \pi/4$ is changed slightly then the answer, $\cos x$ changes only slightly; quantitatively, the absolute condition number is $C(x)|_{x=\pi/4} = \left|-\sin x|_{x=\pi/4}\right| = \sqrt{2}/2$, and the relative condition number is $\kappa(x)|_{x=\pi/4} = \left|-x\sin x/\cos x|_{x=\pi/4}\right| = \pi/4$. When the problem of evaluating the derivative is replaced by one of evaluating the finite difference quotient, then the conditioning becomes bad. Later we will see how to use higher-order-accurate approximations to the derivative to obtain somewhat better results, but we will not achieve the full machine precision using finite difference quotients.

6.3 CHAPTER 6 EXERCISES

1. What are the absolute and relative condition numbers of the following functions? Where are they large?

   (a) $(x-1)^\alpha$
   (b) $1/(1+x^{-1})$
   (c) $\ln x$
   (d) $\log_{10} x$
   (e) $x^{-1}e^x$
   (f) $\arcsin x$

2. Let $f(x) = \sqrt[3]{x}$.

   (a) Find the absolute and relative condition numbers of f.
   (b) Where is f well conditioned in an absolute sense? In a relative sense?
   (c) Suppose $x = 10^{-17}$ is replaced by $\hat{x} = 10^{-16}$ (a small absolute change but a large relative change). Using the absolute condition number of f, how much of a change is expected in f due to this change in the argument?

3. In evaluating the finite difference quotient $(f(x+h) - f(x))/h$, which is supposed to approximate $f'(x)$, suppose that $x = 1$ and $h = 2^{-24}$. What would be the computed result using IEEE single precision? Explain your answer.
4. Use MATLAB or your calculator to compute $\tan(x_j)$ for $x_j = \frac{\pi}{4} + (2\pi)\times 10^j$, $j = 0, 1, 2, \ldots, 20$. In MATLAB, use `format long e` to print out all of the digits in the answers. What is the relative condition number of the problem of evaluating $\tan x$ for $x = x_j$? Suppose the only error that you make in computing the argument $x_j = \frac{\pi}{4} + (2\pi)\times 10^j$ occurs as a result of rounding π to 16 decimal places; that is, assume that the addition and multiplications and division are done exactly. By what absolute amount will your computed argument $\hat{x}_j$ differ from the exact one? Use this to explain your results.
5. **Compound interest.** Suppose a_0 dollars are deposited in a bank that pays 5% interest per year, compounded quarterly. After one quarter the value of the account is

   $$a_0 \times (1 + (0.05)/4)$$

   dollars. At the end of the second quarter, the bank pays interest not only on the original amount a_0, but also on the interest earned in the first quarter;

thus the value of the investment at the end of the second quarter is

$$[a_0 \times (1 + (0.05)/4)] \times (1 + (0.05)/4) = a_0 \times (1 + (0.05)/4)^2$$

dollars. At the end of the third quarter the bank pays interest on this amount, so that the account is now worth $a_0 \times (1 + (0.05)/4)^3$ dollars, and at the end of the whole year the investment is finally worth

$$a_0 \times (1 + (0.05)/4)^4$$

dollars. In general, if a_0 dollars are deposited at an annual interest rate x, compounded n times per year, then the account value after one year is

$$a_0 \times \mathcal{I}_n(x) \quad \text{where} \quad \mathcal{I}_n(x) = \left(1 + \frac{x}{n}\right)^n.$$

This is the *compound interest formula*. It is well known that for fixed x, $\lim_{n\to\infty} \mathcal{I}_n(x) = \exp(x)$.

(a) Determine the relative condition number $\kappa_{\mathcal{I}_n}(x)$ for the problem of evaluating $\mathcal{I}_n(x)$. For $x = 0.05$, would you say that this problem is well conditioned or ill conditioned?

(b) Use MATLAB to compute $\mathcal{I}_n(x)$ for $x = 0.05$ and for $n = 1, 10, 10^2, \ldots, 10^{15}$. Use `format long e` so that you can see if your results are converging to $\exp(x)$, as one might expect. Turn in a table with your results and a listing of the MATLAB command(s) you used to compute these results.

(c) Try to explain the results of part (b). In particular, for $n = 10^{15}$, you probably computed 1 as your answer. Explain why. To see what is happening for other values of n, consider the problem of computing z^n, where $z = (1 + x/n)$, when n is large. What is the relative condition number of this problem? If you make an error of about 10^{-16} in computing z, about what size error would you expect in z^n?

(d) Can you think of a better way than the method you used in part (b) to accurately compute $\mathcal{I}_n(x)$ for $x = 0.05$ and for large values of n? Demonstrate your new method in MATLAB or explain why it should give more accurate results.

7

DIRECT METHODS FOR SOLVING LINEAR SYSTEMS AND LEAST SQUARES PROBLEMS

This is one of the longest chapters in this book. That may be surprising considering the fact that most people learn to solve linear systems $A\mathbf{x} = \mathbf{b}$ in a linear algebra class. How much more can be said about it?!! The answer is quite a lot; this simple sounding problem really lies at the heart of a great many problems in scientific computing, and it has been the focus of a great deal of effort in the numerical analysis community. Whenever one deals with systems of nonlinear equations or optimization problems, ordinary differential equations, or partial differential equations, (and practically every physical phenomenon that can be modeled is modeled by some combination of these), the algorithms that one uses, at their core, almost always solve systems of linear algebraic equations. For this reason it is important that these solvers be *accurate* and *efficient*. This chapter deals with questions of accuracy and efficiency.

At the beginning of the computer age some mathematicians believed that linear systems of a size greater than about 40 by 40 would never be solvable by computers—not because it would take too long, but because rounding errors would accumulate and destroy all accuracy. An early paper in 1943 by Hotelling [53] arrived at the conclusion that errors in Gaussian elimination could grow exponentially with the size of the matrix, making the algorithm all but useless for large linear systems. A later analysis in 1947 by Goldstine and von Neumann [99] showed that the algorithm could solve positive definite systems accurately, and so they suggested replacing a general linear system $A\mathbf{x} = \mathbf{b}$ by $A^T A\mathbf{x} = A^T\mathbf{b}$; we will see in this chapter that this is **not** a good thing to do. Despite these predictions of failure, by the early 1950s engineers were successfully using computers to solve linear systems of a larger size: 100 by 100 and greater [40, 92]. It was not until 1961, when Wilkinson applied the idea of *backward error analysis* to the solution of linear systems, that the algorithm gained a solid theoretical footing and the earlier pessimism was dispelled [107].

7.1 REVIEW OF MATRIX MULTIPLICATION

In this section, we discuss different ways of looking at matrix–vector and matrix–matrix multiplication that often prove useful. For a more comprehensive review of linear algebra in general, see Appendix A.

Let A be an m by n matrix:

$$A = \begin{pmatrix} a_{11} & \dots & a_{1n} \\ \vdots & & \vdots \\ a_{m1} & \dots & a_{mn} \end{pmatrix}.$$

The **transpose** of A, denoted A^T is the matrix whose (i, j)-entry is a_{ji}. If the entries in A are complex numbers, instead of real numbers, then one often uses the **Hermitian transpose** A^*, whose (i, j)-entry is the complex conjugate of a_{ji}, denoted $\bar{a}_{ji}$. The matrix A is said to be **symmetric** (**Hermitian**) if $A = A^T$ $(A = A^*)$. Note that a matrix must be square $(m = n)$ in order to be symmetric or Hermitian and that a real symmetric matrix is also Hermitian.

Let $\mathbf{b}$ be an n-vector:

$$\mathbf{b} = \begin{pmatrix} b_1 \\ \vdots \\ b_n \end{pmatrix}.$$

The matrix–vector product $A\mathbf{b}$ is the m-vector defined by

$$A\mathbf{b} = \begin{pmatrix} a_{11}b_1 + \dots + a_{1n}b_n \\ \vdots \\ a_{m1}b_1 + \dots + a_{mn}b_n \end{pmatrix}.$$

There are other ways to think about the matrix–vector product that often prove useful. The ith entry in $A\mathbf{b}$ is the inner product of the ith row of A with the vector $\mathbf{b}$; that is, if $\mathbf{a}_{i,:}$ [note the pseudo-MATLAB notation; in MATLAB this would be `A(i,:)`] denotes the ith row of A, $(a_{i1}, \dots, a_{in})$, and if $\langle \cdot, \cdot \rangle$ denotes the inner product, then

$$A\mathbf{b} = \begin{pmatrix} \langle \mathbf{a}_{1,:}, \mathbf{b} \rangle \\ \vdots \\ \langle \mathbf{a}_{m,:}, \mathbf{b} \rangle \end{pmatrix}.$$

The product $A\mathbf{b}$ also can be thought of as a linear combination of the columns of A. If $\mathbf{a}_{:,j}$ denotes the jth column of A, $(a_{1j}, \dots, a_{mj})^T$, then

$$A\mathbf{b} = \mathbf{a}_{:,1}b_1 + \dots + \mathbf{a}_{:,n}b_n.$$

Similar interpretations can be given for the product of an m-dimensional row vector $\mathbf{c} = (c_1, \dots, c_m)$ with the m by n matrix A. The result is an n-

dimensional row vector:

$$\mathbf{c}A = \left(\sum_{i=1}^{m} c_i a_{i1}, \ldots, \sum_{i=1}^{m} c_i a_{in} \right).$$

The jth entry in $\mathbf{c}A$ is the inner product of $\mathbf{c}$ with the jth column of A. The vector $\mathbf{c}A$ also can be thought of as a linear combination of the rows $\mathbf{a}_{i,:}$ of A:

$$\mathbf{c}A = \sum_{i=1}^{m} c_i \mathbf{a}_{i,:}.$$

Example 7.1.1. Find $A\mathbf{b}$ and $\mathbf{c}A$, where

$$A = \begin{pmatrix} 1 & 2 & 3 \\ 4 & 5 & 6 \end{pmatrix}, \quad \mathbf{b} = \begin{pmatrix} -1 \\ 0 \\ 2 \end{pmatrix}, \quad \mathbf{c} = (-3, 1).$$

Using the standard rules of multiplication, we find

$$A\mathbf{b} = \begin{pmatrix} 5 \\ 8 \end{pmatrix}, \quad \mathbf{c}A = (1, -1, -3).$$

These results also can be written in the forms:

$$A\mathbf{b} = -1 \cdot \begin{pmatrix} 1 \\ 4 \end{pmatrix} + 0 \cdot \begin{pmatrix} 2 \\ 5 \end{pmatrix} + 2 \cdot \begin{pmatrix} 3 \\ 6 \end{pmatrix},$$

$$\mathbf{c}A = -3 \cdot (1, 2, 3) + 1 \cdot (4, 5, 6).$$

If G is an n by p matrix, then the product AG is defined and has dimensions m by p. The (i, j)-entry in AG is $(AG)_{ij} = \sum_{k=1}^{n} a_{ik} g_{kj}$. The jth column of AG is the product of A with the jth column of G, namely, $\mathbf{a}_{:,1} g_{1j} + \ldots + \mathbf{a}_{:,n} g_{nj}$. The ith row of AG is the product of the ith row of A with G, namely, $a_{i1}\mathbf{g}_{1,:} + \ldots + a_{in}\mathbf{g}_{n,:}$.

7.2 GAUSSIAN ELIMINATION

Suppose A is an n by n nonsingular matrix. Then for any n-vector $\mathbf{b}$, the linear system $A\mathbf{x} = \mathbf{b}$ has a unique solution $\mathbf{x}$, which can be determined by *Gaussian elimination*. This type of algorithm actually appeared in the *Jiuzhang Suanshu* written in approximately 200 BC in China. Gauss developed the method systematically during his study of the orbit of the asteroid Pallas. Using data collected between 1803 and 1809, Gauss derived a set of six linear equations in six unknowns, which he solved using the method that would later bear his name [39, p. 192].

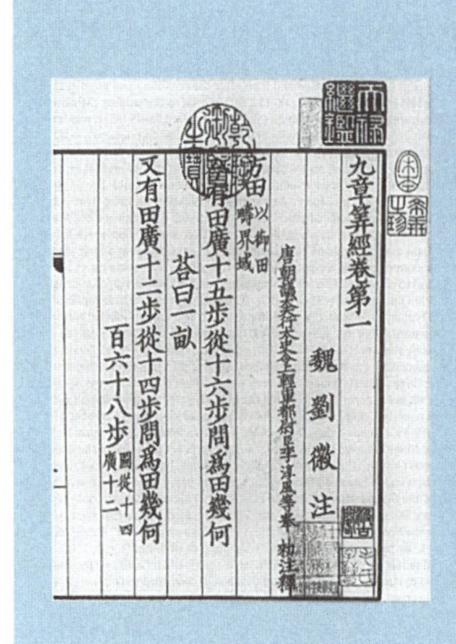
九章筭經卷第一

魏 劉 徽 注

方田 以御田疇界域

今有田廣十五步從十六步問爲田幾何

荅曰一畝

又有田廣十二步從十四步問爲田幾何

百六十八步

Origins of Matrix Methods

Jiuzhang Suanshu is a Chinese manuscript dating from approximately 200 BC and containing 246 problems intended to illustrate methods of solution for everyday problems in areas such as engineering, surveying, and trade. (To the left, we see the opening of chapter 1.) Chapter 8 of this ancient document details the first known example of matrix methods with a method known as *fangcheng*. The method of solution, described in detail, is what would become known centuries later as Gaussian elimination. The coefficients of the system are written as a table on a "counting board." The text also compares two different fangcheng methods by counting the number of counting board operations needed in each method. [2, 57]

Consider the following 3 by 3 linear system:

$$\begin{pmatrix} 1 & 2 & 3 \\ 4 & 5 & 6 \\ 7 & 8 & 0 \end{pmatrix} \begin{pmatrix} x_1 \\ x_2 \\ x_3 \end{pmatrix} = \begin{pmatrix} 1 \\ 0 \\ 2 \end{pmatrix}.$$

Recall that this is shorthand notation for the system of linear equations

$$\begin{aligned} x_1 + 2x_2 + 3x_3 &= 1, \\ 4x_1 + 5x_2 + 6x_3 &= 0, \\ 7x_1 + 8x_2 &= 2. \end{aligned}$$

To solve this linear system using Gaussian elimination, we can append the right-hand side vector to the matrix and then add multiples of one row to another to try to eliminate entries below the diagonal. For example, adding -4 times the first row to the second and -7 times the first row to the third, we obtain

$$\left(\begin{array}{ccc|c} 1 & 2 & 3 & 1 \\ 4 & 5 & 6 & 0 \\ 7 & 8 & 0 & 2 \end{array}\right) \longrightarrow \left(\begin{array}{ccc|c} 1 & 2 & 3 & 1 \\ 0 & -3 & -6 & -4 \\ 0 & -6 & -21 & -5 \end{array}\right).$$

To eliminate the (3, 2)-entry, add -2 times the second row to the third:

$$\left(\begin{array}{ccc|c} 1 & 2 & 3 & 1 \\ 0 & -3 & -6 & -4 \\ 0 & 0 & -9 & 3 \end{array}\right).$$

The system is now *upper triangular* and can be solved easily by *back substitution*. Recalling that the above is shorthand for the system of equations

$$\begin{aligned} x_1 + 2x_2 + 3x_3 &= 1, \\ -3x_2 - 6x_3 &= -4, \\ -9x_3 &= 3, \end{aligned}$$

we solve the third equation to obtain $x_3 = -1/3$; substitute this into the second to get $-3x_2 - 6 \cdot (-1/3) = -4$, so that $x_2 = 2$; substitute each of these into the first equation to get $x_1 + 2 \cdot 2 + 3 \cdot (-1/3) = 1$, so that $x_1 = -2$.

Another way to think about the process of Gaussian elimination is as *factoring* the matrix A into the product of a lower and an upper triangular matrix. When we added -4 times the first row to the second and -7 times the first row to the third, what we actually did was to multiply the original matrix on the left by a certain lower triangular matrix L_1:

$$L_1 A \equiv \begin{pmatrix} 1 & 0 & 0 \\ -4 & 1 & 0 \\ -7 & 0 & 1 \end{pmatrix} \begin{pmatrix} 1 & 2 & 3 \\ 4 & 5 & 6 \\ 7 & 8 & 0 \end{pmatrix} = \begin{pmatrix} 1 & 2 & 3 \\ 0 & -3 & -6 \\ 0 & -6 & -21 \end{pmatrix}.$$

The matrix L_1 is especially easy to invert: Just negate the off-diagonal entries and keep everything else the same:

$$L_1^{-1} = \begin{pmatrix} 1 & 0 & 0 \\ 4 & 1 & 0 \\ 7 & 0 & 1 \end{pmatrix}.$$

Similarly, adding -2 times the second row to the third row in the resulting matrix is equivalent to multiplying on the left by another lower triangular matrix L_2:

$$L_2 L_1 A \equiv \begin{pmatrix} 1 & 0 & 0 \\ 0 & 1 & 0 \\ 0 & -2 & 1 \end{pmatrix} \begin{pmatrix} 1 & 2 & 3 \\ 0 & -3 & -6 \\ 0 & -6 & -21 \end{pmatrix} = \begin{pmatrix} 1 & 2 & 3 \\ 0 & -3 & -6 \\ 0 & 0 & -9 \end{pmatrix}.$$

The inverse of L_2 is also obtained by just negating the off-diagonal entries, and the product $(L_2 L_1)^{-1} = L_1^{-1} L_2^{-1}$ is easy to compute as well: The diagonal entries are 1s and the nonzero off-diagonal entries are those of L_1^{-1} together with those of L_2^{-1}; that is,

$$(L_2 L_1)^{-1} = L_1^{-1} L_2^{-1} = \begin{pmatrix} 1 & 0 & 0 \\ 4 & 1 & 0 \\ 7 & 2 & 1 \end{pmatrix}.$$

Letting U denote the upper triangular matrix obtained from A, we have $L_2 L_1 A = U$, so that $A = LU$, where L is the lower triangular matrix $(L_2 L_1)^{-1}$.

If one wishes to solve several linear systems with the same coefficient matrix but different right-hand side vectors, then one can save the lower and upper triangular factors L and U. Then, to solve $A\mathbf{x} \equiv LU\mathbf{x} = \mathbf{b}$, one first solves the lower triangular system $L\mathbf{y} = \mathbf{b}$ (to obtain $\mathbf{y} = U\mathbf{x}$) and then the upper triangular system $U\mathbf{x} = \mathbf{y}$.

Let us generalize the above to an arbitrary n by n matrix A, and see how a Gaussian elimination routine might be written in MATLAB. The first step is to add multiples of row 1 to each of the other rows in order to eliminate entries below the diagonal in the first column of the matrix. We will apply these same operations to the right-hand side vector **b**:

$$\begin{pmatrix} a_{11} & a_{12} & \dots & a_{1n} & | & b_1 \\ a_{21} & a_{22} & \dots & a_{2n} & | & b_2 \\ \vdots & \vdots & & \vdots & \vdots & \vdots \\ a_{n1} & a_{n2} & \dots & a_{nn} & | & b_n \end{pmatrix} \longrightarrow \begin{pmatrix} a_{11} & a_{12} & \dots & a_{1n} & | & b_1 \\ 0 & a_{22}^{(1)} & \dots & a_{2n}^{(1)} & | & b_2^{(1)} \\ \vdots & \vdots & & \vdots & \vdots & \vdots \\ 0 & a_{n2}^{(1)} & \dots & a_{nn}^{(1)} & | & b_n^{(1)} \end{pmatrix}.$$

The MATLAB code for this step can be written as follows:

```
% Step 1 of Gaussian elimination.
for i=2:n                              % Loop over rows below row 1.
  mult = A(i,1)/A(1,1);           % Subtract this multiple of row 1
                                  % from row i to make A(i,1)=0.
  A(i,:) = A(i,:) - mult*A(1,:);   % This is equivalent to the
                                   % for loop:
                                   % for k=1:n,
                                   % A(i,k) = A(i,k)-mult*A(1,k);
                                   % end;

  b(i) = b(i) - mult*b(1);
end;
```

The next step is to add multiples of row 2 in the modified matrix to rows 3 through n in order to eliminate entries below the diagonal in the second column of the matrix. Again these same operations are applied to **b**. In general, the jth step will operate on rows $j+1$ through n, adding multiples of row j to these rows in order to eliminate entries below the diagonal in column j:

$$\begin{pmatrix} 0 & \dots & 0 & a_{jj}^{(j-1)} & a_{j,j+1}^{(j-1)} & \dots & a_{jn}^{(j-1)} & | & b_j^{(j-1)} \\ 0 & \dots & 0 & a_{j+1,j}^{(j-1)} & a_{j+1,j+1}^{(j-1)} & \dots & a_{j+1,n}^{(j-1)} & | & b_{j+1}^{(j-1)} \\ \vdots & & \vdots & \vdots & \vdots & & \vdots & \vdots & \vdots \\ 0 & \dots & 0 & a_{nj}^{(j-1)} & a_{n,j+1}^{(j-1)} & \dots & a_{nn}^{(j-1)} & | & b_n^{(j-1)} \end{pmatrix} \longrightarrow$$

$$\begin{pmatrix} 0 & \dots & 0 & a_{jj}^{(j-1)} & a_{j,j+1}^{(j-1)} & \dots & a_{jn}^{(j-1)} & | & b_j^{(j-1)} \\ 0 & \dots & 0 & 0 & a_{j+1,j+1}^{(j)} & \dots & a_{j+1,n}^{(j)} & | & b_{j+1}^{(j)} \\ \vdots & & \vdots & \vdots & \vdots & & \vdots & \vdots & \vdots \\ 0 & \dots & 0 & 0 & a_{n,j+1}^{(j)} & \dots & a_{nn}^{(j)} & | & b_n^{(j)} \end{pmatrix}.$$

To accomplish this, the above MATLAB code (slightly modified) must be surrounded by an outer loop over columns $j = 1, \ldots, n-1$. The result is:

```
% Gaussian elimination without pivoting.
for j=1:n-1                          % Loop over columns.
  for i=j+1:n                        % Loop over rows below j.
    mult = A(i,j)/A(j,j);            % Subtract this multiple of
                                     % row j from
                                     % row i to make A(i,j)=0.
    A(i,:) = A(i,:) - mult*A(j,:);  % This does more work than
                                     % necessary;
                                     % do you see why?

    b(i) = b(i) - mult*b(j);
  end;
end;
```

Recall that at the jth stage of Gaussian elimination, the entries in columns 1 through $j-1$ of rows j through n are zero, so that subtracting `mult` times row j from row i ($i > j$) does not alter these zero entries. Therefore the above code could be made more efficient by replacing the line `A(i,:) = A(i,:) - mult*A(j,:);` by

```
A(i,j:n) = A(i,j:n) - mult*A(j,j:n);
```

This operates only on entries j through n of row i.

The above Gaussian elimination code may fail! Can you see what difficulties might arise? One problem is that, in order to use row j to eliminate entries in column j, it must be the case that the (j, j)-entry in the (now modified) matrix is nonzero; that is, the denominator in the expression for `mult` must be nonzero. To ensure this, we will use a procedure called *partial pivoting* that will be described in section 7.2.3.

7.2.1 Operation Counts

The timing of a computer code depends on many things, one of which is the amount of arithmetic that the code does. This used to be the primary factor determining timing, but with faster arithmetic operations available on today's computers, other factors come into play such as the frequency of memory references and the amount of work that can be done in parallel. These other factors will be mentioned in section 7.2.5, but, for now, let us count the number of floating-point operations (additions, subtractions, multiplications, and divisions) performed by our Gaussian elimination code. It should be noted that historically, only multiplications and divisions were counted, because these operations were significantly slower than additions and subtractions. Today, on most computers, the difference in timing between these operations is not great, so we will be concerned with the total number of floating-point operations. Operations on integers, such as incrementing a loop index, are typically less expensive and so are not counted.

In the above Gaussian elimination code, we loop over j going from 1 to $n-1$. Therefore the number of operations performed will be $\sum_{j=1}^{n-1}$ (number of operations performed at stage j). At each stage j, we loop over i going from $j+1$ to n, so the amount of work becomes $\sum_{j=1}^{n-1}\sum_{i=j+1}^{n}$ (number of operations performed on row i at stage j). Now for each row i, we compute `mult`, which requires one division. We then subtract `mult` times row j from row i, which requires n multiplications and n subtractions (using the above code, without the improvement for efficiency). Finally, we subtract `mult` times `b(j)` from `b(i)`, which requires one multiplication and one subtraction. Thus the total number of operations on row i at stage j is $1+2n+2 = 2n+3$. Summing over i and j now, we find the total number of floating-point operations performed by our Gaussian elimination code:

$$\begin{aligned}\sum_{j=1}^{n-1}\sum_{i=j+1}^{n}(2n+3) &= \sum_{j=1}^{n-1}(n-j)(2n+3)\\ &= (2n+3)\sum_{j=1}^{n-1}(n-j) = (2n+3)\sum_{k=1}^{n-1}k\\ &= (2n+3)\frac{n(n-1)}{2} = n^3 + O(n^2).\end{aligned}$$

We are usually interested in the timing of the code when n is large, say $n \geq 1000$, since the code runs very quickly for small n. Therefore, it is customary to determine the highest power of n in the operation count and the constant multiplying that highest power of n and to write the remaining lower-order terms using the $O(\cdot)$ notation, since they will be insignificant for large n. Sometimes the above operation count is written simply as $O(n^3)$, to indicate that it depends on the cube of the matrix size, but it is also important to know the constant multiplying n^3, which in this case is 1.

It was noted above that the efficiency of our Gaussian elimination code could be improved by replacing the line `A(i,:) = A(i,:) - mult*A(j,:)` by one that operates only on entries j through n of row i. Since this requires only $2(n-j+1)$ operations instead of $2n$ operations, the work count becomes

$$\begin{aligned}\sum_{j=1}^{n-1}\sum_{i=j+1}^{n}[2(n-j)+5] &= \sum_{j=1}^{n-1}[2(n-j)^2+5(n-j)]\\ &= 2\sum_{k=1}^{n-1}k^2 + 5\sum_{k=1}^{n-1}k. \qquad (7.1)\end{aligned}$$

Now to compute the operation count, we must use the formula for the sum of the first $n-1$ integers, as well as their squares. These formulas are

$$\sum_{k=1}^{m}k = \frac{m(m+1)}{2}, \qquad \sum_{k=1}^{m}k^2 = \frac{m(m+1)(2m+1)}{6}. \qquad (7.2)$$

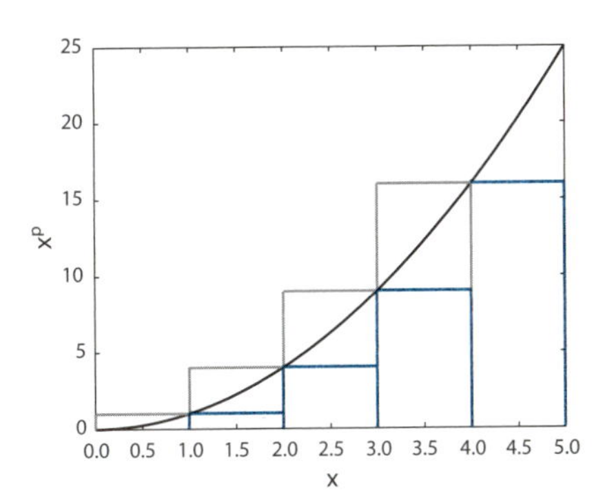

Figure 7.1. The area of the black rectangles is an upper bound on $\int_0^4 x^p\,dx$, while the area of the blue rectangles (which is the same as that of the black rectangles) is a lower bound on $\int_1^5 x^p\,dx$.

Making these substitutions in (7.1) gives the total operation count:

$$2\frac{(n-1)n(2n-1)}{6} + 5\frac{(n-1)n}{2} = \frac{2}{3}n^3 + O(n^2).$$

The work is still $O(n^3)$, but the constant multiplying n^3 has been reduced from 1 to 2/3.

The first formula in (7.2) is easy to remember: it says that the sum of the first m integers is their average value, $(m+1)/2$, times the number of terms, m. The second formula and formulas for sums of higher powers are less easy to remember, but their highest-order terms can be derived as follows. Using rectangles to approximate the area under the curve x^p, for p a positive integer, we find

$$\int_0^m x^p\,dx \leq 1^p + 2^p + \ldots + m^p \leq \int_1^{m+1} x^p\,dx.$$

These inequalities are illustrated in figure 7.1.

Since $\int_0^m x^p\,dx = m^{p+1}/(p+1)$ and $\int_1^{m+1} x^p\,dx = ((m+1)^{p+1}-1)/(p+1) = m^{p+1}/(p+1) + O(m^p)$, and the highest-order term in each of these expressions is $m^{p+1}/(p+1)$, it follows from the above inequalities that

$$\sum_{k=1}^{m} k^p = \frac{m^{p+1}}{p+1} + O(m^p).$$

7.2.2 LU Factorization

We previously interpreted Gaussian elimination as factoring the matrix in the form $A = LU$, where L is lower triangular with 1s on its diagonal (often called a unit lower triangular matrix) and U is upper triangular. It is easy to modify our Gaussian elimination code to save the L and U factors so that they can be used to solve linear systems with the same coefficient matrix but different right-hand side vectors. In fact, this can be done without using any extra storage, because the entries in the strict lower triangle of A (the part of the matrix below the main diagonal) are zeroed out during Gaussian elimination. Therefore this space could be used to store the strict lower triangle of L, and since the diagonal entries in L are always 1, there is no need to store these. The entries in L are simply the multipliers (the variable called `mult` in the previous code) used to eliminate entries in the lower triangle of A. Therefore the Gaussian elimination code can be modified as follows:

```
% LU factorization without pivoting.
for j=1:n-1                            % Loop over columns.
  for i=j+1:n                          % Loop over rows below j.
    mult = A(i,j)/A(j,j);              % Subtract this multiple of
                                       % row j from row i
                                       % to make A(i,j)=0.
    A(i,j+1:n) = A(i,j+1:n) - mult*A(j,j+1:n); % This works on
                                               % columns j+1
                                               % through n.
    A(i,j) = mult;                     % Since A(i,j) becomes 0, use
                                       % the space to store L(i,j).
  end;
end;
```

As noted previously, once A has been factored in the form LU, we can solve a linear system $A\mathbf{x} \equiv LU\mathbf{x} = \mathbf{b}$ by first solving $L\mathbf{y} = \mathbf{b}$ and then solving $U\mathbf{x} = \mathbf{y}$. To solve $L\mathbf{y} = \mathbf{b}$, we solve the first equation for y_1, then substitute this value into the second equation in order to find y_2, etc., as shown below:

$$\begin{pmatrix} \ell_{11} & 0 & \dots & 0 \\ \ell_{21} & \ell_{22} & & 0 \\ \vdots & \vdots & \ddots & \vdots \\ \ell_{n1} & \ell_{2n} & \dots & \ell_{nn} \end{pmatrix} \begin{pmatrix} y_1 \\ y_2 \\ \vdots \\ y_n \end{pmatrix} = \begin{pmatrix} b_1 \\ b_2 \\ \vdots \\ b_n \end{pmatrix} \Longrightarrow$$

$$\begin{aligned} y_1 &= b_1/\ell_{11}, \\ y_2 &= (b_2 - \ell_{21}y_1)/\ell_{22}, \\ \vdots & \quad \vdots \\ y_i &= \left(b_i - \sum_{j=1}^{i-1} \ell_{ij} y_j \right) / \ell_{ii}. \end{aligned}$$

The procedure simplifies slightly when the diagonal entries in L are all 1s. The following MATLAB function solves $L\mathbf{y} = \mathbf{b}$, when L has a unit diagonal and the strict lower triangle of L is stored:

```
function y = lsolve(L, b)
%
% Given a lower triangular matrix L with unit diagonal
% and a vector b,
% this routine solves Ly = b and returns the solution y.
```

```
%
n = length(b);                          % Determine size of problem.
for i=1:n,                              % Loop over equations.
  y(i) = b(i);                          % Solve for y(i) using
  for j=1:i-1,                          % previously computed y(j),
                                        % j=1,...,i-1.
    y(i) = y(i) - L(i,j)*y(j);
  end;
end;
```

Since one subtraction and one multiplication are performed inside the innermost loop of this routine, the total operation count is

$$\sum_{i=1}^{n}\sum_{j=1}^{i-1} 2 = \sum_{i=1}^{n} 2(i-1) = 2\sum_{k=1}^{n-1} k = n(n-1) = n^2 + O(n).$$

Thus, once the matrix has been factored, a triangular solve can be carried out with just $O(n^2)$ operations instead of the $O(n^3)$ required for factorization. This means that if n is large and one saves the LU factors of A, then one can solve with many different right-hand side vectors for almost the same cost as solving with one.

It is left as an exercise to write a MATLAB routine to solve the upper triangular system $U\mathbf{x} = \mathbf{y}$ and to count the number of operations performed.

7.2.3 Pivoting

As noted previously, the codes that we have given so far may fail. If, at the jth step of Gaussian elimination, the (j, j)-entry in the modified matrix is zero, then that row cannot be used to eliminate nonzero entries in the jth column. If the (j, j)-entry is zero, then we could use a different row, say row $k > j$, whose jth element is nonzero in order to eliminate other nonzeros in column j. For convenience, we could interchange rows k and j (remembering also to interchange the right-hand side entries b_k and b_j), so that we maintain the upper triangular form. This process of interchanging rows is called **pivoting,** and the nonzero entry a_{kj} is called the **pivot element.**

Suppose there are no nonzero entries on or below the main diagonal in column j. Then the matrix must be singular. Do you see why? It would mean that after $j-1$ steps of Gaussian elimination, each of the first j columns of the matrix have nonzeros only in their top $j-1$ positions. Now any j vectors of length $j-1$ must be linearly dependent. This implies that the first j columns of the matrix are linearly dependent and so the matrix is singular. Recall that if A is singular then the linear system $A\mathbf{x} = \mathbf{b}$ does not have a unique solution; it has either no solutions (if $\mathbf{b}$ is outside the range of A) or infinitely many solutions (if $\mathbf{b}$ is in the range of A, in which case the solution set consists of any specific solution plus anything from the null space of A). In this case, it would be reasonable for our Gaussian elimination code to return an error message stating that the matrix is singular.

One possible approach to fixing the Gaussian elimination code, then, would be to modify the code to test if a pivot element (the denominator in `mult`) is zero. If it is, then search through that column (below the main diagonal) to find a nonzero entry. When a nonzero entry is found, interchange that row with the pivot row and then do the elimination. If no nonzero entry is found in the column, then return with an error message.

In exact arithmetic, this strategy would be fine, but in finite-precision arithmetic it is not a good idea. Suppose a certain pivot element would be zero in exact arithmetic but because of rounding errors, the computed value is some tiny number, say, the machine precision ϵ. We would use this tiny (and incorrect) number to eliminate entries in other rows by adding huge multiples of the pivot row to these other rows. The results could be completely wrong. Even if there is no error in the pivot entry, adding huge multiples of the pivot row to other rows can result in significant errors, as can be seen with the following linear system:

$$\begin{pmatrix} 10^{-20} & 1 \\ 1 & 1 \end{pmatrix} \begin{pmatrix} x_1 \\ x_2 \end{pmatrix} = \begin{pmatrix} 1 \\ 2 \end{pmatrix}.$$

The exact solution is very nearly $x_1 = x_2 = 1$. (Precisely, it is $x_2 = 1 - 1/(10^{20} - 1)$, $x_1 = 1 + 1/(10^{20} - 1)$.) Using our Gaussian elimination code with double-precision arithmetic, the result after the first stage would be

$$\left(\begin{array}{cc|c} 10^{-20} & 1 & 1 \\ 1 & 1 & 2 \end{array}\right) \longrightarrow \left(\begin{array}{cc|c} 10^{-20} & 1 & 1 \\ 0 & -10^{20} & -10^{20} \end{array}\right),$$

since `mult` would be 10^{20}, and $1 - 10^{20}$ and $2 - 10^{20}$ would each be rounded to -10^{20}. Solving the upper triangular system we would then obtain $x_2 = 1$ and $10^{-20}x_1 + 1 = 1$ so $x_1 = 0$.

This wrong answer was easily avoidable! Had we simply interchanged the two rows and used the second row as the pivot row, we would have obtained

$$\left(\begin{array}{cc|c} 1 & 1 & 2 \\ 10^{-20} & 1 & 1 \end{array}\right) \longrightarrow \left(\begin{array}{cc|c} 1 & 1 & 2 \\ 0 & 1 & 1 \end{array}\right),$$

where `mult` is 10^{-20} and the numbers $1 - 10^{-20}$ and $1 - 2 \times 10^{-20}$ were each rounded to 1. The solution to this upper triangular system is: $x_2 = 1$ and $x_1 + 1 = 2$ so that $x_1 = 1$.

To avoid difficulties with tiny pivot elements, a strategy called **partial pivoting** is generally used. Here one searches for the *largest* entry in the column in absolute value and uses that as the pivot element. This ensures that the multipliers of the pivot row used to eliminate in other rows are all less than or equal to 1 in absolute value. As an example, we consider the 3 by 3 problem given at the beginning of this section:

$$\begin{pmatrix} 1 & 2 & 3 \\ 4 & 5 & 6 \\ 7 & 8 & 0 \end{pmatrix} \begin{pmatrix} x_1 \\ x_2 \\ x_3 \end{pmatrix} = \begin{pmatrix} 1 \\ 0 \\ 2 \end{pmatrix}.$$

Using partial pivoting, we would reduce this problem to upper triangular form as follows:

$$\begin{pmatrix} 1 & 2 & 3 & | & 1 \\ 4 & 5 & 6 & | & 0 \\ 7 & 8 & 0 & | & 2 \end{pmatrix} \xrightarrow{\text{pivot}} \begin{pmatrix} 7 & 8 & 0 & | & 2 \\ 4 & 5 & 6 & | & 0 \\ 1 & 2 & 3 & | & 1 \end{pmatrix} \longrightarrow \begin{pmatrix} 7 & 8 & 0 & | & 2 \\ 0 & 3/7 & 6 & | & -8/7 \\ 0 & 6/7 & 3 & | & 5/7 \end{pmatrix}$$

$$\xrightarrow{\text{pivot}} \begin{pmatrix} 7 & 8 & 0 & | & 2 \\ 0 & 6/7 & 3 & | & 5/7 \\ 0 & 3/7 & 6 & | & -8/7 \end{pmatrix} \longrightarrow \begin{pmatrix} 7 & 8 & 0 & | & 2 \\ 0 & 6/7 & 3 & | & 5/7 \\ 0 & 0 & 9/2 & | & -3/2 \end{pmatrix}.$$

Note that when solving this problem by hand, the arithmetic is easier if one does not pivot, since all of the multipliers are integers. This is a very special case, however, and in general when solving linear systems on a computer one should pivot (if necessary) so that the multipliers are less than or equal to one (or possibly some other moderate size number) in magnitude. There is no additional cost for pivoting in terms of arithmetic, but there is an $O(n^2)$ cost to find the largest entry in each column, and there is the cost of additional data movement.

Our Gaussian elimination code can be modified to incorporate partial pivoting as follows:

```
% Gaussian elimination with partial pivoting.
for j=1:n-1                          % Loop over columns.

  [pivot,k] = max(abs(A(j:n,j)));    % Find the pivot element
                                     % in column j.
                                     % pivot is the largest
                                     % absolute value of an
                                     % entry; k+j-1 is its index.
  if pivot==0,                       % If all entries in the
                                     % column are 0, return with
    disp(' Matrix is singular.')     % an error message.
    break;
  end;

  temp = A(j,:);                     % Otherwise,
  A(j,:) = A(k+j-1,:);               % Interchange rows j and k+j-1.
  A(k+j-1,:) = temp;
  tempb = b(j);
  b(j) = b(k+j-1);
  b(k+j-1) = tempb;

  for i=j+1:n                        % Loop over rows below j.
    mult = A(i,j)/A(j,j);            % Subtract this multiple
                                     % of row j from row
                                     % i to make A(i,j)=0.
```

```
    A(i,j:n) = A(i,j:n) - mult*A(j,j:n);
    b(i) = b(i) - mult*b(j);
  end;
end;
```

In practice, we would need to replace the test if pivot==0 by a test that would reflect the fact that a tiny nonzero value for pivot might be solely due to roundoff error. In this case, we might give the user a warning that the matrix could be singular.

Partial pivoting can also be incorporated into the version of the code that saves the matrix factors. Now, instead of factoring A in the form LU, we factor it in the form PLU, where P is a permutation matrix. A permutation matrix is a matrix that has a 1 in each row and each column and all other elements 0. Exchanging rows can be thought of as applying a permutation matrix on the left. The following theorem guarantees that this factorization exists for any nonsingular matrix A.

Theorem 7.2.1. Every n by n nonsingular matrix A can be factored in the form $A = PLU$, where P is a permutation matrix, L is a unit lower triangular matrix, and U is an upper triangular matrix.

We do not want to leave the reader with the impression that to solve a linear system $A\mathbf{x} = \mathbf{b}$ in MATLAB one must type in the above code! To solve $A\mathbf{x} = \mathbf{b}$, you type `A\b`. The above is (roughly) what MATLAB does when it solves linear systems or computes PLU factorizations.

7.2.4 Banded Matrices and Matrices for Which Pivoting is Not Required

For certain matrices, pivoting in Gaussian elimination is not required. For example, it can be shown that for *symmetric positive definite* matrices, Gaussian elimination can be performed stably without pivoting. A matrix A is **symmetric** if $A = A^T$, and a symmetric matrix is **positive definite** if all of its eigenvalues are positive. An example of a symmetric positive definite matrix is

$$\begin{pmatrix} 2 & -1 & 0 \\ -1 & 2 & -1 \\ 0 & -1 & 2 \end{pmatrix}, \tag{7.3}$$

whose eigenvalues are 2, $2 + \sqrt{2}$, and $2 - \sqrt{2}$.

Instead of factoring a symmetric positive definite matrix in the form LU, where L has unit diagonal, one can factor it in such a way that L and U have the same diagonal elements. With this scaling it turns out that U is just the transpose of L, and so we have factored A in the form $A = LL^T$. This is called the **Cholesky decomposition.**

Gaussian elimination without pivoting is also known to be stable for strictly diagonally dominant matrices. A matrix A is **strictly diagonally dominant** (or

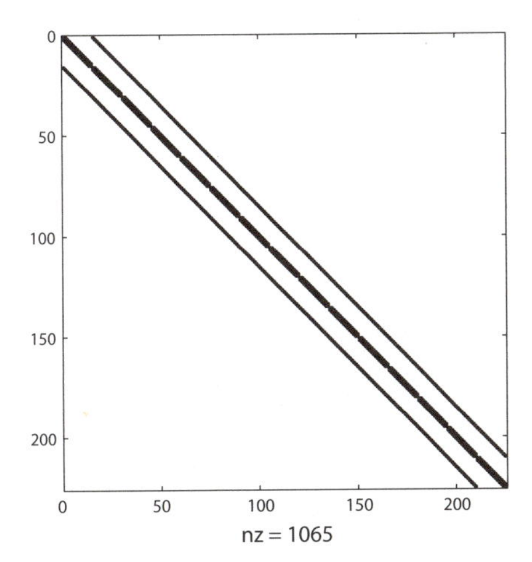

Figure 7.2. Banded structure of a matrix formed from discretizing Poisson's equation on a unit square. Although the matrix has 225×225 entries, only 1065 of these are nonzero.

strictly *row* diagonally dominant) if

$$|a_{ii}| > \sum_{j \neq i} |a_{ij}| \quad \text{for all } i = 1, \ldots, n;$$

that is, the absolute value of each diagonal entry is greater than the sum of the absolute values of all off-diagonal entries in its row.

A number of applications give rise to linear systems that are symmetric positive definite or, sometimes, strictly diagonally dominant, and **banded**; that is, $a_{ij} = 0$, if $|i - j| > m$, where $m \ll n$ is the half bandwidth. The matrix in (7.3) has half bandwidth $m = 1$. The full bandwidth is $2m + 1 = 3$ and so this matrix is called a **tridiagonal** matrix.

Symmetric positive definite and banded matrices often arise in the discretization of partial differential equations, which will be covered later in the text. For example, when Poisson's equation ($\Delta u = f$) on a unit square is discretized using a 16 by 16 grid, the resulting matrix is symmetric and positive definite and has the banded nonzero structure pictured in figure 7.2.

We can save work and storage in the Gaussian elimination algorithm by taking advantage of the banded structure. Consider a matrix with half bandwidth m. To perform Gaussian elimination without pivoting, we use the first row to eliminate entries in column 1 of rows 2 through n. But the entries in column 1 of rows $m+2$ through n are already zero, so we need only operate on the m rows with nonzeros in column 1. Moreover, we need not operate on the entire row. Only columns 1 through $m + 1$ will be affected by adding a multiple of row 1. Thus, the work at the first stage of Gaussian elimination is reduced from $2(n-1)n$ to $2m(m+1)$, and the bandwidth of the matrix has not increased. This is illustrated below for a matrix with half bandwidth 2:

$$\begin{pmatrix} a_{11} & a_{12} & a_{13} & 0 & 0 & 0 \\ a_{21} & a_{22} & a_{23} & a_{24} & 0 & 0 \\ a_{31} & a_{32} & a_{33} & a_{34} & a_{35} & 0 \\ 0 & a_{42} & a_{43} & a_{44} & a_{45} & a_{46} \\ 0 & 0 & a_{53} & a_{54} & a_{55} & a_{56} \\ 0 & 0 & 0 & a_{64} & a_{65} & a_{66} \end{pmatrix} \longrightarrow \begin{pmatrix} a_{11} & a_{12} & a_{13} & 0 & 0 & 0 \\ 0 & \tilde{a}_{22} & \tilde{a}_{23} & a_{24} & 0 & 0 \\ 0 & \tilde{a}_{32} & \tilde{a}_{33} & a_{34} & a_{35} & 0 \\ 0 & a_{42} & a_{43} & a_{44} & a_{45} & a_{46} \\ 0 & 0 & a_{53} & a_{54} & a_{55} & a_{56} \\ 0 & 0 & 0 & a_{64} & a_{65} & a_{66} \end{pmatrix}$$

In general, at stage j of Gaussian elimination, we must eliminate the entry in column j from rows $j+1$ through n. But for a banded matrix, only rows $j+1$ through $\min\{j+m, n\}$ have nonzero entries in column j, and only the entries in columns j through $\min\{j+m, n\}$ of these rows will be affected by adding a multiple of row j. The total work for banded Gaussian elimination without pivoting therefore becomes

$$\sum_{j=1}^{n-1} 2\min\{m, n-j\} \cdot \min\{m+1, n-j+1\} \approx 2m^2 n.$$

For $m << n$, this is a large saving over the $O(n^3)$ operations required for dense Gaussian elimination. Note, however, that while we were able to take advantage of the banded structure in Gaussian elimination, we could not take advantage of any zeros within the band. Frequently, matrices arising from partial differential equations are banded with half bandwidth $m \approx \sqrt{n}$ (for two-dimensional problems), but most of the entries inside the band are also zero. During the process of Gaussian elimination, these entries inside the band *fill in* with nonzeros and so they must be stored and operated on.

To see an example with MATLAB, enter the following commands:

```
n = 15;
A = gallery('poisson',n);
[L,U] = lu(A);
subplot(1,3,1),spy(A), subplot(1,3,2),spy(L),
subplot(1,3,3),spy(U)
```

The banded structure of L and U can be seen in figure 7.3(a) and (b), respectively. Note, the number of nonzeros in L and U is 3389 as compared to 1065 in A, as seen in figure 7.2. Clearly, a large amount of fill-in occurred in the decomposition.

Suppose we need to use partial pivoting. Can we take advantage of a banded structure using Gaussian elimination with partial pivoting? The answer is yes; the half bandwidth of the U factor is at most *doubled* by using partial pivoting. To see this consider, for example, a tridiagonal matrix

$$\begin{pmatrix} x & x & 0 & 0 & 0 \\ x & x & x & 0 & 0 \\ 0 & x & x & x & 0 \\ 0 & 0 & x & x & x \\ 0 & 0 & 0 & x & x \end{pmatrix},$$

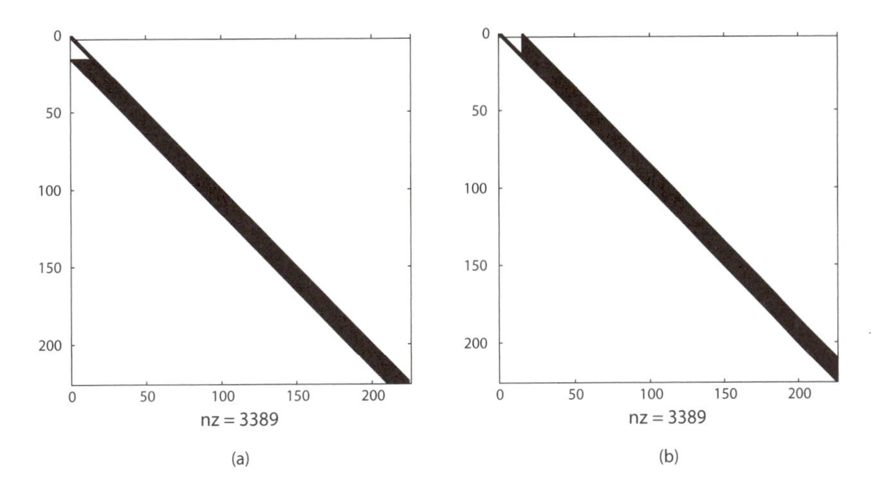

Figure 7.3. Banded structure of L and U for a matrix formed from discretizing Poisson's equation on a unit square.

where the x's represent nonzeros. At the first stage, row 1 or 2 will be chosen as the pivot row. If row 1 is chosen, the bandwidth does not increase, but if row 2 is chosen and moved to the top, the half bandwidth increases from 1 to 2, and the zero entry in column 3 of the new second row (which was row 1) fills in with a nonzero value. The matrix takes the form

$$\begin{pmatrix} x & x & x & 0 & 0 \\ 0 & x & x & 0 & 0 \\ 0 & x & x & x & 0 \\ 0 & 0 & x & x & x \\ 0 & 0 & 0 & x & x \end{pmatrix}.$$

Now row 2 or 3 can be chosen as the pivot row. If row 2 is chosen then no fill-in occurs, but if row 3 is chosen then when it is moved up into the second position, the new row 2 now has two nonzeros above the diagonal, just as row 1 does. During the second stage of elimination, the zero in column 4 of the new third row (which was row 2) becomes nonzero:

$$\begin{pmatrix} x & x & x & 0 & 0 \\ 0 & x & x & x & 0 \\ 0 & 0 & x & x & 0 \\ 0 & 0 & x & x & x \\ 0 & 0 & 0 & x & x \end{pmatrix}.$$

Continuing in this way, it is not difficult to see that each row interchange causes one additional zero to fill in. If row interchanges are made at every step, then the final matrix U will have the form

$$\begin{pmatrix} x & x & x & 0 & 0 \\ 0 & x & x & x & 0 \\ 0 & 0 & x & x & x \\ 0 & 0 & 0 & x & x \\ 0 & 0 & 0 & 0 & x \end{pmatrix}.$$

It is left as an exercise to estimate the number of operations required for Gaussian elimination with partial pivoting for a matrix with half bandwidth m.

7.2.5 Implementation Considerations for High Performance

It was mentioned earlier that *memory references* may be costly on today's supercomputers. Usually there is a memory hierarchy, with larger storage devices such as the main memory or disk being slower to access than smaller storage areas such as local cache and registers. When an arithmetic operation is performed, the operands are fetched from storage, loaded into the registers where the arithmetic is performed, and then the result is stored back into memory. The more work that can be done on a word fetched from memory before storing the result, the better. Also, it is usually more efficient to fetch and store large chunks of data at a time.

Consider the Gaussian elimination code of section 7.2.3. To eliminate the (i, j)-entry at stage j, one fetches the pivot row j and the ith row, $i > j$, from memory and then operates on row i and stores the result. This requires $3(n - j + 1)$ memory accesses and only about $2(n - j + 1)$ arithmetic operations (multiplication of row j and subtraction from row i). If the time for a memory access is greater than that for an arithmetic operation, then most of the time will be spent in fetching operands and storing results.

On the other hand, consider the following basic linear algebra operations:

1. Matrix–vector multiplication: $\mathbf{y} \leftarrow A\mathbf{x} + \mathbf{y}$, where A is an n by n matrix and $\mathbf{x}$ and $\mathbf{y}$ are n-vectors. One must fetch the n^2 entries in A and the $2n$ entries in $\mathbf{x}$ and $\mathbf{y}$ and store the n entries in the result, for a total of $n^2 + 3n$ memory accesses. The arithmetic to compute the product of matrix A with vector $\mathbf{x}$ and add the result to $\mathbf{y}$ is about $2n^2$. Thus the ratio of arithmetic operations to memory fetches is about 2.
2. Matrix–matrix multiplication: $C \leftarrow AB + C$, where A, B, and C are n by n matrices. Here one must fetch the $3n^2$ entries in A, B, and C, and one must store the n^2 entries in the result, for a total of $4n^2$ memory accesses. But the work to compute the product AB and add it to C is about n times the work to perform the matrix–vector multiplication above, namely, $2n^3$. Thus the ratio of arithmetic operations to memory fetches is about $n/2$.

If Gaussian elimination could be expressed in terms of matrix–vector or, better, matrix–matrix multiplication, then the ratio of arithmetic operations to memory accesses would be increased, so that more time would be spent in actually doing the work rather than in fetching the data and storing the results. These basic linear algebra operations are implemented in a package called the BLAS (Basic Linear Algebra Subroutines) [15]. Operations, such as the one performed in our Gaussian elimination code, that act on two vectors (e.g., adding a scalar times one row to another), are labeled BLAS1 and are the least efficient operations when memory references are costly. Matrix–vector operations comprise the BLAS2 routines, which are intermediate in their efficiency, and matrix–matrix operations comprise the BLAS3 routines, which are the most efficient.

Block Algorithms

The Gaussian elimination algorithm can be arranged differently. Suppose we divide the matrix A into a p by p array of m by m blocks A_{ij}, where $n = mp$:

$$A = \begin{pmatrix} A_{11} & A_{12} & \dots & A_{1p} \\ A_{21} & A_{22} & \dots & A_{2p} \\ \vdots & \vdots & & \vdots \\ A_{p1} & A_{p2} & \dots & A_{pp} \end{pmatrix}. \tag{7.4}$$

Assuming for the moment that pivoting is not required, one can eliminate entries below the diagonal in the first block column by computing

$$\begin{pmatrix} A_{22} & \dots & A_{2p} \\ \vdots & & \vdots \\ A_{p2} & \dots & A_{pp} \end{pmatrix} - \begin{pmatrix} A_{21} \\ \vdots \\ A_{p1} \end{pmatrix} A_{11}^{-1}(A_{12}, \dots, A_{1p}).$$

This matrix is called the **Schur complement** of A_{11} in A.

Instead of explicitly computing A_{11}^{-1}, one factors it in the form $A_{11} = L_{11}U_{11}$ and then solves triangular linear systems with multiple right-hand sides. This can be implemented as follows. First bring in the first block column of A and perform Gaussian elimination on that rectangular matrix. This corresponds to a factorization of the form

$$\begin{pmatrix} A_{11} \\ A_{21} \\ \vdots \\ A_{p1} \end{pmatrix} = \begin{pmatrix} L_{11} \\ L_{21} \\ \vdots \\ L_{p1} \end{pmatrix} U_{11}. \tag{7.5}$$

Here L_{11} and U_{11} are the unit lower triangular and upper triangular factors of A_{11}, while $L_{21}, \dots, L_{p1}$ are not lower triangular but contain the multipliers used in the elimination of blocks $A_{21}, \dots, A_{p1}$. This step requires about $2nm$ memory accesses, while the work is about nm^2. To update the rest of the matrix, one then computes

$$\begin{pmatrix} A_{22} & \dots & A_{2p} \\ \vdots & & \vdots \\ A_{p2} & \dots & A_{pp} \end{pmatrix} - \begin{pmatrix} L_{21} \\ \vdots \\ L_{p1} \end{pmatrix} L_{11}^{-1}(A_{12}, \dots, A_{1p}), \tag{7.6}$$

where $L_{11}^{-1}(A_{12}, \dots, A_{1p})$ is computed by solving the lower triangular multiple right-hand side system $L_{11}(U_{12}, \dots, U_{1p}) = (A_{12}, \dots, A_{1p})$. This is a matrix–matrix (BLAS3) operation. It requires about $2n^2$ memory accesses (assuming that $m \ll n$) and about $2n^2m$ arithmetic operations. Elimination in successive block columns is handled similarly.

Supercomputing and Linear Systems

The US national laboratories house many of the world's fastest high performance computers. One of the fastest computers in the world (in fact the fastest in June 2010) is Jaguar (*above*) housed at Oak Ridge National Laboratory, that has performed 1.7 quadrillion floating-point operations per second, or 1.7 petaflops. Such computers are commonly used to solve large linear systems (sometimes containing billions of variables, and solved iteratively using techniques that will be discussed later in this text), necessitating the use of large-scale parallel computing. These massive linear systems often result from differential equations that describe complex physical phenomena. To see the current list of the 500 fastest computers in the world, visit http://www.top500.org. (Image courtesy of the National Center for Computational Science, Oak Ridge National Laboratory.)

A software package that performs Gaussian elimination, as well as many other linear algebra operations, using block algorithms is called LAPACK [1]. It is the software on which MATLAB is based.

Parallelism

A **parallel computer** contains more than one processor, and different processors are capable of working on the same problem simultaneously. Ideally, one might hope that with p processors, an algorithm would run p times as fast. This ideal speedup is seldom achieved, however, even with an algorithm that appears to be highly parallelizable, because of the overhead of communicating between processors. This represents another kind of memory hierarchy. On a **distributed memory** machine, a processor may access data stored in its own memory quickly, but if the data that it needs is on another processor, the time to retrieve it can be much greater.

Gaussian elimination offers many opportunities for parallelism, since elimination in each row (or block row) can be performed independently and in parallel. At a finer level, individual entries in a row could be updated simultaneously. One important reason for using parallel machines, besides improving execution time, is to have more memory. A matrix may be too large to fit into the memory of a single processor, but if it is divided into blocks as in

(7.4), then the different blocks can be distributed among the processors. They should be distributed in such a way as to minimize communication costs.

As an example, but not necessarily the most efficient example, of how Gaussian elimination might be parallelized, suppose that we have p processors and each stores one of the block columns in (7.4). Processor 1 works on its block column, performing the operation in (7.5). It then communicates the results, $L_{11}, \ldots, L_{p1}$, to the other processors, which update their block columns simultaneously according to (7.6). Note that with this arrangement of data, each processor needs all of the blocks $L_{11}, \ldots, L_{p1}$ in order to update its block column. When processor 2 has completed its update, it can begin the analogous process to (7.5) for block column 2. However, all processors must wait until processor 2 completes this work, before starting to do the update analogous to (7.6) for the second stage. In this way, many processors operate simultaneously, but they also must wait for others in order to remain in step.

A library known as ScaLAPACK [14] (for Scalable LAPACK) implements Gaussian elimination and other linear algebra algorithms for distributed memory parallel machines.

7.3 OTHER METHODS FOR SOLVING $A\mathbf{x} = \mathbf{b}$

You might wonder why we have concentrated on Gaussian elimination for solving $A\mathbf{x} = \mathbf{b}$. You probably have learned some other ways. For example, one might compute A^{-1} and set $\mathbf{x} = A^{-1}\mathbf{b}$. You also may have been taught Cramer's rule for solving linear systems. This is a useful way to solve small linear systems by hand, but, as we will see, it is completely inappropriate for large linear systems.

We will concentrate here on operation counts. Recall that Gaussian elimination requires about $\frac{2}{3}n^3$ operations to factor a matrix A in the form $A = PLU$, where P is a permutation matrix, L is a unit lower triangular matrix, and U is an upper triangular matrix. To solve a linear system $A\mathbf{x} = \mathbf{b}$ requires an additional $2n^2$ operations to solve the triangular systems $L\mathbf{y} = \mathbf{b}$ and $U\mathbf{x} = \mathbf{y}$.

Following are operation counts for two other methods.

1. **Compute $\mathbf{A^{-1}}$ and set $\mathbf{x} = \mathbf{A^{-1}b}$.** Sometimes people are taught to solve linear systems this way, but it is more work (and more prone to mistakes!) than Gaussian elimination, even if you are solving a small problem by hand. Consider, for example, the linear system

$$\begin{pmatrix} 1 & 2 & 3 \\ 4 & 5 & 6 \\ 7 & 8 & 0 \end{pmatrix} \begin{pmatrix} x_1 \\ x_2 \\ x_3 \end{pmatrix} = \begin{pmatrix} 1 \\ 0 \\ 2 \end{pmatrix}.$$

A standard procedure for computing the inverse of A involves appending the identity matrix to A, eliminating entries below the diagonal in A, scaling the diagonal elements to be 1, and then eliminating entries above the diagonal in A. The same operations are performed on the appended matrix, and when the procedure is finished, the appended matrix is A^{-1}.

Following is the procedure applied to our matrix:

$$\begin{pmatrix} 1\,2\,3 \mid 1\,0\,0 \\ 4\,5\,6 \mid 0\,1\,0 \\ 7\,8\,0 \mid 0\,0\,1 \end{pmatrix} \rightarrow \begin{pmatrix} 1 & 2 & 3 \mid & 1\,0\,0 \\ 0 & -3 & -6 \mid & -4\,1\,0 \\ 0 & -6 & -21 \mid & -7\,0\,1 \end{pmatrix} \rightarrow$$

$$\begin{pmatrix} 1 & 2 & 3 \mid & 1 & 0\,0 \\ 0 & -3 & -6 \mid & -4 & 1\,0 \\ 0 & 0 & -9 \mid & 1 & -2\,1 \end{pmatrix} \rightarrow \begin{pmatrix} 1\,2\,3 \mid & 1 & 0 & 0 \\ 0\,1\,2 \mid & 4/3 & -1/3 & 0 \\ 0\,0\,1 \mid & -1/9 & 2/9 & -1/9 \end{pmatrix} \rightarrow$$

$$\begin{pmatrix} 1\,2\,0 \mid & 4/3 & -2/3 & 1/3 \\ 0\,1\,0 \mid & 14/9 & -7/9 & 2/9 \\ 0\,0\,1 \mid & -1/9 & 2/9 & -1/9 \end{pmatrix} \rightarrow \begin{pmatrix} 1\,0\,0 \mid & -16/9 & 8/9 & -1/9 \\ 0\,1\,0 \mid & 14/9 & -7/9 & 2/9 \\ 0\,0\,1 \mid & -1/9 & 2/9 & -1/9 \end{pmatrix}.$$

We now apply A^{-1} to the right-hand side vector **b** to obtain

$$\begin{pmatrix} x_1 \\ x_2 \\ x_3 \end{pmatrix} = \frac{1}{9}\begin{pmatrix} -16 & 8 & -1 \\ 14 & -7 & 2 \\ -1 & 2 & -1 \end{pmatrix}\begin{pmatrix} 1 \\ 0 \\ 2 \end{pmatrix} = \frac{1}{9}\begin{pmatrix} -18 \\ 18 \\ -3 \end{pmatrix} = \begin{pmatrix} -2 \\ 2 \\ -1/3 \end{pmatrix}.$$

The procedure we have followed to compute A^{-1} is very much like what one would do in Gaussian elimination to solve n linear systems with coefficient matrix A and with right-hand side vectors $\mathbf{e}_1, \ldots, \mathbf{e}_n$, where $\mathbf{e}_i$ has a 1 in position i and 0s elsewhere. Note that we compute the multipliers necessary to reduce A to upper triangular form only once and then we apply these multipliers to each column of the appended identity matrix; that is, to each of the right-hand side vectors $\mathbf{e}_1, \ldots, \mathbf{e}_n$. We then solve n upper triangular systems with the modified right-hand side vectors. The total work involved is about

$$\frac{2}{3}n^3 + n \times 2n^2 = \frac{8}{3}n^3,$$

which, for large n, is about 4 times as much work as Gaussian elimination. Additionally, we must compute the product of A^{-1} with **b**, but this requires only about $2n^2$ operations, which is of a lower order than the operation count for computing A^{-1}.

2. **Cramer's rule.** This is a useful way to solve small linear systems by hand, but, as we will see, it is *completely inappropriate* for large linear systems. To compute the jth component of the solution, replace column j of A by the right-hand side vector **b**, and compute the ratio of the determinant of this matrix to the determinant of A. For our example problem, expanding determinants by the first row, we find

$$\det(A) = \det\begin{pmatrix} 1\,2\,3 \\ 4\,5\,6 \\ 7\,8\,0 \end{pmatrix} = 1 \cdot \det\begin{pmatrix} 5\,6 \\ 8\,0 \end{pmatrix} - 2 \cdot \det\begin{pmatrix} 4\,6 \\ 7\,0 \end{pmatrix} + 3 \cdot \det\begin{pmatrix} 4\,5 \\ 7\,8 \end{pmatrix}$$

$$= -48 - 2 \cdot (-42) + 3 \cdot (-3) = 27,$$

$$\det\begin{pmatrix}1&2&3\\0&5&6\\2&8&0\end{pmatrix} = 1\cdot\det\begin{pmatrix}5&6\\8&0\end{pmatrix} - 2\cdot\det\begin{pmatrix}0&6\\2&0\end{pmatrix} + 3\cdot\det\begin{pmatrix}0&5\\2&8\end{pmatrix}$$
$$= 1\cdot(-48) - 2\cdot(-12) + 3\cdot(-10) = -54,$$

$$\det\begin{pmatrix}1&1&3\\4&0&6\\7&2&0\end{pmatrix} = 54, \quad \det\begin{pmatrix}1&2&1\\4&5&0\\7&8&2\end{pmatrix} = -9,$$

so that

$$x_1 = \frac{-54}{27} = -2, \quad x_2 = \frac{54}{27} = 2, \quad x_3 = \frac{-9}{27} = -\frac{1}{3}.$$

To apply Cramer's rule to an n by n linear system, we need to compute n determinants. Let us estimate the work for evaluating just one determinant using the expansion procedure outlined here. To evaluate a 2 by 2 determinant

$$\det\begin{pmatrix}a&b\\c&d\end{pmatrix} = ad - bc$$

requires 3 operations: two multiplications and one subtraction. To evaluate a 3 by 3 determinant

$$\det\begin{pmatrix}a&b&c\\d&e&f\\g&h&i\end{pmatrix} = a\cdot\det\begin{pmatrix}e&f\\h&i\end{pmatrix} - b\cdot\det\begin{pmatrix}d&f\\g&i\end{pmatrix} + c\cdot\det\begin{pmatrix}d&e\\g&h\end{pmatrix}$$

requires evaluating three 2 by 2 determinants, multiplying the results by appropriate scalars and adding, for a total of about 3 $\times$ (work for 2 by 2 determinant) + 5 = 14 operations. To evaluate a 4 by 4 determinant one must evaluate four 3 by 3 determinants, multiply by appropriate scalars and add the results, thus doing more than a factor of 4 times as much work as for a 3 by 3 determinant. In general, if W_n is the amount of work required to evaluate an n by n determinant, then

$$W_n > n\cdot W_{n-1} > n\cdot(n-1)\cdot W_{n-2} > \ldots > n\cdot(n-1)\cdot(n-2)\cdots 2\cdot 1 = n!$$

For n as large as 20 or so, a computation of the determinant in this way is out of the question, since $20! \approx 2.4\times 10^{18}$. On a computer that performs, say, 10^9 floating-point operations per second, this would require about 76 years!

There are faster ways than the one given here to compute determinants. In fact, one can find the determinant from the LU factorization of a matrix, since $\det(LU) = \det(L)\cdot\det(U)$, and the determinant of a triangular matrix is just the product of its diagonal entries. To do this with Cramer's rule, however, would defeat the purpose, since we were considering it as an alternative to computing an LU factorization of the matrix. For this reason, as well as accuracy concerns, Cramer's rule is *not* used for solving large linear systems on computers.

TABLE 7.1
Operation counts for different methods for solving linear systems.

Method	Highest-order term in operation count
Gaussian elimination	$(2/3)n^3$
Compute A^{-1}	$(8/3)n^3$
Cramer's rule	$> n!$ if determinants are computed by recursion

Table 7.1 summarizes the operation counts for the three methods discussed for solving linear systems.

7.4 CONDITIONING OF LINEAR SYSTEMS

In chapter 6, we discussed the absolute and relative condition numbers for the problem of evaluating a scalar-valued function of a scalar argument. The problem of solving $A\mathbf{x} = \mathbf{b}$ involves, as input, a matrix A and a right-hand side vector $\mathbf{b}$, while the output is a vector $\mathbf{x}$. To measure the change in output due to a small change in input, we must discuss vector and matrix *norms*.

7.4.1 Norms

Definition. A **norm** for vectors is a function $\|\cdot\|$ satisfying, for all n-vectors $\mathbf{v}$, $\mathbf{w}$:

(i) $\|\mathbf{v}\| \geq 0$, with equality if and only if $\mathbf{v} = 0$;
(ii) $\|\alpha\mathbf{v}\| = |\alpha|\ \|\mathbf{v}\|$ for any scalar α;
(iii) $\|\mathbf{v} + \mathbf{w}\| \leq \|\mathbf{v}\| + \|\mathbf{w}\|$ (triangle inequality).

The most commonly used vector norm on $\mathbf{R}^n$ is the **2-norm** or **Euclidean norm**:

$$\|\mathbf{v}\|_2 \equiv \sqrt{\sum_{i=1}^{n} |v_i|^2}.$$

Other frequently used norms are the **∞-norm**:

$$\|\mathbf{v}\|_\infty \equiv \max_{i=1,\ldots,n} |v_i|,$$

and the **1-norm**:

$$\|\mathbf{v}\|_1 \equiv \sum_{i=1}^{n} |v_i|.$$

More generally, it can be shown that for any $p \geq 1$ (p need not even be an integer), the p**-norm**, defined by

$$\|\mathbf{v}\|_p \equiv \left(\sum_{i=1}^{n} |v_i|^p\right)^{1/p},$$

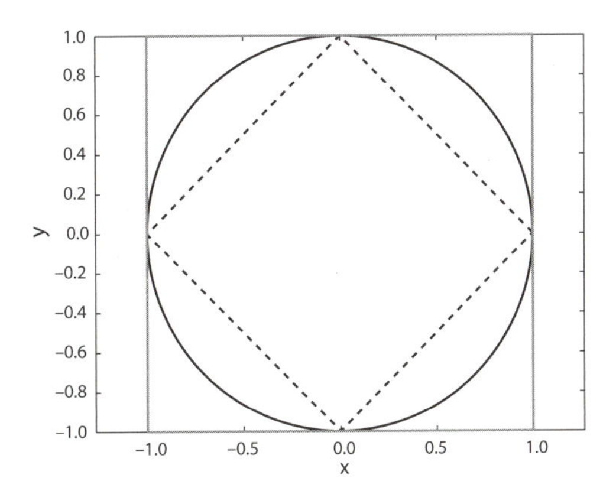

Figure 7.4. Unit circles in the 1-norm, 2-norm, and ∞-norm.

is a norm. We will usually work with the 1-norm, the 2-norm, or the ∞-norm. Of these, only the 2-norm comes from an inner product; that is,

$$\|\mathbf{v}\|_2 = \langle \mathbf{v}, \mathbf{v} \rangle^{1/2}, \quad \text{where } \langle \mathbf{x}, \mathbf{y} \rangle = \sum_{i=1}^{n} x_i y_i.$$

Example 7.4.1. If $\mathbf{v} = (1, 2, -3)^T$, then $\|\mathbf{v}\|_2 = \sqrt{1^2 + 2^2 + (-3)^2} = \sqrt{14}$, $\|\mathbf{v}\|_\infty = \max\{|1|, |2|, |-3|\} = 3$, and $\|\mathbf{v}\|_1 = |1| + |2| + |-3| = 6$. If $\mathbf{w} = (4, 5, 6)^T$, then $\langle \mathbf{v}, \mathbf{w} \rangle = 1 \cdot 4 + 2 \cdot 5 - 3 \cdot 6 = -4$. (You can compute these quantities with MATLAB by entering `v` and `w` and typing `norm(v,1)`, `norm(v,'inf')`, `norm(v,2)` (or simply `norm(v)`), and `w'*v`, respectively.)

Figure 7.4 shows the unit circles in $\mathbf{R}^2$ for the 1-norm (dashed), the 2-norm (black), and the ∞-norm (gray); that is, it shows the set of points (x, y) or vectors $\mathbf{v} \equiv (x, y)^T \in \mathbf{R}^2$ such that $\|\mathbf{v}\|_1 = 1$, such that $\|\mathbf{v}\|_2 = 1$, and such that $\|\mathbf{v}\|_\infty = 1$. The unit circle in the 2-norm is what we normally think of as a circle. It is the set of points (x, y) for which $\sqrt{x^2 + y^2} = 1$. The unit circle in the 1-norm is quite different. It is the set of points (x, y) for which $\|(x, y)\|_1 \equiv |x| + |y| = 1$. It is a square whose corners are the points $(1, 0)$, $(0, 1)$, $(-1, 0)$, and $(0, -1)$. The unit circle in the ∞-norm is again a square, only now aligned with the coordinate axes. It is the set of points (x, y) for which $\|(x, y)\|_\infty \equiv \max\{|x|, |y|\} = 1$. It consists of the line segments where $|y| = 1$ and $-1 \le x \le 1$ (the top and bottom sides) and the line segments where $|x| = 1$ and $-1 \le y \le 1$ (the left and right sides).

One also can define norms for matrices.

Definition. A **matrix norm** is a function $\|\cdot\|$ satisfying, for all m by n matrices A, B:

(i) $\|A\| \ge 0$, with equality if and only if $A = 0$;
(ii) $\|\alpha A\| = |\alpha|\, \|A\|$ for any scalar α;
(iii) $\|A + B\| \le \|A\| + \|B\|$ (triangle inequality).

Note that these three conditions are the same as the requirements for a vector norm. Some definitions require, in addition, that a matrix norm be

submultiplicative; that is, if A is an m by n matrix and C is an n by p matrix, then the product AC must satisfy

(iv) $\|AC\| \leq \|A\| \cdot \|C\|$.

We will always assume that a matrix norm is submultiplicative. There is a potential ambiguity in notation when we talk about a norm of an n by 1 or a 1 by n matrix C, since that "matrix" could also be considered as an n-vector. If thinking of C as a vector, however, we will denote it with a boldface lower case letter **c**; thus $\|C\|$ signifies a matrix norm (that satisfies (iv)) while $\|\mathbf{c}\|$ is a vector norm.

If $\|\cdot\|$ is a vector norm, the *induced* matrix norm is

$$\|A\| \equiv \max_{\|\mathbf{v}\|=1} \|A\mathbf{v}\| = \max_{\mathbf{v}\neq 0} \frac{\|A\mathbf{v}\|}{\|\mathbf{v}\|}.$$

It can be shown that the induced norm satisfies all four properties listed above. Note that if the matrix norm $\|\cdot\|$ (which is denoted in the same way as a vector norm except that its argument is a matrix) is induced by a vector norm $\|\cdot\|$, then for any m by n matrix A and any nonzero n-vector **v**,

$$\|A\| \geq \frac{\|A\mathbf{v}\|}{\|\mathbf{v}\|}, \quad \text{or,} \quad \|A\mathbf{v}\| \leq \|A\| \cdot \|\mathbf{v}\|. \tag{7.7}$$

Any vector norm for which inequality (7.7) holds is said to be *compatible with* or *subordinate to* the matrix norm $\|\cdot\|$. Property (iv) of a matrix norm guarantees that there is some vector norm that is compatible with it, and whenever we mix matrix and vector norms we will assume that they are compatible (i.e., that inequality (7.7) holds).

The following theorems give expressions for the matrix norms induced by the 1-norm, the 2-norm, and the ∞-norm for vectors.

Theorem 7.4.1. Let A be an m by n matrix and let $\|A\|_1$ denote the matrix norm induced by the 1-norm for n-vectors. Then $\|A\|_1$ is the maximum absolute column sum

$$\|A\|_1 = \max_{j=1,\ldots,n} \sum_{i=1}^{m} |a_{ij}|.$$

Proof. Let $A = (\mathbf{a}_{:,1}, \ldots \mathbf{a}_{:,n})$, where $\mathbf{a}_{:,j} = (a_{1j}, \ldots, a_{mj})^T$ denotes the jth column of A. If **v** is an n-vector, then $A\mathbf{v} = \sum_{j=1}^{n} \mathbf{a}_{:,j} v_j$ and so

$$\|A\mathbf{v}\|_1 = \|\sum_{j=1}^{n} \mathbf{a}_{:,j} v_j\|_1 \leq \sum_{j=1}^{n} |v_j| \cdot \|\mathbf{a}_{:,j}\|_1 \text{ (by (iii) and (ii) for vector norms)}$$

$$\leq \max_j \|\mathbf{a}_{:,j}\|_1 \cdot \left(\sum_{j=1}^{n} |v_j| \right) = \max_j \|\mathbf{a}_{:,j}\|_1 \cdot \|\mathbf{v}\|_1.$$

This shows that $\|A\|_1 \leq \max_j \|\mathbf{a}_{:,j}\|_1$. On the other hand, if the index of a column with maximum 1-norm is J and if $v_J = 1$ and all other entries in **v** are 0, then $A\mathbf{v} = \mathbf{a}_{:,J}$, so that $\|\mathbf{v}\|_1 = 1$ and $\|A\mathbf{v}\|_1 = \max_j \|\mathbf{a}_{:,j}\|_1$. This implies that $\|A\|_1 \geq \max_j \|\mathbf{a}_{:,j}\|_1$. Combining the two results we obtain equality. □

Theorem 7.4.2. Let A be an m by n matrix and let $\|A\|_\infty$ denote the matrix norm induced by the ∞-norm for n-vectors. Then $\|A\|_\infty$ is the maximum absolute row sum

$$\|A\|_\infty = \max_{i=1,\ldots,m} \sum_{j=1}^{n} |a_{ij}|.$$

Proof. For any n-vector $\mathbf{v}$, we have

$$\begin{aligned}
\|A\mathbf{v}\|_\infty &= \max_{i=1,\ldots,m} |(A\mathbf{v})_i| = \max_{i=1,\ldots,m} |\sum_{j=1}^{n} a_{ij} v_j| \\
&\leq \max_{i=1,\ldots,m} \sum_{j=1}^{n} |a_{ij}| \cdot |v_j| \\
&\leq \max_{i=1,\ldots,m} (\max_{j=1,\ldots,n} |v_j|) \sum_{j=1}^{n} |a_{ij}| = \|\mathbf{v}\|_\infty \max_{i=1,\ldots,m} \sum_{j=1}^{n} |a_{ij}|.
\end{aligned}$$

This shows that $\|A\|_\infty \leq \max_{i=1,\ldots,m} \sum_{j=1}^{n} |a_{ij}|$. On the other hand, if I is the index of a row with maximal absolute row sum and if $|v_j| = 1$ and $a_{Ij} v_j = |a_{Ij} v_j| = |a_{Ij}|$ for all j (i.e., $v_j = 1$ if $a_{Ij} \geq 0$ and $v_j = -1$ if $a_{Ij} < 0$), then we will have equality above, since for each i, $|\sum_{j=1}^{n} a_{ij} v_j| \leq \sum_{j=1}^{n} |a_{ij}| \leq \sum_{j=1}^{n} |a_{Ij}|$, and for $i = I$ this is an equality. This implies that $\|A\|_\infty \geq \max_{i=1,\ldots,m} \sum_{j=1}^{n} |a_{ij}|$, and combining the two results we obtain equality. □

Theorem 7.4.3. Let A be an m by n matrix and let $\|A\|_2$ denote the matrix norm induced by the 2-norm for n-vectors. Then $\|A\|_2$ is the square root of the largest eigenvalue of $A^T A$.

Proof. The proof of this theorem relies on the *variational characterization of eigenvalues of a symmetric matrix*, which we state here without proof: The largest eigenvalue of the symmetric matrix $A^T A$ is $\max_{\|\mathbf{v}\|_2=1} \langle \mathbf{v}, A^T A\mathbf{v}\rangle$. Since $\|A\|_2^2 = \max_{\|\mathbf{v}\|_2=1} \|A\mathbf{v}\|_2^2 = \max_{\|\mathbf{v}\|_2=1} \langle A\mathbf{v}, A\mathbf{v}\rangle = \max_{\|\mathbf{v}\|_2=1} \langle \mathbf{v}, A^T A\mathbf{v}\rangle$, the result follows. □

Example 7.4.2. Let A be the 3 by 2 matrix

$$A = \begin{pmatrix} 0 & -1 \\ -2 & 3 \\ 1 & 0 \end{pmatrix}.$$

Then $\|A\|_1 = \max\{0 + 2 + 1,\ 1 + 3 + 0\} = 4$ and $\|A\|_\infty = \max\{0 + 1,\ 2 + 3,\ 1 + 0\} = 5$. Forming $A^T A$, we find

$$A^T A = \begin{pmatrix} 0 & -2 & 1 \\ -1 & 3 & 0 \end{pmatrix} \begin{pmatrix} 0 & -1 \\ -2 & 3 \\ 1 & 0 \end{pmatrix} = \begin{pmatrix} 5 & -6 \\ -6 & 10 \end{pmatrix},$$

TABLE 7.2
Errors in solving linear systems with the n by n Hilbert matrix as coefficient matrix.

n	$\|\mathbf{x}_c - \mathbf{x}\|_2$
2	8.95×10^{-16}
5	6.74×10^{-12}
8	2.87×10^{-07}
11	0.0317
14	84.42[1]

[1] MATLAB issues the message: Warning: Matrix is close to singular or badly scaled. Results may be inaccurate.

and the eigenvalues of this matrix satisfy

$$\det \begin{pmatrix} 5-\lambda & -6 \\ -6 & 10-\lambda \end{pmatrix} = (5-\lambda)(10-\lambda) - 36 = \lambda^2 - 15\lambda + 14$$
$$= (\lambda - 1)(\lambda - 14) = 0,$$

or, $\lambda_1 = 1$, $\lambda_2 = 14$. Therefore $\|A\|_2 = \sqrt{14} \approx 3.7417$. (You can compute these quantities with MATLAB by entering `A` and typing `norm(A,1)`, `norm(A,'inf')`, and `norm(A,2)` (or just `norm(A)`), respectively.)

7.4.2 Sensitivity of Solutions of Linear Systems

Consider the following linear system:

$$\underbrace{\begin{pmatrix} 1 & \frac{1}{2} & \cdots & \frac{1}{n} \\ \frac{1}{2} & \frac{1}{3} & \cdots & \frac{1}{n+1} \\ \vdots & \vdots & \ddots & \vdots \\ \frac{1}{n} & \frac{1}{n+1} & \cdots & \frac{1}{2n-1} \end{pmatrix}}_{\text{Hilbert matrix}} \mathbf{x} = \mathbf{b}.$$

This matrix is known as a **Hilbert matrix**, and it is known to be notoriously *ill conditioned*. To see what this means, we will set up a problem where we know the solution and then use MATLAB to solve the linear system; we can compare its computed answer to the one that we know to be right. Letting $\mathbf{x}$ be the vector of all 1s, we form the matrix–vector product $H\mathbf{x}$ (where H is the Hilbert matrix) and set $\mathbf{b}$ equal to this. (Of course, if we compute $\mathbf{b}$ using finite-precision arithmetic, then it will not be the *exact* right-hand side vector corresponding to a solution of all 1s, but it can be expected to differ from this vector by just a moderate multiple of the machine precision.) We

TABLE 7.3
Errors in solving linear systems with a random n by n coefficient matrix.

n	$\|\mathbf{x}_c - \mathbf{x}\|_2$
2	2.48×10^{-16}
5	1.46×10^{-15}
8	9.49×10^{-16}
11	2.15×10^{-15}
14	3.06×10^{-15}

then type `H\b` to compute the solution in MATLAB. Table 7.2 shows results for $n = 2, 5, 8, 11, 14$, where we report the 2-norm of the difference between the computed vector $\mathbf{x}_c$ and the vector $\mathbf{x} = (1, \ldots, 1)^T$.

This would appear to justify the predictions of the early mathematicians, cited at the beginning of this chapter, that as n increased, roundoff errors would accumulate and destroy all accuracy in the computed solution to a linear system! This is a very special linear system, however, and, as we will see, the fact that computed answers differ drastically from true solutions, is not the fault of the algorithm. (In fact, the *true* solution to $A\mathbf{x} = \mathbf{b}$, for the computed $\mathbf{b}$, differs drastically from the vector of all 1s.) For n of even a moderate size, the solution to this linear system is *extremely* sensitive to tiny changes in the matrix entries and entries in the right-hand side vector. Fortunately, most linear systems that arise in practice are not of this extremely ill-conditioned sort. When one does produce such a linear system through modeling some physical process, it is usually because one has not formulated the problem in the best way; perhaps some units are in angstroms while others are in miles, or the quantities that one is solving for are not really the relevant characteristics of the system being modeled.

To illustrate the more usual case, we replaced the Hilbert matrix above with a random n by n matrix and obtained the results in table 7.3.

Suppose we start with a nonsingular n by n linear system $A\mathbf{x} = \mathbf{b}$ and we change the right-hand side vector $\mathbf{b}$ by a small amount. How much does the solution to the linear system change? This question will be important when considering solving linear systems on a computer because, as noted in chapter 5, the entries in $\mathbf{b}$ may not be exactly representable as floating-point numbers. But the question can be asked in general, without reference to any particular computing system or algorithm. This will tell us about the *conditioning* of the linear system, as described in section 6.1.

Suppose, then, that $\hat{\mathbf{b}}$ is a vector such that, in a certain norm, $\|\mathbf{b} - \hat{\mathbf{b}}\|$ is small. Let $\mathbf{x}$ denote the solution to the linear system $A\mathbf{x} = \mathbf{b}$, and let $\hat{\mathbf{x}}$ denote the solution to the linear system $A\hat{\mathbf{x}} = \hat{\mathbf{b}}$. Subtracting these two equations, we find that $A(\mathbf{x} - \hat{\mathbf{x}}) = \mathbf{b} - \hat{\mathbf{b}}$, or, $\mathbf{x} - \hat{\mathbf{x}} = A^{-1}(\mathbf{b} - \hat{\mathbf{b}})$. Taking norms on each side gives the inequality

$$\|\mathbf{x} - \hat{\mathbf{x}}\| \leq \|A^{-1}\| \cdot \|\mathbf{b} - \hat{\mathbf{b}}\|. \tag{7.8}$$

The factor $\|A^{-1}\|$ can be thought of as an absolute condition number for this problem, corresponding to the absolute condition number described in section 6.1 for scalar-valued functions of a scalar argument.

As noted in section 6.1, however, it is usually the relative error rather than the absolute error that is of interest; in this case, we would like to relate $\|\mathbf{x}-\hat{\mathbf{x}}\|/\|\mathbf{x}\|$ to $\|\mathbf{b}-\hat{\mathbf{b}}\|/\|\mathbf{b}\|$. Dividing each side of (7.8) by $\|\mathbf{x}\|$, we find

$$\frac{\|\mathbf{x}-\hat{\mathbf{x}}\|}{\|\mathbf{x}\|} \leq \|A^{-1}\| \cdot \frac{\|\mathbf{b}-\hat{\mathbf{b}}\|}{\|\mathbf{x}\|} = \|A^{-1}\| \cdot \frac{\|\mathbf{b}-\hat{\mathbf{b}}\|}{\|\mathbf{b}\|} \cdot \frac{\|\mathbf{b}\|}{\|\mathbf{x}\|}.$$

Since $\|\mathbf{b}\|/\|\mathbf{x}\| = \|A\mathbf{x}\|/\|\mathbf{x}\| \leq \|A\|$, it follows that

$$\frac{\|\mathbf{x}-\hat{\mathbf{x}}\|}{\|\mathbf{x}\|} \leq \|A^{-1}\| \cdot \|A\| \cdot \frac{\|\mathbf{b}-\hat{\mathbf{b}}\|}{\|\mathbf{b}\|}. \tag{7.9}$$

The number $\|A\| \cdot \|A^{-1}\|$ serves as a sort of relative condition number for the problem of solving $A\mathbf{x} = \mathbf{b}$.

Definition. The number $\kappa(A) \equiv \|A\| \cdot \|A^{-1}\|$ is called the **condition number** of the nonsingular matrix A.

Example 7.4.3. Consider the 2 by 2 matrix

$$A = \begin{pmatrix} 1 & -1 \\ 2 & 2 \end{pmatrix},$$

whose inverse is

$$A^{-1} = \frac{1}{4}\begin{pmatrix} 2 & 1 \\ -2 & 1 \end{pmatrix}.$$

The condition number of A in the 1-norm is $\kappa_1(A) \equiv \|A\|_1 \cdot \|A^{-1}\|_1 = 3 \cdot 1 = 3$. The condition number of A in the ∞-norm is $\kappa_\infty(A) \equiv \|A\|_\infty \cdot \|A^{-1}\|_\infty = 4 \cdot (3/4) = 3$. To find the condition number in the 2-norm, we can compute $A^T A$ and $A^{-T}A^{-1}$ and find their eigenvalues:

$$A^T A = \begin{pmatrix} 5 & 3 \\ 3 & 5 \end{pmatrix}, \quad A^{-T}A^{-1} = \frac{1}{16}\begin{pmatrix} 8 & 0 \\ 0 & 2 \end{pmatrix}.$$

The eigenvalues of $A^T A$ satisfy $(5-\lambda)^2 - 9 = 0$, so $\lambda_1 = 2$ and $\lambda_2 = 8$. Thus $\|A\|_2 = \sqrt{8} = 2\sqrt{2}$. The eigenvalues of $A^{-T}A^{-1}$ are $1/2$ and $1/8$, so $\|A^{-1}\| = 1/\sqrt{2}$. It follows that $\kappa_2(A) \equiv \|A\|_2 \cdot \|A^{-1}\|_2 = 2$.

The 2-norm condition number of a matrix can be viewed geometrically. When an n by n matrix A is applied to a unit vector (corresponding to a point on the unit sphere in $\mathbf{R}^n$), it rotates the vector and stretches or shrinks it by a certain amount. It turns out that A maps the unit sphere in $\mathbf{R}^n$ to a **hyperellipse**, which is just the n-dimensional analog of an ellipse in $\mathbf{R}^2$. For the 2 by 2 matrix in this example, the unit circle in $\mathbf{R}^2$ is mapped to an ellipse, and the ratio of the major axis to the minor axis of this ellipse is the condition number of A, since the unit vector whose image is the major semiaxis is the one that is stretched the most and the unit vector whose image is the minor semiaxis is the one that is stretched the least. It turns out that the unit vectors that map to the major and minor semiaxes of the ellipse are just the eigenvectors of $A^T A$ (also

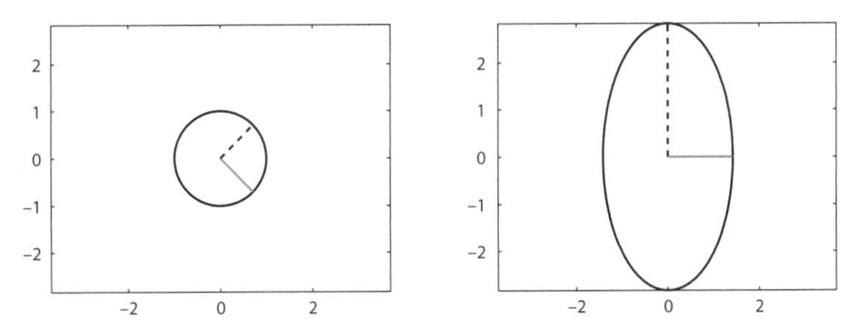

Figure 7.5. A geometric view of the 2-norm condition number of a matrix.

called **right singular vectors** of A), which in this case are $(1/\sqrt{2}, -1/\sqrt{2})^T$ and $(1/\sqrt{2}, 1/\sqrt{2})^T$. The images of these two vectors are $(\sqrt{2}, 0)^T$ and $(0, 2\sqrt{2})^T$, respectively, and the ratio of their lengths is $2 = \kappa_2(A)$. The mapping is pictured in figure 7.5.

Suppose $\mathbf{x}$ is the solution to $A\mathbf{x} = \mathbf{b}$ and $\hat{\mathbf{x}}$ is the solution to $A\hat{\mathbf{x}} = \hat{\mathbf{b}}$, where $\hat{\mathbf{b}} = \mathbf{b} + (\epsilon_1, \epsilon_2)^T$. Then it can be seen using the expression for A^{-1} that

$$\hat{\mathbf{x}} = \mathbf{x} + \frac{1}{4}\begin{pmatrix} 2\epsilon_1 + \epsilon_2 \\ -2\epsilon_1 + \epsilon_2 \end{pmatrix}.$$

If, say, $\epsilon_1 = \epsilon_2 \equiv \epsilon > 0$, then

$$\mathbf{x} - \hat{\mathbf{x}} = -\frac{1}{4}\begin{pmatrix} 3\epsilon \\ -\epsilon \end{pmatrix},$$

and if we consider the ∞-norm, then $\|\mathbf{x} - \hat{\mathbf{x}}\|_\infty = \frac{3}{4}\epsilon = \frac{3}{4}\|\mathbf{b} - \hat{\mathbf{b}}\|_\infty$. Recall that $\frac{3}{4} = \|A^{-1}\|_\infty$ is the absolute condition number for this problem. Suppose $\mathbf{b} = (0, 4)^T$, so that $\mathbf{x} = (1, 1)^T$. Then $\|\mathbf{x}\|_\infty = 1$ and $\|\mathbf{b}\|_\infty = 4$ so that, again with $\epsilon_1 = \epsilon_2 \equiv \epsilon$,

$$\frac{\|\mathbf{x} - \hat{\mathbf{x}}\|_\infty}{\|\mathbf{x}\|_\infty} = \frac{3}{4}\epsilon = 3 \cdot \frac{\|\mathbf{b} - \hat{\mathbf{b}}\|_\infty}{\|\mathbf{b}\|_\infty},$$

where $3 = \kappa_\infty(A)$ is the relative condition number in the ∞-norm.

Example 7.4.4. Use the MATLAB command `cond` to compute the 2-norm condition number of the n by n Hilbert matrix described at the beginning of this section, for $n = 2, 5, 8, 11, 14$.

The results are shown in table 7.4. From this, one can see why the differences between true and computed solutions reported earlier are to be expected.

While inequalities (7.8) and (7.9) give upper bounds on the change in $\mathbf{x}$ due to a small change in $\mathbf{b}$, these upper bounds are not always attained. The size of the individual components of the perturbation vector, as well as its norm, may play an important role. For instance, suppose that the well-conditioned matrix A in Example 7.4.3 is multiplied on the left by a diagonal matrix D with widely varying diagonal elements. This might correspond to changing the units in one of the equations from, say, millimeters to kilometers. The scaled

TABLE 7.4
Condition number of the n by n Hilbert matrix.

n	$\kappa_2(H_n)$
2	19.28
5	$4.77 \times 10^{+05}$
8	$1.53 \times 10^{+10}$
11	$5.23 \times 10^{+14}$
14	$4.69 \times 10^{+17}$

matrix $\mathcal{A} \equiv DA$ may then be ill conditioned in all of the standard norms, but the ill conditioning may be innocuous if we also deal with right-hand side vectors that have been scaled by D and with perturbations whose individual entries are small compared to those of the right-hand side vector.

To be specific, suppose $D = \text{diag}(1, 10^{-6})$. Consider the linear system $\mathcal{A}\mathbf{x} = \boldsymbol{\beta}$, where

$$\mathcal{A} = DA = \begin{pmatrix} 1 & -1 \\ 2 \times 10^{-6} & 2 \times 10^{-6} \end{pmatrix} \quad \text{and} \quad \boldsymbol{\beta} = D\mathbf{b} = \begin{pmatrix} b_1 \\ 10^{-6} b_2 \end{pmatrix}.$$

Suppose $\hat{\mathbf{x}}$ is the solution of $\mathcal{A}\hat{\mathbf{x}} = \hat{\boldsymbol{\beta}}$, where $\hat{\boldsymbol{\beta}} = \boldsymbol{\beta} + (\delta_1, \delta_2)^T$, and $(\delta_1, \delta_2)^T = D(\epsilon_1, \epsilon_2)^T = (\epsilon_1, 10^{-6}\epsilon_2)^T$. Then $\mathbf{x}$ and $\hat{\mathbf{x}}$ are the same as in Example 7.4.3 and hence if, say, $\epsilon_1 = \epsilon_2 \equiv \epsilon$, as in the example, then $\|\mathbf{x} - \hat{\mathbf{x}}\|_\infty = \frac{3}{4}\epsilon = \frac{3}{4}\|\boldsymbol{\beta} - \hat{\boldsymbol{\beta}}\|_\infty$. Yet the absolute condition number, $\|\mathcal{A}^{-1}\|_\infty$, for this problem is 250000.5, since

$$\mathcal{A}^{-1} = \frac{1}{4}\begin{pmatrix} 2 & 10^6 \\ -2 & 10^6 \end{pmatrix}.$$

In this case, the upper bound $\|\mathbf{x} - \hat{\mathbf{x}}\|_\infty \leq \|\mathcal{A}^{-1}\|_\infty \|\boldsymbol{\beta} - \hat{\boldsymbol{\beta}}\|_\infty$ is a large overestimate. For this reason, *componentwise* error bounds are sometimes studied: How much can the solution $\mathbf{x}$ change if each component b_i is changed by a small relative amount to $b_i(1 + \delta_i)$? See, for example, [32]. Here we will continue to work with normwise error bounds but one should keep in mind that in certain special circumstances they may not be tight.

Suppose now that we make a small change not only to $\mathbf{b}$ but also to A. How can we relate the change in the solution to the changes in both A and $\mathbf{b}$? The following theorem shows that, as long as the change in A is small enough so that the modified matrix is still nonsingular and has an inverse whose norm is close to that of A^{-1}, we again obtain an estimate based on the condition number of A.

Theorem 7.4.4. Let A be a nonsingular n by n matrix, let $\mathbf{b}$ be a given n-vector, and let $\mathbf{x}$ satisfy $A\mathbf{x} = \mathbf{b}$. Let $A + E$ be another nonsingular n by n matrix and $\hat{\mathbf{b}}$ another n-vector, and let $\hat{\mathbf{x}}$ satisfy $(A + E)\hat{\mathbf{x}} = \hat{\mathbf{b}}$. Then

$$\frac{\|\mathbf{x} - \hat{\mathbf{x}}\|}{\|\mathbf{x}\|} \leq (\|(A + E)^{-1}\| \cdot \|A\|) \left(\frac{\|\mathbf{b} - \hat{\mathbf{b}}\|}{\|\mathbf{b}\|} + \frac{\|E\|}{\|A\|} \right). \tag{7.10}$$

If $\|E\|$ is small enough so that $\|A^{-1}\| \cdot \|E\| < 1$, then

$$\frac{\|\mathbf{x} - \hat{\mathbf{x}}\|}{\|\mathbf{x}\|} \leq \frac{\kappa(A)}{1 - \kappa(A)\|E\|/\|A\|} \cdot \left(\frac{\|\mathbf{b} - \hat{\mathbf{b}}\|}{\|\mathbf{b}\|} + \frac{\|E\|}{\|A\|} \right). \tag{7.11}$$

Proof. Subtracting the two equations $(A+E)\hat{\mathbf{x}} = \hat{\mathbf{b}}$ and $(A+E)\mathbf{x} = \mathbf{b} + E\mathbf{x}$, we find that $(A+E)(\mathbf{x} - \hat{\mathbf{x}}) = \mathbf{b} - \hat{\mathbf{b}} + E\mathbf{x}$, or, $\mathbf{x} - \hat{\mathbf{x}} = (A+E)^{-1}(\mathbf{b} - \hat{\mathbf{b}} + E\mathbf{x})$. Taking norms on each side we obtain the inequality

$$\|\mathbf{x} - \hat{\mathbf{x}}\| \leq \|(A+E)^{-1}\| \cdot \left(\|\mathbf{b} - \hat{\mathbf{b}}\| + \|E\| \, \|\mathbf{x}\| \right).$$

Dividing each side by $\|\mathbf{x}\|$ gives

$$\frac{\|\mathbf{x} - \hat{\mathbf{x}}\|}{\|\mathbf{x}\|} \leq \|(A+E)^{-1}\| \cdot \left(\frac{\|\mathbf{b} - \hat{\mathbf{b}}\|}{\|\mathbf{b}\|} \cdot \frac{\|\mathbf{b}\|}{\|\mathbf{x}\|} + \|E\| \right).$$

Since $\|\mathbf{b}\|/\|\mathbf{x}\| = \|A\mathbf{x}\|/\|\mathbf{x}\| \leq \|A\|$, this inequality can be replaced by

$$\frac{\|\mathbf{x} - \hat{\mathbf{x}}\|}{\|\mathbf{x}\|} \leq (\|(A+E)^{-1}\| \cdot \|A\|) \cdot \left(\frac{\|\mathbf{b} - \hat{\mathbf{b}}\|}{\|\mathbf{b}\|} + \frac{\|E\|}{\|A\|} \right).$$

This establishes (7.10).

To establish (7.11) we will use a Neumann series expansion of $(A+E)^{-1}$, which will be used again in a different context in chapter 12. Writing $(A+E)^{-1}$ in the form $[A(I + A^{-1}E)]^{-1} = (I + A^{-1}E)^{-1}A^{-1}$, it can be shown that when $\|A^{-1}E\| < 1$, then

$$(I + A^{-1}E)^{-1} = I - A^{-1}E + (A^{-1}E)^2 - \ldots,$$

just as the formula $1/(1+\alpha) = 1 - \alpha + \alpha^2 - \ldots$ holds for scalars α with $|\alpha| < 1$. Thus

$$(A+E)^{-1} = A^{-1} + \left(\sum_{k=1}^{\infty} (-1)^k (A^{-1}E)^k \right) A^{-1},$$

and

$$\|(A+E)^{-1}\| \leq \|A^{-1}\| \left(1 + \sum_{k=1}^{\infty} \|A^{-1}E\|^k \right).$$

Summing the geometric series this becomes

$$\|(A+E)^{-1}\| \leq \|A^{-1}\| \left(1 + \frac{\|A^{-1}E\|}{1 - \|A^{-1}E\|} \right) \leq \|A^{-1}\| \left(1 + \frac{\|A^{-1}\| \, \|E\|}{1 - \|A^{-1}\| \, \|E\|} \right).$$

The right-hand side can be written as

$$\|A^{-1}\| \left(1 + \frac{\kappa(A)\|E\|/\|A\|}{1 - \kappa(A)\|E\|/\|A\|} \right) = \|A^{-1}\| \frac{1}{1 - \kappa(A)\|E\|/\|A\|},$$

and substituting this for $\|(A+E)^{-1}\|$ in (7.10) gives the result (7.11). $\square$

Example 7.4.5. Let $\delta > 0$ be a small number and let

$$A = \begin{pmatrix} 1 & 1+\delta \\ 1-\delta & 1 \end{pmatrix}.$$

Then

$$A^{-1} = \frac{1}{\delta^2} \begin{pmatrix} 1 & -1-\delta \\ -1+\delta & 1 \end{pmatrix}.$$

The condition number of A in the ∞-norm is

$$\kappa_\infty(A) = \|A\|_\infty \|A^{-1}\|_\infty = \frac{(2+\delta)^2}{\delta^2}.$$

If $\delta = 0.01$, then $\kappa_\infty(A) = (201)^2 = 40401$.

This means that if we start with a linear system $A\mathbf{x} = \mathbf{b}$ and perturb $\mathbf{b}$ and/or A by a tiny amount, the solution may change by up to 40401 times as much. For instance, consider the linear system

$$\begin{pmatrix} 1 & 1.01 \\ 0.99 & 1 \end{pmatrix} \begin{pmatrix} x \\ y \end{pmatrix} = \begin{pmatrix} 2.01 \\ 1.99 \end{pmatrix},$$

whose solution is $x = y = 1$. If we replace it by the linear system

$$\begin{pmatrix} 1 & 1.01 \\ 0.99 & 1 \end{pmatrix} \begin{pmatrix} \hat{x} \\ \hat{y} \end{pmatrix} = \begin{pmatrix} 2 \\ 2 \end{pmatrix},$$

so that $\|\mathbf{b} - \hat{\mathbf{b}}\|_\infty / \|\mathbf{b}\|_\infty = 0.01/2.01$, then the solution to the new system is $\hat{x} = -200$, $\hat{y} = 200$, a relative change in ∞-norm of 201.

7.5 STABILITY OF GAUSSIAN ELIMINATION WITH PARTIAL PIVOTING

Suppose we use some algorithm to solve an ill-conditioned linear system on a computer. We cannot necessarily expect to find anything close to the true solution since the matrix and right-hand side vector may have been rounded as they were stored in the computer. Then, even if all other arithmetic operations were performed exactly, we would be solving a slightly different linear system from the one intended and so the solution might be quite different. At best, we could hope to find the exact solution to a nearby problem. An algorithm that does this is called **backward stable**, as mentioned in section 6.2. Thus we can ask the question, "Is Gaussian elimination with partial pivoting backward stable?" What about the other algorithms we discussed; that is, computing A^{-1} and then forming $A^{-1}\mathbf{b}$ or solving a linear system by Cramer's rule?

Consider the following example, where we factor an ill-conditioned matrix in the form LU using Gaussian elimination and retaining only two decimal places:

$$\begin{pmatrix} 1 & 1.01 \\ 0.99 & 1.01 \end{pmatrix} \xrightarrow{L_1} \begin{pmatrix} 1 & 1.01 \\ 0 & 0.01 \end{pmatrix}.$$

Here $1.01-(0.99\times 1.01)$ was computed to be 0.01, since $0.99\times 1.01 = 0.9999$ was rounded to 1. Looking at the product of the L and U factors we find

$$\begin{pmatrix} 1 & 0 \\ 0.99 & 1 \end{pmatrix} \begin{pmatrix} 1 & 1.01 \\ 0 & 0.01 \end{pmatrix} = \begin{pmatrix} 1 & 1.01 \\ 0.99 & 1.0099 \end{pmatrix}.$$

Thus we have computed the exact LU decomposition of a nearby matrix (one whose entries differ from those of the original by only about the machine precision).

It turns out that *for practically all problems, Gaussian elimination with partial pivoting is backward stable*: it finds the true solution to a nearby problem. There are exceptions, but they are *extremely* rare. An open problem in numerical analysis is to further classify and quantify the "unlikelihood" of these examples where Gaussian elimination with partial pivoting is unstable. The classic example, due to Wilkinson [108, p. 212], to illustrate the possible instability has a matrix A with 1s on its main diagonal, -1s throughout its strict lower triangle, 1s in its last column, and 0s everywhere else:

$$A = \begin{pmatrix} 1 & & & 1 \\ -1 & 1 & & 1 \\ \vdots & \ddots & \ddots & \vdots \\ -1 & \dots & -1 & 1 \end{pmatrix}.$$

Take a matrix of this form of size, say, $n = 70$, and try to solve a linear system with this coefficient matrix in MATLAB. You might not see the instability because the entries are integers; in that case, try perturbing the last column with a small random perturbation, say, `A(:,n) = A(:,n) + 1.e-6*randn(n,1)`. You should find that the matrix is well conditioned (type `cond(A)` to see its condition number) but that the computed solution to a linear system with this coefficient matrix is far from correct. (You can set up a problem for which you know the true solution by typing `x=randn(n,1); b=A*x`. Then look at `norm(A\b - x)`.) Despite the existence of such rare examples, Gaussian elimination with partial pivoting is the most-used algorithm for solving linear systems. It is what MATLAB uses and it is the accepted procedure in numerical analysis. Another option is **full pivoting**, where one searches for the largest entry in absolute value in the entire matrix or the remaining submatrix and interchanges both rows and columns in order to move this entry into the pivot position.

How can one tell if an algorithm is backward stable? Short of a detailed analysis, one can do some numerical experiments. If an algorithm for solving $A\mathbf{x} = \mathbf{b}$ is backward stable, then it should produce an approximate solution $\hat{\mathbf{x}}$ for which the *residual* $\mathbf{b} - A\hat{\mathbf{x}}$ is small, even if the problem is ill conditioned so that the *error* $\mathbf{x} - \hat{\mathbf{x}}$ is not small. To see why this is so, let $\mathbf{x}$ satisfy $A\mathbf{x} = \mathbf{b}$ and suppose that the computed solution $\hat{\mathbf{x}}$ satisfies the nearby linear system $(A+E)\hat{\mathbf{x}} = \hat{\mathbf{b}}$. Then $\mathbf{b} - A\hat{\mathbf{x}} = \hat{\mathbf{b}} - A\hat{\mathbf{x}} + \mathbf{b} - \hat{\mathbf{b}} = E\hat{\mathbf{x}} + \mathbf{b} - \hat{\mathbf{b}}$. Taking norms on each side gives the inequality

$$\|\mathbf{b} - A\hat{\mathbf{x}}\| \leq \|E\| \cdot \|\hat{\mathbf{x}}\| + \|\mathbf{b} - \hat{\mathbf{b}}\|.$$

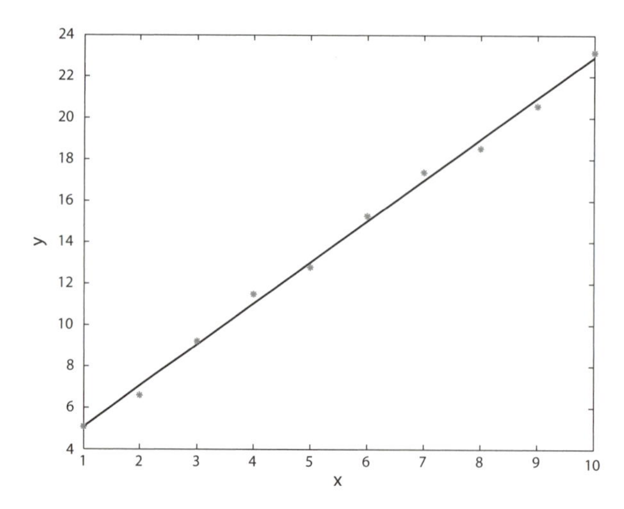

Figure 7.6. Linear least squares fit for a set of data points.

Dividing each side by $\|A\| \cdot \|\hat{\mathbf{x}}\|$ gives

$$\frac{\|\mathbf{b} - A\hat{\mathbf{x}}\|}{\|A\| \cdot \|\hat{\mathbf{x}}\|} \leq \frac{\|E\|}{\|A\|} + \frac{\|\mathbf{b} - \hat{\mathbf{b}}\|}{\|\mathbf{b}\|} \cdot \frac{\|\mathbf{b}\|}{\|A\| \cdot \|\hat{\mathbf{x}}\|}.$$

For $\|E\|$ and $\|\mathbf{b} - \hat{\mathbf{b}}\|$ small, the factor $\|\mathbf{b}\|/(\|A\| \cdot \|\hat{\mathbf{x}}\|)$ will be less than or equal to about 1 (since $\|\mathbf{b}\|/(\|A\| \cdot \|\mathbf{x}\|) \leq 1$), and so for a backward stable algorithm, we should expect the quantity on the left-hand side to be on the order of the machine precision ϵ, *independent* of $\kappa(A)$. Frequently it will be a moderate size constant or, perhaps, a power of the problem size n, times ϵ, but these cases are also judged to be backward stable. In the exercises, you will be asked to do some numerical experiments to verify the backward stability of Gaussian elimination with partial pivoting and to test whether solving a linear system by computing A^{-1} or by Cramer's rule is backward stable.

7.6 LEAST SQUARES PROBLEMS

We have discussed solving a nonsingular system $A\mathbf{x} = \mathbf{b}$ of n linear equations in n unknowns. Such a system has a unique solution. Suppose, on the other hand, that there are more equations than unknowns. Then the system probably has no solution. Instead we might wish to find an *approximate* solution $\mathbf{x}$ satisfying $A\mathbf{x} \approx \mathbf{b}$. This sort of problem arises, for example, in fitting equations to data, which will be discussed in section 7.6.3. If one expects, say, a linear relation between two quantities such as pressure and temperature, then one can test this experimentally by measuring the temperature corresponding to different pressures, plotting the data points and seeing if they lie approximately along a straight line, as illustrated in figure 7.6.

Most likely, the measured data will not lie exactly on a straight line due to measurement errors, but if the measurement errors are small and the conjecture of a linear relationship is correct, then one would expect a "close" fit. Exactly what is meant by a "close" fit may depend on the application, but often one looks at the sum of squares of the differences between the measured values and their predicted values, based on the straight line approximation for which this sum of squares is minimal. This is called a **least squares problem**, and it corresponds to minimizing the square of the 2-norm of the residual, $\|\mathbf{b} - A\mathbf{x}\|_2^2$, in an over-determined linear system.

Let A be an $m \times n$ matrix with $m > n$ and $\mathbf{b}$ a given m-vector. We seek an n-vector $\mathbf{x}$ such that $A\mathbf{x} \approx \mathbf{b}$. For example,

$$\begin{pmatrix} 1 & 1 \\ 1 & -1 \\ 2 & 1 \end{pmatrix} \begin{pmatrix} x_1 \\ x_2 \end{pmatrix} \approx \begin{pmatrix} 2 \\ 0 \\ 4 \end{pmatrix}. \tag{7.12}$$

More precisely, we choose $\mathbf{x}$ to minimize the squared residual norm,

$$\|\mathbf{b} - A\mathbf{x}\|_2^2 = \sum_{i=1}^{m} \left(b_i - \sum_{j=1}^{n} a_{ij} x_j \right)^2. \tag{7.13}$$

This is the *least squares problem* to be addressed in this section, and we will discuss two different approaches.

7.6.1 The Normal Equations

One way to determine the value of $\mathbf{x}$ that achieves the minimum in (7.13) is to differentiate $\|\mathbf{b} - A\mathbf{x}\|_2^2$ with respect to each component x_k and set these derivatives to 0. Because this is a quadratic functional the point where these partial derivatives are 0 will indeed be the minimum. Differentiating, we find

$$\frac{\partial}{\partial x_k} \left(\|\mathbf{b} - A\mathbf{x}\|^2 \right) = \sum_{i=1}^{m} 2 \left(b_i - \sum_{j=1}^{n} a_{ij} x_j \right) (-a_{ik}),$$

and setting these derivatives to 0 for $k = 1, \ldots, n$ gives

$$\sum_{i=1}^{m} a_{ik} \left(\sum_{j=1}^{n} a_{ij} x_j \right) \equiv \sum_{i=1}^{m} a_{ik} (Ax)_i = \sum_{i=1}^{m} a_{ik} b_i.$$

Equivalently, we can write $\sum_{i=1}^{m} (A^T)_{ki} (A\mathbf{x})_i = \sum_{i=1}^{m} (A^T)_{ki} b_i$, $k = 1, \ldots, n$, or,

$$A^T A\mathbf{x} = A^T \mathbf{b}. \tag{7.14}$$

The system (7.14) is called the **normal equations**.

A drawback of this approach is that the 2-norm condition number of $A^T A$ is the square of that of A, since by theorem 7.4.3,

$$\|A^T A\|_2 = \sqrt{\text{largest eigenvalue of } (A^T A)^2} = \text{largest eigenvalue of } (A^T A) = \|A\|_2^2,$$

and similarly $\|(A^T A)^{-1}\|_2 = \|A^{-1}\|_2^2$. As seen in section 7.4, a large condition number can lead to a loss of accuracy in the computed solution.

To solve example (7.12) using the normal equations, we write

$$\begin{pmatrix} 1 & 1 & 2 \\ 1 & -1 & 1 \end{pmatrix} \begin{pmatrix} 1 & 1 \\ 1 & -1 \\ 2 & 1 \end{pmatrix} \begin{pmatrix} x_1 \\ x_2 \end{pmatrix} = \begin{pmatrix} 1 & 1 & 2 \\ 1 & -1 & 1 \end{pmatrix} \begin{pmatrix} 2 \\ 0 \\ 4 \end{pmatrix},$$

which becomes

$$\begin{pmatrix} 6 & 2 \\ 2 & 3 \end{pmatrix} \begin{pmatrix} x_1 \\ x_2 \end{pmatrix} = \begin{pmatrix} 10 \\ 6 \end{pmatrix}.$$

Solving this square linear system by Gaussian elimination, we find

$$\begin{pmatrix} 6 & 2 & | & 10 \\ 2 & 3 & | & 6 \end{pmatrix} \longrightarrow \begin{pmatrix} 6 & 2 & | & 10 \\ 0 & 7/3 & | & 8/3 \end{pmatrix},$$

resulting in

$$x_2 = \frac{8}{7}, \quad x_1 = \frac{9}{7}.$$

It is always a good idea to "check" your answer by computing the residual to make sure that it looks reasonable:

$$\begin{pmatrix} 2 \\ 0 \\ 4 \end{pmatrix} - \begin{pmatrix} 1 & 1 \\ 1 & -1 \\ 2 & 1 \end{pmatrix} \begin{pmatrix} 9/7 \\ 8/7 \end{pmatrix} = \begin{pmatrix} -3/7 \\ -1/7 \\ 2/7 \end{pmatrix}.$$

7.6.2 QR Decomposition

The least squares problem can be approached in a different way. We wish to find a vector $\mathbf{x}$ such that $A\mathbf{x} = \mathbf{b}_*$, where $\mathbf{b}_*$ is the closest vector to $\mathbf{b}$ (in the 2-norm) in the range of A. This is equivalent to minimizing $\|\mathbf{b} - A\mathbf{x}\|_2$ since, by the definition of $\mathbf{b}_*$, $\|\mathbf{b} - \mathbf{b}_*\|_2 \leq \|\mathbf{b} - A\mathbf{y}\|_2$ for all $\mathbf{y}$.

You may recall from linear algebra that the closest vector to a given vector, from a subspace, is the **orthogonal projection** of that vector onto the subspace. This is pictured in figure 7.7.

If $\mathbf{q}_1, \ldots, \mathbf{q}_k$ form an **orthonormal basis** for the subspace then the orthogonal projection of $\mathbf{b}$ onto the subspace is $\sum_{j=1}^{k} \langle \mathbf{b}, \mathbf{q}_j \rangle \mathbf{q}_j$. Assume that the columns of A are linearly independent, so that the range of A (which is the span of its columns) has dimension n. Then the closest vector to $\mathbf{b}$ in range(A) is

$$\mathbf{b}_* = \sum_{j=1}^{n} \langle \mathbf{b}, \mathbf{q}_j \rangle \mathbf{q}_j,$$

where $\mathbf{q}_1, \ldots, \mathbf{q}_n$ form an orthonormal basis for range(A). Let Q be the m by n matrix whose columns are the orthonormal vectors $\mathbf{q}_1, \ldots, \mathbf{q}_n$. Then $Q^T Q = I_{n \times n}$ and the formula for $\mathbf{b}_*$ can be written compactly as

$$\mathbf{b}_* = Q(Q^T \mathbf{b}),$$

since $Q(Q^T\mathbf{b}) = \sum_{j=1}^{n} \mathbf{q}_j (Q^T\mathbf{b})_j = \sum_{j=1}^{n} \mathbf{q}_j (\mathbf{q}_j^T \mathbf{b}) = \sum_{j=1}^{n} \mathbf{q}_j \langle \mathbf{b}, \mathbf{q}_j \rangle$.

GRAM, SCHMIDT, LAPLACE, AND CAUCHY

The Gram–Schmidt method is named for Jorgen Pedersen Gram (1850–1916) and Erhard Schmidt (1876–1959). Schmidt presented the recursive formulas of the process in 1907. In a footnote, Schmidt mentions that his formulas are essentially the same as already given in an 1883 paper of J. P. Gram. In fact, it was neither Gram nor Schmidt who first used the ideas of the method, as the process seems to be a result of Laplace and was essentially used by Cauchy in 1836 [74, 75].

Gram Schmidt Laplace Cauchy

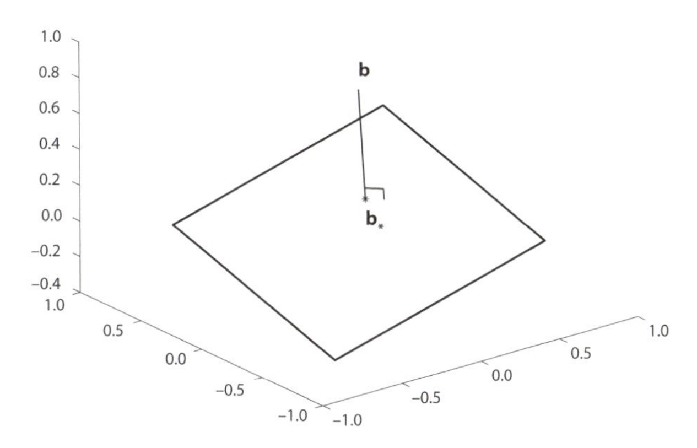

Figure 7.7. The orthogonal projection of a vector onto a subspace.

You may also recall from linear algebra that, given a set of linearly independent vectors such as the columns of A, one can construct an orthonormal set that spans the same space using the *Gram–Schmidt algorithm*:

Given a linearly independent set $\mathbf{v}_1, \mathbf{v}_2, \ldots$, set $\mathbf{q}_1 = \mathbf{v}_1/\|\mathbf{v}_1\|$, and for $j = 2, 3, \ldots$,
  Set $\tilde{\mathbf{q}}_j = \mathbf{v}_j - \sum_{i=1}^{j-1} \langle \mathbf{v}_j, \mathbf{q}_i \rangle \mathbf{q}_i$.
  Set $\mathbf{q}_j = \tilde{\mathbf{q}}_j / \|\tilde{\mathbf{q}}_j\|$.

Note that if $\mathbf{v}_1, \mathbf{v}_2, \ldots, \mathbf{v}_n$ are the columns of A, then the Gram–Schmidt algorithm can be thought of as factoring the m by n matrix A in the form $A = QR$, where $Q = (\mathbf{q}_1, \ldots, \mathbf{q}_n)$ is an m by n matrix with orthonormal columns and R is an n by n upper triangular matrix. To see this, write the equations of the Gram–Schmidt algorithm in the form

$$\mathbf{v}_1 = r_{11}\mathbf{q}_1, \quad r_{11} = \|\mathbf{v}_1\|,$$

$$\mathbf{v}_j = r_{jj}\mathbf{q}_j + \sum_{i=1}^{j-1} r_{ij}\mathbf{q}_i, \quad r_{jj} = \|\tilde{\mathbf{q}}_j\|, \quad r_{ij} = \langle \mathbf{v}_j, \mathbf{q}_i \rangle, \quad j = 2, \ldots, n.$$

These equations can be written using matrices as

$$(\mathbf{v}_1, \ldots, \mathbf{v}_n) = (\mathbf{q}_1, \ldots, \mathbf{q}_n) \begin{pmatrix} r_{11} & r_{12} & \ldots & r_{1n} \\ & r_{22} & \ldots & r_{2n} \\ & & \ddots & \vdots \\ & & & r_{nn} \end{pmatrix}.$$

Thus if $\mathbf{v}_1, \ldots, \mathbf{v}_n$ are the columns of A, then we have written A in the form $A = QR$. Pictorially, this matrix factorization looks like:

$$\boxed{A} = \boxed{Q}\,\boxed{R}$$

This is called the **reduced QR decomposition** of A. (In the **full QR decomposition**, one completes the orthonormal set $\mathbf{q}_1, \ldots, \mathbf{q}_n$ to an orthonormal basis for $\mathbf{R}^m$: $\mathbf{q}_1, \ldots, \mathbf{q}_n, \mathbf{q}_{n+1}, \ldots, \mathbf{q}_m$ so that A is factored in the form $\underbrace{A}_{m\times n} = \underbrace{Q}_{m\times m} \underbrace{\begin{pmatrix} R \\ 0 \end{pmatrix}}_{m\times n}$.)

Having factored A in the form QR, the least squares problem becomes $QR\mathbf{x} = \mathbf{b}_* = QQ^T\mathbf{b}$. We know that this set of equations has a solution since $\mathbf{b}_* = QQ^T\mathbf{b}$ lies in the range of A. Hence if we multiply each side by Q^T, the resulting equation will have the same solution. Since $Q^T Q = I$, we are left with the upper triangular system $R\mathbf{x} = Q^T\mathbf{b}$. Note that the assumption that A has linearly independent columns ensures that the quantities in the Gram–Schmidt algorithm are well defined, so that the diagonal entries in R are nonzero. Hence the n by n linear system has a *unique* solution. If the columns of A are not linearly independent then modifications to this algorithm must be made. We will not discuss this special case here, but see, for example, [32].

Thus the procedure for solving the least squares problem consists of two steps:

1. Compute the reduced QR decomposition of A.
2. Solve the n by n upper triangular system $R\mathbf{x} = Q^T\mathbf{b}$.

This is the way MATLAB solves problems with more equations than unknowns, when you type `A\b`. It does not use the Gram–Schmidt algorithm to compute the QR factorization, however. Different methods are often used for the QR factorization, but these will not be discussed here. See, for example, [96]. You can compute the reduced QR decomposition with MATLAB by typing `[Q,R] = qr(A,0)`, while `[Q,R] = qr(A)` gives the full QR decomposition.

To see how problem (7.12) can be solved using the QR decomposition, we first use the Gram–Schmidt algorithm to construct a pair of orthonormal vectors that span the same space as the columns of the coefficient matrix:

$$\mathbf{q}_1 = \frac{1}{\sqrt{6}} \begin{pmatrix} 1 \\ 1 \\ 2 \end{pmatrix}, \quad r_{11} = \sqrt{6},$$

$$\tilde{\mathbf{q}}_2 = \begin{pmatrix} 1 \\ -1 \\ 1 \end{pmatrix} - \frac{(1 \cdot 1 + (-1) \cdot 1 + 1 \cdot 2)}{\sqrt{6}} \cdot \frac{1}{\sqrt{6}} \begin{pmatrix} 1 \\ 1 \\ 2 \end{pmatrix} = \frac{1}{3} \begin{pmatrix} 2 \\ -4 \\ 1 \end{pmatrix}, \quad r_{12} = \frac{2}{\sqrt{6}},$$

$$\mathbf{q}_2 = \frac{1}{\sqrt{21}} \begin{pmatrix} 2 \\ -4 \\ 1 \end{pmatrix}, \quad r_{22} = \sqrt{\frac{7}{3}}.$$

We then compute the product of Q^T and the right-hand side vector:

$$\begin{pmatrix} 1/\sqrt{6} & 1/\sqrt{6} & 2/\sqrt{6} \\ 2/\sqrt{21} & -4/\sqrt{21} & 1/\sqrt{21} \end{pmatrix} \begin{pmatrix} 2 \\ 0 \\ 4 \end{pmatrix} = \begin{pmatrix} 10/\sqrt{6} \\ 8/\sqrt{21} \end{pmatrix}.$$

Finally we solve the triangular linear system,

$$\begin{pmatrix} \sqrt{6} & 2/\sqrt{6} \\ 0 & \sqrt{7/3} \end{pmatrix} \begin{pmatrix} x_1 \\ x_2 \end{pmatrix} = \begin{pmatrix} 10/\sqrt{6} \\ 8/\sqrt{21} \end{pmatrix},$$

to obtain

$$x_2 = \frac{8}{7}, \quad \sqrt{6}x_1 + \frac{2}{\sqrt{6}} \cdot \frac{8}{7} = \frac{10}{\sqrt{6}} \rightarrow x_1 = \frac{9}{7}.$$

Although the QR decomposition is the preferred method for solving least squares problems on the computer, the arithmetic can become messy when using this algorithm by hand. The normal equations approach is usually easier when solving a small problem by hand.

7.6.3 Fitting Polynomials to Data

Given a set of measured data points (x_i, y_i), $i = 1, \ldots, m$, one sometimes expects a linear relationship of the form $y = c_0 + c_1 x$. As a result of errors in the measurements, the points will not fall exactly on a straight line, but they should be close, and one would like to find the straight line that most closely matches the data; that is, one might look for a least squares solution to the overdetermined linear system

$$\begin{aligned} c_0 + c_1 x_1 &\approx y_1, \\ c_0 + c_1 x_2 &\approx y_2, \\ &\vdots \\ c_0 + c_1 x_m &\approx y_m, \end{aligned}$$

by choosing c_0 and c_1 to minimize

$$\sum_{i=1}^{m} (y_i - (c_0 + c_1 x_i))^2.$$

Writing the above equations with matrices, we have

$$\begin{pmatrix} 1 & x_1 \\ 1 & x_2 \\ \vdots & \vdots \\ 1 & x_m \end{pmatrix} \begin{pmatrix} c_0 \\ c_1 \end{pmatrix} \approx \begin{pmatrix} y_1 \\ y_2 \\ \vdots \\ y_m \end{pmatrix},$$

which can be solved using the normal equations or the QR decomposition, as explained in the previous two subsections.

Suppose we wish to fit a polynomial of degree $n - 1 < m$ to the measured data points; that is, we wish to find coefficients $c_0, c_1, \ldots, c_{n-1}$ such that

$$y_i \approx c_0 + c_1 x_i + c_2 x_i^2 + \ldots + c_{n-1} x_i^{n-1}, \quad i = 1, \ldots, m,$$

or, more precisely, such that $\sum_{i=1}^{m} (y_i - \sum_{j=0}^{n-1} c_j x_i^j)^2$ is minimal. Again writing the system using matrices, we have

$$\underbrace{\begin{pmatrix} 1 & x_1 & x_1^2 & \ldots & x_1^{n-1} \\ 1 & x_2 & x_2^2 & \ldots & x_2^{n-1} \\ \vdots & \vdots & \vdots & & \vdots \\ \vdots & \vdots & \vdots & & \vdots \\ \vdots & \vdots & \vdots & & \vdots \\ 1 & x_m & x_m^2 & \ldots & x_m^{n-1} \end{pmatrix}}_{m \times n} \underbrace{\begin{pmatrix} c_0 \\ c_1 \\ c_2 \\ \vdots \\ c_{n-1} \end{pmatrix}}_{n \times 1} \approx \underbrace{\begin{pmatrix} y_1 \\ y_2 \\ \vdots \\ \vdots \\ \vdots \\ y_m \end{pmatrix}}_{m \times 1}.$$

The m by n system can again be solved using either the normal equations or the QR decomposition of the coefficient matrix.

Example 7.6.1. Consider the following data:

x	y
1	2
2	3
3	6

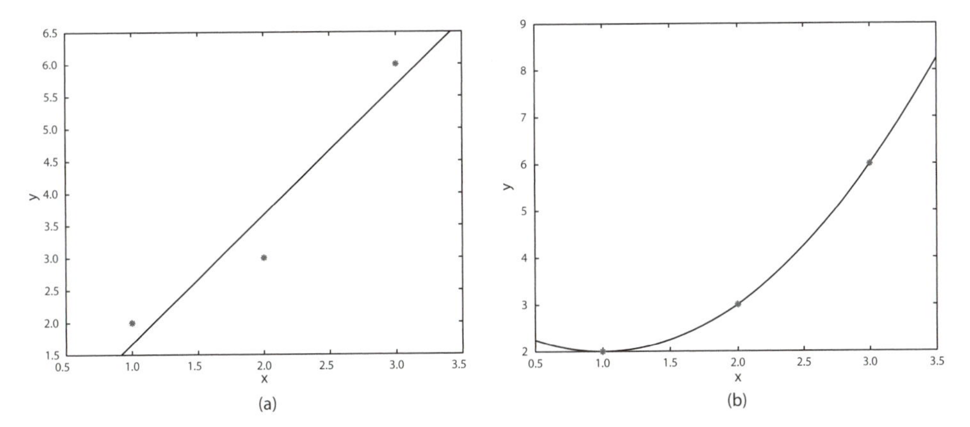

Figure 7.8. Some data and its (a) linear and (b) quadratic least squares fit.

To find the straight line $y = c_0 + c_1 x$ that best fits the data, we must solve, in a least squares sense, the overdetermined system

$$\begin{pmatrix} 1 & 1 \\ 1 & 2 \\ 1 & 3 \end{pmatrix} \begin{pmatrix} c_0 \\ c_1 \end{pmatrix} \approx \begin{pmatrix} 2 \\ 3 \\ 6 \end{pmatrix}.$$

Using the normal equations approach, this becomes

$$\begin{pmatrix} 1 & 1 & 1 \\ 1 & 2 & 3 \end{pmatrix} \begin{pmatrix} 1 & 1 \\ 1 & 2 \\ 1 & 3 \end{pmatrix} \begin{pmatrix} c_0 \\ c_1 \end{pmatrix} = \begin{pmatrix} 1 & 1 & 1 \\ 1 & 2 & 3 \end{pmatrix} \begin{pmatrix} 2 \\ 3 \\ 6 \end{pmatrix},$$

or,

$$\begin{pmatrix} 3 & 6 \\ 6 & 14 \end{pmatrix} \begin{pmatrix} c_0 \\ c_1 \end{pmatrix} = \begin{pmatrix} 11 \\ 26 \end{pmatrix}.$$

Solving this 2 by 2 system, we find $c_0 = -1/3$ and $c_1 = 2$, so the straight line of best fit has the equation $y = -1/3 + 2x$. The data and this line are plotted in figure 7.8(a).

Suppose we fit a quadratic polynomial $y = c_0 + c_1 x + c_2 x^2$ to the above data points. We will see in the next section that there is a unique quadratic that *exactly* fits the three data points. This means that when we solve the least squares problem, the residual will be zero. The equations for c_0, c_1, and c_2 are

$$\begin{pmatrix} 1 & 1 & 1 \\ 1 & 2 & 4 \\ 1 & 3 & 9 \end{pmatrix} \begin{pmatrix} c_0 \\ c_1 \\ c_2 \end{pmatrix} = \begin{pmatrix} 2 \\ 3 \\ 6 \end{pmatrix}.$$

There is no need to form the normal equations here; we can solve this square linear system directly to obtain: $c_0 = 3$, $c_1 = -2$, $c_2 = 1$; that is, $y = 3 - 2x + x^2$. This quadratic is plotted along with the data points in figure 7.8(b).

TABLE 7.5
Data from Galileo's inclined plane experiment.

Release height	Horizontal distance
0.282	0.752
0.564	1.102
0.752	1.248
0.940	1.410

Example 7.6.2. Over 400 years ago Galileo, attempting to find a mathematical description of falling bodies, studied the paths of projectiles. One experiment consisted of rolling a ball down a grooved ramp inclined at a fixed angle to the horizontal, starting the ball at a fixed height h above a table of height 0.778 meters. When the ball left the end of the ramp, it rolled for a short distance along the table and then descended to the floor. Galileo altered the release height h and measured the horizontal distance d that the ball traveled before landing on the floor. Table 7.5 shows data from Galileo's notes (with measurements converted from *puntos* (1 punto $\approx$ 0.94 millimeters) to meters) [35, p. 35].

We can create a least squares linear fit with the following MATLAB code:

```
%%% Set up data
h = [0.282; 0.564; 0.752; 0.940]; d = [0.752; 1.102; 1.248; 1.410];

%%% Form the 4x2 matrix A and solve for the coefficient vector c.
A = [ones(size(h)), h];
c = A\d;
cc = flipud(c);            % order the coefficients in descending
                           % order for polyval

%%% Plot the data points
plot(h,d,'b*'), title('Least Squares Linear Fit'), hold
xlabel('release height'), ylabel('horizontal distance')

%%% Plot the line of best fit
hmin = min(h); hmax = max(h); h1 = [hmin:(hmax-hmin)/100:hmax];
plot(h1,polyval(cc,h1),'r'), axis tight
```

This code produces the line $y = 0.4982 + 0.9926x$. This line is plotted along with the data points in figure 7.9. A measure of the amount by which the line fails to hit the data points is the residual norm:

$$\sqrt{\sum_{i=1:4} (d_i - y(h_i))^2} = 0.0555.$$

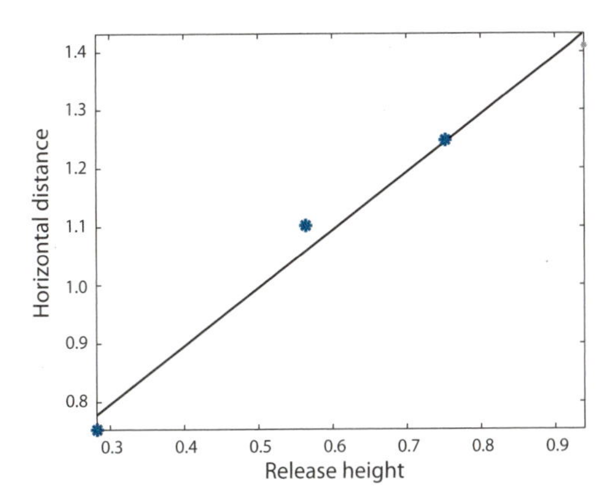

Figure 7.9. Linear least squares fit to data points collected by Galileo.

In the exercises to follow, you will try to derive a mathematical relation between h and d and then do a least squares fit to see if the data obey the mathematical model.

7.7 CHAPTER 7 EXERCISES

1. Write the following matrix in the form LU, where L is a unit lower triangular matrix and U is an upper triangular matrix:

$$\begin{pmatrix} 4 & -1 & -1 \\ -1 & 4 & -1 \\ -1 & -1 & 4 \end{pmatrix}$$

   Write the same matrix in the form LL^T, where L is lower triangular.

2. Write a function `usolve`, analogous to function `lsolve` in section 7.2.2, to solve an upper triangular system $U\mathbf{x} = \mathbf{y}$. [Hint: Loops can be run backwards, say, from `n` down to `1`, by typing in MATLAB: `for i=n:-1:1`. Remember also that the diagonal entries in U are not necessarily 1.]

3. Let

$$A = \begin{pmatrix} 10^{-16} & 1 \\ 1 & 1 \end{pmatrix} \quad \text{and} \quad \mathbf{b} = \begin{pmatrix} 2 \\ 3 \end{pmatrix}.$$

   Here we will see the effect of using a tiny element as a pivot.

   (a) By hand, solve the linear system $A\mathbf{x} = \mathbf{b}$ exactly. Write your answer in a form where it is clear what the approximate values of x_1 and x_2 are.

   (b) In MATLAB, enter the matrix A and type `cond(A)` to determine the 2-norm condition number of A. Would you say that this matrix is well conditioned or ill conditioned in the 2-norm?

   (c) Write a MATLAB code (or use one from the text) that does *not* do partial pivoting to solve the linear system $A\mathbf{x} = \mathbf{b}$. Compare the answer returned by this code to the one that you determined by hand

and to the one that you get in MATLAB by typing `A\b` (which *does* do partial pivoting).

4. Assume that a matrix A satisfies $PA = LU$, where

$$P = \begin{pmatrix} 0 & 0 & 1 \\ 1 & 0 & 0 \\ 0 & 1 & 0 \end{pmatrix}, \quad L = \begin{pmatrix} 1 & 0 & 0 \\ \frac{1}{2} & 1 & 0 \\ \frac{1}{3} & \frac{1}{4} & 1 \end{pmatrix}, \quad \text{and } U = \begin{pmatrix} 2 & 3 & 1 \\ 0 & 1 & 2 \\ 0 & 0 & 2 \end{pmatrix}.$$

Use these factors, without forming the matrix A or taking any inverses, to solve the system of equations $A\mathbf{x} = \mathbf{b}$, where $\mathbf{b} = (2, 10, -12)^T$. Show forward and back substitution by hand.

5. Add partial pivoting to the LU factorization code of section 7.2.2. Use a vector, say, `piv` to keep track of row interchanges. For example, you might initialize `piv` to be `[1:n]`, and then if row `i` is interchanged with row `j`, you would interchange `piv(i)` and `piv(j)`. Now use the vector `piv` along with the L and U factors to solve a linear system $A\mathbf{x} = \mathbf{b}$. To do this, you will first need to perform the same interchanges of entries in $\mathbf{b}$ that were performed on the rows of A. Then you can use function `lsolve` in section 7.2.2 to solve $L\mathbf{y} = \mathbf{b}$, followed by function `usolve` from the previous exercise to solve $U\mathbf{x} = \mathbf{y}$. You can check your answer by computing `norm(b-A*x)` (which should be on the order of the machine precision).

6. How many operations (additions, subtractions, multiplications, and divisions) are required to:

   (a) Compute the sum of two n-vectors?
   (b) Compute the product of an m by n matrix with an n-vector?
   (c) Solve an n by n upper triangular linear system $U\mathbf{x} = \mathbf{y}$?

7. Estimate the number of operations required for Gaussian elimination with partial pivoting on an n by n matrix with half bandwidth m. Use the fact that the half bandwidth of the U factor is at most doubled by using partial pivoting.

8. Compute the 2-norm, the 1-norm, and the ∞-norm of

$$\mathbf{v} = \begin{pmatrix} 4 \\ 5 \\ -6 \end{pmatrix}.$$

9. Compute the 1-norm and the ∞-norm of

$$A = \begin{pmatrix} 5 & 6 \\ 7 & 8 \end{pmatrix}.$$

Also compute the condition number $\kappa(A) \equiv \|A\| \cdot \|A^{-1}\|$ in the 1-norm and the ∞-norm.

10. Show that for all n-vectors $\mathbf{v}$:

   (a) $\|\mathbf{v}\|_\infty \leq \|\mathbf{v}\|_2 \leq \sqrt{n}\|\mathbf{v}\|_\infty$.
   (b) $\|\mathbf{v}\|_2 \leq \|\mathbf{v}\|_1$.
   (c) $\|\mathbf{v}\|_1 \leq n\|\mathbf{v}\|_\infty$.

For each of the above inequalities, identify a nonzero vector $\mathbf{v}$ (with $n > 1$) for which equality holds.

11. Prove that for any nonsingular 2 by 2 matrix, the ∞-norm condition number and the 1-norm condition number are equal. [Hint: Use the fact that the inverse of a 2 by 2 matrix is one over the determinant times the adjoint.]
12. It is shown in the text (inequality (7.9)) that if $A\mathbf{x} = \mathbf{b}$ and $A\hat{\mathbf{x}} = \hat{\mathbf{b}}$, then

$$\frac{\|\mathbf{x} - \hat{\mathbf{x}}\|}{\|\mathbf{x}\|} \leq \kappa(A)\frac{\|\mathbf{b} - \hat{\mathbf{b}}\|}{\|\mathbf{b}\|}.$$

Go through the proof of this inequality and show that for certain nonzero vectors $\mathbf{b}$ and $\hat{\mathbf{b}}$ (with $\hat{\mathbf{b}} \neq \mathbf{b}$), equality will hold.

13. Write a routine to generate an n by n matrix with a given 2-norm condition number. You can make your routine a `function` in MATLAB that takes two input arguments—the matrix size `n` and the desired condition number `condno`—and produces an `n` by `n` matrix `A` with the given condition number as output:

```
function A = matgen(n, condno)
```

Form A by generating two random orthogonal matrices U and V and a diagonal matrix Σ with $\sigma_{ii} = \texttt{condno}^{-(i-1)/(n-1)}$, and setting $A = U\Sigma V^T$. [Note that the largest diagonal entry in Σ is 1 and the smallest is $\texttt{condno}^{-1}$, so the ratio is `condno`.] You can generate a random orthogonal matrix in MATLAB by first generating a random matrix, `Mat = randn(n,n)`, and then computing its QR decomposition, `[Q,R] = qr(Mat)`. The matrix `Q` is then a random orthogonal matrix. You can check the condition number of the matrix you generate by using the function `cond` in MATLAB. Turn in a listing of your code.

For `condno`= $(1,\ 10^4,\ 10^8,\ 10^{12},\ 10^{16})$, use your routine to generate a random matrix `A` with condition number `condno`. Also generate a random vector `xtrue` of length `n`, and compute the product `b = A*xtrue`.

Solve $A\mathbf{x} = \mathbf{b}$ using Gaussian elimination with partial pivoting. This can be done in MATLAB by typing `x = A\b`. Determine the 2-norm of the relative error in your computed solution, $\|\texttt{x} - \texttt{xtrue}\|/\|\texttt{xtrue}\|$. Explain how this is related to the condition number of A. Compute the 2-norm of the relative residual, $\|\mathbf{b} - A\mathbf{x}\|/(\|A\|\|\mathbf{x}\|)$. Does the algorithm for solving $A\mathbf{x} = \mathbf{b}$ appear to be *backward stable*; that is, is the computed solution the exact solution to a nearby problem?

Solve $A\mathbf{x} = \mathbf{b}$ by inverting A: `Ainv = inv(A)`, and then multiplying $\mathbf{b}$ by A^{-1}: `x = Ainv*b`. Again look at relative errors and residuals. Does this algorithm appear to be backward stable?

Finally, solve $A\mathbf{x} = \mathbf{b}$ using Cramer's rule. Use the MATLAB function `det` to compute determinants. (Don't worry—it does not use the $n!$ algorithm discussed in section 7.3.) Again look at relative errors and residuals and determine whether this algorithm is backward stable.

Turn in a table showing the relative errors and residuals for each of the three algorithms and each of the condition numbers tested, along with a brief explanation of the results.

14. Find the polynomial of degree 10 that best fits the function $b(t) = \cos(4t)$ at 50 equally-spaced points t between 0 and 1. Set up the matrix A and right-hand side vector $\mathbf{b}$, and determine the polynomial coefficients in two different ways:

    (a) By using the MATLAB command `x = A\b` (which uses a QR decomposition).
    (b) By solving the normal equations $A^T A\mathbf{x} = A^T\mathbf{b}$. This can be done in MATLAB by typing `x = (A'*A)\(A'*b)`.

    Print the results to 16 digits (using `format long e`) and comment on the differences you see. [Note: You can compute the condition number of A or of $A^T A$ using the MATLAB function `cond`.]
15. In the inclined plane experiment of Galileo, described in Example 7.6.2, if one knows the horizontal speed v of the ball when it leaves the table and enters free fall, then one can determine the time T that it takes for the ball to fall to the ground due to the force of gravity and therefore the horizontal distance vT that it travels before hitting the ground. Assume that the acceleration of the ball on the inclined plane is constant in the direction of the incline: $a(t) = C$, where C will depend on gravity and on the angle of the inclined plane. Then the velocity $v(t)$ will be proportional to t: $v(t) = Ct$, since $v(0) = 0$. The distance $s(t)$ that the ball has moved along the inclined plane is therefore $s(t) = \frac{1}{2}Ct^2$, since $s(0) = 0$. The ball reaches the bottom of the inclined plane when $s(t) = h/\sin\theta$, where θ is the angle of incline.

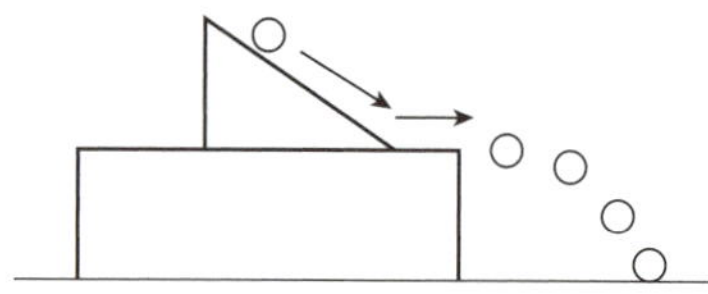

    (a) Write down an expression in terms of C, h, and θ for the time t_b at which the ball reaches the bottom of the inclined plane and for its velocity $v(t_b)$ at that time. Show that the velocity is proportional to $\sqrt{h}$.
    (b) Upon hitting the table, the ball is deflected so that what was its velocity along the incline now becomes its horizontal velocity as it flies off the table. From this observation and part (a), one would expect Galileo's measured distances to be proportional to $\sqrt{h}$. Modify the MATLAB code in Example 7.6.2 to do a least squares fit of the given distances d in table 7.5 to a function of the form $c_0 + c_1\sqrt{h}$. Determine the best coefficients c_0 and c_1 and the residual norm $\sqrt{\sum_{i=1}^{4}(d_i - (c_0 + c_1\sqrt{h_i}))^2}$.

16. Consider the following least squares approach for ranking sports teams. Suppose we have four college football teams, called simply T1, T2, T3, and T4. These four teams play each other with the following outcomes:

    - T1 beats T2 by 4 points: 21 to 17.
    - T3 beats T1 by 9 points: 27 to 18.
    - T1 beats T4 by 6 points: 16 to 10.

- T3 beats T4 by 3 points: 10 to 7.
- T2 beats T4 by 7 points: 17 to 10.

To determine ranking points $r_1, \ldots, r_4$ for each team, we do a least squares fit to the overdetermined linear system:

$$r_1 - r_2 = 4,$$
$$r_3 - r_1 = 9,$$
$$r_1 - r_4 = 6,$$
$$r_3 - r_4 = 3,$$
$$r_2 - r_4 = 7.$$

This system does not have a unique least squares solution, however, since if $(r_1, \ldots, r_4)^T$ is one solution and we add to it any constant vector, such as the vector $(1, 1, 1, 1)^T$, then we obtain another vector for which the residual is exactly the same. Show that if $(r_1, \ldots, r_4)^T$ solves the least squares problem above then so does the vector $(r_1 + c, \ldots, r_4 + c)^T$ for any constant c.

To make the solution unique, we can fix the total number of ranking points, say, at 20. To do this, we add the following equation to those listed above:

$$r_1 + r_2 + r_3 + r_4 = 20.$$

Note that this equation will be satisfied exactly since it will not affect how well the other equalities can be approximated. Use MATLAB to determine the values r_1, r_2, r_3, r_4 that most closely satisfy these equations, and based on your results, rank the four teams. (This method of ranking sports teams is a simplification of one introduced by Ken Massey in 1997 [67]. It has evolved into a part of the famous BCS (Bowl Championship Series) model for ranking college football teams and is one factor in determining which teams play in bowl games.)

17. This exercise uses the MNIST database of handwritten digits, which contains a training set of 60,000 numbers and a test set of 10,000 numbers. Each digit in the database was placed in a 28 by 28 grayscale image such that the center of mass of its pixels is at the center of the picture. To load the database, download it from the book's web page and type `load mnist_all.mat` in MATLAB. Type `who` to see the variables containing training digits (`train0`,...,`train9`) and test digits (`test0`,...,`test9`). You will find digits intended to train an algorithm to recognize a handwritten 0 in the matrix **train0**, which has 5923 rows and 784 columns. Each row corresponds to one handwritten zero. To visualize the first image in this matrix, type

```
digit = train0(1,:);
digitImage = reshape(digit,28,28);
image(rot90(flipud(digitImage),-1)),
colormap(gray(256)), axis square tight off;
```

Note, the **rot90** and **flipud** commands are used so the digits appear as we write them, which is more noticeable with digits like 2 or 3.

(a) Create a 10 by 784 matrix `T` whose ith row contains the *average* pixel values over all the training images of the number $i - 1$. For instance, the first row of T can be formed by typing `T(1,:) = mean(train0);`. Visualize these average digits using the **subplot** command as seen below.

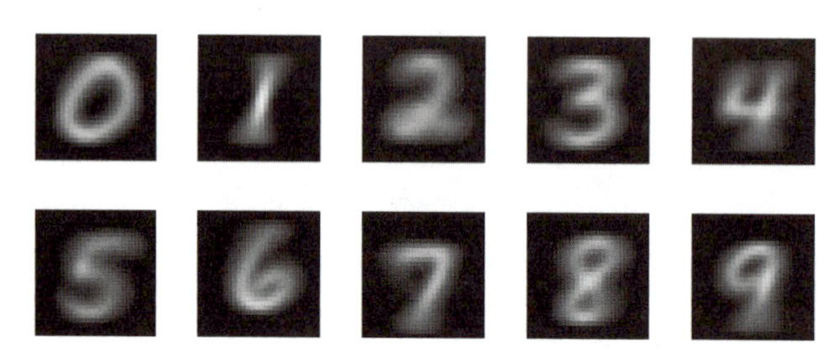

(b) A simple way to identify a test digit is to compare its pixels to those in each row of `T` and determine which row most closely resembles the test digit. Set `d` to be the first test digit in `test0` by typing `d = double(test0(1,:));`. For each row $i = 1, \ldots, 10$, compute `norm(T(i,:) - d)`, and determine for which value of i this is smallest; `d` probably is the digit $i - 1$. Try some other digits as well and report on your results.
(A more sophisticated approach called **Principal Component Analysis** or PCA attempts to identify characteristic properties of each digit, based on the training data, and compares these properties with those of the test digit in order to make an identification.)

POLYNOMIAL AND PIECEWISE POLYNOMIAL INTERPOLATION

8.1 THE VANDERMONDE SYSTEM

Given a set of $n+1$ data points, $(x_0, y_0), (x_1, y_1), \ldots, (x_n, y_n)$, one can find a polynomial of degree at most n that exactly fits these points. One way to do this was demonstrated in the previous section. Since we want

$$y_i = \sum_{j=0}^{n} c_j x_i^j, \quad i = 0, \ldots, n, \tag{8.1}$$

the coefficients $c_0, \ldots, c_n$ must satisfy

$$\begin{pmatrix} 1 & x_0 & x_0^2 & \ldots & x_0^n \\ 1 & x_1 & x_1^2 & \ldots & x_1^n \\ 1 & x_2 & x_2^2 & \ldots & x_2^n \\ \vdots & \vdots & \vdots & & \vdots \\ 1 & x_n & x_n^2 & \ldots & x_n^n \end{pmatrix} \begin{pmatrix} c_0 \\ c_1 \\ c_2 \\ \vdots \\ c_n \end{pmatrix} = \begin{pmatrix} y_0 \\ y_1 \\ y_2 \\ \vdots \\ y_n \end{pmatrix}. \tag{8.2}$$

The matrix here is called a **Vandermonde matrix.** It is named after Alexandre-Théophile Vandermonde (1735–1796), who is generally associated with determinant theory. It can be shown to be nonsingular, provided $x_0, \ldots, x_n$ are distinct, but, unfortunately, it can be very *ill conditioned.* To solve the linear system using Gaussian elimination (or QR factorization) requires $O(n^3)$ operations. Because of the ill conditioning and the large amount of work required, polynomial interpolation problems usually are not solved in this way. In the following sections we give some alternatives.

8.2 THE LAGRANGE FORM OF THE INTERPOLATION POLYNOMIAL

One can simply write down the polynomial p of degree n that satisfies $p(x_i) = y_i, i = 0, \ldots, n$. [Notation: When we refer to a polynomial of degree n, we will

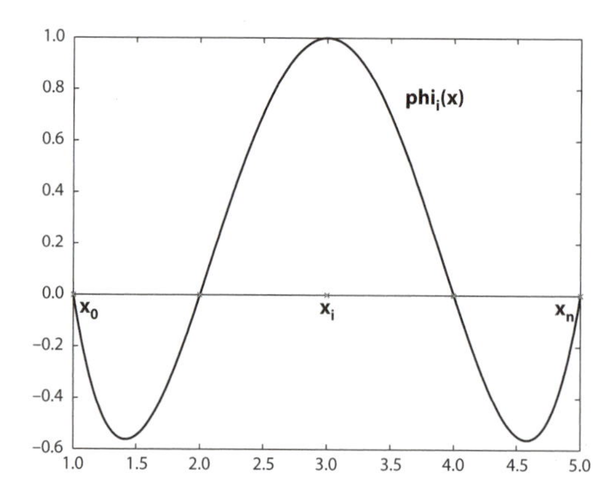

Figure 8.1. A graphical representation of φ_i in the Lagrange form for interpolation.

always mean one whose highest power is *less than or equal to* n; to distinguish this from a polynomial whose highest power is exactly n we will refer to the latter as a polynomial of *exact degree* n.] To do this we will first write down a polynomial $\varphi_i(x)$ that is 1 at x_i and 0 at all of the other nodes:

$$\varphi_i(x) = \prod_{j \neq i} \frac{x - x_j}{x_i - x_j}.$$

Note that if $x = x_i$, then each factor in the product is 1, while if $x = x_j$ for some $j \neq i$, then one of the factors is 0. Note also that $\varphi_i(x)$ is of the desired degree since the product involves n factors: all but one of $j = 0, \ldots, n$. A polynomial of degree n that has value y_i at x_i and value 0 at all other nodes is $y_i \varphi_i(x)$. It follows that a polynomial of degree n that interpolates the points $(x_0, y_0), \ldots, (x_n, y_n)$ is

$$p(x) = \sum_{i=0}^{n} y_i \varphi_i(x) = \sum_{i=0}^{n} y_i \left(\prod_{j \neq i} \frac{x - x_j}{x_i - x_j} \right). \tag{8.3}$$

This is called the **Lagrange form** of the interpolation polynomial.

Example 8.2.1. Consider Example 7.6.1, where we fit a quadratic to the three data points $(1, 2)$, $(2, 3)$, and $(3, 6)$. The Lagrange form of the interpolation polynomial is

$$p(x) = 2 \cdot \frac{(x-2)(x-3)}{(1-2)(1-3)} + 3 \cdot \frac{(x-1)(x-3)}{(2-1)(2-3)} + 6 \cdot \frac{(x-1)(x-2)}{(3-1)(3-2)}.$$

You can check that this is equal to $x^2 - 2x + 3$, the same polynomial computed from the Vandermonde system in section 7.6.3.

Since we simply wrote down the Lagrange form of the interpolation polynomial, there is no computer time involved in determining it, but to evaluate formula (8.3) at a point requires $O(n^2)$ operations since there are $n+1$ terms in the sum (requiring n additions) and each term is a product of $n+1$ factors (requiring n multiplications). In contrast, if one knows p in the form (8.1), then the value of p at a point x can be determined using *Horner's rule* with just $O(n)$ operations: Start by setting $y = c_n$. Then replace y by $yx + c_{n-1}$ so that $y = c_n x + c_{n-1}$. Replace this value by $yx + c_{n-2}$ so that $y = c_n x^2 + c_{n-1} x + c_{n-2}$. Continuing for n steps, we obtain $y = \sum_{j=0}^{n} c_j x^j$, using about $2n$ operations. In the next section we will represent the interpolation polynomial in a form that requires somewhat more work to determine than (8.3) but which requires less work to evaluate.

It was stated in the previous section that the Vandermonde matrix can be shown to be nonsingular. In fact, the above derivation of the Lagrange interpolation polynomial can be used to show this. Since the polynomial in (8.3) clearly goes through the points $(x_0, y_0), \ldots, (x_n, y_n)$, we have shown that the linear system (8.2) has a solution for any right-hand side vector $(y_0, \ldots, y_n)^T$. This implies that the Vandermonde matrix is nonsingular and this, in turn, implies that the solution is unique. We have thus proved the following theorem.

Theorem 8.2.1. If the nodes $x_0, \ldots, x_n$ are distinct, then for any values $y_0, \ldots, y_n$, there is a unique polynomial p of degree n such that $p(x_i) = y_i$, $i = 0, \ldots, n$.

Uniqueness also can be established by using the fact that if a polynomial of degree n vanishes at $n+1$ distinct points, then it must be identically zero. Thus, if there were two nth-degree polynomials p and q such that $p(x_i) = q(x_i) = y_i$, $i = 0, \ldots, n$, then their difference $p - q$ would be an nth-degree polynomial satisfying $(p - q)(x_i) = 0$, $i = 0, \ldots, n$, and would therefore be identically zero.

The Lagrange form (8.3) of the interpolating polynomial can be modified to improve its computational qualities. Often a *barycentric* form of the Lagrange interpolation formula is used. (The barycentric coordinates of a point P in a triangle with vertices A, B, and C are numbers α, β, and γ such that $P = \alpha A + \beta B + \gamma C$, where $\alpha + \beta + \gamma = 1$. Thus P is, in some sense, a "weighted average" of A, B, and C, except that the "weights" α, β, and γ are not necessarily positive. We will write the interpolating polynomial as a "weighted average" of function values, except that the "weights" will themselves be functions that sum to 1.) Define $\varphi(x) = \prod_{j=0}^{n}(x - x_j)$. We can rewrite (8.3) as

$$p(x) = \varphi(x) \sum_{i=0}^{n} \frac{w_i}{x - x_i} y_i, \tag{8.4}$$

where the **barycentric weights** w_i are

$$w_i = \frac{1}{\prod_{j \neq i}(x_i - x_j)}, \quad i = 0, \ldots, n. \tag{8.5}$$

This is sometimes called the first form of the barycentric interpolation formula. An advantage of this form is that, although it still requires $O(n^2)$ operations to compute the weights w_i, $i = 0, \ldots, n$, once these are known, the polynomial can be evaluated at any point x with only $O(n)$ operations.

The formula (8.4) can be further modified using the observation that if each y_i is equal to 1, then $p(x) \equiv 1$, since it is the *unique* interpolating polynomial. Thus

$$1 = \varphi(x) \sum_{i=0}^{n} \frac{w_i}{x - x_i} \quad \text{for all } x.$$

Solving this equation for $\varphi(x)$ and substituting into (8.4) leads to the second form or what is simply called the **barycentric interpolation formula:**

$$p(x) = \left(\sum_{i=0}^{n} \frac{w_i}{x - x_i} y_i \right) \Big/ \left(\sum_{i=0}^{n} \frac{w_i}{x - x_i} \right). \tag{8.6}$$

Note that the coefficient of y_i in this formula is

$$\frac{w_i}{x - x_i} \Big/ \left(\sum_{k=0}^{n} \frac{w_k}{x - x_k} \right),$$

and these coefficients sum to 1. Looking at formula (8.6), it is not at all clear that it actually defines a *polynomial*, but we know that it does based on our derivation.

Example 8.2.2. Compute the quadratic polynomial that goes through the three data points $(1, 2)$, $(2, 3)$, and $(3, 6)$ using the barycentric interpolation formula (8.6) and show that this is the same polynomial determined previously from formula (8.3). We first compute the weights using (8.5):

$$w_0 = \frac{1}{(1-2)(1-3)} = \frac{1}{2}, \quad w_1 = \frac{1}{(2-1)(2-3)} = -1,$$

$$w_2 = \frac{1}{(3-1)(3-2)} = \frac{1}{2}.$$

Substituting these values and the values $y_0 = 2$, $y_1 = 3$, $y_2 = 6$ into formula (8.6) we find

$$p(x) = \left(\frac{2}{2(x-1)} - \frac{3}{x-2} + \frac{6}{2(x-3)} \right) \Big/ \left(\frac{1}{2(x-1)} - \frac{1}{x-2} + \frac{1}{2(x-3)} \right).$$

Multiply numerator and denominator by $\varphi(x) = (x-1)(x-2)(x-3)$ and simplify to verify that this is again the polynomial $x^2 - 2x + 3$.

NEWTON, LAGRANGE, AND INTERPOLATION [20]

Newton (*top*) first touched on the topic of interpolation in a letter of 1675 in which he offered help in the construction of tables of square roots, cube roots, and fourth roots of numbers from 1 to 10,000, calculated to 8 decimal digits. Earlier, Wallis, in *Arithmetica Infinitorum*, had been the first to use the Latin verb *interpolare* (to interpolate) to describe the mathematical procedure, although instances of linear and higher-order interpolation can be traced back to ancient times [72]. In a lecture in 1795, Lagrange (*bottom*) referred to the problem, studied by Newton, of using a "parabolic curve" to interpolate a given curve; that is, interpolating a function with a polynomial. Newton applied his result to the trajectory of a comet, while Lagrange was interested in a problem in surveying in which the position of the observer of three objects must be determined. Lagrange's solution used trial and error which resulted in a curve of the errors that had to be approximated by a polynomial. To solve this problem, Lagrange proposed the interpolating polynomial in the form that now bears his name. Lagrange apparently was unaware that this same formula had been published by Waring in 1779, and rediscovered by Euler in 1783 [72]. It is the same as Newton's polynomial, only expressed differently.

8.3 THE NEWTON FORM OF THE INTERPOLATION POLYNOMIAL

Consider the set of polynomials

$$1,\ x-x_0,\ (x-x_0)(x-x_1),\ldots,\prod_{j=0}^{n-1}(x-x_j).$$

These polynomials form a **basis** for the set of all polynomials of degree n; that is, they are linearly independent and any polynomial of degree n can be written as a linear combination of these. The **Newton form** of the interpolation polynomial is

$$p(x)=a_0+a_1(x-x_0)+a_2(x-x_0)(x-x_1)+\ldots+a_n\prod_{j=0}^{n-1}(x-x_j), \tag{8.7}$$

where $a_0, a_1, \ldots, a_n$ are chosen so that $p(x_i)=y_i$, $i=0,\ldots,n$.

Before seeing how to determine the coefficients, $a_0, \ldots, a_n$, let us estimate the work to evaluate this polynomial at a point x. We can use a procedure somewhat like Horner's rule:

> Start by setting $y = a_n$. For $i = n-1, n-2, \ldots, 0$, replace y by $y(x - x_i) + a_i$.

Note that after the first time through the *for* statement, $y = a_n(x - x_{n-1}) + a_{n-1}$, after the second time $y = a_n(x - x_{n-1})(x - x_{n-2}) + a_{n-1}(x - x_{n-2}) + a_{n-2}$, etc., so that at the end y contains the value of $p(x)$. The work required is about $3n$ operations.

To determine the coefficients in (8.7), we simply apply the conditions $p(x_i) = y_i$ one by one, in order:

$$
\begin{aligned}
p(x_0) &= a_0 = y_0, \\
p(x_1) &= a_0 + a_1(x_1 - x_0) = y_1 \Rightarrow a_1 = \frac{y_1 - a_0}{x_1 - x_0}, \\
p(x_2) &= a_0 + a_1(x_2 - x_0) + a_2(x_2 - x_0)(x_2 - x_1) = y_2 \Rightarrow \\
&\qquad a_2 = \frac{y_2 - a_0 - a_1(x_2 - x_0)}{(x_2 - x_0)(x_2 - x_1)}, \\
&\vdots
\end{aligned}
$$

What we are actually doing here is solving a lower triangular linear system for the coefficients $a_0, \ldots, a_n$:

$$
\begin{pmatrix}
1 & & & & \\
1 & x_1 - x_0 & & & \\
1 & x_2 - x_0 & (x_2 - x_0)(x_2 - x_1) & & \\
\vdots & \vdots & \vdots & \ddots & \\
1 & x_n - x_0 & (x_n - x_0)(x_n - x_1) & \ldots & \prod_{j=0}^{n-1}(x_n - x_j)
\end{pmatrix}
\begin{pmatrix} a_0 \\ a_1 \\ a_2 \\ \vdots \\ a_n \end{pmatrix}
=
\begin{pmatrix} y_0 \\ y_1 \\ y_2 \\ \vdots \\ y_n \end{pmatrix}. \tag{8.8}
$$

We know that the work to solve a lower triangular system is $O(n^2)$, and the work to compute the triangular matrix column by column is also $O(n^2)$. Thus the Newton form of the interpolant requires less work to determine than form (8.1) coming from the Vandermonde system (8.2), and it requires less work to evaluate than the Lagrange form (8.3).

Another advantage of the Newton form over that of Vandermonde or Lagrange is that if you add a new point (x_{n+1}, y_{n+1}), the previously computed coefficients do not change. This amounts to adding one more row at the bottom of the lower triangular matrix, along with a new unknown a_{n+1} and a new right-hand side value y_{n+1}. The previously computed coefficients are then substituted into the last equation to determine a_{n+1}.

It also should be noted that in the barycentric form of the Lagrange interpolation polynomial (8.6), adding a new point is relatively easy. One divides each weight w_i in (8.5) by the new factor $(x_i - x_{n+1})$ and computes a new weight $w_{n+1} = 1/\prod_{j=0}^{n}(x_{n+1} - x_j)$. The work is $O(n)$.

Example 8.3.1. Considering the same problem as in Example 8.2.2, with $x_0 = 1$, $x_1 = 2$, and $x_2 = 3$, we write the Newton form of the interpolation polynomial as

$$p(x) = a_0 + a_1(x-1) + a_2(x-1)(x-2).$$

Since $y_0 = 2$, $y_1 = 3$, and $y_2 = 6$, we determine the coefficients, a_0, a_1, and a_2, by solving the lower triangular system

$$\begin{pmatrix} 1 & & \\ 1 & 1 & \\ 1 & 2 & 2 \end{pmatrix} \begin{pmatrix} a_0 \\ a_1 \\ a_2 \end{pmatrix} = \begin{pmatrix} 2 \\ 3 \\ 6 \end{pmatrix},$$

giving $a_0 = 2$, $a_1 = 1$, and $a_2 = 1$. You should check that this is the same polynomial, $x^2 - 2x + 3$, computed in Example 8.2.2.

Suppose we add another data point $(x_3, y_3) = (5, 7)$. Then the polynomial that interpolates these 4 points is of degree 3 and can be written in the form

$$q(x) = b_0 + b_1(x-1) + b_2(x-1)(x-2) + b_3(x-1)(x-2)(x-3),$$

where b_0, b_1, b_2, and b_3 satisfy

$$\begin{pmatrix} 1 & & & \\ 1 & 1 & & \\ 1 & 2 & 2 & \\ 1 & 4 & 12 & 24 \end{pmatrix} \begin{pmatrix} b_0 \\ b_1 \\ b_2 \\ b_3 \end{pmatrix} = \begin{pmatrix} 2 \\ 3 \\ 6 \\ 7 \end{pmatrix}.$$

Since the first three equations involving b_0, b_1, and b_2 are the same as those for a_0, a_1, and a_2 above, it follows that $b_i = a_i$, $i = 0, 1, 2$. The equation for b_3 becomes $a_0 + 4a_1 + 12a_2 + 24b_3 = 7$, so that $b_3 = -11/24$.

Returning to the quadratic interpolant, it should be noted that the form of the Newton polynomial (though not the polynomial itself) depends on the *ordering* of the data points. Had we labeled the data points as $(x_0, y_0) = (3, 6)$, $(x_1, y_1) = (1, 2)$, and $(x_2, y_2) = (2, 3)$, then we would have written the interpolation polynomial in the form

$$p(x) = \alpha_0 + \alpha_1(x-3) + \alpha_2(x-3)(x-1),$$

and we would have determined α_0, α_1, and α_2 from

$$\begin{aligned} p(3) &= \alpha_0 = 6, \\ p(1) &= \alpha_0 - 2\alpha_1 = 2 \Rightarrow \alpha_1 = 2, \\ p(2) &= \alpha_0 - \alpha_1 - \alpha_2 = 3 \Rightarrow \alpha_2 = 1. \end{aligned}$$

Again you should check that this is the same polynomial $x^2 - 2x + 3$. The ordering of the points can have an effect on the numerical stability of the algorithm.

8.3.1 Divided Differences

A disadvantage of setting up the triangular system (8.8) is that some of the entries may overflow or underflow. If the nodes x_j are far apart, say, the distance between successive nodes is greater than 1, then the product of distances from node n to each of the others is likely to be huge and might result in an overflow. On the other hand, if the nodes are close together, say all are contained in an interval of width less than 1, then the product of the distances from node n to each of the others is likely to be tiny and might underflow. For these reasons a different algorithm, which also requires $O(n^2)$ operations, is often used to determine the coefficients $a_0, \ldots, a_n$.

At this point, it is convenient to change notation slightly. Assume that the ordinates y_i are the values of some function f at x_i: $y_i \equiv f(x_i)$. We will sometimes abbreviate $f(x_i)$ as simply f_i.

Given a collection of nodes $x_{i_0}, x_{i_1}, \ldots, x_{i_k}$, from among the nodes $x_0, \ldots, x_n$, define $f[x_{i_0}, x_{i_1}, \ldots, x_{i_k}]$ (called a **kth-order divided difference**) to be the coefficient of x^k in the polynomial of degree k that interpolates $(x_{i_0}, f_{i_0}), (x_{i_1}, f_{i_1}), \ldots, (x_{i_k}, f_{i_k})$. In the Newton form of the interpolant,

$$\alpha_0 + \alpha_1(x - x_{i_0}) + \alpha_2(x - x_{i_1})(x - x_{i_0}) + \cdots + \alpha_k \prod_{j=0}^{k-1}(x - x_{i_j}),$$

this is the coefficient α_k of $\prod_{j=0}^{k-1}(x - x_{i_j})$. Note that for higher-degree Newton interpolants, interpolating $(x_{i_0}, f_{i_0}), \ldots, (x_{i_k}, f_{i_k}), (x_{i_{k+1}}, f_{i_{k+1}}), \ldots,$ this is again the coefficient of the $(k+1)$st term, $\prod_{j=0}^{k-1}(x - x_{i_j})$, since these coefficients do not change when more data points are added. Thus $f[x_0, \ldots, x_k]$ is the coefficient a_k in (8.7).

With this definition we find

$$f[x_j] = f_j,$$

since $p_{0,j}(x) \equiv f_j$ is the polynomial of degree 0 that interpolates (x_j, f_j). For any two nodes x_i and x_j, $j \neq i$, we have

$$f[x_i, x_j] = \frac{f_j - f_i}{x_j - x_i} = \frac{f[x_j] - f[x_i]}{x_j - x_i},$$

since $p_{1,i,j}(x) \equiv f_i + \frac{f_j - f_i}{x_j - x_i}(x - x_i)$ is the polynomial of degree 1 that interpolates (x_i, f_i) and (x_j, f_j). Proceeding in this way, one finds that

$$f[x_i, x_j, x_k] = \frac{f[x_j, x_k] - f[x_i, x_j]}{x_k - x_i},$$

and more generally:

Theorem 8.3.1. The kth-order divided difference $f[x_0, x_1, \ldots, x_k]$ satisfies

$$f[x_0, x_1, \ldots, x_k] = \frac{f[x_1, \ldots, x_k] - f[x_0, \ldots, x_{k-1}]}{x_k - x_0}. \tag{8.9}$$

Before proving this result, let us see how it can be used to compute the coefficients of the Newton interpolant recursively. We can form a table whose columns are divided differences of order 0, 1, 2, etc.

$$\begin{array}{llll} f[x_0] = f_0 & & & \\ f[x_1] = f_1 & f[x_0, x_1] & & \\ f[x_2] = f_2 & f[x_1, x_2] & f[x_0, x_1, x_2] & \\ f[x_3] = f_3 & f[x_2, x_3] & f[x_1, x_2, x_3] & f[x_0, x_1, x_2, x_3] \end{array}$$

The diagonal entries in this table are the coefficients of the Newton interpolant (8.7). We know the first column of the table. Using theorem 8.3.1, we can compute the entries in each successive column using the entry to the left and the one just above it. If d_{ij} is the (i, j)-entry in the table, then

$$d_{ij} = \frac{d_{i,j-1} - d_{i-1,j-1}}{x_{i-1} - x_{i-j}}.$$

Proof of Theorem. Let p be the polynomial of degree k interpolating $(x_0, f_0), \ldots, (x_k, f_k)$. Let q be the polynomial of degree $k-1$ that interpolates $(x_0, f_0), \ldots, (x_{k-1}, f_{k-1})$, and let r be the polynomial of degree $k-1$ that interpolates $(x_1, f_1), \ldots, (x_k, f_k)$. Then p, q, and r have leading coefficients $f[x_0, \ldots, x_k]$, $f[x_0, \ldots, x_{k-1}]$, and $f[x_1, \ldots, x_k]$, respectively.

We claim that

$$p(x) = q(x) + \frac{x - x_0}{x_k - x_0}(r(x) - q(x)). \tag{8.10}$$

To prove this, we will show that this equation holds at each point x_i, $i = 0, \ldots, k$. Since $p(x)$ is a polynomial of degree k and the right-hand side is a polynomial of degree k, showing that these two polynomials agree at $k+1$ points will imply that they are equal everywhere. For $x = x_0$, we have $p(x_0) = q(x_0) = f_0$, and the second term on the right-hand side of (8.10) is zero, so we have equality. For $i = 1, \ldots, k-1$, $p(x_i) = q(x_i) = r(x_i) = f_i$, and so we again have equality in (8.10) for $x = x_i$. Finally, for $x = x_k$, we have $p(x_k) = r(x_k) = f_k$, and the right-hand side of (8.10) is $q(x_k)+r(x_k)-q(x_k) = f_k$, so we again have equality. This establishes (8.10), and from this it follows that the coefficient of x^k in $p(x)$ is $1/(x_k - x_0)$ times the coefficient of x^{k-1} in $r(x) - q(x)$; that is,

$$f[x_0, \ldots, x_k] = \frac{f[x_1, \ldots, x_k] - f[x_0, \ldots, x_{k-1}]}{x_k - x_0}.$$

□

Example 8.3.2. Looking again at the example from the previous section, with $(x_0, y_0) = (1, 2)$, $(x_1, y_1) = (2, 3)$, and $(x_2, y_2) = (3, 6)$, we compute the following table of divided differences:

$$\begin{array}{lll} f[x_0] = 2 & & \\ f[x_1] = 3 & f[x_0, x_1] = \frac{3-2}{2-1} = 1 & \\ f[x_2] = 6 & f[x_1, x_2] = \frac{6-3}{3-2} = 3 & f[x_0, x_1, x_2] = \frac{3-1}{3-1} = 1 \end{array}$$

Reading off the diagonal entries, we see that the coefficients in the Newton form of the interpolation polynomial are $a_0 = 2$, $a_1 = 1$, and $a_2 = 1$.

8.4 THE ERROR IN POLYNOMIAL INTERPOLATION

Suppose that we interpolate a smooth function f at $n+1$ points using a polynomial p of degree n. How large is the difference between $f(x)$ and $p(x)$ at points x other than the interpolation points (where the difference is 0)? If we use higher- and higher-degree polynomials, interpolating f at more and more points, will the polynomials eventually approximate f well at all points throughout some interval? These questions will be addressed in this section. The answers may surprise you.

Theorem 8.4.1. Assume that $f \in C^{n+1}[a, b]$ and that $x_0, \ldots, x_n$ are in $[a, b]$. Let $p(x)$ be the polynomial of degree n that interpolates f at $x_0, \ldots, x_n$. Then for any point x in $[a, b]$,

$$f(x) - p(x) = \frac{1}{(n+1)!} f^{(n+1)}(\xi_x) \prod_{j=0}^{n} (x - x_j), \tag{8.11}$$

for some point ξ_x in $[a, b]$.

Proof. The result clearly holds if x is one of the nodes $x_0, \ldots, x_n$, so fix $x \in [a, b]$ to be some point other than these. Let q be the polynomial of degree $n+1$ that interpolates f at $x_0, \ldots, x_n$, and x. Then

$$q(t) = p(t) + \lambda \prod_{j=0}^{n} (t - x_j), \quad \text{where } \lambda = \frac{f(x) - p(x)}{\prod_{j=0}^{n}(x - x_j)}. \tag{8.12}$$

Define $\phi(t) \equiv f(t) - q(t)$. Since $\phi(t)$ vanishes at the $n+2$ points $x_0, \ldots, x_n$, and x, Rolle's theorem implies that $\phi'(t)$ vanishes at $n+1$ points between successive pairs. Applying Rolle's theorem again, we find that ϕ'' vanishes at n points, and continuing, that $\phi^{(n+1)}$ must vanish at at least one point ξ_x in the interval $[a, b]$. Hence $0 = \phi^{(n+1)}(\xi_x) = f^{(n+1)}(\xi_x) - q^{(n+1)}(\xi_x)$. Differentiating $q(t)$ in expression (8.12) $n+1$ times, we find that, since $p^{(n+1)}(t) = 0$, $q^{(n+1)}(t) = \lambda(n+1)!$. Thus

$$f^{(n+1)}(\xi_x) = \lambda(n+1)! = \frac{f(x) - p(x)}{\prod_{j=0}^{n}(x - x_j)} \cdot (n+1)!,$$

and the result (8.11) follows. □

The error in polynomial interpolation also can be expressed in terms of *divided differences*. The divided-difference formulas of section 8.3.1 may have reminded you somewhat of *derivatives*. That is because divided differences are approximations to derivatives; the nth-order divided difference approximates $(1/n!)$ times the nth derivative. Consider again the polynomial q in (8.12) that interpolates f at the points $x_0, \ldots, x_n$, and x. The coefficient λ in that formula (the coefficient of the highest power of t) is, by definition, the divided difference

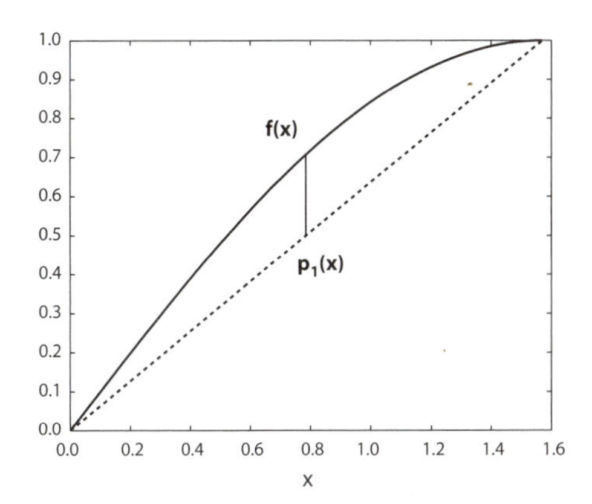

Figure 8.2. The graph of $f(x) = \sin x$ and the first-degree polynomial that interpolates f at 0 and $\pi/2$.

$f[x_0, \ldots, x_n, x]$. Since $q(x) = f(x)$, it follows that

$$f(x) = p(x) + f[x_0, \ldots, x_n, x] \cdot \prod_{j=0}^{n} (x - x_j). \tag{8.13}$$

Comparing this with (8.11), we see that

$$f[x_0, \ldots, x_n, x] = \frac{1}{(n+1)!} f^{(n+1)}(\xi_x). \tag{8.14}$$

Example 8.4.1. Let $f(x) = \sin x$, and let p_1 be the first-degree polynomial that interpolates f at 0 and $\pi/2$. Then $p_1(x) = (2/\pi)x$. The two functions are plotted in figure 8.2 on the interval $[0, \pi/2]$.
Since $|f''(x)| = |\sin x| \le 1$, it follows from theorem 8.4.1 that for any point x,

$$|f(x) - p_1(x)| \le \frac{1}{2!} |(x - 0)(x - \pi/2)|.$$

For $x \in [0, \pi/2]$, this is maximal when $x = \pi/4$, and then

$$|f(x) - p_1(x)| \le \frac{1}{2} (\pi/4)^2.$$

The actual error is $|\sin(\pi/4) - (2/\pi)(\pi/4)| = (\sqrt{2} - 1)/2 \approx 0.207$, and we obtain a reasonably good approximation within the interval $[0, \pi/2]$.

Suppose, however, that we use $p_1(x)$ to *extrapolate* to the value of f at π. The error in this case will be 2, since $\sin \pi = 0$ while $p_1(\pi) = 2$. Theorem 8.4.1 guarantees only that

$$|f(\pi) - p_1(\pi)| \le \frac{1}{2} |(\pi - 0)(\pi - \pi/2)| = \pi^2/4.$$

Suppose we add more and more interpolation points between 0 and $\pi/2$. Will we ever be able to extrapolate to $x = \pi$ and obtain a reasonable approximation to $f(\pi)$? In this case, the answer is yes, since all derivatives of $f(x) = \sin x$ are bounded in absolute value by 1. As the number of interpolation points $n+1$ increases, the factor $1/(n+1)!$ in (8.11) decreases. The factor $|\prod_{j=1}^{n+1} (x - x_j)|$

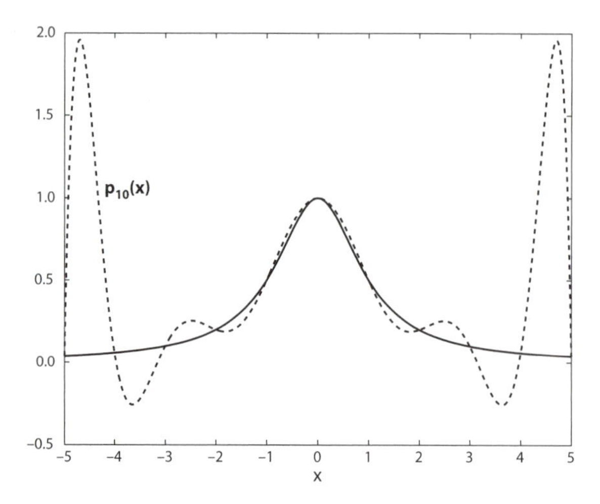

Figure 8.3. The graph of $f(x) = 1/(1 + x^2)$ and its 10th-degree interpolant.

increases if $x = \pi$ and each $x_j \in [0, \pi/2]$, but it increases less rapidly than $(n + 1)!$: $\lim_{n\to\infty} \pi^{n+1}/(n + 1)! = 0$.

Example 8.4.2. (Runge function). Consider the function $f(x) = \frac{1}{1+x^2}$ on the interval $I = [-5, 5]$. Suppose we interpolate f at more and more equally-spaced points throughout the interval I. Will the sequence of interpolants p_n converge to f on I? The answer is *NO*. The function f and its 10th-degree interpolant p_{10} are plotted in figure 8.3.

As the degree n increases, the oscillations near the ends of the interval become larger and larger. This does not contradict theorem 8.4.1, because in this case the quantity $|f^{(n+1)}(\xi_x)\prod_{j=0}^{n}(x - x_j)|$ in (8.11) grows even faster with n than $(n + 1)!$ near $x = \pm 5$. This example is best understood by considering the function $\frac{1}{1+z^2}$, which has poles at $\pm i$, in the complex plane. This goes a bit beyond the scope of this book, but is a topic addressed in complex analysis courses.

8.5 INTERPOLATION AT CHEBYSHEV POINTS AND chebfun

Many of the difficulties associated with high-degree polynomial interpolation can be overcome by clustering the interpolation points near the endpoints of the interval instead of having them equally spaced. Sometimes the **Chebyshev interpolation points** (also called **Gauss–Lobatto points**)

$$x_j = \cos\left(\frac{\pi j}{n}\right), \quad j = 0, \ldots, n, \tag{8.15}$$

(or scaled and translated versions of these points) are used for interpolation. In fact, an entire MATLAB package called `chebfun` [9, 97] has been written to allow users to work with functions that, inside the package, are represented by their polynomial interpolants at the Chebyshev points (8.15). The degree of the polynomial interpolant is chosen automatically to attain a level of accuracy that is close to the machine precision. A key to efficiency in the chebfun package is to convert between different representations of this polynomial interpolant using the *Fast Fourier Transform* (FFT), an algorithm that will be described in section 14.5.1.

It can be shown that when $x_0, \ldots, x_n$ are the Chebyshev points, the factor $\prod_{j=0}^{n}(x - x_j)$ in the error formula (8.11) satisfies $\max_{x\in[-1,1]} |\prod_{j=0}^{n}(x - x_j)| \leq 2^{-n+1}$, which is near optimal [88]. Moreover, the weights in the barycentric interpolation formula (8.6) take on a very simple form:

$$w_j = \frac{2^{n-1}}{n} \begin{cases} (-1)^j/2 & \text{if } j = 0 \text{ or } j = n, \\ (-1)^j & \text{otherwise,} \end{cases} \tag{8.16}$$

which makes this formula especially convenient for evaluating $p(x)$ at points x other than interpolation points. The factor $2^{n-1}/n$ in each w_j can be dropped since it appears as a factor in both the numerator and the denominator of (8.6).

The following theorem [9], which we state without proof, shows that the polynomial interpolant at the Chebyshev points is close to the *best* polynomial approximation in the ∞-norm; that is, its maximum deviation from the function on the interval $[-1, 1]$ is at most a moderate size multiple of $\min_{q_n} \| f - q_n\|_\infty$, where the minimum is taken over all nth-degree polynomials q_n, and $\| f - q_n\|_\infty = \max_{x\in[-1,1]} | f(x) - q_n(x)|$.

Theorem 8.5.1. Let f be a continuous function on $[-1, 1]$, p_n its degree n polynomial interpolant at the Chebyshev points (8.15), and p_n^* its best approximation among all nth-degree polynomials on the interval $[-1, 1]$ in the ∞-norm $\| \cdot \|_\infty$. Then

1. $\| f - p_n\|_\infty \leq (2 + \frac{2}{\pi} \log n)\| f - p_n^*\|_\infty$.
2. If f has a kth derivative of bounded variation (meaning that the total vertical distance along the graph of the function is finite) in $[-1, 1]$ for some $k \geq 1$, then $\| f - p_n\|_\infty = O(n^{-k})$ as $n \to \infty$.
3. If f is analytic in a neighborhood of $[-1, 1]$ (in the complex plane), then $\| f - p_n\|_\infty = O(C^n)$ as $n \to \infty$ for some $C < 1$; in particular, if f is analytic in the closed ellipse with foci ± 1 and semimajor and semiminor axis lengths $M \geq 1$ and $m \geq 0$, then we may take $C = 1/(M + m)$.

Example 8.5.1. (Runge Function Interpolated at Chebyshev Points). Consider again the Runge function $f(x) = \frac{1}{1+x^2}$ on the interval $I = [-5, 5]$. We first make the change of variable $t = x/5$ and define a new function $g(t) = f(x)$, so that g is defined for $t \in [-1, 1]$ and $g(t) = f(5t) = \frac{1}{1+25t^2}$. Figure 8.4 shows the results of interpolating g at $n + 1$ Chebyshev points $t_j = \cos(\pi j/n)$, $j = 0, \ldots, n$, for $n = 5,\ 10,\ 20,$ and 40. For $n \geq 20$, the interpolant is indistinguishable from the function in the plots of figure 8.4. The polynomial interpolant $p_n(t)$ for $g(t)$ is easily transformed back into an interpolant for f on $[-5, 5]$; if $q_n(x) = p_n(x/5)$, then $f(x) - q_n(x) = g(t) - p_n(t)$. Compare these results with the example of the previous section showing interpolation at *equally-spaced* points.

According to theorem 8.5.1, part 2, since g has infinitely many derivatives of bounded variation in the interval $[-1, 1]$, the ∞-norm of the difference between g and its nth-degree polynomial interpolant converges to 0 as $n \to \infty$ faster than any power n^{-k}. Table 8.1 shows the ∞-norm of the error for

TABLE 8.1
Errors in interpolating $g(t) = 1/(1 + 25t^2)$ at $n + 1$ Chebyshev points $t_j = \cos(\pi j/n)$, $j = 0, \ldots, n$.

n	$\|g - p_n\|_\infty$	$\|g - p_n\|_\infty / \|g - p_{n/2}\|_\infty$
5	0.64	
10	0.13	0.21
20	0.018	0.13
40	3.33×10^{-4}	0.019

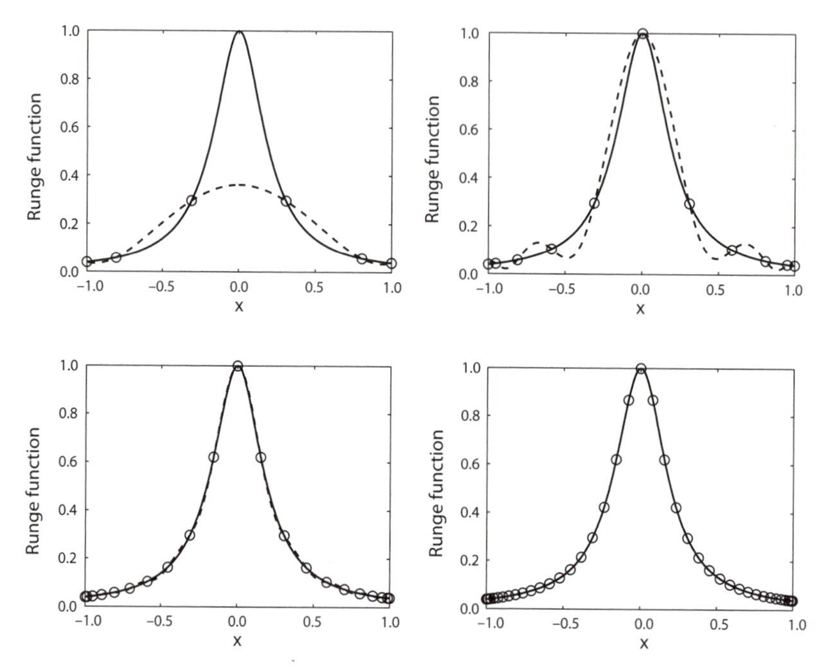

Figure 8.4. Interpolating $g(t) = 1/(1+25t^2)$ at $(n+1)$ Chebyshev points $t_j = \cos(\pi j/n)$, $j = 0, \ldots, n$, for $n = 5$, 10, 20, and 40.

the values of n tested here. Note that the ratios $\|g - p_n\|_\infty / \|g - p_{n/2}\|_\infty$ are *decreasing* with n; if the error were decreasing like $O(n^{-k})$ for some fixed $k \geq 1$ (and *not* like some higher power of n^{-1}), then we would expect these ratios to be approaching the constant 2^{-k}.

The error reduction is explained further by part 3 of theorem 8.5.1. Consider an ellipse in the complex plane centered at the origin with foci ± 1 that does *not* go through or enclose the points $\pm i/5$ (where g has singularities). Such an ellipse must have semiminor axis length $m < 1/5$. Since the distance from the center of an ellipse to either of its foci is $\sqrt{M^2 - m^2}$, it follows that $M^2 - m^2 = 1$ and therefore $M < \sqrt{26}/5$. According to the theorem, this implies that for any constant $C > 5/(1 + \sqrt{26}) \approx 0.82$, the error in the nth-degree polynomial interpolant decreases like $O(C^n)$ as $n \to \infty$.

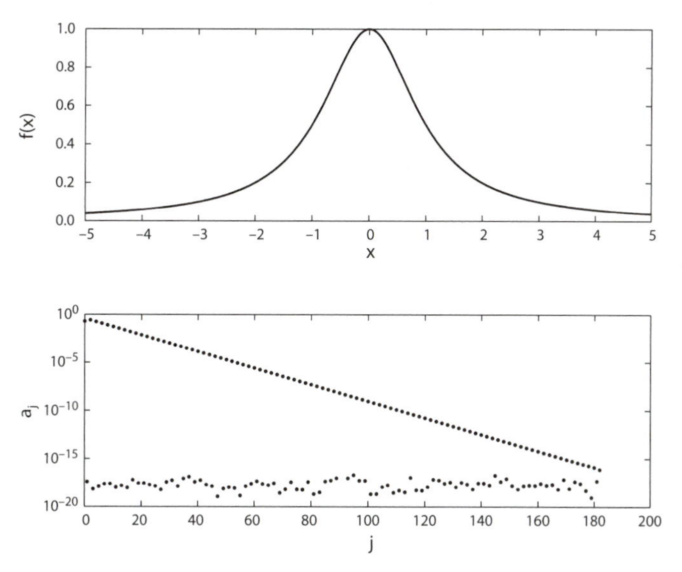

Figure 8.5. Interpolating polynomial (*top*) for the Runge function using the chebfun system and a log plot on the y-axis (*bottom*) of the absolute values of the coefficients of the interpolating polynomial.

Example 8.5.2. (Runge Function in chebfun). If one uses the `chebfun` system to work with the Runge function, then one types in MATLAB

```
f = chebfun('1 ./ (1 + x.^2)',[-5,5])
```

to define the function `f` of the variable `x` on the interval $[-5, 5]$. The package then computes the polynomial interpolant of this function at the Chebyshev points (scaled to the interval $[-5, 5]$) and chooses the degree of the polynomial to achieve a level of accuracy near the machine precision. It returns the following information about `f`:

```
f =
   chebfun column (1 smooth piece)
        interval             length    endpoint values
(        -5,          5)       183       0.038      0.038
vertical scale =   1
```

The fact that `length=183` means that the package chose to represent `f` by a polynomial of degree 182 that interpolates `f` at 183 Chebyshev points. This is a higher-degree polynomial than is usually needed to represent such a benign-looking function, but because of the singularities at $\pm i$ in the complex plane, it turns out to be necessary in this case. One can plot the resulting function by typing `plot(f)`, and one can see from the first plot in figure 8.5 that the interpolating polynomial does indeed look like the Runge function. One can see more about why 183 interpolation points were chosen by typing the commands

```
coeffs = chebpoly(f);
np1 = length(coeffs);
semilogy([0:np1-1],abs(coeffs(np1:-1:1)),'.')
```

The vector coeffs contains the coefficients of the expansion of the interpolation polynomial in Chebyshev polynomials, and we know from above that its length is 183. The Chebyshev polynomials $T_j(x)$ are given by $T_j(x) = \cos(j \arccos x)$, $j = 0, 1, \ldots$, and they also satisfy the recurrence $T_{j+1}(x) = 2xT_j(x) - T_{j-1}(x)$, with $T_0(x) = 1$ and $T_1(x) = x$. They are said to be **orthogonal** polynomials on the interval $[-1, 1]$ with respect to the weight function $(1 - x^2)^{-1/2}$, because $\int_{-1}^{1} T_j(x)T_k(x)(1 - x^2)^{-1/2}\, dx = 0$ for $j \neq k$. Orthogonal polynomials will be discussed further in section 10.3.1. Generally, it is better to express a polynomial p as a linear combination of orthogonal polynomials like the Chebyshev polynomials, than it is to write p in the form $p(x) = \sum_{j=0}^{n} c_j x^j$. Thus, the vector coeffs will contain the values $a_0, \ldots, a_n$ for which $p(x) = \sum_{j=0}^{n} a_j T_j(x)$, where p is the interpolation polynomial. The command semilogy plots the absolute values of these coefficients on a log scale, and we would expect the expansion to be terminated when all remaining coefficients are a moderate size multiple of the machine precision. The second plot in figure 8.5 verifies this. The coefficients of the odd-degree Chebyshev polynomials are all very close to 0, and those of the even-degree Chebyshev polynomials drop below 10^{-15} before the expansion is deemed sufficiently accurate to represent the function.

You can read more about chebfun and download the software from the website www2.maths.ox.ac.uk/chebfun/.

In the past, numerical analysis textbooks have often disparaged high-degree polynomial interpolation and the Lagrange interpolation formula in particular, usually because the interpolation was being carried out at equally-spaced points and the original form (8.3) of the Lagrange formula was being used rather than the barycentric form (8.6). As a result of examples like the above, it now appears that this view was mistaken. In fact, interpolation at Chebyshev points and the advantages of approximating a function by a linear combination of Chebyshev polynomials are expounded upon in [83]. We will see examples later of the use of polynomial interpolation at the Chebyshev points to solve problems of numerical differentiation and integration. Once a function has been accurately approximated by a polynomial, its derivative or integral can be approximated by differentiating or integrating the polynomial *exactly*. Differential equations may be solvable exactly as well if all of the functions involved are replaced by their polynomial interpolants at the Chebyshev points. For more information about polynomial interpolation and, in particular, barycentric Lagrange interpolation, see the excellent survey article [11] by Berrut and Trefethen.

The next section describes an alternative (or complement) to high-degree polynomial approximation, which is *piecewise polynomial interpolation*. Even if high-degree polynomial interpolation (at Chebyshev points) is used on

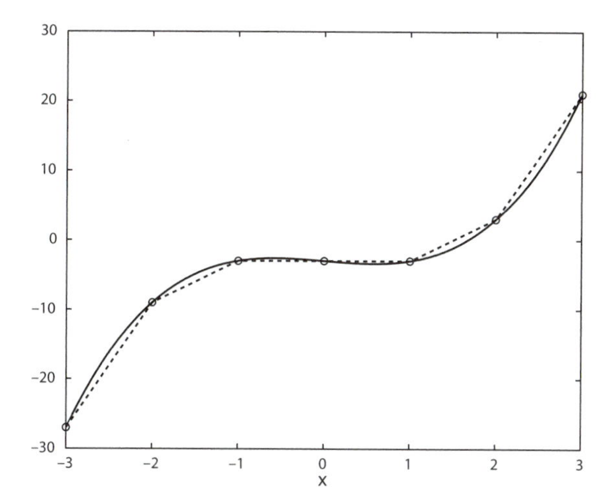

Figure 8.6. A function and its piecewise linear interpolant.

subintervals, it may be necessary to divide a given interval into subintervals to account for *discontinuities* in the function f or some of its derivatives.

8.6 PIECEWISE POLYNOMIAL INTERPOLATION

Another way to approximate a function f on an interval $[a, b]$ is to divide the interval into n subintervals, each of length $h \equiv (b-a)/n$, and to use a low-degree polynomial to approximate f on each subinterval. For example, if the endpoints of these subintervals are $a \equiv x_0, x_1, \ldots, x_{n-1}, x_n \equiv b$, and $\ell(x)$ is the *piecewise linear* interpolant of f, then

$$\ell(x) = f(x_{i-1})\frac{x - x_i}{x_{i-1} - x_i} + f(x_i)\frac{x - x_{i-1}}{x_i - x_{i-1}} \quad \text{in } [x_{i-1}, x_i].$$

[Note that we have written down the Lagrange form of $\ell(x)$.] A function and its piecewise linear interpolant are pictured in figure 8.6.

There are many reasons why one might wish to approximate a function f with piecewise polynomials. For example, suppose one wants to compute $\int_a^b f(x)\,dx$. It might be impossible to integrate f analytically, but it is easy to integrate polynomials. Using the piecewise linear interpolant of f, one would obtain the approximation

$$\begin{aligned}
\int_a^b f(x)\,dx &\approx \sum_{i=1}^{n} \int_{x_{i-1}}^{x_i} \left[f_{i-1}\frac{x - x_i}{x_{i-1} - x_i} + f_i\frac{x - x_{i-1}}{x_i - x_{i-1}} \right] dx \\
&= \sum_{i=1}^{n} \left[f_{i-1}\frac{h}{2} + f_i\frac{h}{2} \right] \\
&= \frac{h}{2}[f_0 + 2f_1 + \cdots + 2f_{n-1} + f_n],
\end{aligned}$$

Creating Models with Piecewise Linear Interpolation

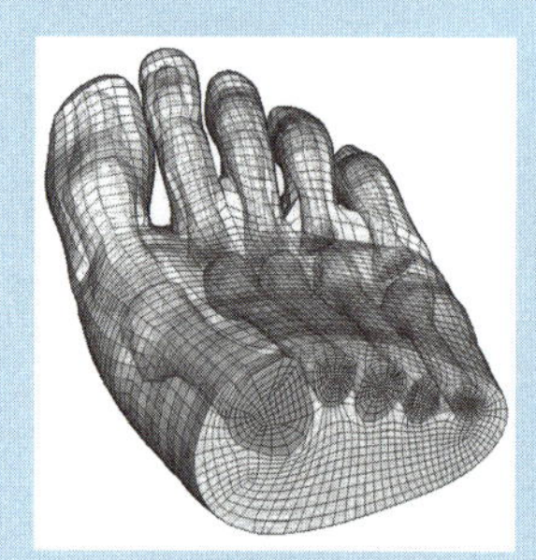

Piecewise linear interpolation is used extensively to create tessellations of objects. The model of the foot to the left includes 17 bones and soft tissue. The Cavanagh lab at the Cleveland Clinic Foundation uses the model for bioengineering and footwear design purposes. Such models are often used in conjunction with numerical methods for partial differential equations to model complex physical phenomena. (Image courtesy of Peter Cavanagh, Ph.D. D.Sc. University of Washington. cavanagh@uw.edu)

where we have again used the abbreviation $f_j \equiv f(x_j)$. You may recognize this as the *trapezoid rule*. Numerical integration methods will be discussed in chapter 10.

Another application is interpolation in a table (although this is becoming less and less prevalent with the availability of calculators). For instance, suppose you know that $\sin(\pi/6) = 1/2$ and $\sin(\pi/4) = \sqrt{2}/2$. How can you estimate, say, $\sin(\pi/5)$? One possibility would be to use the linear interpolant of $\sin x$ on the interval $[\pi/6, \pi/4]$. This is

$$\ell(x) = \frac{1}{2} \cdot \frac{x - \pi/4}{\pi/6 - \pi/4} + \frac{\sqrt{2}}{2} \cdot \frac{x - \pi/6}{\pi/4 - \pi/6}.$$

Hence $\ell(\pi/5) = (1/2) \cdot (3/5) + (\sqrt{2}/2) \cdot (2/5) = (3 + 2\sqrt{2})/10 \approx 0.58$.

What can be said about the *error* in the piecewise linear interpolant? We know from theorem 8.4.1 that in the subinterval $[x_{i-1}, x_i]$,

$$f(x) - \ell(x) = \frac{f''(\xi_x)}{2!}(x - x_{i-1})(x - x_i),$$

for some $\xi_x \in [x_{i-1}, x_i]$. If $|f''(x)| \le M$ for all $x \in [x_{i-1}, x_i]$, then

$$|f(x) - \ell(x)| \le \frac{M}{2} \max_{x \in [x_{i-1}, x_i]} |(x - x_{i-1})(x - x_i)| = \frac{M}{2}\left(\frac{h}{2}\right)^2 = \frac{Mh^2}{8},$$

and if $|f''(x)| \le M$ for all $x \in [a, b]$, then this same estimate will hold for each subinterval. Hence

$$|f(x) - \ell(x)| \le \frac{Mh^2}{8} \quad \text{for all } x \in [a, b].$$

To make the error less than some desired tolerance δ one can choose h so that $Mh^2/8 < \delta$; that is, $h < \sqrt{8\delta/M}$. Thus the piecewise linear interpolant always converges to the function $f \in C^2[a, b]$ as the subinterval size h goes to 0.

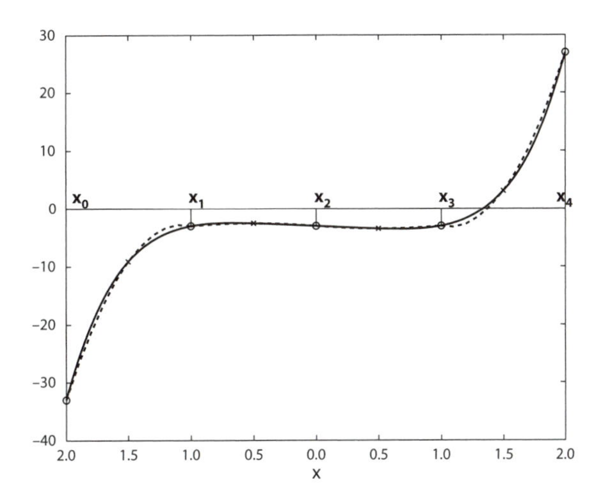

Figure 8.7. A function and its piecewise quadratic interpolant.

For convenience, we used a fixed mesh spacing $h \equiv (b-a)/n = x_i - x_{i-1}$, $i = 1, \ldots, n$. To obtain an accurate approximation with fewer subintervals, one might put more of the nodes x_i in regions where f is highly nonlinear and fewer nodes in regions where the function can be well approximated by a straight line. This can be done adaptively. One might start with fairly large subintervals, in each subinterval $[x_{i-1}, x_i]$ test the difference between $f((x_i + x_{i-1})/2)$ and $\ell((x_i + x_{i-1})/2)$, and if this difference is greater than δ, then replace that subinterval by two subintervals, each of width $(x_i - x_{i-1})/2$. This adaptive strategy is not foolproof; it may happen that f and ℓ are in close agreement at the midpoint of the subinterval but not at other points, and then the subinterval will not be divided as it should be. This strategy often works well, however, and gives a good piecewise linear approximation to f at a lower cost than using many equally-sized subintervals.

Another way to increase the accuracy of the interpolant is to use higher-degree piecewise polynomials; for example, quadratics that interpolate f at the midpoints as well as the endpoints of the subintervals. A function and its piecewise quadratic interpolant are pictured in figure 8.7.

We know from theorem 8.4.1 that the error in the quadratic interpolant $q(x)$ in interval $[x_{i-1}, x_i]$ is

$$f(x) - q(x) = \frac{f'''(\xi_x)}{3!}(x - x_{i-1})\left(x - \frac{x_i + x_{i-1}}{2}\right)(x - x_i),$$

for some point $\xi_x \in [x_{i-1}, x_i]$. If $|f'''(x)/3!|$ is bounded by a constant M for all $x \in [a, b]$, then the error in the quadratic interpolant decreases as $O(h^3)$, since the factor $|(x - x_{i-1})(x - \frac{x_i + x_{i-1}}{2})(x - x_i)|$ is less than h^3 for any $x \in [x_{i-1}, x_i]$.

This piecewise quadratic interpolant, like the piecewise linear interpolant, is *continuous*, but its derivative, in general, is *not* continuous. Another idea might be to require: $q(x_i) = f(x_i)$ *and* $q'(x_i) = f'(x_i)$, $i = 0, 1, \ldots, n$. But if we try to achieve this with a piecewise quadratic q, then within each subinterval

$[x_{i-1}, x_i]$, q' must be the linear interpolant of f':

$$q'(x) = f'(x_{i-1})\frac{x - x_i}{x_{i-1} - x_i} + f'(x_i)\frac{x - x_{i-1}}{x_i - x_{i-1}}, \quad x \in [x_{i-1}, x_i].$$

Integrating q', we find that $q(x) = \int_{x_{i-1}}^{x} q'(t)\,dt + C$, for some constant C. In order that $q(x_{i-1})$ be equal to $f(x_{i-1})$, we must have $C = f(x_{i-1})$. But now that C is fixed, there is no way to force $q(x_i) = f(x_i)$. Thus the goal of matching function values and first derivatives *cannot* be achieved with piecewise quadratics.

8.6.1 Piecewise Cubic Hermite Interpolation

Suppose instead that we use piecewise cubics p and try to match f and f' at the nodes. A quadratic p' that matches f' at x_{i-1} and x_i can be written in the form

$$p'(x) = f'(x_{i-1})\frac{x - x_i}{x_{i-1} - x_i} + f'(x_i)\frac{x - x_{i-1}}{x_i - x_{i-1}} + \alpha(x - x_{i-1})(x - x_i)$$

for $x \in [x_{i-1}, x_i]$. Here we have introduced an additional parameter α that can be varied in order to match function values. Integrating, we find

$$p(x) = -\frac{f'(x_{i-1})}{h}\int_{x_{i-1}}^{x}(t - x_i)\,dt + \frac{f'(x_i)}{h}\int_{x_{i-1}}^{x}(t - x_{i-1})\,dt$$
$$+\alpha\int_{x_{i-1}}^{x}(t - x_{i-1})(t - x_i)\,dt + C.$$

The condition $p(x_{i-1}) = f(x_{i-1})$ implies that $C = f(x_{i-1})$. Making this substitution and performing the necessary integration then gives

$$p(x) = -\frac{f'(x_{i-1})}{h}\left(\frac{(x - x_i)^2}{2} - \frac{h^2}{2}\right) + \frac{f'(x_i)}{h}\frac{(x - x_{i-1})^2}{2}$$
$$+\alpha(x - x_{i-1})^2\left(\frac{x - x_{i-1}}{3} - \frac{h}{2}\right) + f(x_{i-1}). \tag{8.17}$$

The condition $p(x_i) = f(x_i)$ then implies that

$$p(x_i) = f'(x_{i-1})\frac{h}{2} + f'(x_i)\frac{h}{2} - \alpha\frac{h^3}{6} + f(x_{i-1}) = f(x_i) \Longrightarrow$$

$$\alpha = \frac{3}{h^2}(f'(x_{i-1}) + f'(x_i)) + \frac{6}{h^3}(f(x_{i-1}) - f(x_i)). \tag{8.18}$$

Example 8.6.1. Let $f(x) = x^4$ on $[0, 2]$. Find the piecewise cubic Hermite interpolant of f using the two subintervals $[0, 1]$ and $[1, 2]$.

For this problem $x_0 = 0$, $x_1 = 1$, $x_2 = 2$, and $h = 1$. Substituting these values and the values of f and its derivative into (8.17) and (8.18), we

Splines at Boeing

During World War II, the British aircraft industry constructed templates for airplanes by passing thin wooden strips (called "splines") through key interpolation points in a design laid out on the floor of a large design loft. In the 1960s, work by Boeing researchers allowed numerical splines to replace these "lofting" techniques. By the mid-1990s, computations at Boeing routinely involved 30–50,000 data points, a feat not possible by hand. The use of splines by Boeing extended beyond design of aircraft. They were also integrated into the navigation and guidance systems. In 2005, Boeing airplanes made more than 42,000 flights, and within the company more than 10,000 design applications and 20,000 engineering applications were run [31]. Thus the Boeing Company was responsible for roughly 500 million spline evaluations each day, and this number was expected to rise each year [47]. Splines are used extensively at Boeing and throughout the manufacturing and engineering sector.

find that

$$p(x) = \begin{cases} 2x^3 - x^2 & \text{in } [0, 1], \\ 6x^3 - 13x^2 + 12x - 4 & \text{in } [1, 2]. \end{cases}$$

You should check that p and p' match f and f' at the nodes 0, 1, and 2.

8.6.2 Cubic Spline Interpolation

The piecewise cubic Hermite interpolant has one continuous derivative. If we give up the requirement that the derivative of our interpolant match that of f at the nodes, then we can obtain an even smoother piecewise cubic interpolant. A **cubic spline** interpolant s is a piecewise cubic that interpolates f at the nodes (also called **knots**) $x_0, \ldots, x_n$ and has *two* continuous derivatives.

We will not be able to write down a formula for $s(x)$ in $[x_{i-1}, x_i]$ that involves only values of f and its derivatives in this subinterval. Instead, we will have to solve a global linear system, involving the values of f at all of the nodes, to determine the formula for $s(x)$ in all of the subintervals simultaneously. Before doing this, let us count the number of equations and unknowns (also called **degrees of freedom**). A cubic polynomial is determined by 4 parameters; since s is cubic in each subinterval and there are n subintervals, the number of unknowns is $4n$. The conditions $s(x_i) = f_i$, $i = 0, 1, \ldots, n$ provide $2n$ equations, since the cubic in each of the n subintervals must match f at 2 points. The conditions that s' and s'' be continuous at $x_1, \ldots, x_{n-1}$ provide an additional $2n - 2$ equations, for a total of $4n - 2$ equations in $4n$ unknowns. This leaves 2 degrees of freedom. These can be used to enforce various conditions on the endpoints of the spline, resulting in different types of cubic splines.

To derive the equations of a cubic spline interpolant, let us start by setting $z_i = s''(x_i)$, $i = 1, \ldots, n-1$. Suppose now that the values z_i were known. Since

s'' is linear within each subinterval, it must satisfy

$$s''(x) = z_{i-1}\frac{x - x_i}{x_{i-1} - x_i} + z_i\frac{x - x_{i-1}}{x_i - x_{i-1}} \quad \text{in } [x_{i-1}, x_i].$$

We will simplify the notation somewhat by assuming a uniform mesh spacing, $x_i - x_{i-1} = h$ for all i; this is not necessary but it does make the presentation simpler. With this assumption, then, and letting s_i denote the restriction of s to subinterval i, we can write

$$s_i''(x) = \frac{1}{h}z_{i-1}(x_i - x) + \frac{1}{h}z_i(x - x_{i-1}). \tag{8.19}$$

Integrating, we find

$$s_i'(x) = -\frac{1}{h}z_{i-1}\frac{(x_i - x)^2}{2} + \frac{1}{h}z_i\frac{(x - x_{i-1})^2}{2} + C_i, \tag{8.20}$$

and

$$s_i(x) = \frac{1}{h}z_{i-1}\frac{(x_i - x)^3}{6} + \frac{1}{h}z_i\frac{(x - x_{i-1})^3}{6} + C_i(x - x_{i-1}) + D_i, \tag{8.21}$$

for some constants C_i and D_i.

Now we begin applying the necessary conditions to s_i. From (8.21), the condition $s_i(x_{i-1}) = f_{i-1}$ implies that

$$\frac{h^2}{6}z_{i-1} + D_i = f_{i-1}, \quad \text{or} \quad D_i = f_{i-1} - \frac{h^2}{6}z_{i-1}. \tag{8.22}$$

The condition $s_i(x_i) = f_i$ implies, with (8.21) and (8.22), that

$$\frac{h^2}{6}z_i + C_i h + f_{i-1} - \frac{h^2}{6}z_{i-1} = f_i,$$

or,

$$C_i = \frac{1}{h}\left[f_i - f_{i-1} + \frac{h^2}{6}(z_{i-1} - z_i)\right]. \tag{8.23}$$

Making these substitutions in (8.21) gives

$$s_i(x) = \frac{1}{h}z_{i-1}\frac{(x_i - x)^3}{6} + \frac{1}{h}z_i\frac{(x - x_{i-1})^3}{6} + \frac{1}{h}\left[f_i - f_{i-1} + \frac{h^2}{6}(z_{i-1} - z_i)\right]$$
$$\times(x - x_{i-1}) + f_{i-1} - \frac{h^2}{6}z_{i-1}. \tag{8.24}$$

Once the parameters $z_1, \ldots, z_{n-1}$ are known, one can use this formula to evaluate s anywhere.

We will use the continuity of s' to determine $z_1, \ldots, z_{n-1}$. The condition $s_i'(x_i) = s_{i+1}'(x_i)$, together with (8.20) and (8.23), implies that

$$\frac{h}{2}z_i + \frac{1}{h}(f_i - f_{i-1}) + \frac{h}{6}(z_{i-1} - z_i) = -\frac{h}{2}z_i + \frac{1}{h}(f_{i+1} - f_i) + \frac{h}{6}(z_i - z_{i+1}),$$
$$i = 1, \ldots, n-1.$$

Moving the unknown z_j's to the left-hand side and the known f_j's to the right-hand side, this becomes

$$\frac{2h}{3}z_i + \frac{h}{6}z_{i-1} + \frac{h}{6}z_{i+1} = -\frac{2}{h}f_i + \frac{1}{h}f_{i-1} + \frac{1}{h}f_{i+1}, \quad i = 1, \ldots, n-1.$$

This is a symmetric tridiagonal system of linear equations for the values $z_1, \ldots, z_{n-1}$:

$$\begin{pmatrix} \alpha_1 & \beta_1 & & \\ \beta_1 & \ddots & \ddots & \\ & \ddots & \ddots & \beta_{n-2} \\ & & \beta_{n-2} & \alpha_{n-1} \end{pmatrix} \begin{pmatrix} z_1 \\ \vdots \\ \vdots \\ z_{n-1} \end{pmatrix} = \begin{pmatrix} b_1 \\ \vdots \\ \vdots \\ b_{n-1} \end{pmatrix}, \tag{8.25}$$

where

$$\alpha_i = \frac{2h}{3}, \quad \beta_i = \frac{h}{6} \quad \text{for all } i,$$

$$b_i = \frac{1}{h}(f_{i+1} - 2f_i + f_{i-1}), \quad i = 2, \ldots, n-2,$$

$$b_1 = \frac{1}{h}(f_2 - 2f_1 + f_0) - \frac{h}{6}z_0,$$

$$b_{n-1} = \frac{1}{h}(f_n - 2f_{n-1} + f_{n-2}) - \frac{h}{6}z_n.$$

We are free to choose z_0 and z_n, and different choices give rise to different kinds of splines. The choice $z_0 = z_n = 0$ gives what is called the **natural cubic spline**. This can be shown to be the C^2 piecewise cubic interpolant of f whose curvature is (approximately) minimal. The **complete spline** is obtained by choosing z_0 and z_n so that the first derivative of the spline matches that of f at the endpoints (or takes on some given values at the endpoints). The **not-a-knot spline** is obtained by choosing z_0 and z_n to ensure third derivative continuity at both x_1 and x_{n-1}. This choice may be appropriate if no endpoint derivative information is available.

Example 8.6.2. Let $f(x) = x^4$ on $[0, 2]$. Find the natural cubic spline interpolant of f using the two subintervals $[0, 1]$ and $[1, 2]$.

Since this problem has only two subintervals ($n = 2$), the tridiagonal system (8.25) consists of just one equation in one unknown. Thus $\alpha_1 = 2/3$ and $b_1 = f_2 - 2f_1 + f_0 = 14$, so $z_1 = 21$. Using this value, together with $z_0 = z_2 = 0$, in (8.24) we can find s_1 and s_2:

$$s_1(x) = \frac{7}{2}x^3 - \frac{5}{2}x,$$

$$s_2(x) = -\frac{7}{2}x^3 + 21x^2 - \frac{47}{2}x + 7.$$

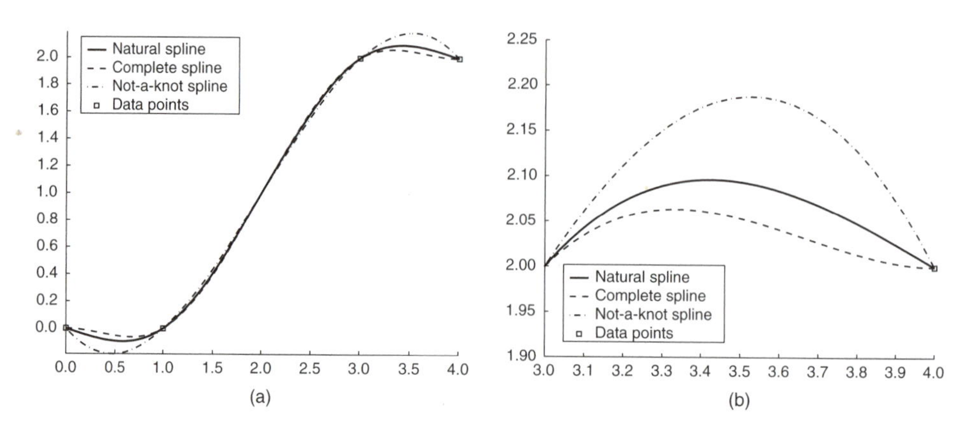

Figure 8.8. A complete and natural cubic spline interpolant, along with (a) the not-a-knot spline and (b) a close-up of the splines.

You should check that all of the conditions

$$s_1(0) = f(0), \quad s_1(1) = s_2(1) = f(1), \quad s_2(2) = f(2),$$

$$s_1'(1) = s_2'(1), \quad s_1''(1) = s_2''(1),$$

and

$$s_1''(0) = s_2''(2) = 0,$$

are satisfied.

Example 8.6.3. Use MATLAB to find a cubic spline that goes through the points $(0, 0)$, $(1, 0)$, $(3, 2)$, and $(4, 2)$.

Note that the spacing between nodes in this problem is not uniform. We will use the `spline` function in MATLAB which, by default, computes the not-a-knot spline. Letting `X = [0 1 3 4]` and `Y = [0 0 2 2]`, the not-a-knot spline can be computed and plotted with the command `plot(x, ppval (spline(X,Y), x))`, where `x` is an array of values at which to evaluate and plot the spline, such as `x = linspace(0, 4, 200)`. For the not-a-knot spline, the vectors `X` and `Y` have the same length. To specify slopes at the endpoints of the interval, and thus obtain a complete cubic spline interpolant, append these values at the beginning and end of the vector `Y`. For example, the command `plot(x, ppval (spline(X, [0 Y 0]), x))` produces the complete spline plotted in figure 8.8 with $f'(0) = f'(4) = 0$. Figure 8.8 also shows the natural cubic spline interpolant, and to the right is a close-up of the splines between $x = 3$ and $x = 4$ to highlight the difference in the end conditions. It can be seen that the complete cubic spline has zero slope at $x = 4$; it is true but less easy to see that the natural cubic spline has zero curvature at $x = 4$.

8.7 SOME APPLICATIONS

Numerical interpolation techniques, of the type discussed in this chapter, are widely used throughout industry, government, and academia. In this section

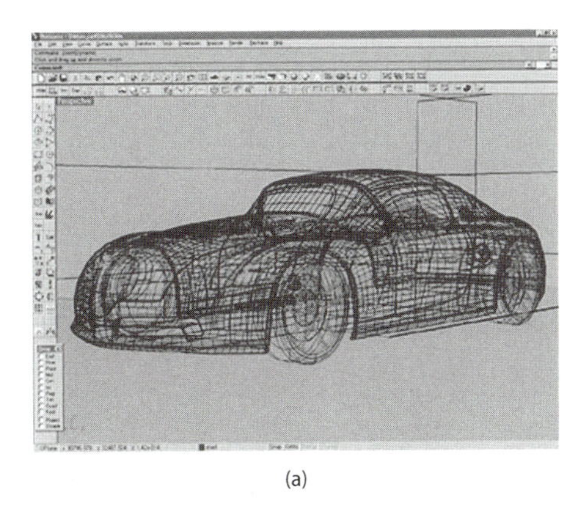

(a)

(b)

Figure 8.9. (a) A NURBS model of a car and (b) the same car after it is fully rendered.

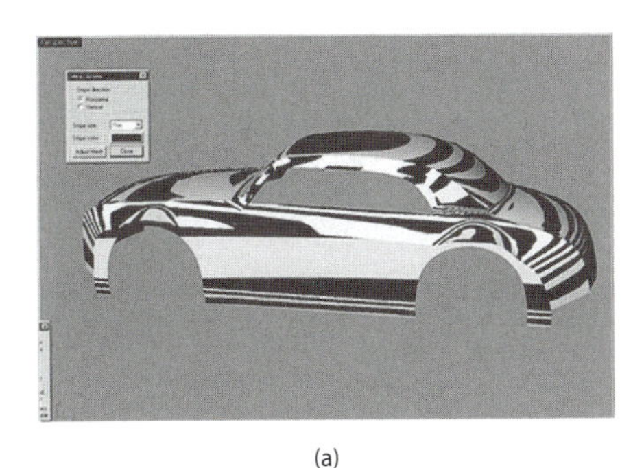

(a)

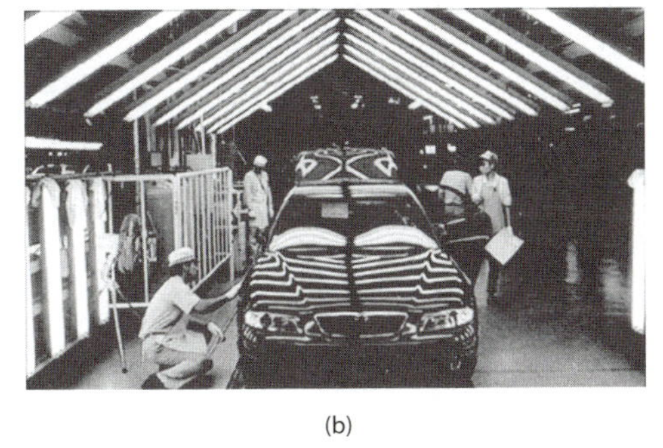

(b)

Figure 8.10. (a) Simulated reflection lines and (b) inspecting reflection lines for quality assurance. (Photograph courtesy of Gerald Farin and Dianne Hansford.)

we show examples of such uses, without going into detail on the exact algorithms applied.

NURBS. NonUniform Rational B-Splines (NURBS) are a set of mathematical functions that can accurately model complex geometrical shapes in 2 or 3 dimensions. They are used in such fields as computer graphics and manufacturing, where they are often combined with computer aided design (CAD) software. In figure 8.9(a), we see a NURBS model of a car created in the CAD software, Rhinoceros, which is NURBS modeling software for Windows. The company's web page http://www.rhino3d.com has a gallery of models that include jewelry, architecture, automobiles and cartoon characters. In figure 8.9(b), we see the same car after it is fully rendered.

Note that the reflections on the car are smooth curves. Acquiring smooth *reflection lines* is a goal in automotive design. A dent in the car's body, on the other hand, produces a discontinuity in the reflection line through the deformity. In the early days of car design, prototype cars were molded in clay and then painted to be reflective. Today, many aspects of automobile design occur within a computer. For example, computers can simulate reflection lines as seen in figure 8.10(a). In figure 8.10(b), reflection lines are inspected as a quality assurance tool, something that many high-end automakers do just before the car leaves the factory [38].

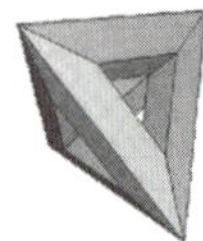 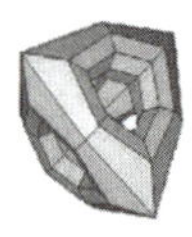  

Figure 8.11. Successive subdivision of a surface.

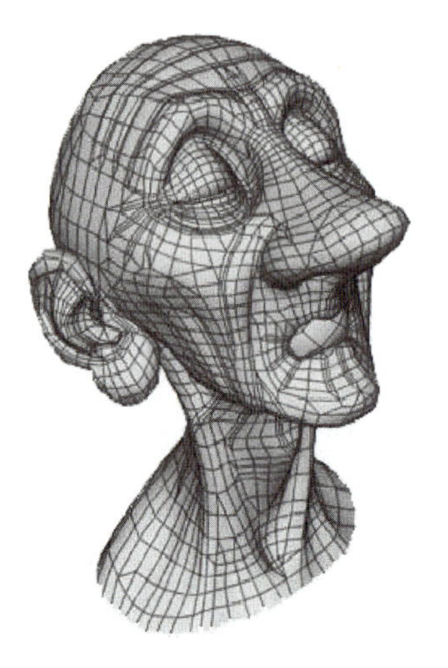 

Figure 8.12. Pixar's animated short *Geri's Game* used surface subdivision. (© Pixar 1997 Animation Studios)

Subdivision. Unlike NURBS, surface subdivision is not based on spline surfaces. The algorithm uses a sequence of two simple geometric operations: splitting and averaging. The idea is that a smooth object can be approximated through successive subdivisions of a surface as seen in figure 8.11.
While the technique has been known for some time, Pixar took the algorithm to new heights in its animated short *Geri's Game*, which won the 1998 Oscar for Best Short Animation. Pixar used surface subdivision to approximate complicated surfaces such as Geri's head and hands as seen in figure 8.12.

8.8 CHAPTER 8 EXERCISES

1. Following is census data showing the population of the US between 1900 and 2000:

Years after 1900	Population in millions
0	76.0
20	105.7
40	131.7
60	179.3
80	226.5
100	281.4

(a) Plot population versus years after 1900 using MATLAB by entering the years after 1900 into a vector `x` and the populations into a vector `y` and giving the command `plot(x,y,'o')`. Find the fifth-degree polynomial that passes through each data point, and determine its value in the year 2020. Plot this polynomial on the same graph as the population data (but with the x-axis now extending from 0 through 120). Do you think that this a reasonable way to estimate the population of the US in future years? (To see a demonstration of other methods of extrapolation, type `census` in MATLAB.)
Turn in your plot and the value that you estimated for the population in the year 2020 along with your explanation as to why this is or is not a good way to estimate the population.

(b) Write down the *Lagrange form* of the second-degree polynomial that interpolates the population in the years 1900, 1920, and 1940.

(c) Determine the coefficients of the *Newton form* of the interpolants of degrees 0, 1, and 2, that interpolate the first one, two, and three data points, respectively. Verify that the second-degree polynomial that you construct here is identical to that in part (b).

2. The secant method for finding a root of a function $f(x)$ fits a first-degree polynomial (i.e., a straight line) through the points $(x_{k-1}, f(x_{k-1}))$ and $(x_k, f(x_k))$ and takes the root of this polynomial as the next approximation x_{k+1}. Another root-finding algorithm, called *Muller's method*, fits a quadratic through the three points, $(x_{k-2}, f(x_{k-2}))$, $(x_{k-1}, f(x_{k-1}))$, and $(x_k, f(x_k))$, and takes the root of this quadratic that is closest to x_k as the next approximation x_{k+1}. Write down a formula for this quadratic. Suppose $f(x) = x^3 - 2$, $x_0 = 0$, $x_1 = 1$, and $x_2 = 2$. Find x_3.

3. Write down a divided-difference table and the Newton form of the interpolating polynomial for the following set of data.

x	1	3/2	0	2
$f(x)$	2	6	0	14

Show how this polynomial can be evaluated at a given point x by using four steps of a Horner's rule-like method.

4. (a) Use MATLAB to fit a polynomial of degree 12 to the Runge function

$$f(x) = \frac{1}{1+x^2},$$

interpolating the function at 13 equally-spaced points between -5 and 5. (You can set the points with the command `x = [-5:5/6:5];`.) You may use the MATLAB routine `polyfit` to do this computation or you may use your own routine. (To find out how to use `polyfit` type `help polyfit` in MATLAB.) Plot the function and your 12th-degree interpolant on the same graph. You can evaluate the polynomial at points throughout the interval $[-5, 5]$ using MATLAB's `polyval` routine, or you may write your own routine. Turn in your plot and a listing of your code.

(b) Repeat part (a) with a polynomial of degree 12 that interpolates f at 13 scaled Chebyshev points,

$$x_j = 5 \cos \frac{\pi j}{12}, \quad j = 0, \ldots, 12.$$

Again you may use MATLAB's `polyfit` and `polyval` routines to fit the polynomial to the function at these points and evaluate the result at points throughout the interval, or you may use the barycentric interpolation formula (8.6) with weights defined by (8.5) or (8.16) to evaluate the interpolation polynomial at points other than the interpolation points in the interval $[-5, 5]$. Turn in your plot and a listing of your code.

(c) If one attempts to fit a much higher-degree interpolation polynomial to the Runge function using MATLAB's `polyfit` routine, MATLAB will issue a warning that the problem is ill conditioned and answers may not be accurate. Still, the barycentric interpolation formula (8.6) with the special weights (8.16) for the values of a polynomial that interpolates f at the Chebyshev points may be used. Use this formula with 21 scaled Chebyshev interpolation points, $x_j = 5 \cos \frac{\pi j}{20}$, $j = 0, \ldots, 20$, to compute a more accurate approximation to this function. Turn in a plot of the interpolation polynomial $p_{20}(x)$ and the errors $|f(x) - p_{20}(x)|$ over the interval $[-5, 5]$.

(d) Download the `chebfun` package from the website http://www2.maths.ox.ac.uk/chebfun/ and follow the directions there to install the package. Enter the Runge function into **chebfun** as in example 8.5.2:

```
f = chebfun('1 ./ (1 + x.^2)',[-5,5])
```

and then plot the result by typing `plot(f)`. Also plot the actual Runge function on the same graph. The two should be indistinguishable.

5. The Chebyshev interpolation points in (8.15) are defined for the interval $[-1, 1]$. Suppose we wish to approximate a function on an interval $[a, b]$. Write down the linear transformation ℓ that maps the interval $[-1, 1]$ to $[a, b]$, with $\ell(a) = -1$ and $\ell(b) = 1$.

6. It was stated in the text that the Chebyshev polynomials $T_j(x)$, $j = 0, 1, \ldots$, could be defined in two ways: (1) $T_j(x) = \cos(j \arccos x)$, and (2) $T_{j+1}(x) = 2xT_j(x) - T_{j-1}(x)$, where $T_0(x) = 1$ and $T_1(x) = x$. Note that $\cos(0 \cdot \arccos x) = 1$ and $\cos(1 \cdot \arccos x) = x$, so the two expressions for $T_0(x)$ and $T_1(x)$ are the same. Show that the two definitions of $T_{j+1}(x)$, $j = 1, 2, \ldots$ are the same, by using formulas for the cosine of a sum and difference of two angles to show that

$$\cos((j + 1) \arccos x) = 2x \cos(j \arccos x) - \cos((j - 1) \arccos x).$$

7. Use MATLAB to fit various piecewise polynomials to the Runge function, using the same nodes as in part (a) of exercise 4.

(a) Find the piecewise linear interpolant of $f(x)$. (You may use MATLAB routine `interp1` or write your own routine. Type `help interp1` in MATLAB to find out how to use `interp1`.)

(b) Find the piecewise cubic Hermite interpolant of $f(x)$. (Write your own routine for this, using formulas (8.17) and (8.18).)
(c) Find a cubic spline interpolant of $f(x)$. (You may use either the `interp1` command or the `spline` command in MATLAB.)

In each case above, plot $f(x)$ and the interpolant on the same graph. If you wish, you can put all of your graphs on the same page by using the `subplot` command.

8. Computer libraries often use tables of function values together with piecewise linear interpolation to evaluate elementary functions such as $\sin x$, because table lookup and interpolation can be faster than, say, using a Taylor series expansion.

(a) In MATLAB, create a vector `x` of 1000 uniformly-spaced values between 0 and π. Then create a vector `y` with the values of the sine function at each of these points.
(b) Next create a vector `r` of 100 randomly distributed values between 0 and π. (This can be done in MATLAB by typing `r = pi*rand(100,1);`.) Estimate `sin r` as follows: For each value `r(j)`, find the two consecutive `x` entries, `x(i)` and `x(i+1)` that satisfy `x(i)` $\leq$ `r(j)` $\leq$ `x(i+1)`. Having identified the subinterval that contains `r(j)`, use linear interpolation to estimate `sin (r(j))`. Compare your results with those returned by MATLAB when you type `sin(r)`. Find the maximum absolute error and the maximum relative error in your results.

9. Determine the piecewise polynomial function

$$P(x) = \begin{cases} P_1(x) & \text{if } 0 \leq x \leq 1, \\ P_2(x) & \text{if } 1 \leq x \leq 2, \end{cases}$$

that is defined by the conditions

- $P_1(x)$ is linear.
- $P_2(x)$ is quadratic.
- $P(x)$ and $P'(x)$ are continuous at $x = 1$.
- $P(0) = 1$, $P(1) = -1$, and $P(2) = 0$.

Plot this function.

10. Let f be a given function satisfying $f(0) = 1$, $f(1) = 2$, and $f(2) = 0$. A **quadratic spline** interpolant $r(x)$ is defined as a piecewise quadratic that interpolates f at the nodes ($x_0 = 0$, $x_1 = 1$, and $x_2 = 2$) and whose first derivative is continuous throughout the interval. Find the quadratic spline interpolant of f that also satisfies $r'(0) = 0$. [Hint: Start from the left subinterval.]
11. Let $f(x) = x^2(x-1)^2(x-2)^2(x-3)^2$. What is the piecewise cubic Hermite interpolant of f on the grid $x_0 = 0$, $x_1 = 1$, $x_2 = 2$, $x_3 = 3$? Let $g(x) = ax^3 + bx^2 + cx + d$ for some parameters a, b, c, and d. Write down the piecewise cubic Hermite interpolant of g on the same grid. [Note: You should not need to do any arithmetic for either of these problems.]

12. Show that the following function is a natural cubic spline through the points (0, 1), (1, 1), (2, 0), and (3, 10):

$$s(x) = \begin{cases} 1 + x - x^3 & \text{if } 0 \le x < 1, \\ 1 - 2(x-1) - 3(x-1)^2 + 4(x-1)^3 & \text{if } 1 \le x < 2, \\ 4(x-2) + 9(x-2)^2 - 3(x-2)^3 & \text{if } 2 \le x \le 3. \end{cases}$$

13. Determine all the values of a, b, c, d, e, and f for which the following function is a cubic spline:

$$s(x) = \begin{cases} ax^2 + b(x-1)^3 & \text{if } x \in (-\infty, 1], \\ cx^2 + d & \text{if } x \in [1, 2], \\ ex^2 + f(x-2)^3 & \text{if } x \in [2, \infty). \end{cases}$$

14. Use MATLAB's `spline` function to find the not-a-knot cubic spline interpolant of $f(x) = e^{-2x}\sin(10\pi x)$, using $n = 20$ subintervals of equal size between $x_0 = 0$ and $x_{21} = 1$. (To learn how to use the spline routine, type `help spline` in MATLAB.) Plot the function along with its cubic spline interpolant, marking the knots on the curves.
15. A sampling of altitudes on a hillside is given below:

$y \uparrow$	Height(x, y)						
6	0	0	0	1	0	0	0
5	0	0	0	2	0	0	0
4	0	0	2	4	2	0	0
3	1	2	4	8	4	2	2
2	0	0	2	4	2	0	0
1	0	0	0	2	0	0	0
0	0	0	0	1	0	0	0
	0	1	2	3	4	5	6 $\rightarrow x$

In this exercise, we will fit a two-dimensional cubic spline (bicubic spline) through this data and plot the result.

(a) Using the `spline` command, fit a not-a-knot cubic spline to the data in each row, evaluating the spline at the points $x = 0, 0.1, 0.2, \ldots, 5.9, 6.0$. This will produce a 7 by 61 array of values.
(b) Fit a not-a-knot cubic spline to the data in each column of this array, evaluating each of these splines at $y = 0.0, 0.1, 0.2, \ldots, 5.9, 6.0$. This will produce a 61 by 61 array of values. Call this matrix M.
(c) Plot the values in M with the command `mesh(0:.1:6,0:.1:6,M)`.
(d) Alternatively, plot the data in M with the commands

```
surfl(0:.1:6,0:.1:6,M);
colormap(copper), shading interp
```

(e) Suppose the data above were samplings of the amount of light reflected off a shiny object. To produce a plot for this type of effect, use the commands

```
imagesc(0:.1:6,0:.1:6,M);
colormap(hot), colorbar
```

16. PostScript and TrueType letters are created with splines using only a few points for each letter. The MATLAB code below creates a parametric spline with the following data.

t	0	1	2	3	4	5	6
x	1	2	3	2	1.2	2	2.7
y	1	0	1	2.5	3.4	4	3.2

Note that in the resulting curve, y is not a "function" of x since the curve may take on more than one y value at a given x value; hence a parametric spline is constructed.

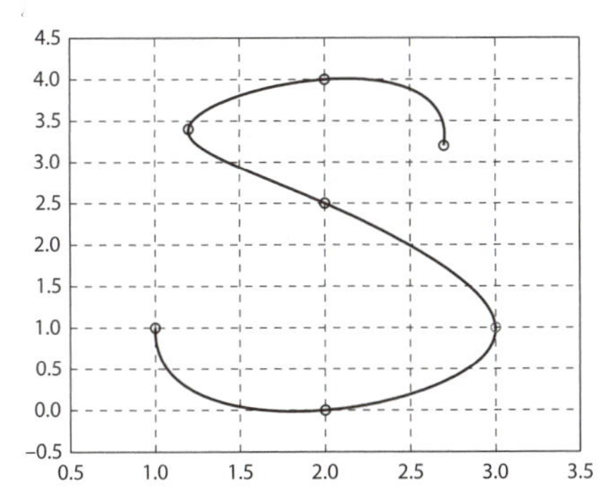

MATLAB code:

```
x = [1 2 3 2 1.2 2 2.7]; y = [1 0 1 2.5 3.4 4 3.2];
n = length(x);
t = 0:1:n-1;
tt = 0:.01:n-1;    % Finer mesh to sample spline
xx = spline(t,x,tt); yy = spline(t,y,tt); hold on
plot(xx,yy','LineWidth',2), plot(x,y,'o'), grid on
```

(a) Create and print the script letter defined by the following data.

t	0	1	2	3	4	5	6	7	8	9	10	11
x	3	1.75	0.90	0	0.50	1.50	3.25	4.25	4.25	3	3.75	6.00
y	4	1.60	0.50	0	1.00	0.50	0.50	2.25	4.00	4	3.25	4.25

(b) On the same axes, plot this letter along with the letter doubled in size. (The command `2*x` will double the size of the font in the x direction.) Note the ease of scaling the size of a font that is stored in such a way.
(c) Edit your MATLAB code to use the command `comet(xx,yy)` which will animate the drawing of your letter.
(d) Create and print another script letter.

9

NUMERICAL DIFFERENTIATION AND RICHARDSON EXTRAPOLATION

Suppose a scientist collects a set of data relating two properties such as temperature and pressure, or distance that an object has fallen and time since release, or amount that a plant has grown and amount of water that it has received. The scientist postulates that there is some smooth function that describes this relationship, but does not know exactly what that function is. One or more derivatives of the function might be needed, but without an analytic expression, the derivatives can only be approximated numerically. Sometimes a function is defined by running a computer program. It might be the solution to a design problem that depends on a parameter, and the scientist might wish to optimize some aspect of the design with respect to this parameter. This requires determining the derivative of the function to be optimized. While an analytic expression for this derivative might be determined by carefully going through every line of the code, the resulting expression might be so complicated that it would be more efficient to simply estimate the derivative numerically.

Another application of numerical differentiation is in approximating solutions of differential equations, a topic that will be discussed in later chapters of this text. Using *finite difference methods*, derivatives are approximated by linear combinations of function values, and the differential equation is thereby replaced by a system of linear or nonlinear equations involving function values. These systems of equations may involve millions or billions of unknowns. Some of the world's fastest supercomputers are commonly used to solve systems that result from discretizing differential equations with numerical approximations to derivatives.

9.1 NUMERICAL DIFFERENTIATION

We have already seen (in section 6.2) one way to approximate the derivative of a function f:

$$f'(x) \approx \frac{f(x+h) - f(x)}{h}, \tag{9.1}$$

for some small number h. To determine the accuracy of this approximation, we use Taylor's theorem, assuming that $f \in C^2$:

$$f(x+h) = f(x) + hf'(x) + \frac{h^2}{2} f''(\xi), \quad \xi \in [x, x+h] \quad \Longrightarrow$$

$$f'(x) = \frac{f(x+h) - f(x)}{h} - \frac{h}{2} f''(\xi).$$

The term $\frac{h}{2} f''(\xi)$ is called the **truncation error**, or, the **discretization error**, and the approximation is said to be first-order accurate since the truncation error is $O(h)$.

As noted in section 6.2, however, roundoff also plays a role in the evaluation of the finite difference quotient (9.1). For example, if h is so small that $x+h$ is rounded to x, then the computed difference quotient will be 0. More generally, even if the only error made is in rounding the values $f(x+h)$ and $f(x)$, then the computed difference quotient will be

$$\frac{f(x+h)(1+\delta_1) - f(x)(1+\delta_2)}{h} = \frac{f(x+h) - f(x)}{h} + \frac{\delta_1 f(x+h) - \delta_2 f(x)}{h}.$$

Since each $|\delta_i|$ is less than the machine precision ϵ, this implies that the rounding error is less than or equal to

$$\frac{\epsilon(|f(x)| + |f(x+h)|)}{h}.$$

Since the truncation error is proportional to h and the rounding error is proportional to $1/h$, the best accuracy is achieved when these two quantities are approximately equal. Ignoring the constants $|f''(\xi)/2|$ and $(|f(x)| + |f(x+h)|)$, this means that

$$h \approx \frac{\epsilon}{h} \Rightarrow h \approx \sqrt{\epsilon},$$

and in this case the error (truncation error or rounding error) is about $\sqrt{\epsilon}$. Thus, with formula (9.1), we can approximate a derivative to only about the square root of the machine precision.

Example 9.1.1. Let $f(x) = \sin x$ and use (9.1) to approximate $f'(\pi/3) = 0.5$.

Table 9.1 shows the results produced by MATLAB. At first the error decreases linearly with h, but as we decrease h below $\sqrt{\epsilon} \approx 10^{-8}$, this changes and the error starts to grow due to roundoff.

TABLE 9.1
Results of approximating the first derivative of $f(x) = \sin x$ at $x = \pi/3$ using formula (9.1).

h	$(\sin(\pi/3+h) - \sin(\pi/3))/h$	Error
1.0e−01	4.559018854107610e−01	4.4e−02
1.0e−02	4.956615757736871e−01	4.3e−03
1.0e−03	4.995669040007700e−01	4.3e−04
1.0e−04	4.999566978958203e−01	4.3e−05
1.0e−05	4.999956698670261e−01	4.3e−06
1.0e−06	4.999995669718871e−01	4.3e−07
1.0e−07	4.999999569932356e−01	4.3e−08
1.0e−08	4.999999969612645e−01	3.0e−09
1.0e−09	5.000000413701855e−01	−4.1e−08
1.0e−10	5.000000413701855e−01	−4.1e−08
1.0e−11	5.000000413701855e−01	−4.1e−08
1.0e−12	5.000444502911705e−01	−4.4e−05
1.0e−13	4.996003610813204e−01	4.0e−04
1.0e−14	4.996003610813204e−01	4.0e−04
1.0e−15	5.551115123125783e−01	−5.5e−02
1.0e−16	0	5.0e−01

Another way to approximate the derivative is to use a *centered-difference formula*:

$$f'(x) \approx \frac{f(x+h) - f(x-h)}{2h}. \tag{9.2}$$

We can again determine the truncation error by using Taylor's theorem. (In this and other chapters throughout this book, unless otherwise stated, we will always assume that functions expanded in Taylor series are sufficiently smooth for the Taylor expansion to exist. The amount of smoothness required will be clear from the expansion used. (See theorem 4.2.1.) In this case, we need $f \in C^3$.) Expanding $f(x+h)$ and $f(x-h)$ about the point x, we find

$$f(x+h) = f(x) + hf'(x) + \frac{h^2}{2} f''(x) + \frac{h^3}{6} f'''(\xi), \quad \xi \in [x, x+h],$$

$$f(x-h) = f(x) - hf'(x) + \frac{h^2}{2} f''(x) - \frac{h^3}{6} f'''(\eta), \quad \eta \in [x-h, x].$$

Subtracting the two equations and solving for $f'(x)$ gives

$$f'(x) = \frac{f(x+h) - f(x-h)}{2h} - \frac{h^2}{12}(f'''(\xi) + f'''(\eta)).$$

Thus the truncation error is $O(h^2)$, and this difference formula is second-order accurate.

To study the effects of roundoff, we again make the simplifying assumption that the only roundoff that occurs is in rounding the values $f(x+h)$ and

TABLE 9.2
Results of approximating the first derivative of $f(x) = \sin x$ at $x = \pi/3$ using formula (9.2).

h	$(\sin(\pi/3+h) - \sin(\pi/3-h))/(2h)$	Error
1.0e−01	4.991670832341411e−01	8.3e−04
1.0e−02	4.999916667083382e−01	8.3e−06
1.0e−03	4.999999166666047e−01	8.3e−08
1.0e−04	4.999999991661674e−01	8.3e−10
1.0e−05	4.999999999921734e−01	7.8e−12
1.0e−06	4.999999999588667e−01	4.1e−11
1.0e−07	5.000000002919336e−01	−2.9e−10
1.0e−08	4.999999969612645e−01	3.0e−09
1.0e−09	5.000000413701855e−01	−4.1e−08
1.0e−10	5.000000413701855e−01	−4.1e−08
1.0e−11	5.000000413701855e−01	−4.1e−08
1.0e−12	5.000444502911705e−01	−4.4e−05
1.0e−13	4.996003610813204e−01	4.0e−04
1.0e−14	4.996003610813204e−01	4.0e−04
1.0e−15	5.551115123125783e−01	−5.5e−02
1.0e−16	0	5.0e−01

$f(x-h)$. Then the computed difference quotient is

$$\frac{f(x+h)(1+\delta_1) - f(x-h)(1+\delta_2)}{2h} = \frac{f(x+h) - f(x-h)}{2h} + \frac{\delta_1 f(x+h) - \delta_2 f(x-h)}{2h},$$

and the roundoff term $(\delta_1 f(x+h) - \delta_2 f(x-h))/2h$ is bounded in absolute value by $\epsilon(|f(x+h)| + |f(x-h)|)/(2h)$. Once again ignoring constant terms involving f and its derivatives, the greatest accuracy is now achieved when

$$h^2 \approx \frac{\epsilon}{h} \Rightarrow h \approx \epsilon^{1/3},$$

and then the error (truncation error or rounding error) is $\epsilon^{2/3}$. With this formula we can obtain greater accuracy, to about the 2/3 power of the machine precision.

Example 9.1.2. Let $f(x) = \sin x$ and use (9.2) to approximate $f'(\pi/3) = 0.5$.

Table 9.2 shows the results produced by MATLAB. At first the error decreases quadratically with h, but as we decrease h below $\epsilon^{1/3} \approx 10^{-5}$, this changes and the error starts to grow due to roundoff. A graph of the error is shown in figure 9.1. The best achievable accuracy is greater than that with the forward-difference formula (9.1), however.

One can approximate higher derivatives similarly. To derive a second-order-accurate approximation to the second derivative, we again use Taylor's

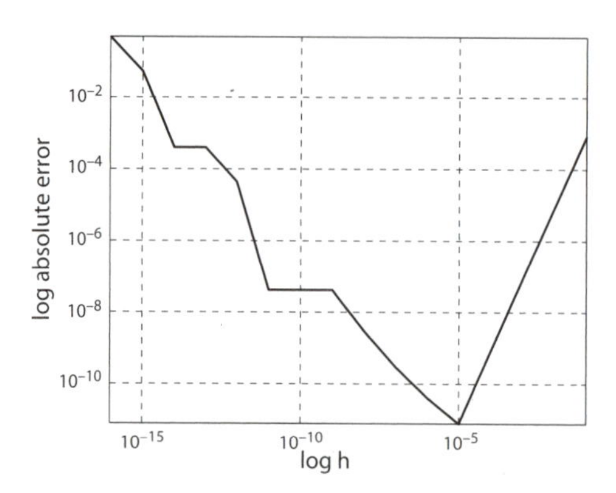

Figure 9.1. Log–log plot of the error in a computed centered-difference approximation to the first derivative of $\sin x$ at $x = \pi/3$.

theorem:

$$f(x+h) = f(x) + hf'(x) + \frac{h^2}{2} f''(x) + \frac{h^3}{6} f'''(x) + \frac{h^4}{4!} f''''(\xi), \quad \xi \in [x, x+h],$$

$$f(x-h) = f(x) - hf'(x) + \frac{h^2}{2} f''(x) - \frac{h^3}{6} f'''(x) + \frac{h^4}{4!} f''''(\eta), \quad \eta \in [x-h, x].$$

Adding these two equations gives

$$f(x+h) + f(x-h) = 2f(x) + h^2 f''(x) + \frac{h^4}{12} f''''(\nu), \quad \nu \in [\eta, \xi].$$

Solving for $f''(x)$, we obtain the formula

$$f''(x) = \frac{f(x+h) - 2f(x) + f(x-h)}{h^2} - \frac{h^2}{12} f''''(\nu).$$

Using the approximation

$$f''(x) \approx \frac{f(x+h) - 2f(x) + f(x-h)}{h^2}, \tag{9.3}$$

the truncation error is $O(h^2)$. Note, however, that a similar rounding error analysis predicts rounding errors of size about ϵ/h^2, so the smallest total error occurs when h is about $\epsilon^{1/4}$ and then the truncation error and the rounding error are each about $\sqrt{\epsilon}$. With machine precision $\epsilon \approx 10^{-16}$, this means that h should not be taken to be less than about 10^{-4}. Evaluation of standard finite difference quotients for higher derivatives is even more sensitive to the effects of roundoff.

Example 9.1.3. Let $f(x) = \sin x$ and use (9.3) to approximate $f''(\pi/3) = -\frac{\sqrt{3}}{2}$. Create a plot of the error versus h. As expected, we see from figure 9.2 that the smallest error of about 10^{-8} occurs when $h \approx 10^{-4}$.

Suppose the function f cannot be evaluated at arbitrary points but is known only at a fixed set of measurement points $x_0, \ldots, x_n$. Let the measured values of f at these points be $y_0, \ldots, y_n$. If the distance between the measurement

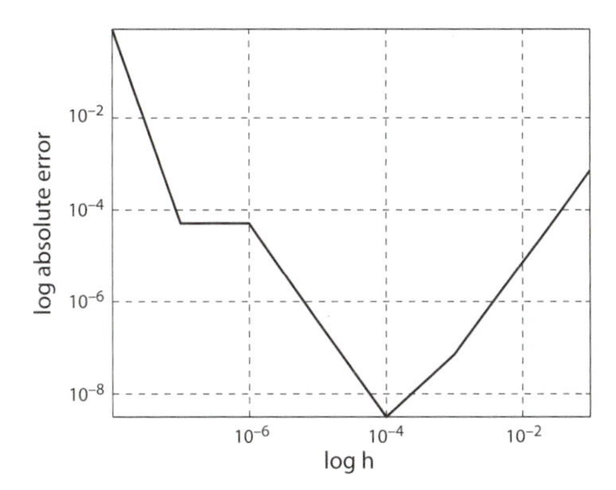

Figure 9.2. Log–log plot of the error in a computed centered difference approximation to the second derivative of $\sin x$ at $x = \pi/3$.

points is small (but not so small that roundoff destroys our results), then we might expect to obtain reasonable approximations to $f'(x)$ or $f''(x)$ by using approximations that involve measured function values at points near x. For example, if x lies between x_{i-1} and x_i, then we can approximate $f'(x)$ by using the values of f at x_{i-1} and x_i. Expanding $f(x_{i-1})$ and $f(x_i)$ in Taylor series about x, we find:

$$f(x_{i-1}) = f(x) + (x_{i-1} - x) f'(x) + \frac{(x_{i-1} - x)^2}{2!} f''(\xi), \quad \xi \in [x_{i-1}, x],$$

$$f(x_i) = f(x) + (x_i - x) f'(x) + \frac{(x_i - x)^2}{2!} f''(\eta), \quad \eta \in [x, x_i].$$

Subtracting these two equations and dividing by $x_i - x_{i-1}$ gives

$$f'(x) = \frac{f(x_i) - f(x_{i-1})}{x_i - x_{i-1}} - \frac{(x_i - x)^2}{2(x_i - x_{i-1})} f''(\eta) + \frac{(x_{i-1} - x)^2}{2(x_i - x_{i-1})} f''(\xi).$$

The approximation

$$f'(x) \approx \frac{f(x_i) - f(x_{i-1})}{x_i - x_{i-1}}$$

has truncation error less than or equal to $(x_i - x_{i-1})$ times the maximum absolute value of f'' in this interval. (If x happens to be the midpoint of the interval, then the truncation error is actually proportional to $(x_i - x_{i-1})^2$, which can be seen by retaining one more term in the Taylor series expansions.)

Suppose, however, that the measured function values y_{i-1} and y_i have errors; errors in measurements are usually *much* larger than computer rounding errors. The same analysis that was used to see the effects of rounding errors can be used to see the effects of errors in the measured data. If $y_i = f(x_i)(1 + \delta_1)$ and $y_{i-1} = f(x_{i-1})(1 + \delta_2)$, then the difference between the value $(y_i - y_{i-1})/(x_i - x_{i-1})$ and the intended derivative approximation

$(f(x_i) - f(x_{i-1}))/(x_i - x_{i-1})$ is

$$\frac{\delta_1 f(x_i) - \delta_2 f(x_{i-1})}{x_i - x_{i-1}}.$$

In order to obtain a reasonable approximation to $f'(x)$, the distance between x_i and x_{i-1} must be *small* enough so that the truncation error $(x_i - x_{i-1})$ times the maximum value of f'' in the interval is small, but *large* enough so that $\delta(|f(x_i)| + |f(x_{i-1})|)/(x_i - x_{i-1})$ is still small, where δ is the largest error in the measurements. This combination can be difficult (or impossible!) to achieve. For higher-order derivatives the restrictions on the measurement points and the accuracy of measurements are even more severe. If one has measured values of f at x, $x+h$, and $x-h$, and one uses these measured values in the difference formula (9.3) for the second derivative, then the measurement errors δ must be smaller than h^2 in order for the difference quotient using measured values to be close to the intended difference quotient. For these reasons, approximating derivatives of a function from measured data can be fraught with difficulties.

Another possible approach to computing derivatives is to interpolate f with a polynomial and to differentiate the polynomial. Given the value of f at $x_0, \ldots, x_n$, one can write down the nth-degree polynomial that interpolates f at these points. Considering some of the difficulties that were pointed out with high-degree polynomial interpolation at equally-spaced points, however, this approach is not recommended for n very large, *unless* the points $x_0, \ldots, x_n$ are distributed like the Chebyshev points (8.15) (i.e., clustered near the endpoints of the interval).

The barycentric formula (8.6) for $p(x)$ can be differentiated directly to obtain

$$p'(x) = \frac{-s_1(x)s_2(x) + s_3(x)s_4(x)}{(s_1(x))^2}, \tag{9.4}$$

where

$$\begin{aligned} s_1(x) &= \sum_{i=0}^{n} \frac{w_i}{x - x_i}, \quad s_2(x) = \sum_{i=0}^{n} \frac{w_i}{(x - x_i)^2} y_i, \\ s_3(x) &= \sum_{i=0}^{n} \frac{w_i}{x - x_i} y_i, \; s_4(x) = \sum_{i=0}^{n} \frac{w_i}{(x - x_i)^2}, \end{aligned} \tag{9.5}$$

but other formulas can be derived that may require less work to evaluate and have better numerical properties; see, for example, [11]. In the `chebfun` package, introduced in section 8.5, $p(x)$ is expressed as a linear combination of Chebyshev polynomials, $p(x) = \sum_{j=0}^{n} a_j T_j(x)$, and then an easy computation can be done to determine the coefficients $b_0, \ldots, b_{n-1}$ in the expansion $p'(x) = \sum_{j=0}^{n-1} b_j T_j(x)$ [9].

Example 9.1.4. Interpolate $\sin x$ at the scaled Chebyshev points $x_j = \pi \cos \frac{\pi j}{n}$, $j = 0, \ldots, n$, and then differentiate the resulting interpolation polynomial $p(x)$ using (9.4) and (9.5).

Using $n = 5$, 10, 20, and 40 interpolation points and computing the maximum error on the interval $[-\pi, \pi]$, one obtains the results in table 9.3.

TABLE 9.3
Results of approximating the first derivative of $f(x) = \sin x$ by interpolating the function at the scaled Chebyshev points $x_j = \pi \cos(\pi j/n)$, $j = 0, \ldots, n$, and differentiating the interpolation polynomial $p(x)$.

n	$\max_{x\in[-\pi,\pi]} \lvert p'(x) - \cos x \rvert$
5	0.07
10	3.8e−05
20	1.5e−13
40	3.4e−13

We appear to be limited to about 13 decimal places of accuracy using this approach.

This same computation, as well as the computation of higher derivatives, can be done using the `chebfun` package. To compute the first two derivatives of $\sin x$, type

```
f = chebfun('sin(x)',[-pi,pi])
fp = diff(f)
fpp = diff(f,2)
```

Chebfun reports that the length of `f` is 22, indicating that it has represented the function $\sin x$ on the interval $[-\pi, \pi]$ using a polynomial of degree 21 that interpolates $\sin x$ at 22 scaled Chebyshev points $x_j = \pi \cos \frac{\pi j}{21}$, $j = 0, \ldots, 21$. The length of `fp` is 21 since it is the derivative of the interpolation polynomial for `f` and hence one degree lower. Similarly, the length of `fpp` is 20. Evaluating `fp` and `fpp` at many points throughout the interval $[-\pi, \pi]$ and computing the maximum absolute value of the difference between `fp` and $(d/dx) \sin x = \cos x$ and the maximum absolute value of the difference between `fpp` and $(d^2/dx^2) \sin x = -\sin x$, we find that the error in `fp` is 4.2×10^{-14}, while that in `fpp` is 1.9×10^{-12}.

Example 9.1.5. (Computing derivatives of the Runge function in chebfun). In section 8.4, we introduced the Runge function as an example of a function for which high-degree polynomial interpolation at *equally-spaced* nodes was extremely bad. In the next section we showed, however, that interpolation at the Chebyshev points (8.15) could be quite effective and that this could be carried out with the `chebfun` package. We defined the Runge function in `chebfun` by typing

```
f = chebfun('1 ./ (1 + x.^2)',[-5,5]);
```

We can easily differentiate this function by typing `fp = diff(f)`, or we can differentiate it any number of times k by typing `fpp = diff(f,k)`. Recall that the function itself was represented by a polynomial of degree 182, so that the first derivative of this polynomial has degree 181 and its kth derivative has degree $182 - k$. This is reflected in the `length` of the `chebfuns` `f`, `fp`, and `fpp`.

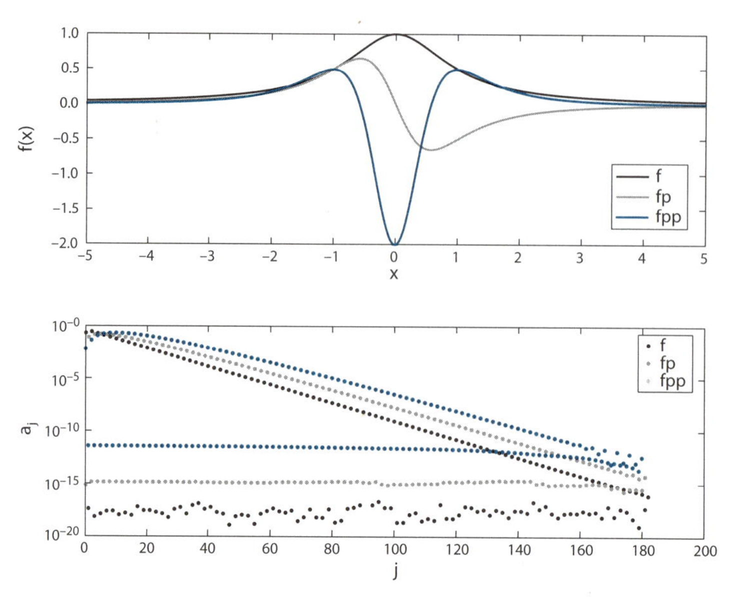

Figure 9.3. The polynomial interpolant of the Runge function produced by chebfun and its first and second derivatives (*top*). Coefficients of the highest degree Chebyshev polynomials in their expansion (*bottom*).

In the top plot of figure 9.3 we have plotted the polynomial interpolant of the Runge function produced by `chebfun` and its first and second derivatives.

We can get an idea of the accuracy of the first and second derivatives by looking at the size of the coefficients of the highest-degree Chebyshev polynomials in their expansions, just as we looked at the coefficients in the expansion of f in section 8.5. To do this, we type

```
coeffs = chebpoly(f); coeffsfp = chebpoly(fp);
coeffsfpp = chebpoly(fpp); np1 = length(coeffs);
semilogy([0:np1-1],abs(coeffs(np1:-1:1)),'.', ...
         [0:np1-2],abs(coeffsfp(np1-1:-1:1)),'.r', ...
         [0:np1-3],abs(coeffsfpp(np1-2:-1:1)),'.g')
```

The results are shown in the bottom plot of figure 9.3. Note that some accuracy is inevitably lost in higher derivatives, but this is a *very* good result. The coefficients in both `fp` and `fpp` drop well below 10^{-10} before the expansion is terminated. Had we differentiated the polynomial that interpolated this function at equally-spaced points, the results would have been disastrous.

Earlier in this section we discussed the effects of a small change in function values, say, $f(x)$ and $f(x+h)$, on a corresponding finite difference quotient such as $(f(x+h) - f(x))/h$. We found that since changes in f were multiplied by $1/h$, the problem of computing a finite difference quotient is *ill conditioned* if h is very small. However, the goal is not to compute a finite difference quotient but to compute the actual *derivative* of the function and, as illustrated in the previous two examples, this can be accomplished in different ways.

What is the conditioning of the problem of computing a *derivative* $f'(x)|_{x=a}$, given a and the values of f? Figure 9.4 should convince you that

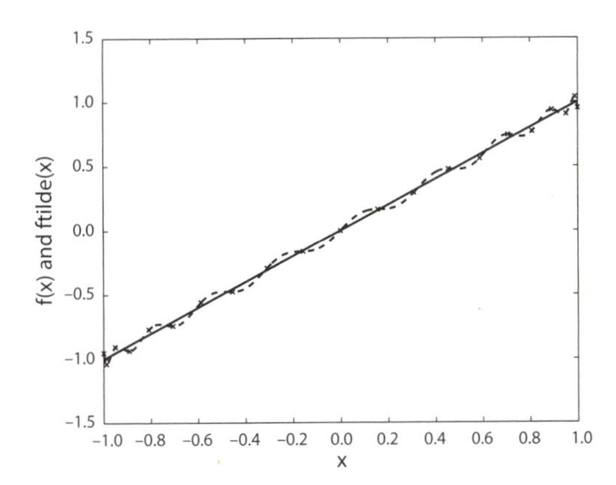

Figure 9.4. Two functions with very close values but very different derivatives.

this problem too can be ill conditioned; two functions $f(x)$ and $\tilde{f}(x)$ can have very close values throughout an interval and yet have very different derivatives. In the figure, the solid line has slope 1, while the derivative of the dashed curve varies significantly over the interval. The dashed curve might be obtained as an "interpolation polynomial" for the solid line, if the function values that it interpolated differed slightly from those on the line. This example differs from the usual case of numerical interpolation, however, in that the distance between interpolation points is of the same order of magnitude as the errors in function values. When errors in function values come from computer roundoff, they are usually *much* smaller than the distance between the points at which the function values are used, and for this reason one can often obtain reasonable accuracy with numerical differentiation.

9.2 RICHARDSON EXTRAPOLATION

Another way to obtain higher-order accuracy in the approximation of derivatives is called **Richardson extrapolation.** Tradition credits Huygens with proposing this idea, which he applied to Archimedes' method for finding the circumference of a circle. In 1910, Lewis Fry Richardson suggested the use of results obtained with several stepsizes to eliminate the first error term in the central difference formula. In 1927, he called the technique the *deferred approach to the limit* and used it to solve a 6th order differential eigenvalue problem [17, 19].

Richardson extrapolation is a general procedure that can be applied to many types of problems, such as approximating integrals and solutions to differential equations, as well as derivatives. Consider again the centered-difference formula (9.2). That was derived from a Taylor series, and, had we saved a few more terms in the Taylor series, we would have found:

$$f(x+h) = f(x) + hf'(x) + \frac{h^2}{2} f''(x) + \frac{h^3}{6} f'''(x) + \frac{h^4}{24} f''''(x) + O(h^5),$$

$$f(x-h) = f(x) - hf'(x) + \frac{h^2}{2} f''(x) - \frac{h^3}{6} f'''(x) + \frac{h^4}{24} f''''(x) + O(h^5),$$

Lewis Fry Richardson

Lewis Fry Richardson (1881–1953) made numerous mathematical contributions including being one of the first to apply mathematics to predict weather. This work (which occurred during the 1910s and 1920s) predated the computer age. Richardson calculated that 60,000 people would be needed in order to perform the hand calculations necessary to produce the prediction of the next day's weather before the weather arrived. Richardson, a pacifist, modeled the causes of war, set up equations governing arms buildup by nations, and researched factors that would reduce the frequency of wars. Richardson's work produced a power law distribution relating the size of a war to its frequency. Richardson's study of the causes of war led him to measure coastlines, as he searched for a relation between the probability of two countries going to war and the length of their common border. This work later influenced Benoit Mandelbrot and his development of fractal theory.

so that subtracting and solving for $f'(x)$ gives

$$f'(x) = \frac{f(x+h) - f(x-h)}{2h} - \frac{h^2}{6} f'''(x) + O(h^4). \tag{9.6}$$

Let $\varphi_0(h) \equiv \frac{f(x+h)-f(x-h)}{2h}$ denote the centered-difference approximation to $f'(x)$, so that equation (9.6) can be written in the form

$$f'(x) = \varphi_0(h) - \frac{h^2}{6} f'''(x) + O(h^4). \tag{9.7}$$

Then $\varphi_0(h/2)$, the centered-difference approximation to $f'(x)$ using spacing $h/2$, satisfies

$$f'(x) = \varphi_0(h/2) - \frac{(h/2)^2}{6} f'''(x) + O(h^4). \tag{9.8}$$

If we multiply equation (9.8) by 4 and subtract (9.7), then the terms involving h^2 will cancel. Dividing the result by 3, we have

$$f'(x) = \frac{4}{3}\varphi_0(h/2) - \frac{1}{3}\varphi_0(h) + O(h^4). \tag{9.9}$$

Thus, we have derived a difference formula, $\varphi_1(h) \equiv \frac{4}{3}\varphi_0(h/2) - \frac{1}{3}\varphi_0(h)$, whose error is $O(h^4)$. It requires roughly twice as much work as the second-order centered-difference formula, since we must evaluate φ_0 at h and $h/2$, but the truncation error now decreases *much* faster with h. Moreover, the rounding error can be expected to be on the order of ϵ/h, as it was for the centered-difference formula, so the greatest accuracy will be achieved for $h^4 \approx \epsilon/h$, or, $h \approx \epsilon^{1/5}$, and then the error will be about $\epsilon^{4/5}$.

TABLE 9.4
Results of approximating the first derivative of $f(x) = \sin x$ at $x = \pi/3$ using centered differences and one step of Richardson extrapolation.

h	$\varphi_1(h) \equiv \frac{4}{3}\varphi_0(h/2) - \frac{1}{3}\varphi_0(h)$	Error
1.0e−01	4.999998958643311e−01	1.0e−07
1.0e−02	4.999999999895810e−01	1.0e−11
1.0e−03	4.999999999999449e−01	5.5e−14

This process can be repeated. Equation (9.9) can be written in the form

$$f'(x) = \varphi_1(h) + Ch^4 + O(h^5), \tag{9.10}$$

where we have written the $O(h^4)$ term as $Ch^4 + O(h^5)$. The analogous equation using spacing $h/2$ is

$$f'(x) = \varphi_1(h/2) + C(h/2)^4 + O(h^5).$$

If we take 2^4 times this equation and subtract (9.10), then the $O(h^4)$ terms will cancel. Dividing the result by $2^4 - 1 = 15$ will give a difference formula for $f'(x)$ whose truncation error is at most $O(h^5)$. (A more careful analysis shows that it is actually $O(h^6)$, since the $O(h^5)$ terms also cancel.)

$$f'(x) = \frac{16}{15}\varphi_1(h/2) - \frac{1}{15}\varphi_1(h) + O(h^6).$$

Example 9.2.1. Let $f(x) = \sin x$. We will use centered differences and Richardson extrapolation to approximate $f'(\pi/3) = 0.5$.

Let $\varphi_0(h)$ denote the centered-difference formula for $f'(\pi/3)$: $\varphi_0(h) = (\sin(\pi/3 + h) - \sin(\pi/3 - h))/(2h)$. Using this formula with one step of Richardson extrapolation in MATLAB produced the results in table 9.4. We can see the fourth-order convergence at first, but by the time we reach $h = 1.0\text{e}{-03}$, we are near machine precision and roundoff begins to play a role. Smaller values of h give less accurate results for this reason. If we do a second step of Richardson extrapolation, defining $\varphi_2(h) = (16/15)\varphi_1(h/2) - (1/15)\varphi_1(h)$, then for $h = 1.0\text{e}{-01}$ we obtain an approximation whose error is 1.6e−12.

To apply this procedure generally, suppose that we wish to approximate a value L using an approximation $\varphi_0(h)$ that depends on a parameter h and whose error can be expressed in terms of powers of h:

$$L = \varphi_0(h) + a_1 h + a_2 h^2 + a_3 h^3 + \dots.$$

If we use the same approximation procedure but with h replaced by $h/2$, then we have

$$L = \varphi_0(h/2) + \frac{a_1}{2}h + \frac{a_2}{4}h^2 + \frac{a_3}{8}h^3 + \dots.$$

Aiding Archimedes with Extrapolation

Archimedes (287–212 BC) approximated π by determining the length of the perimeter of a polygon inscribed within a circle. Assuming the circle has unit radius, the value of π lies between half of the value of this perimeter and half of the length of the perimeter of a polygon circumscribed about the circle. Let $\{C_n\}$ denote the circumference of an n-sided polygon inscribed in the circle. Then, $\lim_{n\to\infty} C_n = 2\pi$. By using polygons with 96 sides, Archimedes determined that $\frac{223}{71} < \pi < \frac{22}{7}$, which translates to $3.1408 < \pi < 3.1429$ in our modern decimal notation.

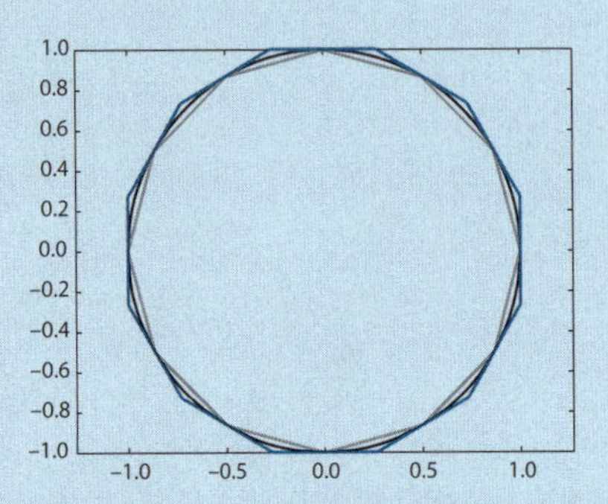

Let us consider using Richardson extrapolation. Half the perimeter of an inscribed square is $2\sqrt{2} \approx 2.8284$, and that of an inscribed hexagon is 3.0 exactly. The error in these approximations can be shown to be of order $(1/n)^2$, where n is the number of sides. Thus the ratio of the errors in these two estimates is 9 : 4. Therefore, the extrapolated value is

$$\frac{9}{5} \cdot 3 - \frac{4}{5} \cdot 2.8284 = 3.13728.$$

To obtain an estimate of this accuracy without using Richardson extrapolation would require a polygon of 35 sides.

The approximation $2\varphi_0(h/2) - \varphi_0(h)$ will have error $O(h^2)$, since twice the second equation minus the first gives

$$L = 2\varphi_0(h/2) - \varphi_0(h) - \frac{1}{2}a_2h^2 - \frac{3}{4}a_3h^3 + \dots.$$

Defining $\varphi_1(h) \equiv 2\varphi_0(h/2) - \varphi_0(h)$, this can be written in the form

$$L = \varphi_1(h) + b_2h^2 + b_3h^3 + \dots,$$

for certain constants b_2, b_3, etc. Analogously, approximating L by $\varphi_1(h/2)$, we find

$$L = \varphi_1(h/2) + \frac{b_2}{4}h^2 + \frac{b_3}{8}h^3 + \dots.$$

Multiplying this equation by 4, subtracting the previous one, and dividing by 3 gives

$$L = \frac{4}{3}\varphi_1(h/2) - \frac{1}{3}\varphi_1(h) - \frac{1}{6}b_3h^3 + \dots .$$

Continuing in this way, we can define an approximation $\varphi_2(h) \equiv \frac{4}{3}\varphi_1(h/2) - \frac{1}{3}\varphi_1(h)$ whose error is $O(h^3)$, etc. In general, each application of Richardson extrapolation increases the order of the error by one, but the procedure works especially well when some powers are not present in the error formula. For example, if the error involves only even powers of h, then each extrapolation step increases the order of the error by two.

9.3 CHAPTER 9 EXERCISES

1. Use MATLAB to evaluate the second-order-accurate approximation (9.3) to $f''(x)$ for $f(x) = \sin x$ and $x = \pi/6$. Try $h = 10^{-1}, 10^{-2}, \dots, 10^{-16}$, and make a table of values of h, the computed finite difference quotient, and the error. Explain your results.
2. Use formula (9.3) with $h = 0.2$, $h = 0.1$, and $h = 0.05$ to approximate $f''(x)$, where $f(x) = \sin x$ and $x = \pi/6$. Use one step of Richardson extrapolation, combining the results from $h = 0.2$ and $h = 0.1$, to obtain a higher-order-accurate approximation. Do the same with the results from $h = 0.1$ and $h = 0.05$. Finally do a second step of Richardson extrapolation, combining the two previously extrapolated values, to obtain a still higher-order-accurate approximation. Make a table of the computed results and their errors. What do you think is the order of accuracy after one step of Richardson extrapolation? How about after two?
3. Use `chebfun` to evaluate $f''(x)$ for $f(x) = \sin x$ and $x = \pi/6$. What is the degree of the interpolation polynomial that it produces for f, and what is the error in its approximation to $f''(\pi/6)$?
4. (a) Use the formula $T_j(x) = \cos(j \arccos x)$ for the Chebyshev polynomials to show that

$$\int T_j(x)\,dx = \frac{1}{2}\left[\frac{\cos((j+1)\arccos x)}{j+1} - \frac{\cos((j-1)\arccos x)}{j-1}\right] + C$$
$$= \frac{1}{2}\left[\frac{T_{j+1}(x)}{j+1} - \frac{T_{j-1}(x)}{j-1}\right] + C, \quad j = 2, 3, \dots,$$

where C is an arbitrary constant. [Hint: First make the change of variable $x = \cos\theta$, so that $dx = -\sin\theta\,d\theta$ and the integral becomes $\int T_j(x)\,dx = -\int \cos(j\theta)\sin\theta\,d\theta$. Now use the identity $\frac{1}{2}[\sin((j+1)\theta) - \sin((j-1)\theta)] = \cos(j\theta)\sin\theta$ to evaluate the integral.]

(b) Suppose $p(x) = \sum_{j=0}^{n} a_j T_j(x)$. Use part (a) together with the fact that $T_0(x) = 1$, $T_1(x) = x$, $T_2(x) = 2x^2 - 1$ to determine coefficients $A_0, \dots, A_n, A_{n+1}$ such that $\int p(x)\,dx = \sum_{j=0}^{n+1} A_j T_j(x)$; that is, express $A_0, \dots, A_n, A_{n+1}$ in terms of $a_0, \dots, a_n$. [Note: The coefficient A_0 can be arbitrary to account for the arbitrary constant in the indefinite integral.]

(c) Now suppose $q(x) = \sum_{j=0}^{n+1} A_j T_j(x)$. Reversing the process in part (b), determine coefficients $a_0, \ldots, a_n$ such that $q'(x) = \sum_{j=0}^{n} a_j T_j(x)$; that is, express $a_0, \ldots, a_n$ in terms of $A_0, \ldots, A_n, A_{n+1}$. [Hint: Work backwards, first expressing a_n in terms of A_{n+1}, then expressing a_{n-1} in terms of A_n, then expressing a_{j-1} in terms of A_j and a_{j+1}, $j = n-1, \ldots, 1$.]

5. Using Taylor series, derive the error term for the approximation

$$f'(x) \approx \frac{1}{2h}[-3f(x) + 4f(x+h) - f(x+2h)].$$

6. Consider a forward-difference approximation for the second derivative of the form

$$f''(x) \approx Af(x) + Bf(x+h) + Cf(x+2h).$$

Use Taylor's theorem to determine the coefficients A, B, and C that give the maximal order of accuracy and determine what this order is.

7. Suppose you are given the values of f and f' at points $x_0 + h$ and $x_0 - h$ and you wish to approximate $f'(x_0)$. Find coefficients α and β that make the following approximation accurate to $O(h^4)$:

$$f'(x_0) \approx \alpha \frac{f'(x_0+h) + f'(x_0-h)}{2} + \beta \frac{f(x_0+h) - f(x_0-h)}{2h}.$$

Compute the coefficients by combining the Taylor series expansions of $f(x)$ and $f'(x)$ about the point x_0:

$$f(x) = f(x_0) + (x-x_0)f'(x_0) + \frac{(x-x_0)^2}{2!}f''(x_0) + \frac{(x-x_0)^3}{3!}f'''(x_0) + \frac{(x-x_0)^4}{4!}f^{(4)}(x_0) + \frac{(x-x_0)^5}{5!}f^{(5)}(c_1),$$

$$f'(x) = f'(x_0) + (x-x_0)f''(x_0) + \frac{(x-x_0)^2}{2!}f'''(x_0) + \frac{(x-x_0)^3}{3!}f^{(4)}(x_0) + \frac{(x-x_0)^4}{4!}f^{(5)}(c_2).$$

[Hint: Combine the Taylor expansions into $(f(x_0+h) - f(x_0-h))$ and $(f'(x_0+h) + f'(x_0-h))$ and then combine these two to cancel the leading order error term (in this case $O(h^2)$).]

10

NUMERICAL INTEGRATION

Many integrals of seemingly simple functions cannot be evaluated analytically; for example,

$$\int e^{-x^2}\,dx.$$

This integral is so important in probability and statistics that it has been given a special name—the **error function** or **erf**:

$$\text{erf}(x) \equiv \frac{2}{\sqrt{\pi}} \int_0^x e^{-t^2}\,dt,$$

and tables of values of the function have been compiled. While the indefinite integral cannot be expressed in terms of elementary algebraic and transcendental functions, the definite integral can certainly be approximated numerically. To handle such problems we need good numerical methods for the approximation of definite integrals. The formulas for doing this are often called **quadrature** formulas. The name comes from the Latin word *quadratus* meaning square, as ancient mathematicians attempted to find the area under a curve by finding a square with approximately the same area.

You probably learned some quadrature rules in a calculus course—the trapezoid rule and Simpson's rule for example. These can be thought of as low-order *Newton–Cotes formulas*, which will be described in section 10.1. Usually, one approximates the area under a curve by the sum of areas of trapezoids or other shapes that intersect the curve at certain points. These are examples of *composite* rules, which will be described in section 10.2. More sophisticated quadrature formulas can be derived by interpolating the integrand at carefully chosen points with a polynomial and integrating the polynomial exactly. In section 10.3 we describe Gauss quadrature and in section 10.4 Clenshaw–Curtis quadrature.

10.1 NEWTON–COTES FORMULAS

One way to approximate $\int_a^b f(x)\,dx$ is to replace f by its polynomial interpolant and integrate the polynomial. Writing the polynomial interpolant

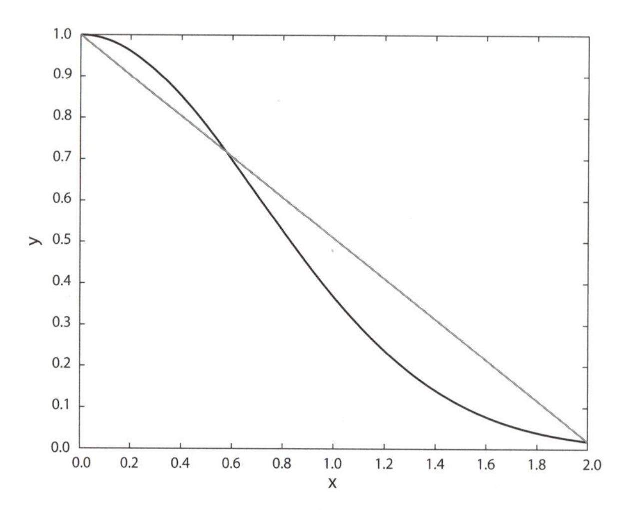

Figure 10.1. The area under a curve from a to b is approximated by the area of a trapezoid.

in the Lagrange form, this becomes

$$\int_a^b f(x)\,dx \approx \sum_{i=0}^{n} f(x_i) \int_a^b \left(\prod_{\substack{j=0 \\ j \neq i}}^{n} \frac{x - x_j}{x_i - x_j} \right) dx.$$

Formulas derived in this way, using equally-spaced nodes, are called **Newton–Cotes formulas**, named for Isaac Newton and Roger Cotes.

For $n = 1$, the Newton–Cotes formula is the **trapezoid rule.** Since the first-degree polynomial interpolating f at a and b is

$$p_1(x) = f(a)\frac{x-b}{a-b} + f(b)\frac{x-a}{b-a},$$

when we integrate $p_1(x)$ we obtain

$$\begin{aligned}\int_a^b f(x)\,dx \approx \int_a^b p_1(x)\,dx &= \frac{f(a)}{a-b}\frac{(x-b)^2}{2}\bigg|_a^b + \frac{f(b)}{b-a}\frac{(x-a)^2}{2}\bigg|_a^b \\ &= \frac{b-a}{2}[f(a)+f(b)].\end{aligned}$$

This is the area of the trapezoid with height $f(a)$ at a and $f(b)$ at b, as pictured in figure 10.1.

To find the error in the Newton–Cotes formulas, we use equation (8.11) for the difference between $f(x)$ and $p(x)$ and integrate over x:

$$\int_a^b f(x)\,dx - \int_a^b p(x)\,dx = \frac{1}{(n+1)!}\int_a^b f^{(n+1)}(\xi_x)\left(\prod_{\ell=0}^{n}(x - x_\ell)\right) dx.$$

ROGER COTES

Roger Cotes (1682–1716) worked with Newton in the preparation of the second edition of the *Principia*. His treatise on Newton's differential method was published posthumously. He comments that his theorems were developed prior to reading Newton's text. While giving little more explanation for the method, he provides formulas for numerically integrating with 3 to 11 equally-spaced points.

For the trapezoid rule ($n = 1$), this becomes

$$\begin{aligned}\int_a^b f(x)\,dx - \int_a^b p_1(x)\,dx &= \frac{1}{2}\int_a^b f''(\xi_x)(x-a)(x-b)\,dx \\ &= \frac{1}{2}f''(\eta)\int_a^b (x-a)(x-b)\,dx, \quad \eta \in [a,b] \\ &= -\frac{1}{12}(b-a)^3 f''(\eta). \qquad (10.1)\end{aligned}$$

Note that the second line is justified by the Mean Value Theorem for integrals, since $(x-a)(x-b)$ has the same sign (i.e., negative) throughout the interval $[a, b]$.

Example 10.1.1. Using the trapezoid rule to approximate $\int_0^2 e^{-x^2}\,dx$, we find

$$\int_0^2 e^{-x^2}\,dx \approx e^0 + e^{-4} \approx 1.0183.$$

The error in this approximation is $-\frac{2}{3}f''(\eta)$, for some $\eta \in [0, 2]$. In this case, $f''(x) = (4x^2 - 2)e^{-x^2}$, and the absolute value of f'' is maximal on the interval at $x = 0$, where $|f''(0)| = 2$. Therefore the absolute error in this approximation is less than or equal to $4/3$.

What is the Newton–Cotes formula for $n = 2$? Using the interpolation points $x_0 = a$, $x_1 = (a+b)/2$, and $x_2 = b$, we can write the second-degree interpolation polynomial in the form

$$\begin{aligned}p_2(x) = {} & f(a)\frac{(x-(a+b)/2)(x-b)}{(a-(a+b)/2)(a-b)} + f\left(\frac{a+b}{2}\right)\frac{(x-a)(x-b)}{((a+b)/2-a)((a+b)/2-b)} \\ & + f(b)\frac{(x-a)(x-(a+b)/2)}{(b-a)(b-(a+b)/2)},\end{aligned}$$

and then the formula is determined by integrating each of these quadratic terms. This can be messy.

Another way to determine the formula is called the **method of undetermined coefficients.** We know that the formula must be exact for constants, linear functions, and quadratic functions. Moreover, it is the *only* formula of the form

$$\int_a^b f(x)\,dx \approx A_1 f(a) + A_2 f\left(\frac{a+b}{2}\right) + A_3 f(b)$$

that is exact for all polynomials of degree 2 or less. This follows from the fact that the formula must integrate each of the quadratic terms in the formula for $p_2(x)$ correctly. Thus, for instance, if we define $\varphi_1(x) \equiv (x - (a+b)/2)(x-b)/[(a-(a+b)/2)(a-b)]$, then

$$\int_a^b \varphi_1(x)\,dx = A_1\varphi_1(a) + A_2\varphi_1\left(\frac{a+b}{2}\right) + A_3\varphi_1(b) \Longrightarrow$$

$$A_1 = \int_a^b \varphi_1(x)\,dx.$$

Similar formulas hold for A_2 and A_3. To determine A_1, A_2, and A_3, without evaluating these complicated integrals, however, note that

$$\int_a^b 1\,dx = b-a \Rightarrow A_1 + A_2 + A_3 = b - a,$$

$$\int_a^b x\,dx = \frac{b^2-a^2}{2} \Rightarrow A_1 a + A_2\frac{a+b}{2} + A_3 b = \frac{b^2-a^2}{2},$$

$$\int_a^b x^2\,dx = \frac{b^3-a^3}{3} \Rightarrow A_1 a^2 + A_2\left(\frac{a+b}{2}\right)^2 + A_3 b^2 = \frac{b^3-a^3}{3}.$$

This gives a system of 3 linear equations for the 3 unknowns A_1, A_2, and A_3. Solving these linear equations, we find

$$A_1 = A_3 = \frac{b-a}{6}, \quad A_2 = \frac{4(b-a)}{6}.$$

This gives *Simpson's rule*:

$$\int_a^b f(x)\,dx \approx \frac{b-a}{6}\left[f(a) + 4f\left(\frac{a+b}{2}\right) + f(b)\right]. \tag{10.2}$$

This is the area under the quadratic that interpolates f at a, b, and $(a+b)/2$, as pictured in figure 10.2.

From the way it was derived, it is clear that Simpson's rule is exact for all polynomials of degree 2 or less. It turns out that it is also exact for polynomials of degree 3. To see this, note that

$$\int_a^b x^3\,dx = \frac{1}{4}(b^4 - a^4) = \frac{1}{4}(b-a)(b^3 + ab^2 + a^2 b + a^3),$$

and

$$\frac{b-a}{6}\left[a^3 + 4\left(\frac{a+b}{2}\right)^3 + b^3\right] = \frac{1}{4}(b-a)(b^3 + ab^2 + a^2 b + a^3).$$

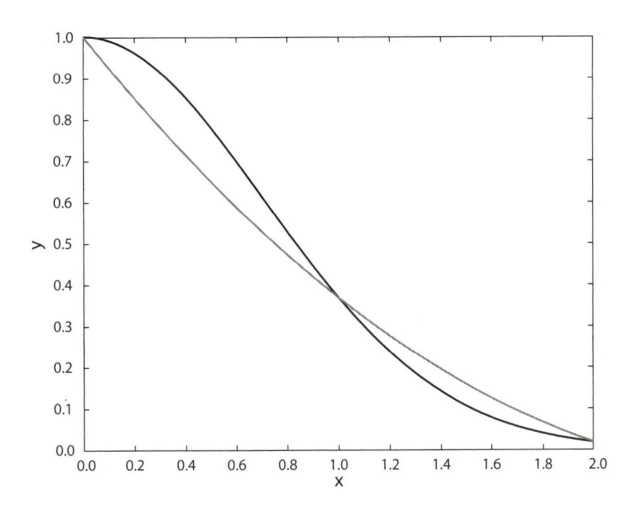

Figure 10.2. The area under a curve from a to b is approximated by the area under a quadratic.

Instead of using (8.11) to analyze the error in Simpson's rule, we will use a Taylor series expansion of both the integral and its approximation about the point a, and see how many terms match. Expanding $f((a+b)/2)$ and $f(b)$ about a, we find

$$f\left(\frac{a+b}{2}\right) = f(a) + \frac{b-a}{2} f'(a) + \frac{(b-a)^2}{8} f''(a) + \frac{(b-a)^3}{48} f'''(a) + \frac{(b-a)^4}{384} f^{(4)}(a) + \dots,$$

$$f(b) = f(a) + (b-a) f'(a) + \frac{(b-a)^2}{2} f''(a) + \frac{(b-a)^3}{6} f'''(a) + \frac{(b-a)^4}{24} f^{(4)}(a) + \dots.$$

From this it follows that the right-hand side of (10.2) satisfies

$$\frac{b-a}{6}\left[f(a) + 4f\left(\frac{a+b}{2}\right) + f(b)\right] = (b-a) f(a) + \frac{(b-a)^2}{2} f'(a) + \frac{(b-a)^3}{6} f''(a) + \frac{(b-a)^4}{24} f'''(a) + \frac{5(b-a)^5}{576} f^{(4)}(a) + \dots. \quad (10.3)$$

Next define $F(x) \equiv \int_a^x f(t)\,dt$, and note, by the Fundamental Theorem of Calculus, that $F'(x) = f(x)$. Expanding $F(b)$ in a Taylor series about a, and using the fact that $F(a) = 0$, we find

$$F(b) = (b-a)f(a) + \frac{(b-a)^2}{2} f'(a) + \frac{(b-a)^3}{6} f''(a) + \frac{(b-a)^4}{24} f'''(a)$$
$$+ \frac{(b-a)^5}{120} f^{(4)}(a) + \dots. \tag{10.4}$$

Comparing (10.3) and (10.4), we see that the error in Simpson's rule is $(1/2880)(b-a)^5 f^{(4)}(a) + O((b-a)^6)$. It can be shown that this error can be written as $(1/2880)(b-a)^5 f^{(4)}(\xi)$ for some $\xi \in [a, b]$.

Example 10.1.2. Using Simpson's rule to approximate $\int_0^2 e^{-x^2}\,dx$, we find

$$\int_0^2 e^{-x^2}\,dx \approx \frac{1}{3}\left[e^0 + 4e^{-1} + e^{-4}\right] \approx 0.8299.$$

The error is $(2^5/2880) f^{(4)}(\xi)$, for some $\xi \in [0, 2]$. A tedious calculation shows that $|f^{(4)}(x)| \le 12$ for $x \in [0, 2]$, and so the error in this approximation is less than or equal to 0.1333.

One could obtain even higher-order accuracy using higher-degree interpolation polynomials but this is not recommended when the interpolation points are equally spaced; it is usually better to use piecewise polynomials instead.

10.2 FORMULAS BASED ON PIECEWISE POLYNOMIAL INTERPOLATION

The **composite trapezoid rule** is obtained by dividing the interval $[a, b]$ into subintervals and using the trapezoid rule to approximate the integral over each subinterval, as pictured in figure 10.3.

Here we use the approximation

$$\int_a^b f(x)\,dx = \sum_{i=1}^n \int_{x_{i-1}}^{x_i} f(x)\,dx \approx \sum_{i=1}^n \int_{x_{i-1}}^{x_i} p_{1,i}(x)\,dx,$$

where $p_{1,i}(x)$ is the linear interpolant of f over the subinterval $[x_{i-1}, x_i]$, and

$$\int_{x_{i-1}}^{x_i} p_{1,i}(x) = f(x_{i-1}) \int_{x_{i-1}}^{x_i} \frac{x - x_i}{x_{i-1} - x_i}\,dx + f(x_i) \int_{x_{i-1}}^{x_i} \frac{x - x_{i-1}}{x_i - x_{i-1}}\,dx = (x_i - x_{i-1})$$
$$\times \frac{f(x_i) + f(x_{i-1})}{2}.$$

Thus we have the formula

$$\int_a^b f(x)\,dx \approx \sum_{i=1}^n (x_i - x_{i-1}) \frac{f(x_i) + f(x_{i-1})}{2}.$$

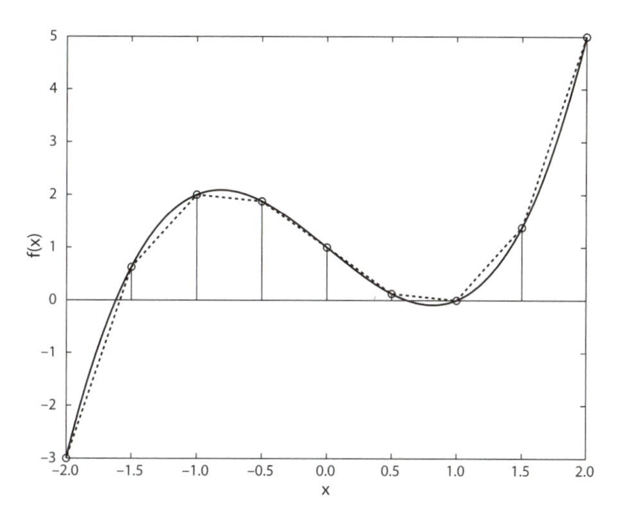

Figure 10.3. The area under a curve from a to b is approximated using the composite trapezoid rule.

With a uniform mesh spacing, $x_i - x_{i-1} = h$ for all i, this becomes

$$\int_a^b f(x)\,dx \approx \frac{h}{2}\sum_{i=1}^{n}(f(x_i)+f(x_{i-1})) = \frac{h}{2}[f_0+2f_1+\ldots+2f_{n-1}+f_n], \quad (10.5)$$

where we have again used the symbol f_j for $f(xj)$.

From (10.1) it follows that the error over each subinterval is $O(h^3)$ for the composite trapezoid rule; more precisely, for the ith subinterval $[x_{i-1}, x_i]$, the error is $-\frac{1}{12}h^3 f''(\xi_i)$ for some point $\xi_i \in [x_{i-1}, x_i]$. Summing the errors over all of the $(b-a)/h$ subintervals, we have a total error of $O(h^2)$.

Example 10.2.1. Using the composite trapezoid rule to approximate $\int_0^2 e^{-x^2}\,dx$, and letting T_h denote the composite trapezoid rule approximation with spacing h, we obtain the results in table 10.1. Note that the error is reduced by about a factor of 4, each time h is cut in half.

TABLE 10.1
Results of using the composite trapezoid rule to approximate $\int_0^2 e^{-x^2}\,dx$.

h	T_h	Error
1.0000	0.877037	5.0×10^{-3}
0.5000	0.880619	1.5×10^{-3}
0.2500	0.881704	3.8×10^{-4}
0.1250	0.881986	9.5×10^{-5}
0.0625	0.882058	2.4×10^{-5}

The **composite Simpson's rule** is derived similarly. Using Simpson's rule to approximate the integral over each subinterval, where we interpolate the function f at the midpoint as well as the endpoints of each subinterval, we obtain the formula

$$\int_a^b f(x)\,dx \approx \sum_{i=1}^{n} \frac{x_i - x_{i-1}}{6}\left[f(x_{i-1}) + 4f\left(\frac{x_i + x_{i-1}}{2}\right) + f(x_i)\right].$$

Simpson's Rule

Simpson's rule is named after the English mathematician Thomas Simpson (*bottom*). Simpson published textbooks on algebra, geometry, and trigonometry, all of which went through multiple editions [3]. Among his accomplishments was not the invention of Simpson's rule, however. The method was already well known and had been published by James Gregory (*top*), a Scottish mathematician who died 35 years prior to Simpson's birth. Simpson, however, did contribute a (geometric) proof of the convergence of the composite form of the method [106].

TABLE 10.2
Results of using the composite Simpson's rule to approximate $\int_0^2 e^{-x^2}\,dx$.

h	S_h	Error
1.0000	0.88181243	2.7×10^{-4}
0.5000	0.88206551	1.6×10^{-5}
0.2500	0.88208040	9.9×10^{-7}
0.1250	0.88208133	6.2×10^{-8}
0.0625	0.88208139	3.9×10^{-9}

For a uniform mesh spacing $x_i - x_{i-1} = h$, for all $i = 1, 2, \ldots, n$, this becomes

$$\int_a^b f(x)\,dx \approx \frac{h}{6}[f_0 + 4f_{1/2} + 2f_1 + \ldots + 2f_{n-1} + 4f_{n-1/2} + f_n]. \tag{10.6}$$

Since it was shown, using (10.3) and (10.4), that the error in Simpson's rule for the interval $[a, b]$ is $O((b-a)^5)$, it follows that the error over each subinterval in the composite Simpson's rule is $O(h^5)$ and hence the total error is $O(h^4)$.

Example 10.2.2. Using the composite Simpson's rule to approximate $\int_0^2 e^{-x^2}\,dx$, and letting S_h denote the composite Simpson's rule approximation with spacing h, we obtain the results in table 10.2. Note that the error is reduced by about a factor of 16, each time h is cut in half.

Table 10.3 summarizes the formulas and error terms in the basic and composite trapezoid rule and Simpson's rule, using a uniform mesh spacing $h = (b-a)/n$.

10.3 GAUSS QUADRATURE

Suppose we wish to approximate $\int_a^b f(x)\,dx$ by a sum of the form

$$\sum_{i=0}^{n} A_i f(x_i).$$

TABLE 10.3
Quadrature formulas and their errors.

Method	Approximation to $\int_a^b f(x)\,dx$	Error
Trapezoid rule	$\frac{b-a}{2}[f(a)+f(b)]$	$-\frac{1}{12}(b-a)^3 f''(\eta),\ \eta \in [a,b]$
Simpson's rule	$\frac{b-a}{6}\left[f(a)+4f\left(\frac{a+b}{2}\right)+f(b)\right]$	$\frac{1}{2880}(b-a)^5 f^{(4)}(\xi),\ \xi \in [a,b]$
Composite trapezoid rule	$\frac{h}{2}[f_0+2f_1+\ldots+2f_{n-1}+f_n]$	$O(h^2)$
Composite Simpson's rule	$\frac{h}{6}[f_0+4f_{1/2}+2f_1+\ldots +2f_{n-1}+4f_{n-1/2}+f_n]$	$O(h^4)$

We have seen how, once the x_i's are fixed, we can choose the A_i's so that this formula integrates all polynomials of degree n or less exactly. Suppose we are also free to choose the nodes x_i. How can one choose both the A_i's and the x_i's to make this formula exact for as high-degree polynomials as possible? A formula in which the A_i's and x_i's are chosen in this way is called a **Gauss quadrature formula**. These methods were developed by Gauss in a communication of 1816 to the Göttingen Society in which Gauss began by re-proving the Cotes formulas.

Let us start by considering the case $n = 0$. Here we use the approximation

$$\int_a^b f(x)\,dx \approx A_0 f(x_0).$$

In order that this formula be exact for constants, it must be the case that

$$\int_a^b 1\,dx = b - a = A_0 \Rightarrow A_0 = b - a.$$

In order that the formula be exact for polynomials of degree 1 as well, it must be the case that

$$\int_a^b x\,dx = \frac{b^2 - a^2}{2} = (b - a)x_0 \Rightarrow x_0 = \frac{a + b}{2}.$$

Choosing A_0 and x_0 in this way, the formula will not be exact for polynomials of degree 2, since

$$\int_a^b x^2\,dx = \frac{b^3 - a^3}{3} \neq (b - a)\left(\frac{a + b}{2}\right)^2.$$

Thus the one-point Gauss quadrature formula,

$$\int_a^b f(x)\,dx \approx (b - a) f\left(\frac{a + b}{2}\right),$$

is exact for polynomials of degree 1 or less, but not for higher-degree polynomials.

If we try the same approach for $n = 1$, we quickly become stuck. For $n = 1$, we use an approximation of the form

$$\int_a^b f(x)\,dx \approx A_0 f(x_0) + A_1 f(x_1).$$

The requirements are

$$\int_a^b 1\,dx = b - a \Rightarrow A_0 + A_1 = b - a,$$

$$\int_a^b x\,dx = \frac{b^2 - a^2}{2} \Rightarrow A_0 x_0 + A_1 x_1 = \frac{b^2 - a^2}{2},$$

$$\int_a^b x^2\,dx = \frac{b^3 - a^3}{3} \Rightarrow A_0 x_0^2 + A_1 x_1^2 = \frac{b^3 - a^3}{3},$$

$$\vdots$$

It is difficult to know whether this system of nonlinear equations has a solution, or how many, if any, additional equations can be added.

We will see that an $(n + 1)$-point Gauss quadrature formula is exact for polynomials of degree $2n + 1$ or less. The approach that we take is quite different from that above, however.

10.3.1 Orthogonal Polynomials

Definition. Two polynomials p and q are said to be **orthogonal** on the interval $[a, b]$ if

$$\langle p, q\rangle \equiv \int_a^b p(x)q(x)\,dx = 0.$$

They are said to be **orthonormal** if, in addition, $\langle p, p\rangle = \langle q, q\rangle = 1$.

From the linearly independent polynomials $1,\ x,\ x^2, \ldots,$ one can construct an orthonormal set $q_0,\ q_1,\ q_2, \ldots,$ using the *Gram–Schmidt algorithm* (just as was done for vectors in section 7.6.2):

Set $q_0 = 1/[\int_a^b 1^2\,dx]^{1/2} = 1/\sqrt{b-a}$. For $j = 1, 2, \ldots,$

Set $\tilde{q}_j(x) = xq_{j-1}(x) - \sum_{i=0}^{j-1} \langle xq_{j-1}(x), q_i(x)\rangle q_i(x)$,

where $\langle xq_{j-1}(x), q_i(x)\rangle \equiv \int_a^b xq_{j-1}(x)q_i(x)\,dx$.

Set $q_j(x) = \tilde{q}_j(x)/\|\tilde{q}_j(x)\|$, where $\|\tilde{q}_j(x)\| \equiv [\int_a^b (\tilde{q}_j(x))^2\,dx]^{1/2}$.

Note that since q_{j-1} is orthogonal to all polynomials of degree $j-2$ or less, $\langle xq_{j-1}(x), q_i(x)\rangle = \langle q_{j-1}(x), xq_i(x)\rangle = 0$, if $i \le j - 3$. Thus the formula for $\tilde{q}_j$ can be replaced by

$$\tilde{q}_j(x) = xq_{j-1}(x) - \langle xq_{j-1}(x), q_{j-1}(x)\rangle q_{j-1}(x) - \langle xq_{j-1}(x), q_{j-2}(x)\rangle q_{j-2}(x),$$

and the orthonormal polynomials satisfy a *three-term recurrence*; that is, q_j is obtained from q_{j-1} and q_{j-2}.

The following theorem shows the relation between orthogonal polynomials and Gauss quadrature formulas.

Theorem 10.3.1. If $x_0, x_1, \ldots, x_n$ are the zeros of $q_{n+1}(x)$, the $(n+1)$st orthogonal polynomial on $[a, b]$, then the formula

$$\int_a^b f(x)\,dx \approx \sum_{i=0}^{n} A_i f(x_i), \tag{10.7}$$

where

$$A_i = \int_a^b \varphi_i(x)\,dx, \quad \varphi_i(x) \equiv \prod_{\substack{j=0 \\ j\neq i}}^{n} \frac{x - x_j}{x_i - x_j}, \tag{10.8}$$

is exact for polynomials of degree $2n+1$ or less.

Proof. Suppose f is a polynomial of degree $2n+1$ or less. If we divide f by q_{n+1}, using ordinary long division for polynomials, then the result will be a polynomial p_n of degree n and a remainder term r_n of degree n; that is, we can write f in the form $f = q_{n+1} p_n + r_n$ for certain nth-degree polynomials p_n and r_n. Then $f(x_i) = r_n(x_i)$, $i = 0, 1, \ldots, n$, since $q_{n+1}(x_i) = 0$. Integrating f, we find

$$\int_a^b f(x)\,dx = \int_a^b q_{n+1}(x) p_n(x)\,dx + \int_a^b r_n(x)\,dx = \int_a^b r_n(x)\,dx,$$

since q_{n+1} is orthogonal to polynomials of degree n or less.

By the choice of the A_i's, it follows that formula (10.7) is exact for polynomials of degree n or less. Hence

$$\int_a^b f(x)\,dx = \int_a^b r_n(x)\,dx = \sum_{i=0}^{n} A_i r_n(x_i) = \sum_{i=0}^{n} A_i f(x_i).$$

□

Example 10.3.1. Let $[a, b] = [-1, 1]$, and find the Gauss quadrature formula for $n = 1$.

We are looking for a formula of the form

$$\int_{-1}^{1} f(x)\,dx \approx A_0 f(x_0) + A_1 f(x_1),$$

which, according to the theorem, must be exact for polynomials of degree 3 or less.

We start by constructing orthogonal polynomials $\tilde{q}_0$, $\tilde{q}_1$, $\tilde{q}_2$ on $[-1, 1]$. (These polynomials are called **Legendre polynomials.**) The roots of $\tilde{q}_2$ will be the nodes x_0 and x_1. Note that we do not need to normalize these polynomials; their roots will be the same regardless of normalization. We need only use the correct formulas for orthogonalizing, when the polynomials are not normalized. When doing things by hand, the algebra is often easier without normalization. Thus, let $\tilde{q}_0 = 1$. Then

$$\tilde{q}_1(x) = x - \frac{\langle x, 1 \rangle}{\langle 1, 1 \rangle} \cdot 1 = x - \frac{\int_{-1}^{1} x\, dx}{\int_{-1}^{1} 1\, dx} \cdot 1 = x,$$

and

$$\begin{aligned}
\tilde{q}_2(x) &= x^2 - \frac{\langle x^2, 1 \rangle}{\langle 1, 1 \rangle} \cdot 1 - \frac{\langle x^2, x \rangle}{\langle x, x \rangle} \cdot x \\
&= x^2 - \frac{\int_{-1}^{1} x^2\, dx}{\int_{-1}^{1} 1\, dx} \cdot 1 - \frac{\int_{-1}^{1} x^3\, dx}{\int_{-1}^{1} x^2\, dx} \cdot x \\
&= x^2 - \frac{1}{3}.
\end{aligned}$$

The roots of $\tilde{q}_2$ are $\pm\frac{1}{\sqrt{3}}$.

Thus our formula will have the form

$$\int_{-1}^{1} f(x)\, dx \approx A_0 f\left(-\frac{1}{\sqrt{3}}\right) + A_1 f\left(\frac{1}{\sqrt{3}}\right).$$

We can use the method of undetermined coefficients to find A_0 and A_1:

$$\int_{-1}^{1} 1\, dx = 2 = A_0 + A_1, \quad \int_{-1}^{1} x\, dx = 0 = A_0\left(-\tfrac{1}{\sqrt{3}}\right) + A_1\left(\tfrac{1}{\sqrt{3}}\right) \Longrightarrow$$

$$A_0 = A_1 = 1.$$

Although the coefficients A_0 and A_1 were determined by forcing the formula to be exact only for polynomials of degree 0 and 1, we know, because of the way the nodes were chosen, that the formula will actually be exact for polynomials of degree 3 or less. To check this, we note that

$$\int_{-1}^{1} x^2\, dx = \frac{2}{3} = \left(-\frac{1}{\sqrt{3}}\right)^2 + \left(\frac{1}{\sqrt{3}}\right)^2,$$

$$\int_{-1}^{1} x^3\, dx = 0 = \left(-\frac{1}{\sqrt{3}}\right)^3 + \left(\frac{1}{\sqrt{3}}\right)^3,$$

$$\int_{-1}^{1} x^4\, dx = \frac{2}{5} \neq \left(-\frac{1}{\sqrt{3}}\right)^4 + \left(\frac{1}{\sqrt{3}}\right)^4 = \frac{2}{9}.$$

One can also discuss **weighted orthogonal polynomials**. Given a nonnegative weight function w on an interval $[a, b]$, one can define the **weighted inner product** of two polynomials p and q to be

$$\langle p, q \rangle_w \equiv \int_a^b p(x)q(x)w(x)\, dx.$$

The polynomials p and q are said to be orthogonal with respect to the weight function w, or w-orthogonal, if $\langle p, q\rangle_w = 0$.

Suppose one wishes to approximate $\int_a^b f(x)w(x)\,dx$. Then the roots of the $(n+1)$st w-orthogonal polynomial can be used as nodes in an $(n+1)$-point **weighted Gauss quadrature formula** that will be exact for polynomials of degree $2n+1$ or less, just as was shown previously for the case $w(x) \equiv 1$.

Theorem 10.3.2. If w is a nonnegative weight function on $[a, b]$ and if $x_0, x_1, \ldots, x_n$ are the zeros of the $(n+1)$st w-orthogonal polynomial on $[a, b]$, then the formula

$$\int_a^b f(x)w(x)\,dx \approx \sum_{i=0}^{n} A_i f(x_i), \tag{10.9}$$

where

$$A_i = \int_a^b \varphi_i(x)w(x)\,dx, \quad \varphi_i(x) \equiv \prod_{\substack{j=0 \\ j\neq i}}^{n} \frac{x - x_j}{x_i - x_j}, \tag{10.10}$$

is exact for polynomials of degree $2n+1$ or less.

Example 10.3.2. Let $[a, b] = [-1, 1]$, and find the weighted Gauss quadrature formula for $n = 1$, with weight function $w(x) = (1-x^2)^{-1/2}$.

We are looking for a formula of the form

$$\int_{-1}^{1} f(x)w(x)\,dx \approx A_0 f(x_0) + A_1 f(x_1),$$

which, according to the theorem, must be exact for polynomials of degree 3 or less.

We start by constructing w-orthogonal polynomials $\tilde{q}_0$, $\tilde{q}_1$, $\tilde{q}_2$ on $[-1, 1]$. (These polynomials are called **Chebyshev polynomials.** We will see later that they are the same Chebyshev polynomials defined earlier by the formula $T_j(x) = \cos(j \arccos x)$.) The roots of $\tilde{q}_2$ will be the nodes x_0 and x_1. Let $\tilde{q}_0 = 1$; then

$$\tilde{q}_1(x) = x - \frac{\langle x, 1\rangle_w}{\langle 1, 1\rangle_w} \cdot 1 = x - \frac{\int_{-1}^{1} x(1-x^2)^{-1/2}\,dx}{\int_{-1}^{1}(1-x^2)^{-1/2}\,dx} \cdot 1 = x,$$

and

$$\begin{aligned}\tilde{q}_2(x) &= x^2 - \frac{\langle x^2, 1\rangle_w}{\langle 1, 1\rangle_w} \cdot 1 - \frac{\langle x^2, x\rangle_w}{\langle x, x\rangle_w} \cdot x \\ &= x^2 - \frac{\int_{-1}^{1} x^2(1-x^2)^{-1/2}\,dx}{\int_{-1}^{1}(1-x^2)^{-1/2}\,dx} \cdot 1 - \frac{\int_{-1}^{1} x^3(1-x^2)^{-1/2}\,dx}{\int_{-1}^{1} x^2(1-x^2)^{-1/2}\,dx} \cdot x \\ &= x^2 - \frac{\pi/2}{\pi} \cdot 1 - \frac{0}{\pi/2} \cdot x = x^2 - \frac{1}{2}.\end{aligned}$$

The roots of $\tilde{q}_2$ are $\pm\frac{1}{\sqrt{2}}$.

Thus our formula will have the form

$$\int_{-1}^{1} f(x)(1-x^2)^{-1/2}\,dx \approx A_0 f\left(-\frac{1}{\sqrt{2}}\right) + A_1 f\left(\frac{1}{\sqrt{2}}\right).$$

We can use the method of undetermined coefficients to find A_0 and A_1:

$$\int_{-1}^{1} (1-x^2)^{-1/2}\,dx = \pi = A_0 + A_1, \quad \int_{-1}^{1} x(1-x^2)^{-1/2}\,dx = 0$$

$$= A_0\left(-\frac{1}{\sqrt{2}}\right) + A_1\left(\frac{1}{\sqrt{2}}\right) \Longrightarrow$$

$$A_0 = A_1 = \frac{\pi}{2}.$$

Although the coefficients A_0 and A_1 were determined by forcing the formula to be exact only for polynomials of degree 0 and 1, we know, because of the way the nodes were chosen, that the formula will actually be exact for polynomials of degree 3 or less. To check this, we note that

$$\int_{-1}^{1} x^2(1-x^2)^{-1/2}\,dx = \frac{\pi}{2} = \frac{\pi}{2}\left(-\frac{1}{\sqrt{2}}\right)^2 + \frac{\pi}{2}\left(\frac{1}{\sqrt{2}}\right)^2,$$

$$\int_{-1}^{1} x^3(1-x^2)^{-1/2}\,dx = 0 = \frac{\pi}{2}\left(-\frac{1}{\sqrt{2}}\right)^3 + \frac{\pi}{2}\left(\frac{1}{\sqrt{2}}\right)^3,$$

$$\int_{-1}^{1} x^4(1-x^2)^{-1/2}\,dx = \frac{3\pi}{8} \neq \frac{\pi}{2}\left(-\frac{1}{\sqrt{2}}\right)^4 + \frac{\pi}{2}\left(\frac{1}{\sqrt{2}}\right)^4 = \frac{\pi}{4}.$$

10.4 CLENSHAW–CURTIS QUADRATURE

In the Newton–Cotes quadrature rules, a polynomial that interpolates the integrand at equally-spaced points is integrated exactly, in order to approximate the definite integral. In Gauss quadrature, the interpolation points are chosen so that the resulting formula is exact for polynomials of as high degree as possible—degree $2n+1$ using $n+1$ Gauss quadrature points. The *Clenshaw–Curtis quadrature formula* is based on the interpolating polynomial at the Chebyshev points (8.15) and offers advantages over each of the other two.

We saw earlier that interpolating a function with a high-degree polynomial that matches the function at equally-spaced points is *not* a good idea, as illustrated by the Runge example in section 8.4. Interpolating at the Chebyshev points did a much better job, as illustrated in section 8.5. Hence, one might expect that integrating the polynomial that interpolates the function at the Chebyshev points would provide a better approximation to the actual integral than is produced by the corresponding Newton–Cotes formula. This is, indeed, the case. While the Gauss quadrature formula is *optimal* in the sense of integrating exactly polynomials of the highest possible degree, if it could be replaced by a method that required less work and produced similarly accurate

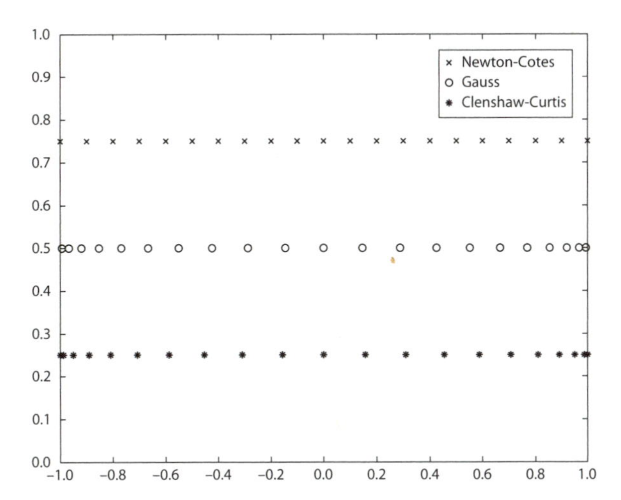

Figure 10.4. Interpolation points that are used for 20-point Newton–Cotes, Gauss, and Clenshaw–Curtis quadrature on the interval $[-1, 1]$.

results, even if the degree of the polynomials that were integrated exactly decreased, then this would be an improvement. The *Clenshaw–Curtis formula* accomplishes this. While the $(n + 1)$-point Clenshaw–Curtis formula is exact only for polynomials of degree n or less, it can be shown to provide a level of accuracy close to that of the corresponding Gauss quadrature formula with significantly less work [95]. While the best methods for computing Gauss quadrature points and weights require $O(n^2)$ operations, the Clenshaw–Curtis formula can be implemented using the FFT (to be described in section 14.5.1) with only $O(n \log n)$ operations.

For comparison, we illustrate in figure 10.4 the interpolation points that are used for 20-point Newton–Cotes, Gauss, and Clenshaw–Curtis quadrature on the interval $[-1, 1]$. Note that both the Gauss quadrature points and those of Clenshaw–Curtis are clustered towards the endpoints of the interval.

Example 10.4.1. Using the `chebfun` package in MATLAB to approximate $\int_0^2 e^{-x^2}\, dx$, we type

```
f = chebfun('exp(-x.^2)',[0,2]), intf = sum(f)
```

The package informs us that the length of `f` is 25, indicating that it has approximated the integrand with a polynomial of degree 24 that matches the integrand at 25 shifted and scaled Chebyshev points in the interval $[0, 2]$. [Note that to translate the Chebyshev points, which are defined on $[-1, 1]$, to an interval $[a, b]$, one uses the transformation $x \rightarrow \frac{b+a}{2} + \frac{b-a}{2}x$. In this case, then, each $x_j = \cos(j\pi/n)$ is replaced by $1 + \cos(j\pi/n)$, $j = 0, \ldots, n$.] Integration in the `chebfun` package is done using the `sum` command. (Just as the `sum` command in MATLAB sums the entries in a vector, the `sum` command in `chebfun` "sums" (i.e., integrates) the values of a function.) It returns the value `intf = 0.882081390762422`, which is accurate to the machine precision.

10.5 ROMBERG INTEGRATION

The idea of *Richardson extrapolation* described in the previous chapter can be applied to the approximation of integrals, as well as derivatives. Consider again the composite trapezoid rule (10.5). The error in this approximation is $O(h^2)$; that is, it is of the form Ch^2 plus higher powers of h. We will show later that if $f \in C^\infty$, then the error term consists only of *even* powers of h, and thus

$$\int_a^b f(x)\,dx = \frac{h}{2}[f_0 + 2f_1 + \ldots + 2f_{n-1} + f_n] + Ch^2 + O(h^4).$$

Define $T_h \equiv \frac{h}{2}[f_0+2f_1+\ldots+2f_{n-1}+f_n]$, and consider the two approximations T_h and $T_{h/2}$. Since the error term for $T_{h/2}$ is $C(h/2)^2 + O(h^4)$, that is,

$$\int_a^b f(x)\,dx = \frac{h}{4}[f_0 + 2f_{1/2} + 2f_1 + \ldots + 2f_{n-1} + 2f_{n-1/2} + f_n] + \frac{C}{4}h^2 + O(h^4),$$

we can eliminate the $O(h^2)$ error term by forming the combination $T_{h/2}^{(1)} \equiv (4/3)T_{h/2} - (1/3)T_h$. Thus

$$\begin{aligned}
\int_a^b f(x)\,dx &= \frac{4}{3}T_{h/2} - \frac{1}{3}T_h + O(h^4) \\
&= \frac{h}{3}[f_0 + 2f_{1/2} + 2f_1 + \ldots + 2f_{n-1} + 2f_{n-1/2} + f_n] \\
&\quad - \frac{h}{6}[f_0 + 2f_1 + \ldots + 2f_{n-1} + f_n] + O(h^4) \\
&= \frac{h}{6}[f_0 + 4f_{1/2} + 2f_1 + \ldots + 2f_{n-1} + 4f_{n-1/2} + f_n] + O(h^4).
\end{aligned}$$

Note that this is the composite Simpson's rule (10.6). Thus the composite trapezoid rule, with one step of Richardson extrapolation, gives the composite Simpson's rule.

As noted in Chapter 9, this process can be repeated to obtain even higher-order accuracy. Since the error in the composite trapezoid rule consists of only even powers of h, each step of Richardson extrapolation increases the order of the error by 2. Using the trapezoid rule with repeated application of Richardson extrapolation is called **Romberg integration**. Figure 10.5 shows a MATLAB routine for Romberg integration. It estimates the error and adjusts the number of subintervals to achieve a desired error tolerance.

This code requires fewer function evaluations to achieve a high level of accuracy for a very smooth function f than the standard composite trapezoid or Simpson's rule algorithms. Because it forms approximations with different values of h and different levels of extrapolation, the difference between these approximations provides a reasonable estimate of the error.

Example 10.5.1. Using the Romberg integration routine of figure 10.5 to approximate $\int_0^2 e^{-x^2}\,dx$, and setting `tol=1.e-9`, we obtain the approximation 0.88208139076230, for which the error is about 1.2e−13. Because the error estimate is somewhat pessimistic, the routine often obtains more accuracy than

```
function [q,cnt] = romberg(a, b, tol)

% Approximates the integral from a to b of f(x)dx to absolute tolerance
% tol by using the trapezoid rule with repeated Richardson extrapolation.
% Returns number of function evaluations in cnt.

% Make first estimate using one interval.

n = 1;  h = b-a;
fv = [f(a); f(b)];
est(1,1) = .5*h*(fv(1)+fv(2));
cnt = 2;                          % Count no. of function evaluations.

% Keep doubling the number of subintervals until desired tolerance is
% achieved or max no. of subintervals (2^10 = 1024) is reached.

err = 100*abs(tol);  k = 0;       % Initialize err to something > tol.
while err > tol & k < 10,
  k = k+1;  n = 2*n;  h = h/2;
  fvnew = zeros(n+1,1);           % Store computed values of f to reuse
  fvnew(1:2:n+1) = fv(1:n/2+1);   % when h is cut in half.
  for i=2:2:n,                    % Compute f at midpoints of previous intervals
    fvnew(i) = f(a+(i-1)*h);
  end;
  cnt = cnt + (n/2);              % Update no. of function evaluations.
  fv = fvnew;
  trap = .5*(fv(1)+fv(n+1));      % Use trapezoid rule with new h value
  for i=2:n,                      % to estimate integral.
    trap = trap + fv(i);
  end;
  est(k+1,1) = h*trap;          % Store new estimate in first column of tableau.

%   Perform Richardson extrapolations.
  for j=2:k+1,
    est(k+1,j) = ((4^(j-1))*est(k+1,j-1) - est(k,j-1))/(4^(j-1)-1);
  end;
  q = est(k+1,k+1);

%   Estimate error.
  err = max([abs(q - est(k,k)); abs(q-est(k+1,k))]);
end;
```

Figure 10.5. MATLAB code for Romberg integration.

is requested. The routine used 65 function evaluations. This is about the same number of function evaluations required by the composite Simpson's rule with $h = 0.0625$, where the error was 3.9e−9.

10.6 PERIODIC FUNCTIONS AND THE EULER–MACLAURIN FORMULA

It turns out that for certain functions the composite trapezoid rule is *far* more accurate than the $O(h^2)$ error estimate derived in section 10.2 would suggest.

This is established by the *Euler–Maclaurin formula*, proved independently by Leonhard Euler and Colin Maclaurin around 1735 [16]. Despite the fact that this result is quite old, many a numerical analyst today has been amazed to observe the level of accuracy achieved when using the composite trapezoid rule with equally-spaced nodes, for integrating periodic functions.

Let us further examine the error in the composite trapezoid rule:

$$\int_a^b f(x)\,dx = \sum_{i=1}^{n} \int_{x_{i-1}}^{x_i} f(x)\,dx \approx \sum_{i=1}^{n} (x_i - x_{i-1}) \frac{f(x_i) + f(x_{i-1})}{2}.$$

Expanding $f(x)$ in Taylor series about x_{i-1} and x_i, we find

$$\begin{aligned} f(x) = {} & f(x_{i-1}) + (x - x_{i-1}) f'(x_{i-1}) + \frac{(x - x_{i-1})^2}{2!} f''(x_{i-1}) + \frac{(x - x_{i-1})^3}{3!} \\ & \times f'''(x_{i-1}) + \frac{(x - x_{i-1})^4}{4!} f^{(4)}(x_{i-1}) + \ldots, \\ f(x) = {} & f(x_i) + (x - x_i) f'(x_i) + \frac{(x - x_i)^2}{2!} f''(x_i) + \frac{(x - x_i)^3}{3!} f'''(x_i) \\ & + \frac{(x - x_i)^4}{4!} f^{(4)}(x_i) + \ldots . \end{aligned}$$

Add one half times the first equation and one half times the second equation and integrate from x_{i-1} to x_i to obtain

$$\begin{aligned} \int_{x_{i-1}}^{x_i} f(x)\,dx = {} & (x_i - x_{i-1}) \frac{f(x_i) + f(x_{i-1})}{2} + \frac{(x_i - x_{i-1})^2}{4} (f'(x_{i-1}) - f'(x_i)) \\ & + \frac{(x_i - x_{i-1})^3}{12} (f''(x_{i-1}) + f''(x_i)) + \frac{(x_i - x_{i-1})^4}{48} (f'''(x_{i-1}) - f'''(x_i)) \\ & + \frac{(x_i - x_{i-1})^5}{240} (f^{(4)}(x_{i-1}) + f^{(4)}(x_i)) + \ldots . \end{aligned}$$

Now assume that $h = x_i - x_{i-1}$ is constant and sum from $i = 1$ to n:

$$\begin{aligned} \int_a^b f(x)\,dx = {} & \frac{h}{2} [f_0 + 2 f_1 + \ldots + 2 f_{n-1} + f_n] + \frac{h^2}{4} [f'(a) - f'(b)] \\ & + \frac{h^3}{12} [f_0'' + 2 f_1'' + \ldots + 2 f_{n-1}'' + f_n''] + \frac{h^4}{48} [f'''(a) - f'''(b)] \\ & + \frac{h^5}{240} [f_0^{(4)} + 2 f_1^{(4)} + \ldots + 2 f_{n-1}^{(4)} + f_n^{(4)}] + \ldots . \end{aligned} \tag{10.11}$$

Note that the error term $(h^3/12)[f_0'' + 2 f_1'' + \ldots + 2 f_{n-1}'' + f_n'']$ is the composite trapezoid rule approximation to $(h^2/6) \int_a^b f''(x)\,dx = (h^2/6)[f'(b) - f'(a)]$ and similarly, the error term $(h^5/240)[f_0^{(4)} + 2 f_1^{(4)} + \ldots + 2 f_{n-1}^{(4)} + f_n^{(4)}]$ is the composite trapezoid rule approximation to $(h^4/120) \int_a^b f^{(4)}(x) dx$

$= (h^4/120)[f'''(b) - f'''(a)]$. Using (10.11) with f replaced by f'', we find

$$\int_a^b f''(x)\,dx = \frac{h}{2}[f_0'' + 2f_1'' + \ldots + 2f_{n-1}'' + f_n''] + \frac{h^2}{4}[f'''(a) - f'''(b)] + \frac{h^3}{12}[f_0^{(4)} + 2f_1^{(4)} + \ldots + 2f_{n-1}^{(4)} + f_n^{(4)}] + O(h^4).$$

Thus the error term $(h^3/12)[f_0'' + 2f_1'' + \ldots + 2f_{n-1}'' + f_n'']$ in (10.11) can be written as

$$\frac{h^3}{12}[f_0'' + 2f_1'' + \ldots + 2f_{n-1}'' + f_n''] = \frac{h^2}{6}[f'(b) - f'(a)] - \frac{h^4}{24}[f'''(a) - f'''(b)] - \frac{h^5}{72}[f_0^{(4)} + 2f_1^{(4)} + \ldots + 2f_{n-1}^{(4)} + f_n^{(4)}] + O(h^6).$$

Making this substitution in (10.11) gives

$$\int_a^b f(x)\,dx = \frac{h}{2}[f_0 + 2f_1 + \ldots + 2f_{n-1} + f_n] - \frac{h^2}{12}[f'(b) - f'(a)] + \frac{h^4}{48}[f'''(b) - f'''(a)] - \frac{7h^5}{720}[f_0^{(4)} + 2f_1^{(4)} + \ldots + 2f_{n-1}^{(4)} + f_n^{(4)}] + \ldots.$$

Continuing in this way, we obtain the **Euler–Maclaurin summation formula** given in the following theorem.

Theorem 10.6.1 (Euler-Maclaurin theorem). If $f \in C^{2n}[a,b]$ and T_h is the composite trapezoid rule approximation (10.5) to $\int_a^b f(x)\,dx$, then

$$T_h - \int_a^b f(x)\,dx = \frac{h^2}{12}[f'(b) - f'(a)] - \frac{h^4}{720}[f^{(3)}(b) - f^{(3)}(a)] + \frac{h^6}{30240} \times [f^{(5)}(b) - f^{(5)}(a)] - \ldots + (-1)^{n-2}\frac{b_{2n-2}}{(2n-2)!}h^{2n-2}[f^{(2n-1)}(b) - f^{(2n-1)}(a)] + (-1)^{n-1}\frac{b_{2n}}{(2n)!}h^{2n}f^{(2n)}(\xi), \quad \xi \in [a,b]. \tag{10.12}$$

The numbers $(-1)^{j-1}b_{2j}$ in this formula are called **Bernoulli numbers**.

This formula shows that the error terms in the trapezoid rule are all even powers of h, as stated in the previous section. But it shows even more. Suppose that the function f is *periodic* with period $b-a$ or $(b-a)/m$ for some positive integer m. Then $f(b) = f(a)$, $f'(b) = f'(a)$, etc. Thus all of the terms in (10.12) except the last are zero. Moreover, if f is infinitely differentiable, then n in (10.12) can be taken to be arbitrarily large. This means that the error in the trapezoid rule, using equally-spaced nodes, decreases faster than any power of h. Such a convergence rate is said to be **superalgebraic**. Hence for periodic

TABLE 10.4
Results of using the composite trapezoid rule with equally-spaced nodes to approximate $\int_0^{2\pi} \exp(\sin x)\, dx$.

h	T_h	Error
π	6.28318530717959	$1.7 \times 10^{+00}$
$\pi/2$	7.98932343982204	3.4×10^{-02}
$\pi/4$	7.95492777270178	1.3×10^{-06}
$\pi/8$	7.95492652101285	5.3×10^{-15}
$\pi/16$	7.95492652101284	
$\pi/32$	7.95492652101284	

TABLE 10.5
Results of using the composite trapezoid rule with equally-spaced nodes to approximate $\int_1^{1+4\pi} \exp(\sin x)\, dx$.

h	T_h	Error
π	17.28411774036522	$1.4 \times 10^{+00}$
$\pi/2$	15.86488766100667	4.5×10^{-02}
$\pi/4$	15.90985267778412	3.6×10^{-07}
$\pi/8$	15.90985304202569	
$\pi/16$	15.90985304202569	

functions, when the integral is taken over an integer multiple of the period, the trapezoid rule is *extremely* accurate. When implemented on a computer or calculator using finite-precision arithmetic, it usually attains the machine precision with just a moderate number of nodes.

Example 10.6.1. Compute $\int_0^{2\pi} \exp(\sin x)\, dx$ using the composite trapezoid rule with equally-spaced nodes.

Since this function is periodic with period 2π, we expect extremely good answers using only a small number of subintervals. Table 10.4 shows the results. With 16 subintervals ($h = \pi/8$), essentially the machine precision has been reached. In exact arithmetic, the addition of more nodes would continue to reduce the error dramatically, but we have reached the limits of the computer's arithmetic.

Example 10.6.2. Compute $\int_1^{1+4\pi} \exp(\sin x)\, dx$ using the composite trapezoid rule with equally-spaced nodes.

Note that it does not matter where the integral starts, as long as it extends over a region whose length is an integer multiple of the period. Thus we again expect extremely accurate answers using a modest number of subintervals. Table 10.5 shows the results. With 32 subintervals ($h = \pi/8$), the machine precision has been reached.

For comparison purposes, using the composite trapezoid rule with subinterval sizes distributed *randomly* between $\pi/16$ and $\pi/8$, the error was about 3.5×10^{-03}. The Euler–Maclaurin formula holds only for subintervals of equal size; when they are unequal the error is again $O(h^2)$, where h is the largest subinterval width.

It should be noted that this phenomenon of superalgebraic convergence is *not* unique to the composite trapezoid rule. Other methods, such as Clenshaw–Curtis quadrature (and Gauss quadrature) achieve superalgebraic convergence even for *nonperiodic* smooth functions. A key to understanding this is to note that the coefficients in the Chebyshev series for a smooth function $f(x)$ defined in the interval $[-1, 1]$, or the coefficients in the Fourier cosine series for a smooth function $f(\cos\theta)$ defined in $[0, \pi]$, decay extremely rapidly. Hence if a quadrature rule integrates the first n terms in these series exactly, then the size of the remaining terms and hence the error in the quadrature formula will decrease faster than any power of n.

10.7 SINGULARITIES

We have discussed some *very* powerful integration formulas for smooth functions. Suppose that the integrand f is not so smooth; perhaps it approaches $\pm\infty$ at one of the endpoints of the interval, while the integral remains finite. Or perhaps the integral that we wish to compute has one or both endpoints equal to $\pm\infty$.

In the simple case where f has a finite jump discontinuity inside the interval of integration, most of the error estimates presented previously will fail because they assume that f and some of its derivatives are continuous. If we know the point of discontinuity, say c, however, then we can break the integral into two pieces:

$$\int_a^b f(x)\,dx = \int_a^c f(x)\,dx + \int_c^b f(x)\,dx. \tag{10.13}$$

Provided f is smooth in each piece, we can apply the analysis of the previous sections to quadrature rules used for each piece separately.

Suppose f becomes infinite as $x \to a$, the lower limit of integration. The upper limit can be treated similarly, and if f has a singularity within the interval of integration, it can be treated as an endpoint by dividing the interval into pieces, as in (10.13). Specifically, suppose that the integral has the form

$$\int_a^b \frac{g(x)}{(x-a)^\theta}\,dx, \quad 0 < \theta < 1, \tag{10.14}$$

where g has sufficiently many continuous derivatives on $[a, b]$ in order for the previous analysis to be applicable to g. It can be shown that for fairly general functions g (i.e., ones with $g(a) \neq 0$), the integral in (10.14) is finite even though the integrand may approach ∞ at $x = a$. There are several approaches that one might take to approximate such an integral. One is to divide the integral into two pieces:

$$\int_a^{a+\delta} \frac{g(x)}{(x-a)^\theta}\,dx + \int_{a+\delta}^b \frac{g(x)}{(x-a)^\theta}\,dx.$$

The second integral can be approximated using any of the standard techniques, since the integrand is smooth and its derivatives bounded on this interval.

To approximate the first integral, one might first expand g in a Taylor series about a:

$$g(x) = g(a) + (x-a)g'(a) + \frac{(x-a)^2}{2}g''(a) + \ldots.$$

Then the first integral becomes

$$\int_a^{a+\delta} \left[\frac{g(a)}{(x-a)^\theta} + g'(a)(x-a)^{1-\theta} + \frac{g''(a)}{2}(x-a)^{2-\theta} + \ldots \right] dx$$

$$= g(a) \left. \frac{(x-a)^{1-\theta}}{1-\theta} \right|_a^{a+\delta} + g'(a) \left. \frac{(x-a)^{2-\theta}}{2-\theta} \right|_a^{a+\delta} + \frac{g''(a)}{2} \left. \frac{(x-a)^{3-\theta}}{3-\theta} \right|_a^{a+\delta} + \ldots$$

$$= g(a)\frac{\delta^{1-\theta}}{1-\theta} + g'(a)\frac{\delta^{2-\theta}}{2-\theta} + \frac{g''(a)}{2}\frac{\delta^{3-\theta}}{3-\theta} + \ldots.$$

This sum can be terminated at an appropriate point and used as an approximation to the first integral.

Suppose that the integrand f is smooth but the upper endpoint of integration is infinite. One can again break the integral into two pieces:

$$\int_a^\infty f(x)\,dx = \int_a^R f(x)\,dx + \int_R^\infty f(x)\,dx,$$

for some large number R. If R is sufficiently large, it may be possible, by analytical means, to show that the second integral is negligible. Another possible approach is to make a change of variable $\xi = 1/x$ in the second integral, so that it becomes

$$\int_0^{1/R} f(1/\xi)\xi^{-2}\,d\xi.$$

If $f(1/\xi)\xi^{-2}$ approaches a finite limit as $\xi \to 0$, then this limit can be used in standard quadrature formulas to approximate this integral. If $f(1/\xi)\xi^{-2}$ becomes infinite as $\xi \to 0$, then the integral might be handled as above, by finding a value $\theta \in (0,1)$ such that $f(1/\xi)\xi^{-2}$ behaves like $g(\xi)/\xi^\theta$ as $\xi \to 0$, where g is smooth.

Devising quadrature formulas for functions with singularities remains a topic of current research.

10.8 CHAPTER 10 EXERCISES

1. Derive the Newton–Cotes formula for $\int_0^1 f(x)\,dx$ using the nodes 0, $\frac{1}{3}$, $\frac{2}{3}$, and 1.
2. Find the formula of the form

$$\int_0^1 f(x)\,dx \approx A_0 f(0) + A_1 f(1)$$

that is exact for all functions of the form $f(x) = ae^x + b\cos(\pi x/2)$, where a and b are constants.

3. Consider the integration formula,

$$\int_{-1}^{1} f(x)\,dx \approx f(\alpha) + f(-\alpha).$$

(a) For what value(s) of α, if any, will this formula be exact for all polynomials of degree 1 or less?
(b) For what value(s) of α, if any, will this formula be exact for all polynomials of degree 3 or less?
(c) For what value(s) of α, if any, will this formula be exact for all polynomials of the form $a + bx + cx^3 + dx^4$, where a, b, c, and d are constants?

4. Find a formula of the form

$$\int_{0}^{1} x f(x)\,dx \approx A_0 f(x_0) + A_1 f(x_1),$$

that is exact for all polynomials of degree 3 or less.

5. The Chebyshev polynomials $T_j(x) = \cos(j \arccos x)$, $j = 0, 1, \ldots,$ are orthogonal with respect to the weight function $(1 - x^2)^{-1/2}$ on $[-1, 1]$; that is, $\int_{-1}^{1} T_j(x) T_k(x)(1 - x^2)^{-1/2}\,dx = 0$, if $j \neq k$. Use the Gram–Schmidt process applied to the linearly independent set $\{1, x, x^2\}$ to construct the first three *orthonormal* polynomials for this weight function and show that they are indeed scalar multiples of T_0, T_1, and T_2.

6. Consider the composite *midpoint rule* for approximating an integral

$$\int_{a}^{b} f(x)\,dx \approx h \sum_{i=1}^{n} f\left(\frac{x_i + x_{i-1}}{2}\right),$$

where $h = (b - a)/n$ and $x_i = a + ih$, $i = 0, 1, \ldots, n$.

(a) Draw a graph to show geometrically what area is being computed by this formula.
(b) Show that this formula is exact if f is either constant or linear in each subinterval.
(c) Assuming that $f \in C^2[a, b]$, show that the midpoint rule is second-order accurate; that is, the error is less than or equal to a constant times h^2. To do this, you will first need to show that the error in each subinterval is of order h^3. To see this, expand f in a Taylor series about the midpoint $x_{i-1/2} = (x_i + x_{i-1})/2$ of the subinterval:

$$f(x) = f(x_{i-1/2}) + (x - x_{i-1/2}) f'(x_{i-1/2}) + \frac{(x - x_{i-1/2})^2}{2} f''(\xi_{i-1/2}),$$
$$\xi_{i-1/2} \in [x_{i-1}, x_i].$$

By integrating each term, show that the difference between the true value $\int_{x_{i-1}}^{x_i} f(x)\,dx$ and the approximation $h f(x_{i-1/2})$ is of order h^3. Finally, combine the results from all subintervals to show that the total error is of order h^2.

7. Write a MATLAB code to approximate

$$\int_0^1 \cos(x^2)\,dx$$

using the composite trapezoid rule and one to approximate the integral using the composite Simpson's rule, with equally-spaced nodes. The number of intervals $n = 1/h$ should be an input to each code. Turn in listings of your codes.
Do a convergence study to verify the second-order accuracy of the composite trapezoid rule and the fourth-order accuracy of the composite Simpson's rule; that is, run your code with several different h values and make a table showing the error E_h with each value of h and the ratios E_h/h^2 for the composite trapezoid rule and E_h/h^4 for the composite Simpson's rule. These ratios should be nearly constant for small values of h. You can determine the error in your computed integral by comparing your results with those of MATLAB routine `quad`. To learn about routine `quad`, type `help quad` in MATLAB. When you run `quad`, ask for a high level of accuracy, say,
```
q = quad('cos(x.^2)',0,1,[1.e-12 1.e-12])
```
where the last argument `[1.e-12 1.e-12]` indicates that you want an answer that is accurate to 10^{-12} in both a relative and an absolute sense. [Note that when you use routine `quad` you must define a function, either inline or in a separate file, that evaluates the integrand $\cos(x^2)$ at a *vector* of values of x; hence you need to write `cos(x.^2)`, instead of `cos(x^2)`.]
8. Download routine `romberg.m` from the book's web page and use it to compute $\int_0^1 \cos(x^2)\,dx$, using an error tolerance of 10^{-12}. (To use `romberg` you will need to call your function `f` and have it defined in a separate file `f.m` or as an inline function.) Have it return the number of function evaluations required by typing `[q,cnt] = romberg(0,1,1.e-12)`. The output argument `cnt` will contain the number of function evaluations required. Based on the results obtained in the previous exercise, estimate how many function evaluations would be required by the composite trapezoid rule to reach an error of 10^{-12}. Explain why the difference is so large.
9. Use `chebfun` to evaluate $\int_0^1 \cos(x^2)\,dx$. How many function evaluations does it use? (This is the length of the function `f` defined by, `f = chebfun('cos(x.^2)',[0,1])`.)

11

NUMERICAL SOLUTION OF THE INITIAL VALUE PROBLEM FOR ORDINARY DIFFERENTIAL EQUATIONS

In this chapter we study problems of the form

$$y'(t) = f(t, y(t)), \quad t \geq t_0,$$

$$y(t_0) = y_0, \tag{11.1}$$

where the independent variable t usually represents time, $y \equiv y(t)$ is the unknown function that we seek, and y_0 is a given initial value. This is called an **initial value problem** (IVP) for the ordinary differential equation (ODE) $y' = f(t, y)$. (Sometimes the differential equation holds for $t \leq t_0$ or for t in some interval about t_0. Such problems are still referred to as initial value problems.) For now, we will assume that y is a scalar-valued function of t. We will later consider *systems* of ODEs. We start with some examples.

Example 11.0.1. The rate of growth of a population with no external limits is proportional to the size of the population. If $y(t)$ represents the population at time t, then $y' = ky$ for some positive constant k. We can solve this equation analytically: $y(t) = Ce^{kt}$ for some constant C. Knowing the initial size of the population, say, $y(0) = 100$, we can determine C; in this case $C = 100$. Thus a population with *no* external limits grows exponentially (but such populations do not exist for long before food, space, etc. begin to impose limits).

This model of population growth is known as the **Malthusian** growth model, named after Thomas Malthus who developed the model in his anonymously authored book *An Essay on the Principle of Population* published in 1798. The constant k is sometimes called the Malthusian parameter. Malthus's book was influential, being cited by Charles Darwin for its key impact on his ideas of natural selection. Based on this model, Malthus warned that population growth would outpace agricultural production and might lead to an oversupply of labor, lower wages, and eventual widespread poverty.

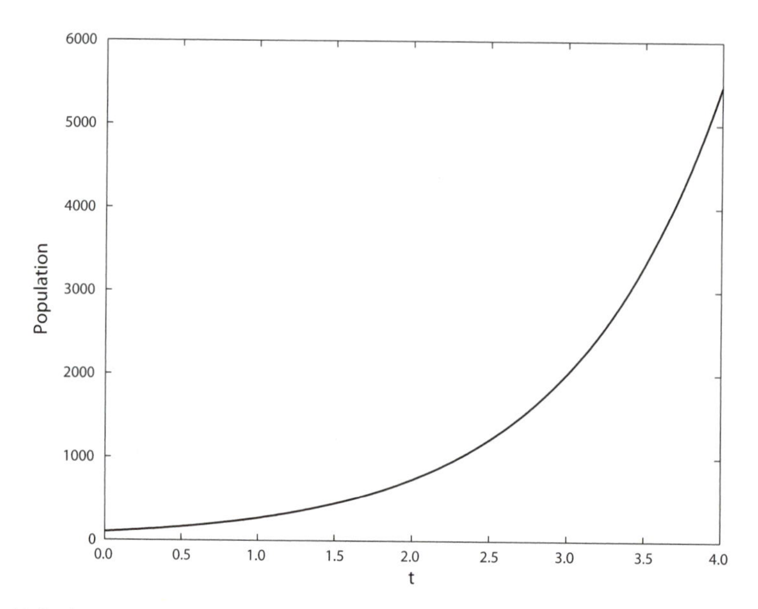

Figure 11.1. An example of exponential growth.

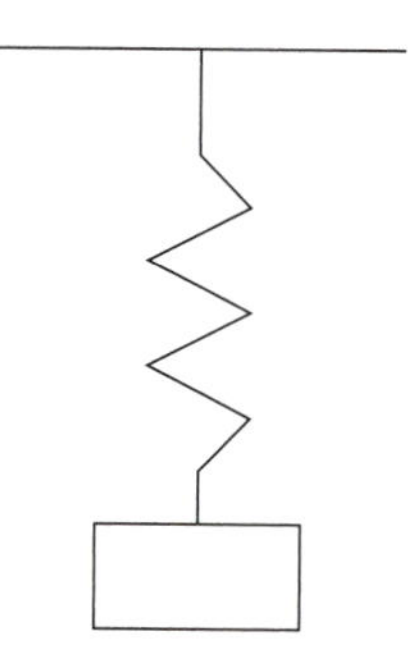

Figure 11.2. Hooke's law models the motion of a spring.

Example 11.0.2. Hooke's law says that the acceleration of an object on a spring is proportional to the distance of the object from equilibrium. If $y(t)$ represents the distance from equilibrium at time t, this can be written as $y'' = -ky$, where $k > 0$ is the spring constant and the negative sign means that the acceleration is back towards the equilibrium point. This equation has solutions of the form $y(t) = C_1 \sin(\sqrt{k}t) + C_2 \cos(\sqrt{k}t)$ for certain constants C_1 and C_2. It is easy to check by differentiating that these are indeed solutions. To specify a unique solution, however, we will need more than one initial value.

This *second-order* differential equation (involving a second derivative) can be written as a system of two first-order equations. Letting $z = y'$, we have the system

$$\begin{aligned} y' &= z, \\ z' &= -y, \end{aligned} \quad \text{or,} \quad \begin{pmatrix} y \\ z \end{pmatrix}' = \begin{pmatrix} 0 & 1 \\ -1 & 0 \end{pmatrix} \begin{pmatrix} y \\ z \end{pmatrix}.$$

We need initial values for both y and z (i.e., we need to know the initial displacement and speed of the object) in order to determine a unique solution.

Note that if, instead of being given *initial values*, we were given some sort of *boundary values* for this problem, questions of the existence and uniqueness of solutions would become more difficult. For example, suppose we were given $y(0) = 0$ and $y(\pi)$. Solutions with $y(0) = 0$ are of the form $C_1 \sin t$, and hence have value 0 at $t = \pi$. Thus the boundary value problem has infinitely many solutions if the given value is $y(\pi) = 0$ and it has no solutions if $y(\pi) \neq 0$.

Hooke's law is named after physicist Robert Hooke who published his discovery in 1676 as a Latin anagram `ceiiinosssttuv`. Before the advent of patents and intellectual property rights, published anagrams often announced a new discovery without unveiling its details. It was in 1678 that Hooke published the solution to the anagram `Ut tensio sic vis`, which means "of the extension, so the force."

Example 11.0.3. In general, it is difficult to find a formula for the solution of a differential equation, and often analytic formulas do not exist. Consider, for example, the *Lotka–Volterra* predator–prey equations. Letting $F(t)$ denote the number of foxes (predator) at time t and $R(t)$ the number of rabbits (prey) at time t, these equations have the form

$$R' = (\alpha - \beta F)R,$$
$$F' = (\gamma R - \delta)F,$$

where α, β, γ, and δ are positive constants. The reasoning behind these equations is that the rate of growth of the rabbit population increases with the number of rabbits but decreases with the number of foxes who eat rabbits; the rate of growth of the fox population decreases with the number of foxes because food becomes scarce, but it increases with increasing food supply in the form of rabbits. These equations cannot be solved analytically, but given initial values for F and R, their solution can be approximated numerically. For a demonstration, type `lotkademo` in MATLAB.

The Lotka–Volterra equations are named for Alfred Lotka and Vito Volterra who independently derived the equations in 1925 and 1926, respectively. Lotka wrote one of the first books on mathematical biology, entitled *Elements of Mathematical Biology*, in 1924.

11.1 EXISTENCE AND UNIQUENESS OF SOLUTIONS

Before we attempt to solve an initial value problem numerically, we must consider the question of whether or not a solution exists and, if so, whether it is the only solution. Sometimes solutions of an initial value problem exist only locally, as the following example illustrates.

$$y' = y \tan t,$$

$$y(0) = 1.$$

The solution is $y(t) = \sec t$, but the secant becomes infinite when $t = \pm\pi/2$, so the solution is valid only for $-\pi/2 < t < \pi/2$. As another example, consider

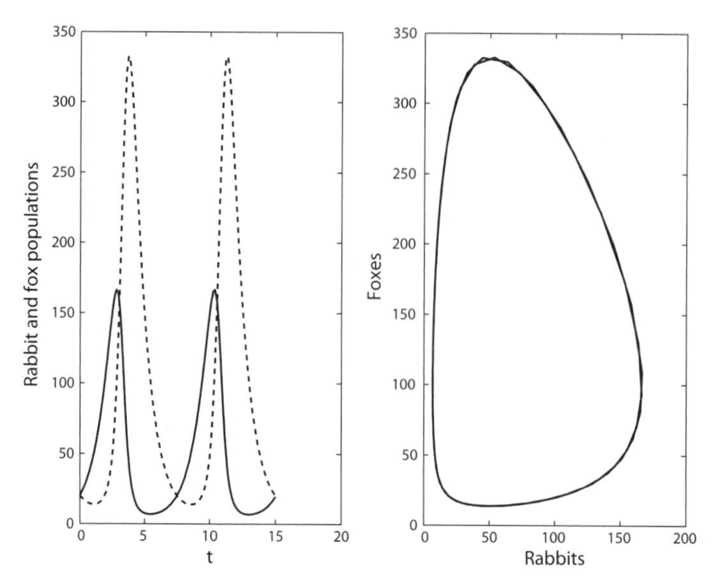

Figure 11.3. The Lotka–Volterra equations describe predator–prey systems (*left*: Time history; *right*: Phase plane plot).

the initial value problem

$$y' = y^2,$$

$$y(0) = 1.$$

You can check that the solution is $y(t) = 1/(1-t)$, $t \in [0, 1)$, but this becomes infinite at $t = 1$, so no solution exists beyond that point.

In studying questions of the existence and uniqueness of solutions to (11.1), we consider the right-hand side function f to be a function of two *independent* variables t and y and make assumptions about its behavior as a function of each of these variables. The following theorem gives sufficient conditions for the initial value problem to have a solution *locally*.

Theorem 11.1.1. If f is continuous in a rectangle R centered at (t_0, y_0),

$$R = \{(t, y) : \ |t - t_0| \le \alpha, \ \ |y - y_0| \le \beta\}, \tag{11.2}$$

then the IVP (11.1) has a solution $y(t)$ for $|t - t_0| \le \min(\alpha, \beta/M)$, where $M = \max_R |f(t, y)|$.

Even when a solution exists, it may not be unique, as the following example shows:

$$y' = y^{2/3},$$

$$y(0) = 0.$$

One solution is $y(t) \equiv 0$ and another is $y(t) = \frac{1}{27}t^3$.

The following theorem gives sufficient conditions for local existence *and* uniqueness.

CAUCHY'S EXISTENCE

Augustin Cauchy (1789–1857) [105] was a prolific French mathematician who, among many other achievements, invented the name of the determinant, founded complex analysis, and supplied a rigorous foundation for differential equations. Paul Painlevé, a mathematician and later prime minister of France, stated in an article *Encyclopédie des Sciences Mathématiques* of 1910, "It is A.-L. Cauchy who has set the general theory of differential equations on an indestructible base."[20] While others, namely Lagrange, Laplace and Poisson had done significant earlier work on differential equations, it was Cauchy who first posed and resolved the problem of the existence of a solution to a general first-order differential equation.

Theorem 11.1.2. If f *and* $\partial f/\partial y$ are continuous in the rectangle R in (11.2), then the IVP (11.1) has a *unique* solution $y(t)$ for $|t - t_0| \leq \min(\alpha, \beta/M)$, where $M = \max_R |f(t, y)|$.

Note that in the example of nonuniqueness, $f(t, y) = y^{2/3}$, so that $\partial f/\partial y = (2/3)y^{-1/3}$, which is not continuous in any interval about $y = 0$.

Finally, we state a theorem guaranteeing *global* existence and uniqueness of solutions.

Theorem 11.1.3. Assume that t_0 lies in the interval $[a, b]$. If f is continuous in the strip $a \leq t \leq b$, $-\infty < y < \infty$, and uniformly *Lipschitz continuous* in y—that is, there exists a number L such that for all y_1, y_2, and $t \in [a, b]$,

$$|f(t, y_2) - f(t, y_1)| \leq L|y_2 - y_1| \tag{11.3}$$

—then the IVP (11.1) has a unique solution in $[a, b]$.

Note that if f is differentiable with respect to y, then f satisfies the Lipschitz condition in the theorem if there exists a constant L such that for all y and all $t \in [a, b]$, $|(\partial f/\partial y)(t, y)| \leq L$. This follows from Taylor's theorem since

$$f(t, y_2) = f(t, y_1) + (y_2 - y_1)\frac{\partial f}{\partial y}(t, \xi),$$

for some ξ between y_1 and y_2. Hence

$$|f(t, y_2) - f(t, y_1)| \leq \max\left|\frac{\partial f}{\partial y}\right| \cdot |y_2 - y_1|.$$

This theorem is stated for a single ordinary differential equation, but it holds for systems as well. Just take y in the theorem to be a vector and replace the absolute value signs $|\cdot|$ by any vector norm; for example, the Euclidean norm $\|\mathbf{v}\|_2 \equiv \sqrt{\sum_j |v_j|^2}$ or the ∞-norm $\|\mathbf{v}\|_\infty = \max_j |v_j|$.

In addition to the existence and uniqueness of solutions, we need to know that the problem is **well posed**, meaning that the solution depends continuously on the initial data. More specifically, if we expect to compute a good approximate solution numerically, then it should be the case that "small" changes in the initial data lead to correspondingly "small" changes in the solution, since initial values will, inevitably, be rounded. (Compare this to the *conditioning* of a system of linear algebraic equations.)

Theorem 11.1.4. Suppose $y(t)$ and $z(t)$ satisfy

$$y' = f(t, y), \; z' = f(t, z),$$
$$y(t_0) = y_0, \quad z(t_0) = z_0 \equiv y_0 + \delta_0,$$

where f satisfies the hypotheses of theorem 11.1.3, and $t_0 \in [a, b]$. Then for any $t \in [a, b]$,

$$|z(t) - y(t)| \leq e^{L|t-t_0|} \cdot |\delta_0|. \tag{11.4}$$

Proof. Integrating the differential equations, we find

$$y(t) = y_0 + \int_{t_0}^{t} f(s, y(s))\, ds,$$

$$z(t) = z_0 + \int_{t_0}^{t} f(s, z(s))\, ds.$$

Subtracting these two equations gives

$$z(t) - y(t) = \delta_0 + \int_{t_0}^{t} (f(s, z(s)) - f(s, y(s)))\, ds.$$

Assume first that $t > t_0$. Take absolute values on each side and use the Lipschitz condition to find

$$|z(t) - y(t)| \leq |\delta_0| + L \int_{t_0}^{t} |z(s) - y(s)|\, ds.$$

Define $\Phi(t) \equiv \int_{t_0}^{t} |z(s) - y(s)|\, ds$. Then $\Phi'(t) = |z(t) - y(t)|$, and the above inequality becomes

$$\Phi'(t) \leq L\Phi(t) + |\delta_0|, \quad \Phi(t_0) = 0. \tag{11.5}$$

If the differential inequality in (11.5) were an equality, the solution to this initial value problem would be

$$\frac{|\delta_0|}{L} \left(e^{L(t-t_0)} - 1\right). \tag{11.6}$$

Because it is an inequality, it can be shown that $\Phi(t)$ is less than or equal to this expression (Gronwall's inequality). Hence using (11.5) and (11.6) we have

$$|z(t) - y(t)| = \Phi'(t) \leq L\Phi(t) + |\delta_0| \leq |\delta_0| e^{L(t-t_0)}.$$

If $t < t_0$, define $\tilde{y}(t) = y(a + t_0 - t)$ and $\tilde{z}(t) = z(a + t_0 - t)$, $a \leq t \leq t_0$, so that $\tilde{z}(a) - \tilde{y}(a) = z(t_0) - y(t_0) = \delta_0$. Then $\tilde{y}'(t) = -y'(a + t_0 - t) = -f(a + t_0 - t, \tilde{y}(t))$, and similarly for $\tilde{z}'(t)$. If we define $\tilde{f}(t,x) = -f(a+t_0-t,x)$, then $\tilde{y}'(t) = \tilde{f}(t,\tilde{y}(t))$ and $\tilde{z}'(t) = \tilde{f}(t,\tilde{z}(t))$, $a \le t \le t_0$. Apply the first part of the proof to $\tilde{z}(t) - \tilde{y}(t)$ to obtain

$$|\tilde{z}(t) - \tilde{y}(t)| \le |\delta_0| e^{L(t-a)},$$

which is equivalent to $|z(a+t_0-t) - y(a+t_0-t)| \le |\delta_0| e^{L(t-a)}$, or, letting $s = a + t_0 - t$, $|z(s) - y(s)| \le |\delta_0| e^{L(t_0-s)}$. □

11.2 ONE-STEP METHODS

Assuming that the initial value problem (11.1) is well posed, we will approximate its solution at time T by dividing the interval $[t_0, T]$ into small subintervals and replacing the time derivative over each subinterval by a finite difference quotient, of the sort discussed in chapter 9. This will result in a system of algebraic equations involving the values of the approximate solution at the endpoints of the subintervals. (For notational simplicity, we will assume from here on that the differential equation holds for $t > t_0$, but the methods could also be used to compute a solution for $t < t_0$, provided that the problem is well posed.)

Let the endpoints of the subintervals (called **nodes** or **mesh points**) be denoted $t_0, t_1, \ldots, t_N$, where $t_N = T$, and let the approximate solution at time t_j be denoted as y_j. A *one-step method* is one in which the approximate solution at time t_{k+1} is determined from that at time t_k. *Multistep methods*, which will be discussed in the next section, use approximate solution values at earlier time steps $t_{k-1}, t_{k-2}, \ldots$, as well, in the calculation of y_{k+1}. We will assume for simplicity that the subinterval width $h = t_{k+1} - t_k$ is uniform for all k. We will also assume that the solution $y(t)$ has as many continuous derivatives as are needed in order to analyze the accuracy of the method using Taylor's theorem.

11.2.1 Euler's Method

In his three-volume textbook *Institutionum Calculi Integralis*, in a chapter entitled "On the integration of differential equations by approximation," Leonhard Euler commented on the unsatisfactory nature of using series to solve differential equations and proposed the following procedure instead [20].

Starting with $y_0 = y(t_0)$, *Euler's method* sets

$$y_{k+1} = y_k + hf(t_k, y_k), \quad k = 0, 1, \ldots. \tag{11.7}$$

According to Taylor's theorem, the true solution satisfies

$$y(t_{k+1}) = y(t_k) + hy'(t_k) + \frac{h^2}{2} y''(\xi_k) = y(t_k) + hf(t_k, y(t_k)) + \frac{h^2}{2} y''(\xi_k), \quad \xi_k \in [t_k, t_{k+1}], \tag{11.8}$$

so Euler's method is obtained by dropping the $O(h^2)$ term in this formula.

If one plots the solutions to the ODE $y' = f(t,y)$ corresponding to different initial values $y(t_0)$, then Euler's method can be thought of as moving from one solution curve to another, along the tangent line to the curve, at each time

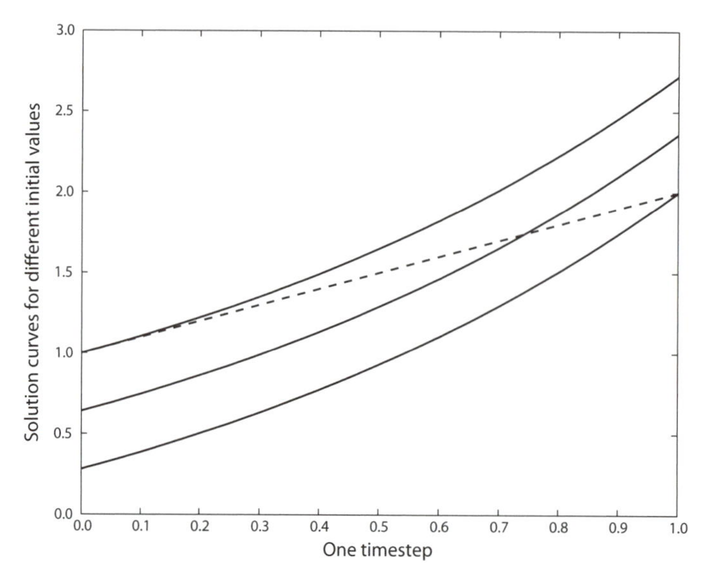

Figure 11.4. Euler's method estimates $y(t_{k+1})$ by following the tangent line at (t_k, y_k). Note how this moves the approximation from one solution curve to another (corresponding to a different initial value).

step, as pictured in figure 11.4. Hence the accuracy of the approximation will certainly be tied to the well-posedness of the initial value problem.

Example 11.2.1. Consider the initial value problem

$$y' = y, \quad y(0) = 1,$$

whose solution is $y(t) = e^t$. This problem is well posed, since $f(t, y) = y$ and $\partial f/\partial y = 1$. Starting with $y_0 = 1$, Euler's method sets

$$y_{k+1} = y_k + hy_k = (1+h)y_k, \quad k = 0, 1, \ldots.$$

If $h = T/n$ for some fixed T, then $y_n = (1+h)^n = (1+T/n)^n \to e^T$, as $n \to \infty$.

Example 11.2.2. If Euler's method is applied to the Lotka–Volterra predator–prey system in Example 11.0.3, then

$$\begin{aligned} R_{k+1} &= R_k + h(\alpha - \beta F_k)R_k, \\ F_{k+1} &= F_k + h(\gamma R_k - \delta)F_k. \end{aligned}$$

If we define the vector $\mathbf{y} \equiv (R, F)^T$ and the vector-valued function $\mathbf{f}(t, \mathbf{y}) = ((\alpha - \beta F)R, (\gamma R - \delta)F)^T$, then Euler's method again has the form

$$\mathbf{y}_{k+1} = \mathbf{y}_k + h\mathbf{f}(t_k, \mathbf{y}_k).$$

In discussing this method, Euler noted in his text that the error at each step can be made arbitrarily small by choosing sufficiently small time steps. But he also noted that errors can accumulate as one progressively steps through an interval. It was Cauchy in 1820 who provided a rigorous proof of convergence for the method.

To analyze the accuracy of a method, we will discuss *local* and *global* errors. The **local truncation error** is defined roughly as the amount by which the true solution fails to satisfy the difference equations. But with this definition, one must be careful about the form in which the difference equations are written. The approximation to $y'(t_k)$ in Euler's method is $(y_{k+1} - y_k)/h$, and the difference equations could be written in the form

$$\frac{y_{k+1} - y_k}{h} = f(t_k, y_k).$$

Substituting the true solution $y(t_j)$ for y_j everywhere in this formula, we find, using (11.8), that

$$\frac{y(t_{k+1}) - y(t_k)}{h} = f(t_k, y(t_k)) + \frac{h}{2} y''(\xi_k),$$

so the local truncation error in Euler's method is $O(h)$, and the method is said to be **first order**. (Unfortunately, the definition of local truncation error is inconsistent in the literature. Some sources describe the local error in Euler's method as $O(h^2)$ since the first neglected term in the Taylor series is $O(h^2)$. We will always use the definition that the local truncation error is the amount by which the true solution fails to satisfy the difference equations *when these equations are written in a form where the left-hand side approximates the derivative* y'; this is $1/h$ times the first neglected term in a Taylor series expansion for $y(t_{k+1})$ about t_k.)

For a reasonable method, one expects that the local truncation error will approach 0 as $h \to 0$. This holds for Euler's method since

$$\lim_{h \to 0} \frac{y(t_{k+1}) - y(t_k)}{h} = y'(t_k) = f(t_k, y(t_k)).$$

A method is called **consistent** if the local truncation error approaches 0 as $h \to 0$.

We also will be concerned with the **global error** for a method; that is, the difference between the true solution $y(t_k)$ at mesh points t_k and the approximate solution y_k. The following theorem shows that Euler's method, applied to a well-posed problem, is **convergent**: As $h \to 0$, the maximum difference between y_k and $y(t_k)$ for t_k a mesh point in a fixed interval $[t_0, T]$ approaches 0.

Theorem 11.2.1. Let $y(t)$ be the solution to $y' = f(t, y)$ with initial value $y(t_0)$ given, where $f(t, y)$ satisfies the hypotheses of theorem 11.1.3. Let $T \in [a, b]$ be fixed, with $T > t_0$, and let $h = (T - t_0)/N$. Define

$$y_{k+1} = y_k + hf(t_k, y_k), \quad k = 0, 1, \ldots, N - 1. \tag{11.9}$$

Assume that $y_0 \to y(t_0)$ as $h \to 0$. Then for all k such that $t_k \in [t_0, T]$, $y_k \to y(t_k)$ as $h \to 0$ and the convergence is uniform; that is, $\max_k |y(t_k) - y_k| \to 0$.

Proof. Subtracting (11.9) from (11.8) and letting d_j denote the difference $y(t_j) - y_j$ gives

$$d_{k+1} = d_k + h[f(t_k, y(t_k)) - f(t_k, y_k)] + \frac{h^2}{2} y''(\xi_k), \quad \xi_k \in [t_k, t_{k+1}].$$

Taking absolute values on each side and using the Lipschitz condition (11.3), we can write

$$|d_{k+1}| \le |d_k| + hL|d_k| + \frac{h^2}{2}M = (1 + hL)|d_k| + \frac{h^2}{2}M, \tag{11.10}$$

where $M \equiv \max_{t\in[t_0,T]} |y''(t)|$.

At this point in the proof, we need the following fact.

Fact. Suppose

$$\gamma_{k+1} \le (1+\alpha)\gamma_k + \beta, \quad k = 0, 1, \ldots, \tag{11.11}$$

where $\alpha > 0$ and $\beta \ge 0$. Then

$$\gamma_n \le e^{n\alpha}\gamma_0 + \frac{e^{n\alpha} - 1}{\alpha}\beta.$$

To see that this inequality holds, note that by repeatedly applying inequality (11.11), we can write

$$\begin{aligned}\gamma_n &\le (1+\alpha)\gamma_{n-1} + \beta \le (1+\alpha)^2\gamma_{n-2} + [(1+\alpha)+1]\beta \le \ldots \\ &\le (1+\alpha)^n\gamma_0 + \left[\sum_{j=0}^{n-1}(1+\alpha)^j\right]\beta.\end{aligned}$$

The sum of the geometric series is

$$\sum_{j=0}^{n-1}(1+\alpha)^j = \frac{(1+\alpha)^n - 1}{\alpha},$$

and so we have

$$\gamma_n \le (1+\alpha)^n\gamma_0 + \frac{(1+\alpha)^n - 1}{\alpha}\beta.$$

Since $(1+\alpha)^n \le e^{n\alpha}$ for all $\alpha > 0$ (because $1 + \alpha \le e^\alpha = 1 + \alpha + (\alpha^2/2)e^\xi$, ξ between 0 and α), it follows that

$$\gamma_n \le e^{n\alpha}\gamma_0 + \frac{e^{n\alpha} - 1}{\alpha}\beta.$$

Continuing with the proof of the theorem and using the fact above, with $\alpha = hL$ and $\beta = (h^2/2)M$, to bound $|d_{k+1}|$ in (11.10) gives

$$|d_{k+1}| \le e^{(k+1)hL}|d_0| + \frac{e^{(k+1)hL} - 1}{L}\frac{h}{2}M,$$

and

$$\max_{\{k:\ t_k\in[t_0,T]\}} |d_k| \le e^{L(T-t_0)}|d_0| + \frac{e^{L(T-t_0)} - 1}{L}\frac{h}{2}M,$$

since $kh \le T - t_0$. Since both terms on the right-hand side approach 0 as $h \to 0$, the theorem is proved. □

LEONHARD EULER

Leonhard Euler (1707–1783) was a Swiss mathematician who made far-reaching contributions in many areas of mathematics including analytic geometry, trigonometry, calculus and number theory. Euler introduced the symbols e, i and $f(x)$ for a function. His abilities in memorization were well known. He once mentally performed a calculation to settle an argument among students whose calculations differed in the fiftieth decimal place. Such skill served him well when he lost sight in both eyes in 1766. He continued to publish by dictating his results, producing over 800 papers during his lifetime.

We have proved not only that Euler's method converges but that its global error is of the same order, $O(h)$, as its local truncation error. We will later give a general theorem about the convergence and order of accuracy of one-step methods, whose proof is almost identical to that for Euler's method.

Before proceeding to other one-step methods, let us consider the effect of rounding errors on Euler's method. It was noted in chapter 9 that roundoff can make it difficult to approximate derivatives using finite difference quotients. Yet this is what we do in solving differential equations, so one would expect the same sorts of difficulties here. Suppose the computed solution to the difference equations (11.7) satisfies

$$\tilde{y}_{k+1} = \tilde{y}_k + hf(t_k, \tilde{y}_k) + \delta_k, \tag{11.12}$$

where δ_k accounts for the roundoff in computing $f(t_k, \tilde{y}_k)$, multiplying it by h, and adding the result to $\tilde{y}_k$. Assume that $\max_k |\delta_k| \leq \delta$. If $\tilde{d}_j \equiv \tilde{y}_j - y_j$ denotes the difference between the computed approximation and the approximation that would be obtained with exact arithmetic, then subtracting (11.7) from (11.12) gives

$$\tilde{d}_{k+1} = \tilde{d}_k + h[f(t_k, \tilde{y}_k) - f(t_k, y_k)] + \delta_k.$$

Taking absolute values on each side and assuming that f satisfies the Lipschitz condition (11.3), we find

$$|\tilde{d}_{k+1}| \leq (1 + hL)|\tilde{d}_k| + \delta.$$

Applying the fact proved in theorem 11.2.1, with $\alpha = hL$ and $\beta = \delta$, we find

$$|\tilde{d}_{k+1}| \leq e^{(k+1)hL}|\tilde{d}_0| + \frac{e^{(k+1)hL} - 1}{hL}\,\delta \leq e^{L(T-t_0)}|\tilde{d}_0| + \frac{e^{L(T-t_0)} - 1}{hL}\,\delta,$$

where $\tilde{d}_0$ represents the roundoff (if any) in evaluating the initial value y_0. As in the discussion in chapter 9, we see that if h is too small then rounding errors, being proportional to $1/h$ will dominate. To obtain the most accurate approximate solution, one should choose h small enough so that the global error in Euler's method (which is $O(h)$) is small, but not so small that the error due to roundoff (which is $O(1/h)$) is too large.

11.2.2 Higher-Order Methods Based on Taylor Series

Euler's method was derived by dropping the $O(h^2)$ term in the Taylor series expansion (11.8). One can derive methods with higher-order local truncation error by retaining more terms in the Taylor series.

Since y satisfies

$$
\begin{aligned}
y(t+h) &= y(t) + hy'(t) + \frac{h^2}{2}y''(t) + O(h^3) \\
&= y(t) + hf(t, y(t)) + \frac{h^2}{2}\left(\frac{\partial f}{\partial t} + \frac{\partial f}{\partial y} f\right)(t, y(t)) + O(h^3),
\end{aligned}
$$

dropping the $O(h^3)$ term gives the approximate formula

$$
y_{k+1} = y_k + hf(t_k, y_k) + \frac{h^2}{2}\left(\frac{\partial f}{\partial t} + \frac{\partial f}{\partial y} f\right)(t_k, y_k). \tag{11.13}
$$

Note that we have used the fact that since $y'(t) = f(t, y(t))$, the second derivative is

$$
y'' = \frac{\partial f}{\partial t} + \frac{\partial f}{\partial y}\frac{dy}{dt} = \frac{\partial f}{\partial t} + \frac{\partial f}{\partial y} f.
$$

Formula (11.13) is called the **second-order Taylor method**, and the local truncation error is $O(h^2)$, since writing the difference equations in the form

$$
\frac{y_{k+1} - y_k}{h} = f(t_k, y_k) + \frac{h}{2}\left(\frac{\partial f}{\partial t} + \frac{\partial f}{\partial y} f\right)(t_k, y_k),
$$

and substituting $y(t_j)$ for y_j everywhere we find

$$
\frac{y(t_{k+1}) - y(t_k)}{h} = f(t_k, y(t_k)) + \frac{h}{2}\left(\frac{\partial f}{\partial t} + \frac{\partial f}{\partial y} f\right)(t_k, y(t_k)) + O(h^2).
$$

While the local truncation error in this method is of higher order than that in Euler's method, this method requires computing partial derivatives of the right-hand side function f. Such derivatives are usually difficult or costly to compute. One could use even higher-order Taylor methods, requiring even more partial derivatives of f, but because of the difficulty of evaluating these derivatives, such methods are seldom used.

11.2.3 Midpoint Method

The **midpoint method** is defined by first taking a half step with Euler's method to approximate the solution at time $t_{k+1/2} \equiv (t_k + t_{k+1})/2$, and then taking a

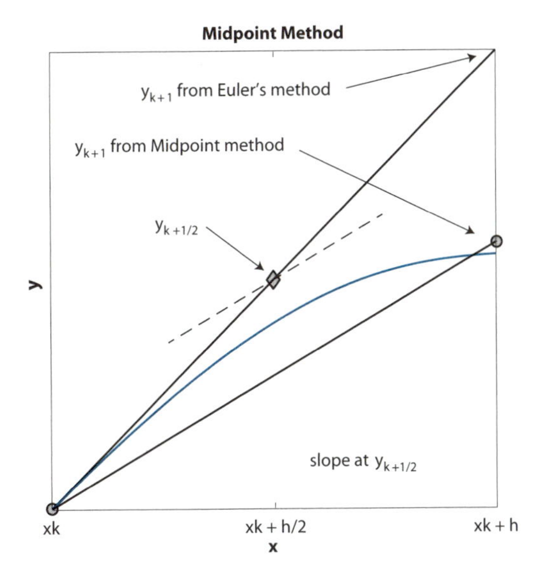

Figure 11.5. A graphical depiction of the midpoint method.

full step using the value of f at $t_{k+1/2}$ and the approximate solution $y_{k+1/2}$:

$$y_{k+1/2} = y_k + \frac{h}{2} f(t_k, y_k), \tag{11.14}$$

$$y_{k+1} = y_k + hf(t_{k+1/2}, y_{k+1/2}). \tag{11.15}$$

Note that this method requires two function evaluations per step instead of one as in Euler's method. Hence each step will be more expensive, but the extra expense may be worth it if we are able to use a significantly larger time step.

Figure 11.5 gives a graphical depiction of the midpoint method. Note how the first half step follows the tangent line to a solution curve at (t_k, y_k) and generates a value $y_{k+1/2}$. Then the full step starts from t_k again but follows the tangent line to a solution curve through the point $(t_{k+1/2}, y_{k+1/2})$. As with Euler's method, the midpoint method can be thought of as jumping between solution curves for the ODE corresponding to different initial values, or, different values at the previous time step. In the figure, we see that the midpoint method lands on a solution curve whose value at the previous time step is closer to the given value than that of the solution curve on which Euler's method lands.

To determine the local truncation error for this method, expand the true solution in a Taylor series about $t_{k+1/2} = t_k + h/2$:

$$y(t_{k+1}) = y(t_{k+1/2}) + (h/2) f(t_{k+1/2}, y(t_{k+1/2})) + \frac{(h/2)^2}{2} y''(t_{k+1/2}) + O(h^3),$$

$$y(t_k) = y(t_{k+1/2}) - (h/2) f(t_{k+1/2}, y(t_{k+1/2})) + \frac{(h/2)^2}{2} y''(t_{k+1/2}) + O(h^3).$$

Subtracting these two equations gives

$$y(t_{k+1}) - y(t_k) = hf(t_{k+1/2}, y(t_{k+1/2})) + O(h^3).$$

Now expanding $y(t_{k+1/2})$ about t_k gives

$$y(t_{k+1/2}) = y(t_k) + (h/2)f(t_k, y(t_k)) + O(h^2),$$

and making this substitution we have

$$\begin{aligned} y(t_{k+1}) - y(t_k) &= hf\left(t_{k+1/2}, y(t_k) + (h/2)f(t_k, y(t_k)) + O(h^2)\right) + O(h^3) \\ &= hf\left(t_{k+1/2}, y(t_k) + (h/2)f(t_k, y(t_k))\right) + O(h^3), \end{aligned} \quad (11.16)$$

where the second equality follows because f is Lipschitz in its second argument so that $hf(t, y + O(h^2)) = hf(t, y) + O(h^3)$. Since, from (11.14) and (11.15), the approximate solution satisfies

$$y_{k+1} = y_k + hf\left(t_{k+1/2}, y_k + (h/2)f(t_k, y_k)\right),$$

or, equivalently,

$$\frac{y_{k+1} - y_k}{h} = f\left(t_{k+1/2}, y_k + (h/2)f(t_k, y_k)\right),$$

and, from (11.16), the true solution satisfies

$$\frac{y(t_{k+1}) - y(t_k)}{h} = f\left(t_{k+1/2}, y(t_k) + (h/2)f(t_k, y(t_k))\right) + O(h^2),$$

the local truncation error in the midpoint method is $O(h^2)$; that is, the method is second-order accurate.

11.2.4 Methods Based on Quadrature Formulas

Integrating the differential equation (11.1) from t to $t + h$ gives

$$y(t + h) = y(t) + \int_t^{t+h} f(s, y(s))\, ds. \quad (11.17)$$

The integral on the right-hand side of this equation can be approximated with any of the quadrature rules described in chapter 10 to obtain an approximate solution method.

For example, using the trapezoid rule to approximate the integral,

$$\int_t^{t+h} f(s, y(s))\, ds = \frac{h}{2}[f(t, y(t)) + f(t + h, y(t + h))] + O(h^3),$$

leads to the **trapezoidal method** for solving the IVP (11.1),

$$y_{k+1} = y_k + \frac{h}{2}[f(t_k, y_k) + f(t_{k+1}, y_{k+1})]. \quad (11.18)$$

This is called an **implicit** method, since the new value y_{k+1} appears on both the left- and right-hand sides of the equation. To determine y_{k+1}, one must solve a *nonlinear* equation. The previous methods that we have discussed have been **explicit**, meaning that the value of y at the new time step can be computed explicitly from its value at the previous step. Since the error in the trapezoid rule approximation to the integral is $O(h^3)$, the local truncation error for this method is $O(h^2)$.

To avoid solving the nonlinear equation in the trapezoidal method, one can use **Heun's method**, which first estimates y_{k+1} using Euler's method and then uses that estimate in the right-hand side of (11.18):

$$\tilde{y}_{k+1} = y_k + hf(t_k, y_k),$$

$$y_{k+1} = y_k + (h/2)[f(t_k, y_k) + f(t_{k+1}, \tilde{y}_{k+1})].$$

Heun's method follows a line whose slope is the average of the slope of a solution curve at (t_k, y_k) and the slope of a solution curve at $(t_{k+1}, \tilde{y}_{k+1})$, where $\tilde{y}_{k+1}$ is the result of a step with Euler's method.

11.2.5 Classical Fourth-Order Runge–Kutta and Runge–Kutta–Fehlberg Methods

Heun's method is also known as a *second-order Runge–Kutta method*. It can be derived as follows. Introduce an intermediate value $\tilde{y}_{k+\alpha}$ which will serve as an approximation to $y(t_k + \alpha h)$, for a certain parameter α. Set

$$\tilde{y}_{k+\alpha} = y_k + \alpha hf(t_k, y_k),$$

$$y_{k+1} = y_k + \beta hf(t_k, y_k) + \gamma hf(t_k + \alpha h, \tilde{y}_{k+\alpha}), \tag{11.19}$$

where the parameters α, β, and γ will be chosen to match as many terms as possible in the Taylor series expansion of $y(t_{k+1})$ about t_k.

To see how this can be done, we will need to use a *multivariate* Taylor series expansion. (Multivariate Taylor series expansions are derived in Appendix B, but for readers who have not seen this material before, the following paragraph can be skipped and the statements about order of accuracy simply taken as facts.) To determine α, β, and γ, first expand $f(t_k + \alpha h, \tilde{y}_{k+\alpha})$ about (t_k, y_k):

$$\begin{aligned} f(t_k+\alpha h, \tilde{y}_{k+\alpha}) &= f(t_k + \alpha h, y_k+\alpha hf(t_k, y_k)) \\ &= f(t_k, y_k)+\alpha h[f_t + ff_y]+\frac{\alpha^2h^2}{2}[f_{tt}+2ff_{ty}+f^2 f_{yy}] + O(h^3), \end{aligned}$$

where, unless otherwise stated, f and its derivatives (denoted here as f_t, f_y, etc.) are evaluated at (t_k, y_k). Making this substitution in (11.19) gives

$$y_{k+1} = y_k + (\beta + \gamma)hf + \alpha\gamma h^2[f_t + ff_y] + \frac{\alpha^2\gamma h^3}{2}[f_{tt} + 2ff_{ty} + f^2 f_{yy}] + O(h^4). \tag{11.20}$$

Now, the Taylor expansion for the true solution $y(t_{k+1})$ about t_k is

$$\begin{aligned} y(t_{k+1}) &= y(t_k) + hy'(t_k) + \frac{h^2}{2}y''(t_k) + \frac{h^3}{6}y'''(t_k) + O(h^4) \\ &= y(t_k) + hf + \frac{h^2}{2}[f_t + ff_y] + \frac{h^3}{6}[f_{tt} + 2ff_{ty} + f_t f_y \\ &\quad + f^2 f_{yy} + ff_y^2] + O(h^4). \end{aligned} \tag{11.21}$$

Comparing (11.20) and (11.21), we see that by choosing α, β, and γ properly, we can match the $O(h)$ term and the $O(h^2)$ term, but not the $O(h^3)$ term. In order to do this, we need

$$\beta + \gamma = 1, \quad \alpha\gamma = \frac{1}{2}. \tag{11.22}$$

Methods of the form (11.19) that satisfy (11.22) are called **second-order Runge–Kutta methods**, and their local truncation error is $O(h^2)$. Heun's method is of this form with $\alpha = 1$, $\beta = \gamma = \frac{1}{2}$.

One can take this idea further and introduce two intermediate values and additional parameters, and it can be shown that by choosing the parameters properly one can achieve fourth-order accuracy. The algebra quickly becomes complicated, so we will omit the derivation.

The most common fourth-order Runge–Kutta method, called the *classical fourth-order Runge–Kutta method*, can be written in the form

$$\begin{aligned}
q_1 &= f(t_k, y_k), \\
q_2 &= f(t_k + \frac{h}{2}, y_k + \frac{h}{2}q_1), \\
q_3 &= f(t_k + \frac{h}{2}, y_k + \frac{h}{2}q_2), \\
q_4 &= f(t_k + h, y_k + hq_3), \\
y_{k+1} &= y_k + \frac{h}{6}\left[q_1 + 2q_2 + 2q_3 + q_4\right].
\end{aligned}$$

Note that if f does not depend on y, then this corresponds to *Simpson's rule* for quadrature:

$$y(t+h) - y(t) = \int_t^{t+h} f(s)\,ds \approx \frac{h}{6}\left[f(t) + 4f(t+\frac{h}{2}) + f(t+h)\right].$$

Recall that the error in Simpson's rule is $O(h^5)$, and from this the fourth-order accuracy of the classical Runge–Kutta method follows, assuming that f is independent of y.

In implementing methods for solving initial value problems, it is important to be able to estimate the error in the computed solution and adjust the step size accordingly. In an effort to devise such a procedure, Fehlberg in 1969 looked at fourth-order Runge–Kutta methods with five function evaluations (unlike the classical fourth-order method which requires only four function evaluations), and fifth-order Runge–Kutta methods with six function evaluations. Since function evaluations are usually the most expensive part of the solution procedure, one wants to minimize the number of function evaluations required to achieve a desired level of accuracy. Fehlberg found that by carefully choosing the parameters, he could derive methods of fourth and fifth order that used the same evaluation points; that is, the fourth-order method required function evaluations at five points, and the fifth-order method required function evaluations at the same five points plus one more. Thus, in total, only six function evaluations were required for a fifth-order-accurate method, with an associated fourth-order method used to estimate the error and vary the step size if necessary. This is called the *Runge–Kutta–Fehlberg method*, and it is among the most powerful one-step explicit methods. The MATLAB ODE solver, ode45, is based on this idea, using a pair of Runge–Kutta methods devised by Dormand and Prince.

Example 11.2.3. The following example is part of the MATLAB demonstration routine odedemo. It uses ode45 to solve a pair of ODEs modeling the van der Pol equations, and illustrates nicely how the time step is adjusted dynamically to obtain an accurate solution. The solution changes rapidly over certain time intervals, and here ode45 uses a small time step. It changes very slowly over others, where ode45 can use a larger time step without sacrificing accuracy. We will see later that when this phenomenon of very rapid change in certain time intervals and slow change in others becomes extreme, then different methods designed for *stiff* problems may be needed, but for this example the Runge–Kutta method in ode45 is adequate.

The pair of ODEs to be solved is

$$y_1' = y_2,$$
$$y_2' = (1 - y_1^2)y_2 - y_1.$$

We first create a function vanderpol that returns the values of y_1' and y_2', given y_1, y_2, and t.

```
function [dydt] = vanderpol(t,y)
dydt = [y(2); (1 - y(1)^2)*y(2) - y(1)];
```

We will solve this system of ODEs over the time interval from $t = 0$ to $t = 20$, with the initial values $y_1(0) = 2$, $y_2(0) = 0$.

```
[T,Y] = ode45(@vanderpol, [0, 20], [2, 0]);
```

The vector T that is returned contains the times at which the solution was computed, and the array Y contains in its ith row, the values of y_1 and y_2 at T(i). We can plot the first component of the solution and see the time steps created for the computation with the following commands.

```
plot(T,Y(:,1),'-', T,Y(:,1),'o')
xlabel('t'), ylabel('y_1(t)'), title('van der Pol equation')
```

The result is shown in figure 11.6.

11.2.6 An Example Using MATLAB's ODE Solver

To further illustrate the use of ode45 in MATLAB, we consider a model of romantic love from the aptly entitled article "The Lighter Side of Differential

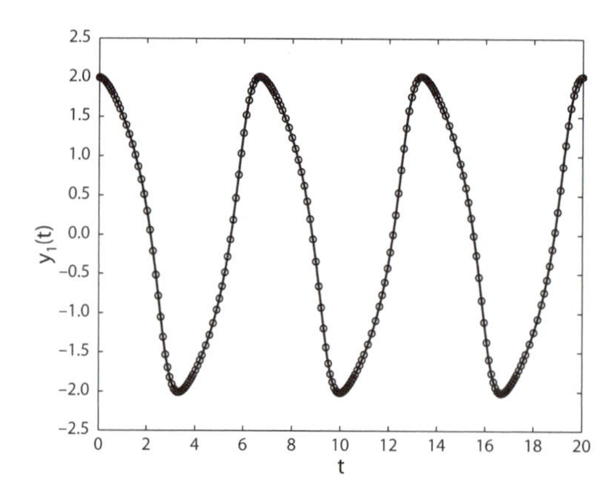

Figure 11.6. Numerical solution to the van der Pol equations.

Equations" [71]. In this model, there are two characters, Romeo and Juliet, whose affection is quantified on a scale from -5 to 5 described below.

hysterical hatred	*disgust*	*indifference*	*sweet affection*	*ecstatic love*
−5	−2.5	0	2.5	5

The characters struggle with frustrated love due to the lack of reciprocity of their feelings. Mathematically, they might say:

> ROMEO: "My feelings for Juliet decrease in proportion to her love for me."
> JULIET: "My love for Romeo grows in proportion to his love for me."

We will measure time in days, beginning at $t = 0$ and ending at $t = 60$. This ill-fated love affair might be modeled by the equations

$$\frac{dx}{dt} = -0.2y,$$
$$\frac{dy}{dt} = 0.8x, \tag{11.23}$$

where x and y are Romeo's and Juliet's love, respectively. Note that affection from Romeo increases Juliet's love, while Juliet's affection prompts Romeo to pull away. To begin the story, we will take $x(0) = 2$, indicating that Romeo is somewhat smitten upon first seeing Juliet, while we will take $y(0) = 0$, to reflect Juliet's indifference upon meeting her eventual beau. While some may say that their fate was in the stars, in this model it is undeniably cast within these equations.

To numerically solve the equations using MATLAB, we define a function, love, which corresponds to the ODE in (11.23). This is accomplished with the following code:

```
function dvdt = love(t,v)
dvdt = zeros(2,1)
dvdt(1) = -0.2*v(2);
dvdt(2) = 0.8*v(1)
```

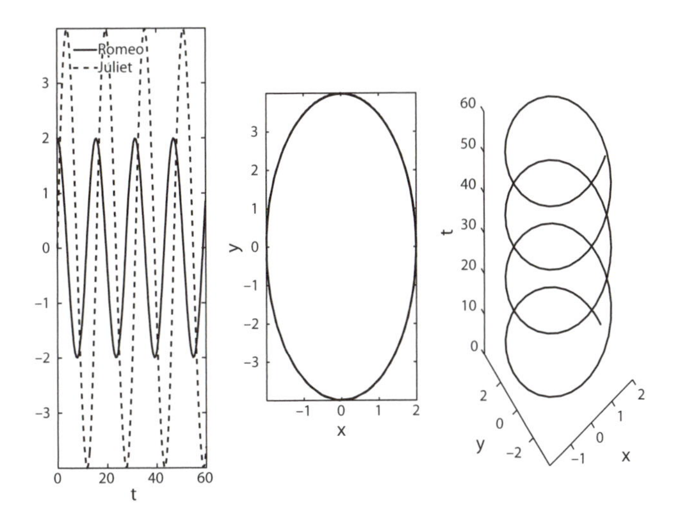

Figure 11.7. Three graphs of numerical results from the differential equations in (11.23) describing the affections of Romeo and Juliet.

Now, we can use **ode45** to solve the initial value problem (with the first entry in vector v initially set to 2 and the second entry initially set to 0) by making the call

```
[T,V] = ode45(@love,[0,60],[2,0]);
```

There are a variety of ways to graphically display the results. We demonstrate three ways with the following code.

```
subplot(1,3,1)
plot(T,V(:,1),'-',T,V(:,2),'-.');
h = legend('Romeo','Juliet',2); axis tight
xlabel('t')
subplot(1,3,2)
plot(V(:,1),V(:,2)), xlabel('x'), ylabel('y')
axis equal tight
subplot(1,3,3)
plot3(V(:,1),V(:,2),T)
xlabel('x'),ylabel('y'),zlabel('t')
axis tight
```

The graphs created from this code can be seen in figure 11.7. From these plots, we see that Juliet's emotions swing to higher and lower levels than the cooler Romeo's. Both lovers, however, are fated to be on an emotional roller coaster, as their feelings for one another swing back and forth. In an exercise at the end of the chapter, we consider how a change in the differential equation alters their fate.

11.2.7 Analysis of One-Step Methods

A general, explicit one-step method can be written in the form

$$y_{k+1} = y_k + h\psi(t_k, y_k, h). \tag{11.24}$$

- Euler's method:

$$y_{k+1} = y_k + hf(t_k, y_k) \Rightarrow \psi(t, y, h) = f(t, y).$$

- Heun's method:

$$y_{k+1} = y_k + \frac{h}{2}[f(t_k, y_k) + f(t_{k+1}, y_k + hf(t_k, y_k))] \Rightarrow$$

$$\psi(t, y, h) = \frac{1}{2}[f(t, y) + f(t + h, y + hf(t, y))].$$

- Midpoint method:

$$y_{k+1} = y_k + hf(t_{k+1/2}, y_k + (h/2)f(t_k, y_k)) \Rightarrow$$

$$\psi(t, y, h) = f(t + h/2, y + (h/2)f(t, y)).$$

We will now formalize several of the definitions given in section 11.2.1.

Definition. The one-step method (11.24) is **consistent** if $\lim_{h\to 0} \psi(t, y, h) = f(t, y)$.

This is equivalent to the definition stated earlier that

$$\lim_{h\to 0} \left[\frac{y(t_{k+1}) - y(t_k)}{h} - \psi(t_k, y(t_k), h)\right] = 0,$$

since $\lim_{h\to 0}(y(t_{k+1}) - y(t_k))/h = f(t_k, y(t_k))$. All of the one-step methods listed above are consistent.

Definition. The one-step method (11.24) is **stable** if there is a constant K and a step size $h_0 > 0$ such that the difference between two solutions y_n and $\tilde{y}_n$ with initial values y_0 and $\tilde{y}_0$, respectively, satisfies

$$|y_n - \tilde{y}_n| \le K|y_0 - \tilde{y}_0|,$$

whenever $h \le h_0$ and $nh \le T - t_0$.

Fact. If $\psi(t, y, h)$ is Lipschitz continuous in y; that is, if there exists a constant L such that for all y and $\tilde{y}$ (and for all t in an interval about t_0),

$$|\psi(t, y, h) - \psi(t, \tilde{y}, h)| \le L|y - \tilde{y}|,$$

then (11.24) is stable.

The one-step methods listed above are stable, assuming that $f(t, y)$ is uniformly Lipschitz in y. In fact, it would be difficult to construct a consistent one-step method that is not stable when $f(t, y)$ is Lipschitz in y, since $\lim_{h\to 0} \psi(t, y, h) = f(t, y)$. In the next section, we will see that this is not the case for multistep methods. Very reasonable looking consistent multistep methods are sometimes highly unstable.

Definition. The **local truncation error** is

$$\tau(t, h) = \frac{y(t+h) - y(t)}{h} - \psi(t, y(t), h).$$

We have seen that for Euler's method the local truncation error is $O(h)$ and for Heun's method and the midpoint method the local truncation error is $O(h^2)$.

With these definitions, we can now prove a general theorem about explicit one-step methods that is analogous to theorem 11.2.1 for Euler's method.

Theorem 11.2.2. If a one-step method of the form (11.24) is stable and consistent and if $|\tau(t, h)| \leq Ch^p$, then the global error is bounded by

$$\max_{\{k:\ t_0+kh \leq T\}} |y_k - y(t_k)| \leq Ch^p \frac{e^{L(T-t_0)} - 1}{L} + e^{L(T-t_0)}|y_0 - y(t_0)|,$$

where L is the Lipschitz constant for ψ.

Proof. The proof is similar to that for Euler's method. The approximate solution satisfies

$$y_{k+1} = y_k + h\psi(t_k, y_k, h),$$

while the true solution satisfies

$$y(t_{k+1}) = y(t_k) + h\psi(t_k, y(t_k), h) + h\tau(t_k, h).$$

Subtracting these two equations and letting $d_j = y(t_j) - y_j$, gives

$$d_{k+1} = d_k + h[\psi(t_k, y(t_k), h) - \psi(t_k, y_k, h)] + h\tau(t_k, h).$$

Taking absolute values on each side and using the Lipschitz condition on ψ and the bound on $|\tau(t, h)|$ we find

$$|d_{k+1}| \leq (1 + hL)|d_k| + Ch^{p+1}.$$

Substituting the analogous inequality for $|d_k|$, then $|d_{k-1}|$, etc., gives the inequality

$$|d_{k+1}| \leq (1 + hL)^{k+1}|d_0| + \left[\sum_{j=0}^{k}(1 + hL)^j\right] Ch^{p+1}.$$

Summing the geometric series, this becomes

$$|d_{k+1}| \leq (1 + hL)^{k+1}|d_0| + \frac{(1 + hL)^{k+1} - 1}{hL} Ch^{p+1} = (1 + hL)^{k+1}|d_0|$$

$$+ \frac{(1 + hL)^{k+1} - 1}{L} Ch^p.$$

Now using the fact that $(1 + hL)^{k+1} \le e^{(k+1)hL}$ (since, for any x, $1 + x \le e^x = 1 + x + (x^2/2)e^{\xi}$, ξ between 0 and x), the above inequality can be replaced by

$$|d_{k+1}| \le e^{(k+1)hL}|d_0| + \frac{e^{(k+1)hL} - 1}{L} Ch^p.$$

Since $kh \le T - t_0$ if $t_k \in [t_0, T]$, it follows that

$$\max_{\{k:\ t_k \in [t_0, T]\}} |d_k| \le e^{L(T-t_0)}|d_0| + \frac{e^{L(T-t_0)} - 1}{L} Ch^p. \qquad \square$$

We have shown that a stable one-step method whose local truncation error is $O(h^p)$ also has global error of order h^p. This is a *very* powerful result. It is easy to determine the local truncation error just by comparing terms in a Taylor series expansion. To directly analyze global error can be much more difficult.

We give one more definition, in order to state the result of this theorem in a form that will be repeated later for multistep methods.

Definition. The one-step method (11.24) is **convergent** if, for all well-posed IVPs, $\max_{\{k:\ t_k \in [t_0, T]\}} |y(t_k) - y_k| \to 0$ as $y_0 \to y(t_0)$ and $h \to 0$.

Theorem 11.2.2 implies that if a one-step method is *stable* and *consistent* then it is *convergent*.

11.2.8 Practical Implementation Considerations

As noted previously, packages designed to solve initial value problems almost always use a variable step size and/or variable-order methods for different parts of the problem. This requires an estimate of the error at each step and corresponding adjustment of the step size or order so as to meet the user's requested error tolerance, but not to greatly exceed it since this would require extra work.

One simple way to estimate the error in a step of size h is to replace that step by two steps, each of size $h/2$, and compare the results. Using Euler's method, for example, one might set

$$y_{k+1} = y_k + hf(t_k, y_k),$$

and

$$\hat{y}_{k+1/2} = y_k + (h/2) f(t_k, y_k)$$

$$\hat{y}_{k+1} = \hat{y}_{k+1/2} + (h/2) f(t_{k+1/2}, \hat{y}_{k+1/2}).$$

The difference between y_{k+1} and $\hat{y}_{k+1}$ might be used as an estimate of the error in this step. Moreover, if $y(t_{k+1}; y_k)$ denotes the exact solution to the ODE, *with initial value* $y(t_k) = y_k$, then a Taylor series expansion of $y(t_{k+1}; y_k)$ shows that

$$y(t_{k+1}; y_k) = y_k + hf(t_k, y_k) + \frac{h^2}{2}y''(t_k) + O(h^3) = y_{k+1} + \frac{h^2}{2}y''(t_k) + O(h^3), \tag{11.25}$$

and expanding $y(t_{k+1/2}; y_k)$ about t_k and $y(t_{k+1}; y_k)$ about $t_{k+1/2}$ gives

$$\begin{aligned}
y(t_{k+1/2}; y_k) &= y_k + (h/2) f(t_k, y_k) + \frac{(h/2)^2}{2}y''(t_k) + O(h^3) \\
&= \hat{y}_{k+1/2} + \frac{(h/2)^2}{2}y''(t_k) + O(h^3), \\
y(t_{k+1}; y_k) &= y(t_{k+1/2}; y_k) + (h/2) f(t_{k+1/2}, y(t_{k+1/2}; y_k)) \\
&\quad + \frac{(h/2)^2}{2}y''(t_{k+1/2}) + O(h^3) \\
&= \hat{y}_{k+1/2} + \frac{(h/2)^2}{2}y''(t_k) + (h/2) f(t_{k+1/2}, \hat{y}_{k+1/2}) \\
&\quad + \frac{(h/2)^2}{2}y''(t_k) + O(h^3) \\
&= \hat{y}_{k+1} + \frac{h^2}{4}y''(t_k) + O(h^3).
\end{aligned} \tag{11.26}$$

The next-to-last line was obtained by noting that

$$y(t_{k+1/2}; y_k) = \hat{y}_{k+1/2} + ((h/2)^2/2)y''(t_k) + O(h^3),$$

$$(h/2) f(t_{k+1/2}, y(t_{k+1/2}; y_k)) = (h/2) f(t_{k+1/2}, \hat{y}_{k+1/2}) + O(h^3),$$

$$((h/2)^2/2)y''(t_{k+1/2}) = ((h/2)^2/2)y''(t_k) + O(h^3).$$

It follows that one can improve the *local error* in the approximation by using a step of Richardson extrapolation. From (11.25) and (11.26), it follows that the approximation $2\hat{y}_{k+1} - y_{k+1}$ differs from $y(t_{k+1}; y_k)$ by $O(h^3)$ instead of $O(h^2)$. Thus, for the price of doing two additional half steps, one can obtain a (locally) more accurate approximation, $2\hat{y}_{k+1} - y_{k+1}$, as well as an error estimate, $|(2\hat{y}_{k+1} - y_{k+1}) - \hat{y}_{k+1}| = |\hat{y}_{k+1} - y_{k+1}|$.

As noted earlier, another way to estimate errors is to use a higher-order method with the same step size. For example, Runge–Kutta methods of orders 2 and 3 are sometimes used, as are the Runge–Kutta–Fehlberg pairs of orders 4 and 5. Which method works best is, of course, problem dependent, and the efficiency of a software package often depends more on how good the error estimator and the step size adjustment are than on which underlying method is actually used.

ANIMATING A SYSTEM OF POINTS

As discussed in section 1.1, the movement of cloth in computer animation results from models of spring–mass systems. To animate the robe on Yoda to the left, a numerical ODE solver is used to solve for the position of each point on Yoda's robe. Hair on digitally created characters is also modeled as a spring–mass system to create the movement. Thus numerical ODE solvers play an important role in the special effects of movies. (Image courtesyof Lucasfilm Ltd. *Star Wars: Episode II - Attack of the Clones* ™ & © 2002 Lucasfilm Ltd. All rights reserved. Used under authorization. Unauthorized duplication is a violationof applicable law. Model created by Kecskemeti B. Zoltan. Digital Work by Industrial Light & Magic.)

11.2.9 Systems of Equations

Everything that we have done so far, and most of what we will do throughout this chapter, applies equally well to systems of ODEs. Let $\mathbf{y}(t)$ denote a vector of values $\mathbf{y}(t) = (y_1(t), \ldots, y_n(t))^T$, and let $\mathbf{f}(t, \mathbf{y}(t))$ denote a vector-valued function of its arguments: $\mathbf{f}(t, \mathbf{y}(t)) = (f_1(t, \mathbf{y}(t)), \ldots, f_n(t, \mathbf{y}(t)))^T$. The initial value problem

$$\mathbf{y}' = \mathbf{f}(t, \mathbf{y}), \quad \mathbf{y}(t_0) = \mathbf{y}_0,$$

represents the system of equations

$$\begin{aligned} y_1{}' &= f_1(t, y_1, \ldots, y_n), \\ y_2{}' &= f_2(t, y_1, \ldots, y_n), \\ &\vdots \\ y_n{}' &= f_n(t, y_1, \ldots, y_n), \end{aligned}$$

with initial conditions

$$y_1(t_0) = y_{10}, \quad y_2(t_0) = y_{20}, \ldots, y_n(t_0) = y_{n0}.$$

Here we denote the time step with a second subscript, since the first subscript denotes the index within the vector $\mathbf{y}_0$.

All of the methods discussed so far can be applied to systems. For example, Euler's method still has the form

$$\mathbf{y}_{k+1} = \mathbf{y}_k + h\mathbf{f}(t_k, \mathbf{y}_k),$$

which componentwise means that

$$y_{i,k+1} = y_{ik} + hf_i(t_k, y_{1k}, \dots, y_{nk}), \quad i = 1, \dots, n.$$

All of the theorems proved so far apply to systems when $|\cdot|$ is replaced by $\|\cdot\|$, where $\|\cdot\|$ can be any vector norm, such as the 2-norm or the ∞-norm. The proofs are virtually identical as well.

11.3 MULTISTEP METHODS

By the time we reach time step $k+1$ in an IVP solver, we have computed approximations to the solution at time steps $k, k-1, k-2, \dots$. It seems a shame to throw away the values $y_{k-1}, y_{k-2}, \dots$, and compute y_{k+1} using only y_k. On the other hand, if one knows the solution at time step k, then this uniquely determines the solution at all future times (theorem 11.1.3), so one might think of it as overkill to use computed values at two or more time steps to determine the approximation at the next time step. Nevertheless, this is what is sometimes done, and such methods are called **multistep methods**.

11.3.1 Adams–Bashforth and Adams–Moulton Methods

Integrating the ODE in (11.1) from t_k to t_{k+1}, we find

$$y(t_{k+1}) = y(t_k) + \int_{t_k}^{t_{k+1}} f(s, y(s))\, ds. \tag{11.27}$$

Using the computed values $y_k, y_{k-1}, \dots, y_{k-m+1}$, let $p_{m-1}(s)$ be the $(m-1)$st-degree polynomial that interpolates the values $f(t_k, y_k)$, $f(t_{k-1}, y_{k-1}), \dots, f(t_{k-m+1}, y_{k-m+1})$ at the points $t_k, t_{k-1}, \dots, t_{k-m+1}$. The Lagrange form of the interpolant p_{m-1} is

$$p_{m-1}(s) = \sum_{\ell=0}^{m-1} \left(\prod_{\substack{j=0 \\ j\neq\ell}}^{m-1} \frac{s - t_{k-j}}{t_{k-\ell} - t_{k-j}} \right) f(t_{k-\ell}, y_{k-\ell}).$$

Suppose we replace $f(s, y(s))$ in the integral on the right-hand side of (11.27) by $p_{m-1}(s)$. The result is

$$y_{k+1} = y_k + \int_{t_k}^{t_{k+1}} p_{m-1}(s)\, ds = y_k + h \sum_{\ell=0}^{m-1} b_\ell f(t_{k-\ell}, y_{k-\ell}),$$

where

$$b_\ell = \frac{1}{h} \int_{t_k}^{t_{k+1}} \left(\prod_{\substack{j=0 \\ j\neq\ell}}^{m-1} \frac{s - t_{k-j}}{t_{k-\ell} - t_{k-j}} \right) ds.$$

This is known as the m-**step Adams–Bashforth method**. Note that in order to start the method we need, in addition to the initial value y_0, the $m-1$ values $y_1, \dots, y_{m-1}$. These are often computed using a one-step method.

Adams and Bashforth

The Adams–Bashforth methods were designed by John Couch Adams (*left*) to solve a differential equation modeling capillary action devised by Francis Bashforth. Introducing his theory and Adams's method, Bashforth published *The Theories of Capillary Action* [8] in 1883 that compared theoretical and measured forms of drops of fluid. Adams is probably most well known for predicting the existence and position of Neptune (which was discovered a year later by Leverrier), using only mathematics. Discrepancies between Uranus's orbit and the laws of Kepler and Newton led Adams to this prediction. Bashforth was a professor of applied mathematics for the advanced class of artillery officers at Woolwich. Stemming from his interest in ballistics, Bashforth did a series of experiments between 1864 and 1880 that laid the groundwork for our current understanding of air resistance.

For $m = 1$, this is Euler's method: $y_{k+1} = y_k + hf(t_k, y_k)$. For $m = 2$, we have

$$y_{k+1} = y_k + h[b_0 f(t_k, y_k) + b_1 f(t_{k-1}, y_{k-1})],$$

where

$$b_0 = \frac{1}{h}\int_{t_k}^{t_{k+1}} \frac{s - t_{k-1}}{t_k - t_{k-1}}\, ds = \frac{3}{2}, \quad b_1 = \frac{1}{h}\int_{t_k}^{t_{k+1}} \frac{s - t_k}{t_{k-1} - t_k}\, ds = -\frac{1}{2}.$$

Thus the two-step Adams–Bashforth method is

$$y_{k+1} = y_k + h\left[\frac{3}{2} f(t_k, y_k) - \frac{1}{2} f(t_{k-1}, y_{k-1})\right]. \tag{11.28}$$

The Adams–Bashforth methods are *explicit*. Since the error in approximating the integral is $O(h^{m+1})$, the local truncation error, defined as

$$\frac{y(t_{k+1}) - y(t_k)}{h} - \sum_{\ell=0}^{m-1} b_\ell f(t_{k-\ell}, y(t_{k-\ell})),$$

is $O(h^m)$.

The *Adams–Moulton methods* are *implicit* methods and, like the Adams–Bashforth methods, are due solely to Adams. Forest Ray Moulton's name is associated with the methods because he realized that these methods could be used with the Adams–Bashforth methods as a predictor–corrector pair [46]. The Adams–Moulton methods are obtained by using the polynomial q_m that interpolates the values $f(t_{k+1}, y_{k+1}), f(t_k, y_k), \ldots, f(t_{k-m+1}, y_{k-m+1})$ at the points $t_{k+1}, t_k, \ldots, t_{k-m+1}$, in place of $f(s, y(s))$ in the integral on the right-hand side of (11.27). Thus,

$$q_m(s) = \sum_{\ell=0}^{m} \left(\prod_{\substack{j=0 \\ j \neq \ell}}^{m} \frac{s - t_{k+1-j}}{t_{k+1-\ell} - t_{k+1-j}} \right) f(t_{k+1-\ell}, y_{k+1-\ell}),$$

and

$$y_{k+1} = y_k + h \sum_{\ell=0}^{m} c_\ell f(t_{k+1-\ell}, y_{k+1-\ell}),$$

where

$$c_\ell = \frac{1}{h} \int_{t_k}^{t_{k+1}} \left(\prod_{\substack{j=0 \\ j \neq \ell}}^{m} \frac{s - t_{k+1-j}}{t_{k+1-\ell} - t_{k+1-j}} \right) ds.$$

For $m = 0$, this becomes

$$y_{k+1} = y_k + hf(t_{k+1}, y_{k+1}). \tag{11.29}$$

This is called the *backward Euler method*. For $m = 1$, we have

$$y_{k+1} = y_k + h[c_0 f(t_{k+1}, y_{k+1}) + c_1 f(t_k, y_k)],$$

where

$$c_0 = \frac{1}{h} \int_{t_k}^{t_{k+1}} \frac{s - t_k}{t_{k+1} - t_k} ds = \frac{1}{2},$$

$$c_1 = \frac{1}{h} \int_{t_k}^{t_{k+1}} \frac{s - t_{k+1}}{t_k - t_{k+1}} ds = \frac{1}{2}.$$

Thus the Adams–Moulton method corresponding to $m = 1$ is

$$y_{k+1} = y_k + \frac{h}{2}[f(t_{k+1}, y_{k+1}) + f(t_k, y_k)]. \tag{11.30}$$

We have seen this method before; it is the *trapezoidal method*. The local truncation error for an Adams–Moulton method using an mth-degree polynomial interpolant is $O(h^{m+1})$.

11.3.2 General Linear m-Step Methods

A general linear m-step method can be written in the form

$$\sum_{\ell=0}^{m} a_\ell y_{k+\ell} = h \sum_{\ell=0}^{m} b_\ell f(t_{k+\ell}, y_{k+\ell}), \tag{11.31}$$

where $a_m = 1$ and the other a_ℓ's and the b_ℓ's are constants. If $b_m = 0$ the method is explicit; otherwise it is implicit.

Example 11.3.1. The two-step Adams–Bashforth method (11.28) can be written in the form (11.31) by using the formula for y_{k+2}:

$$y_{k+2} - y_{k+1} = h\left[\frac{3}{2} f(t_{k+1}, y_{k+1}) - \frac{1}{2} f(t_k, y_k)\right].$$

For this method, $a_2 = 1$, $a_1 = -1$, $a_0 = 0$, $b_2 = 0$, $b_1 = 3/2$, and $b_0 = -1/2$.

Example 11.3.2. The one-step Adams–Moulton (or trapezoidal) method (11.30) can be written in the form (11.31) as

$$y_{k+1} - y_k = h\left[\frac{1}{2} f(t_{k+1}, y_{k+1}) + \frac{1}{2} f(t_k, y_k)\right].$$

Here $a_1 = 1$, $a_0 = -1$, $b_1 = 1/2$, and $b_0 = 1/2$.

The local truncation error is

$$\tau(t, y, h) \equiv \frac{1}{h}\sum_{\ell=0}^{m} a_\ell y(t + \ell h) - \sum_{\ell=0}^{m} b_\ell f(t + \ell h, y(t + \ell h)). \tag{11.32}$$

Theorem 11.3.1. The local truncation error of the multistep method (11.31) is of order $p \geq 1$ if and only if

$$\sum_{\ell=0}^{m} a_\ell = 0 \quad \text{and} \quad \sum_{\ell=0}^{m} \ell^j a_\ell = j\sum_{\ell=0}^{m} \ell^{j-1} b_\ell, \quad j = 1, \ldots, p. \tag{11.33}$$

Proof. Expand $\tau(t, y, h)$ in a Taylor series. (We assume, as always, that y is sufficiently differentiable for the Taylor series expansion to exist. In this case, y must be p times differentiable.)

$$\begin{aligned}
\tau(t, y, h) &= \frac{1}{h}\sum_{\ell=0}^{m} a_\ell y(t + \ell h) - \sum_{\ell=0}^{m} b_\ell y'(t + \ell h) \\
&= \frac{1}{h}\sum_{\ell=0}^{m} a_\ell \sum_{j=0}^{p} \frac{(\ell h)^j}{j!} y^{(j)}(t) - \sum_{\ell=0}^{m} b_\ell \sum_{j=0}^{p-1} \frac{(\ell h)^j}{j!} y^{(j+1)}(t) + O(h^p) \\
&= \frac{1}{h}\left(\sum_{\ell=0}^{m} a_\ell\right) y(t) + \sum_{j=1}^{p} \frac{h^{j-1}}{j!}\left(\sum_{\ell=0}^{m} \ell^j a_\ell\right) y^{(j)}(t) \\
&\quad - \sum_{j=0}^{p-1} \frac{h^j}{j!}\left(\sum_{\ell=0}^{m} \ell^j b_\ell\right) y^{(j+1)}(t) + O(h^p) \\
&= \frac{1}{h}\left(\sum_{\ell=0}^{m} a_\ell\right) y(t) + \sum_{j=1}^{p} \frac{h^{j-1}}{j!}\left(\sum_{\ell=0}^{m} \ell^j a_\ell - j\sum_{\ell=0}^{m} \ell^{j-1} b_\ell\right) y^{(j)}(t) + O(h^p).
\end{aligned}$$

Thus the conditions in (11.32) are necessary and sufficient for τ to be $O(h^p)$. □

Example 11.3.3. The two-step Adams–Bashforth method is second order, since

$$\begin{aligned} a_0 + a_1 + a_2 &= 0 + (-1) + 1 = 0, \\ 0 \cdot a_0 + 1 \cdot a_1 + 2 \cdot a_2 &= 1 = b_0 + b_1 + b_2 = -1/2 + 3/2 + 0, \\ 0^2 \cdot a_0 + 1^2 \cdot a_1 + 2^2 \cdot a_2 &= 3 = 2 \cdot (0 \cdot b_0 + 1 \cdot b_1 + 2 \cdot b_2) = 2 \cdot (3/2), \quad \text{but} \\ 0^3 \cdot a_0 + 1^3 \cdot a_1 + 2^3 \cdot a_2 &= 7 \neq 3 \cdot (0^2 \cdot b_0 + 1^2 \cdot b_1 + 2^2 \cdot b_2) = 9/2. \end{aligned}$$

Example 11.3.4. The one-step Adams–Moulton method is second order, since

$$\begin{aligned} a_0 + a_1 &= -1 + 1 = 0, \\ 0 \cdot a_0 + 1 \cdot a_1 &= 1 = b_0 + b_1 = 1/2 + 1/2, \\ 0^2 \cdot a_0 + 1^2 \cdot a_1 &= 1 = 2 \cdot (0 \cdot b_0 + 1 \cdot b_1) = 2 \cdot (1/2), \quad \text{but} \\ 0^3 \cdot a_0 + 1^3 \cdot a_1 &= 1 \neq 3 \cdot (0^2 \cdot b_0 + 1^2 \cdot b_1) = 3/2. \end{aligned}$$

For one-step methods, high-order local truncation error generally means high-order global error. We had only to establish that ψ in (11.24) is uniformly Lipschitz in y in order to prove convergence. This is not the case for multistep methods. There are many reasonable looking multistep methods that are *unstable*! Consider the following simple example:

$$y_{k+2} - 3y_{k+1} + 2y_k = h\left[\frac{13}{12} f(t_{k+2}, y_{k+2}) - \frac{5}{3} f(t_{k+1}, y_{k+1}) - \frac{5}{12} f(t_k, y_k)\right]. \tag{11.34}$$

It is left as an exercise to show that the local truncation error is $O(h^2)$. Suppose, however, that this method is applied to the trivial problem $y' = 0$, $y(0) = 1$, whose solution is $y(t) \equiv 1$. In order to start the method, we must first compute y_1, perhaps using a one-step method. Suppose we compute $y_1 = 1 + \delta$, with some tiny error δ. Then

$$\begin{aligned} y_2 &= 3y_1 - 2y_0 = 1 + 3\delta, \\ y_3 &= 3y_2 - 2y_1 = 1 + 7\delta, \\ y_4 &= 3y_3 - 2y_2 = 1 + 15\delta, \\ &\vdots \\ y_k &= 3y_{k-1} - 2y_{k-2} = 1 + (2^k - 1)\delta. \end{aligned}$$

If δ is about 2^{-53}, like the roundoff error in double precision, then after 53 steps the error has grown to about size 1, and after 100 steps it has grown to 2^{47}!

In order to understand this instability and how to avoid it, we need to study linear difference equations.

11.3.3 Linear Difference Equations

Linear multistep methods take the form of linear difference equations:

$$\sum_{\ell=0}^{m} \alpha_\ell y_{k+\ell} = \beta_k, \quad k = 0, 1, \ldots, \tag{11.35}$$

where $\alpha_m = 1$. Given values for $y_0, \ldots, y_{m-1}$ and knowing the β_k's, $k = 0, 1, \ldots$, this equation determines $y_m, y_{m+1}, \ldots$.

Consider the corresponding *homogeneous* linear difference equation

$$\sum_{\ell=0}^{m} \alpha_\ell x_{k+\ell} = 0, \quad k = 0, 1, \ldots. \tag{11.36}$$

If a sequence $y_0, y_1, \ldots$ satisfies (11.35) and if the sequence $x_0, x_1, \ldots$ satisfies (11.36), then the sequence $y_0 + x_0, y_1 + x_1, \ldots$ also satisfies (11.35). Moreover, if $\hat{y}_0, \hat{y}_1, \ldots$ is any *particular* solution to (11.35), then *all* solutions are of the form $\hat{y}_0 + x_0, \hat{y}_1 + x_1, \ldots$, for some sequence $x_0, x_1, \ldots$ that solves (11.36).

Let us look for solutions of (11.36) in the form $x_j = \lambda^j$. This will be a solution provided

$$\sum_{\ell=0}^{m} \alpha_\ell \lambda^{k+\ell} = \lambda^k \left(\sum_{\ell=0}^{m} \alpha_\ell \lambda^\ell \right) = 0.$$

Define

$$\chi(\lambda) \equiv \sum_{\ell=0}^{m} \alpha_\ell \lambda^\ell \tag{11.37}$$

to be the **characteristic polynomial** of (11.36).

If the roots $\lambda_1, \ldots, \lambda_m$ of $\chi(\lambda)$ are distinct, then the general solution of (11.36) is

$$x_j = \sum_{i=1}^{m} c_i \lambda_i^j, \quad j = 0, 1, \ldots.$$

The constants $c_1, \ldots, c_m$ are determined by imposing initial conditions $x_0 = \hat{x}_0, \ldots, x_{m-1} = \hat{x}_{m-1}$.

Example 11.3.5. The Fibonacci numbers are defined by the difference equation

$$F_{k+2} - F_{k+1} - F_k = 0, \quad k = 0, 1, \ldots,$$

with initial conditions

$$F_0 = 0, \quad F_1 = 1.$$

The characteristic polynomial is $\chi(\lambda) = \lambda^2 - \lambda - 1$, and its roots are $\lambda_1 = (1+\sqrt{5})/2$ and $\lambda_2 = (1-\sqrt{5})/2$. Therefore the general solution to the difference equation is

$$F_j = c_1 \left(\frac{1+\sqrt{5}}{2} \right)^j + c_2 \left(\frac{1-\sqrt{5}}{2} \right)^j, \quad j = 0, 1, \ldots.$$

The initial conditions imply that

$$F_0 = c_1 + c_2 = 0, \quad F_1 = c_1 \left(\frac{1+\sqrt{5}}{2} \right) + c_2 \left(\frac{1-\sqrt{5}}{2} \right) = 1,$$

from which it follows that $c_1 = 1/\sqrt{5}$ and $c_2 = -1/\sqrt{5}$.

Example 11.3.6. The unstable method (11.34) applied to the problem $y' = 0$ gave rise to the difference equation

$$y_{k+2} - 3y_{k+1} + 2y_k = 0,$$

with initial conditions

$$y_0 = 1, \quad y_1 = 1 + \delta.$$

The characteristic polynomial for this equation is $\chi(\lambda) = \lambda^2 - 3\lambda + 2$, and its roots are $\lambda_1 = 1$ and $\lambda_2 = 2$. Therefore the general solution to the difference equation is

$$y_j = c_1 \cdot 1^j + c_2 \cdot 2^j = c_1 + 2^j c_2.$$

The initial conditions imply that

$$y_0 = c_1 + c_2 = 1, \quad y_1 = c_1 + 2c_2 = 1 + \delta,$$

from which it follows that $c_1 = 1 - \delta$ and $c_2 = \delta$. If δ were 0, this would give the desired solution, $y_j \equiv 1$. But because δ is nonzero, the result contains a small multiple of the other solution, 2^j, which quickly overshadows the desired solution as j increases.

If some of the roots of the characteristic polynomial have multiplicity greater than one—say, $\lambda_1, \ldots, \lambda_s$ are the distinct roots with multiplicities $q_1, \ldots, q_s$—then the general solution of (11.36) is a linear combination of

$$\{x_j = j^\ell \lambda_i^j : \ \ell = 0, \ldots, q_i - 1, \ i = 1, \ldots, s\}.$$

To illustrate this, suppose, for example, that λ_i is a double root of χ in (11.37). This means that

$$\chi(\lambda_i) = \sum_{\ell=0}^{m} \alpha_\ell \lambda_i^\ell = 0, \quad \chi'(\lambda_i) = \sum_{\ell=1}^{m} \alpha_\ell \ell \lambda_i^{\ell-1} = 0.$$

Setting $x_j = j\lambda_i^j$ in the left-hand side of (11.36), we find

$$\sum_{\ell=0}^{m} \alpha_\ell (k+\ell) \lambda_i^{(k+\ell)} = k\lambda_i^k \sum_{\ell=0}^{m} \alpha_\ell \lambda_i^\ell + \lambda_i^{k+1} \sum_{\ell=1}^{m} \alpha_\ell \ell \lambda_i^{\ell-1} = k\lambda_i^k \chi(\lambda_i) + \lambda_i^{k+1} \chi'(\lambda_i) = 0.$$

Thus $x_j = j\lambda_i^j$, is a solution of (11.36), just as $x_j = \lambda_i^j$ is, when λ_i is a double root of χ.

This leads to a result known as the *root condition*.

Root Condition. If all roots of χ are less than or equal to 1 in absolute value and any root with absolute value 1 is a simple root, then solutions of (11.36) remain bounded as $j \to \infty$.

This follows because if $|\lambda_i| \leq 1$, then λ_i^j is bounded by 1 for all j. If $|\lambda_i| < 1$ and ℓ is a fixed positive integer, then while $j^\ell \lambda_i^j$ may grow for j in a certain range, it approaches 0 as $j \to \infty$, since λ_i^j decreases at a faster rate than j^ℓ grows.

Another way to look at linear difference equations is as follows. Define a vector

$$\mathbf{x}_{k+1} \equiv \begin{pmatrix} x_{k+1} \\ x_{k+2} \\ \vdots \\ x_{k+m} \end{pmatrix}.$$

Then the difference equation (11.36) can be written in the form

$$\mathbf{x}_{k+1} = A\mathbf{x}_k, \quad \text{where} \quad A = \begin{pmatrix} 0 & 1 & & \\ & \ddots & \ddots & \\ & & 0 & 1 \\ -\alpha_0 & \dots & -\alpha_{m-2} & -\alpha_{m-1} \end{pmatrix}. \tag{11.38}$$

The matrix A in (11.38) is called a **companion matrix**, and it can be shown that the characteristic polynomial χ in (11.37) is the characteristic polynomial of this matrix; that is, $\chi(\lambda) = (-1)^m \det(A - \lambda I)$. It follows from (11.38) that $\mathbf{x}_k = A^k \mathbf{x}_0$, where $\mathbf{x}_0$ is the vector of initial values $(x_0, \dots, x_{m-1})^T$. The root condition says that the eigenvalues of A are less than or equal to 1 in absolute value and that any eigenvalue with absolute value 1 is a simple eigenvalue. Under these assumptions it can be shown that the norms of powers of A remain bounded: $\sup_{k \geq 0} \|A^k\|$ is finite, for any matrix norm $\|\cdot\|$.

Having analyzed solutions to the linear homogeneous difference equation (11.36), we must now relate these to solutions of the linear inhomogeneous equation (11.35). If we define

$$\mathbf{y}_{k+1} \equiv \begin{pmatrix} y_{k+1} \\ y_{k+2} \\ \vdots \\ y_{k+m} \end{pmatrix}, \quad \mathbf{b}_k \equiv \begin{pmatrix} 0 \\ \vdots \\ 0 \\ \beta_k \end{pmatrix},$$

then the inhomogeneous system can be written in the form $\mathbf{y}_{k+1} = A\mathbf{y}_k + \mathbf{b}_k$. It follows by induction that

$$\mathbf{y}_{k+1} = A^{k+1}\mathbf{y}_0 + \sum_{\nu=0}^{k} A^\nu \mathbf{b}_{k-\nu}.$$

Taking the infinity norm of the vectors on each side gives the inequality

$$\|\mathbf{y}_{k+1}\|_\infty \le \|A^{k+1}\|_\infty \|\mathbf{y}_0\|_\infty + \sum_{\nu=0}^{k} \|A^\nu\|_\infty \|\mathbf{b}_{k-\nu}\|_\infty.$$

As noted previously, if χ satisfies the root condition, then $M \equiv \sup_{k\ge 0} \|A^k\|_\infty$ is finite. Using this with the above inequality we have

$$|y_{k+m}| \le \|\mathbf{y}_{k+1}\|_\infty \le M\left(\|\mathbf{y}_0\|_\infty + \sum_{\nu=0}^{k} |\beta_{k-\nu}|\right).$$

Thus, when χ satisfies the root condition, not only do solutions of the homogeneous difference equation (11.36) remain bounded, but so do solutions of the inhomogeneous equation (11.35), *provided* that $\sum_{\nu=0}^{k} |\beta_\nu|$ remains bounded as $k \to \infty$. If $\beta_\nu = h\sum_{\ell=0}^{m} b_\ell f(t_{\nu+\ell}, y_{\nu+\ell})$, as in the linear multistep method (11.31), then this will be the case if $h = (T - t_0)/N$ for some fixed time T and $k \le N \to \infty$. This is the basis of the *Dahlquist equivalence theorem*.

11.3.4 The Dahlquist Equivalence Theorem

Germund Dahlquist was a mathematician and numerical analyst at the Royal Institute of Technology in Stockholm, Sweden. Beginning in 1951, Dahlquist made groundbreaking contributions to the theory of numerical methods for initial value problems in differential equations. In the early 1950s numerical methods existed, but their behavior was not well understood; anomalies arose due to the complexity of problems that could now be attacked via digital computers, and understanding of these anomalies was lacking. Dahlquist introduced stability and convergence analysis to numerical methods for differential equations and thereby put the subject on a sound theoretical footing [13].

Definition. The linear multistep method (11.31) is **convergent** if for every initial value problem $y' = f(t, y)$, $y(t_0) = \hat{y}_0$, where f satisfies the hypotheses of theorem 11.1.3, and any choice of initial values $y_0(h), \ldots, y_{m-1}(h)$ such that $\lim_{h\to 0} |y(t_0 + ih) - y_i(h)| = 0$, $i = 0, 1, \ldots, m-1$, we have

$$\lim_{h\to 0} \max_{\{k:\ t_0+kh\in[t_0,T]\}} |y(t_0 + kh) - y_k| = 0.$$

Theorem 11.3.2 (Dahlquist equivalence theorem). The multistep method (11.31) is *convergent* if and only if the local truncation error is of order $p \ge 1$ and $\chi(\lambda)$ in (11.37) satisfies the root condition.

A linear multistep method whose associated characteristic polynomial satisfies the root condition is called **zero-stable**. The Dahlquist equivalence theorem can be restated using this terminology as: A linear multistep method is convergent if and only if it is *consistent* ($\tau(t, y, h) = O(h)$ or higher order) and *zero-stable*. It also can be shown that if the method is zero-stable and if the local truncation error is $O(h^p)$, $p \ge 1$, then the global error is $O(h^p)$.

11.4 STIFF EQUATIONS

Up to now we have been concerned with what happens in the limit as $h \to 0$. In order for a method to be useful, it certainly should converge to the true solution as $h \to 0$. But in practice, we use a fixed nonzero step size h, or at least, there is a lower bound on h based on allowable time for the computation and also perhaps on the machine precision, since below a certain point roundoff will begin to cause inaccuracy. Therefore we would like to understand how different methods behave with a fixed time step h. This, of course, will be problem dependent but we would like to use methods that give reasonable results with a fixed time step h for as wide a class of problems as possible.

Consider the following system of two ODEs in two unknowns:

$$\begin{aligned} y_1' &= -100y_1 + y_2, \\ y_2' &= -(1/10)y_2. \end{aligned} \tag{11.39}$$

If $\mathbf{y} = (y_1, y_2)^T$, then this system also can be written in the form

$$\mathbf{y}' = A\mathbf{y}, \quad A = \begin{pmatrix} -100 & 1 \\ 0 & -1/10 \end{pmatrix}. \tag{11.40}$$

We can solve this system analytically. From the second equation it follows that $y_2(t) = e^{-t/10}y_2(0)$. Substituting this expression into the first equation and solving for $y_1(t)$, we find

$$y_1(t) = e^{-100t}\left(y_1(0) - \frac{10}{999}y_2(0)\right) + e^{-t/10}\frac{10}{999}y_2(0).$$

The first term decays to 0 *very* rapidly. The other term in the expression for $y_1(t)$ and the expression for $y_2(t)$ behave very benignly, decaying to 0 over time, but at a fairly slow pace. Hence, while a very small time step might be required for the initial transient, one would expect to be able to use a much larger time step once the e^{-100t} term has decayed to near 0.

Suppose Euler's method is used to solve this problem. The equations are

$$y_{2,k+1} = (1 - h/10)y_{2k} \implies y_{2k} = (1 - h/10)^k y_2(0),$$

and

$$\begin{aligned} y_{1,k+1} &= (1 - 100h)y_{1k} + hy_{2k} \\ &= (1 - 100h)y_{1k} + h(1 - h/10)^k y_2(0) \\ &= (1 - 100h)^2 y_{1,k-1} + h[(1 - 100h)(1 - h/10)^{k-1} + (1 - h/10)^k]y_2(0) \\ &\vdots \\ &= (1 - 100h)^{k+1}y_1(0) + h(1 - h/10)^k\left[\sum_{\ell=0}^{k}\left(\frac{1 - 100h}{1 - h/10}\right)^\ell\right]y_2(0). \end{aligned}$$

Summing the geometric series and doing some algebra, we find

$$y_{1,k+1} = (1 - 100h)^{k+1}\left[y_1(0) - \frac{10}{999}y_2(0)\right] + (1 - h/10)^{k+1}\frac{10}{999}y_2(0).$$

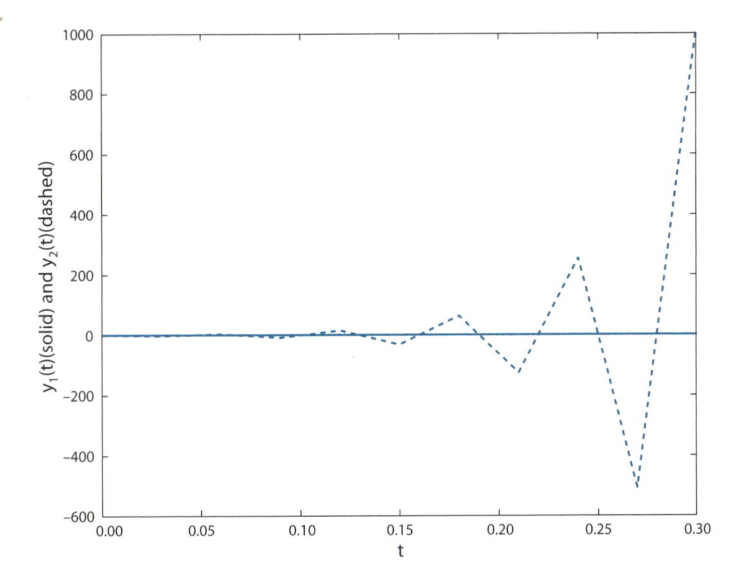

Figure 11.8. Euler's method's poor performance for a system of stiff equations.

If $h > 1/50$, then $|1 - 100h| > 1$, and the first term *grows* geometrically, in contrast to the true solution. Even if the initial values are such that $y_1(0) - (10/999)y_2(0) = 0$, roundoff will cause the approximation in Euler's method to grow unless $h < 1/50$. The local truncation error for Euler's method is fairly small even for $h > 1/50$, but the algorithm will fail because of *instability* (but a different kind of instability than discussed previously, because this one occurs only for $h > 1/50$). Figure 11.8 is a plot of results from Euler's method applied to this problem, using $h = 0.03$.

The appearance of wild oscillations, as in figure 11.8, is typical of the behavior of unstable algorithms. Often one sees only one large jump in the plot because the last jump is so much larger than the others that it completely overshadows them; upon changing the scaling of the axes, one begins to see that the oscillations begin much earlier.

Equations like the system (11.39) are called *stiff* equations, and they are characterized by having components evolving on very different time scales. They are very common in practice—arising in areas such as chemical kinetics, control theory, reactor kinetics, weather prediction, electronics, mathematical biology, cosmology, etc.

11.4.1 Absolute Stability

To analyze the behavior of a method with a particular time step h, one might consider a very simple test equation:

$$y' = \lambda y, \tag{11.41}$$

where λ is a complex constant. The solution is $y(t) = e^{\lambda t} y(0)$. Sample solutions are plotted in figure 11.9 for the real part of λ greater than 0, equal to 0, and

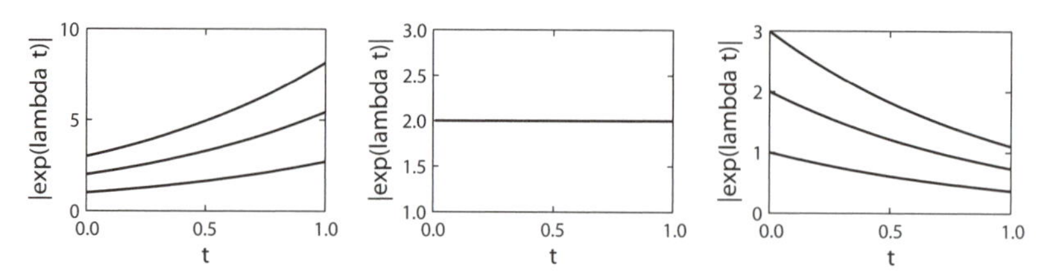

Figure 11.9. Sample solutions to $y' = \lambda y$ (*left*: Re(lambda) $>$ 0; *middle*: Re(lambda)$=$ 0; *right*: Re(lambda)$<$ 0).

less than 0. Note that $y(t) \to 0$ as $t \to \infty$ if and only if $\mathcal{R}(\lambda) < 0$, where $\mathcal{R}(\cdot)$ denotes the real part.

Equation (11.41) is really much simpler than the problems one is likely to want to solve numerically, but certainly if a method performs poorly on this test equation then it cannot be expected to do well in practice. The idea behind this test equation is the following. First, suppose one wishes to solve a system of the form

$$\mathbf{y}' = A\mathbf{y}, \tag{11.42}$$

where A is an n by n matrix. Assume that A is *diagonalizable*, meaning that A can be written in the form $A = V\Lambda V^{-1}$, where Λ is a diagonal matrix of eigenvalues and the columns of V are eigenvectors of A. Multiplying equation (11.42) by V^{-1} on the left, we find

$$(V^{-1}\mathbf{y})' = \Lambda(V^{-1}\mathbf{y}),$$

so that the equations for the components of $V^{-1}\mathbf{y}$ separate:

$$(V^{-1}\mathbf{y})'_j = \lambda_j(V^{-1}\mathbf{y})_j, \quad j = 1, \ldots, n.$$

Thus, once we look at the equation in a basis where A is diagonal (i.e., we look at $V^{-1}\mathbf{y}$ instead of $\mathbf{y}$), then it looks like a system of independent equations, each of the form (11.41). Now, a drawback of this approach is that it does not take into account how different the vector $V^{-1}\mathbf{y}$ may be from $\mathbf{y}$, and for some problems this difference can be quite significant. Nevertheless, if we understand the behavior of a method applied to the test equation (11.41) for different values of λ, then that gives us some information about how it behaves for a system of linear ODEs like (11.42). A system of nonlinear ODEs behaves locally like a linear system, and so we also will gain some information about the behavior of the method when applied to a nonlinear system of ODEs.

With this in mind, we make the following definition.

Definition. The **region of absolute stability** of a method is the set of all numbers $h\lambda \in \mathbf{C}$ such that $y_k \to 0$ as $k \to \infty$, when the method is applied to the test equation (11.41) using step size h.

Example 11.4.1. When Euler's method is applied to the test equation (11.41), the resulting formula for y_{k+1} is:

$$y_{k+1} = y_k + h\lambda y_k = (1 + h\lambda)y_k = \ldots = (1 + h\lambda)^{k+1}y_0.$$

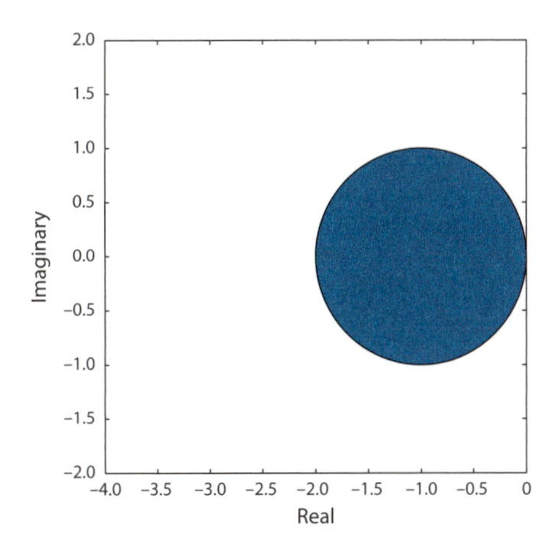

Figure 11.10. Region of absolute stability for Euler's method.

It follows that the region of absolute stability is

$$\{h\lambda : \ |1 + h\lambda| < 1\}.$$

This is the interior of a disk in the complex plane centered at -1 and of radius 1, as pictured in figure 11.10.

This implies, for example, that if one is solving a system of two ODEs, $\mathbf{y}' = A\mathbf{y}$, where A is diagonalizable and has eigenvalues, say, $-1 + 10i$ and $-10 - 10i$, then one must choose h to satisfy

$$|1 + h(-1 + 10i)| < 1 \quad \text{and} \quad |1 + h(-10 - 10i)| < 1,$$

or, equivalently,

$$(1 - h)^2 + 100h^2 < 1 \quad \text{and} \quad (1 - 10h)^2 + 100h^2 < 1.$$

To satisfy the first inequality, one must have $h < 2/101$ and to satisfy the second, $h < 1/10$. In order to achieve stability in both eigencomponents, one needs $h < \min\{2/101, 1/10\} = 2/101$.

Note that in the region of absolute stability, differences at one step become smaller at later steps; that is, suppose z_{k+1} also satisfies $z_{k+1} = (1 + h\lambda)z_k$, but $z_k \neq y_k$. Then the difference between z_{k+1} and y_{k+1} is $z_{k+1} - y_{k+1} = (1 + h\lambda)(z_k - y_k)$, and hence $|z_{k+1} - y_{k+1}| < |z_k - y_k|$, since $|1 + h\lambda| < 1$.

Ideally, one would like the region of absolute stability to consist of the entire left half plane. This would mean that for any complex number λ with $\mathcal{R}(\lambda) < 0$, the approximate solution generated by the method for the test problem (11.41), using any step size $h > 0$, would decay over time, as the true solution does. A method with this property is called *A-stable*.

Definition. A method is **A-stable** if its region of absolute stability contains the entire left half plane.

Example 11.4.2. When the backward Euler method is applied to the test equation (11.41), the resulting formula for y_{k+1} is

$$y_{k+1} = y_k + h\lambda y_{k+1} \Longrightarrow y_{k+1} = \frac{1}{1 - h\lambda}\, y_k = \ldots = \frac{1}{(1 - h\lambda)^{k+1}}\, y_0.$$

It follows that the region of absolute stability is

$$\{h\lambda : \; |1 - h\lambda| > 1\}.$$

Since h is real and λ is complex, the absolute value of $1 - h\lambda$ is

$$\sqrt{(1 - h\mathcal{R}(\lambda))^2 + (h\mathcal{I}(\lambda))^2}.$$

If $\mathcal{R}(\lambda) < 0$, this is always greater than 1, since $1 - h\mathcal{R}(\lambda) > 1$. Therefore the region of absolute stability of the backward Euler method contains the entire left half plane, and the method is A-stable.

Following are some more examples of difference methods and their regions of absolute stability.

Example 11.4.3. When the classical fourth-order Runge–Kutta method is applied to the test equation (11.41), the resulting formulas are

$$\begin{aligned}
q_1 &= \lambda y_k, \\
q_2 &= \lambda\big(y_k + \frac{h}{2} q_1\big) = \big(\lambda + \frac{1}{2} h\lambda^2\big) y_k, \\
q_3 &= \lambda\big(y_k + \frac{h}{2} q_2\big) = \big(\lambda + \frac{1}{2} h\lambda^2 + \frac{1}{4} h^2\lambda^3\big) y_k, \\
q_4 &= \lambda\big(y_k + h q_3\big) = \big(\lambda + h\lambda^2 + \frac{1}{2} h^2\lambda^3 + \frac{1}{4} h^3\lambda^4\big) y_k, \\
y_{k+1} &= y_k + \frac{h}{6}\big[q_1 + 2q_2 + 2q_3 + q_4\big] \\
&= y_k + \frac{h}{6}\big[6\lambda + 3h\lambda^2 + h^2\lambda^3 + \frac{1}{4} h^3\lambda^4\big] y_k \\
&= \big[1 + h\lambda + \frac{1}{2}(h\lambda)^2 + \frac{1}{6}(h\lambda)^3 + \frac{1}{24}(h\lambda)^4\big] y_k.
\end{aligned}$$

It follows that the region of absolute stability is

$$\big\{z \in \mathbf{C} : \; \big|1 + z + \frac{z^2}{2} + \frac{z^3}{6} + \frac{z^4}{24}\big| < 1\big\}.$$

This region is pictured in figure 11.11.

Example 11.4.4. When the trapezoidal method is applied to the test equation (11.41), the resulting formula for y_{k+1} is

$$y_{k+1} = y_k + \frac{h}{2}(\lambda y_k + \lambda y_{k+1}) \Longrightarrow y_{k+1} = \left(\frac{1 + h\lambda/2}{1 - h\lambda/2}\right) y_k.$$

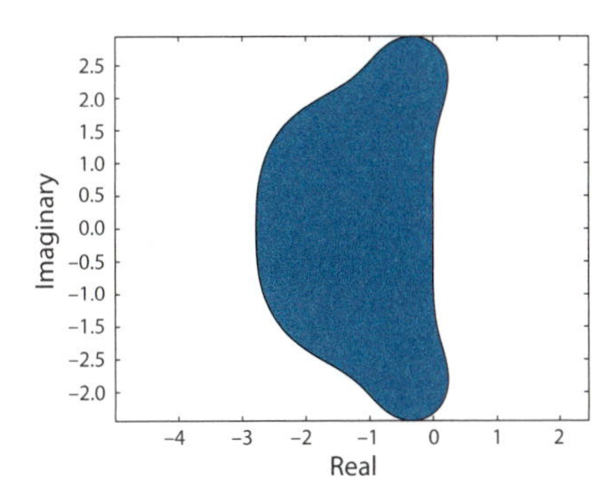

Figure 11.11. Region of absolute stability for the fourth-order Runge–Kutta method.

It follows that $y_k \to 0$ as $k \to \infty$ if and only if

$$\left|\frac{1 + h\lambda/2}{1 - h\lambda/2}\right| < 1 \Longleftrightarrow |1 + h\lambda/2| < |1 - h\lambda/2| \Longleftrightarrow \mathcal{R}(h\lambda) < 0.$$

Thus the region of absolute stability is precisely the open left half plane $\{z \in \mathbf{C} : \mathcal{R}(z) < 0\}$; therefore the trapezoidal method is *A-stable.*

You may have noticed in the examples that the only methods that were A-stable were implicit methods. This is not an accident. It can be shown that *there are no explicit A-stable methods of the form (11.31).* It can be shown further that:

Theorem 11.4.1 (Dahlquist). The highest order of an A-stable linear multistep method of the form (11.31) is two.

Despite this negative result, one can look for high-order methods with large regions of absolute stability, even if the region of absolute stability does not contain the entire left half plane. The *backward differentiation formulas*, or, *BDF methods*, are a class of implicit methods designed to have large regions of absolute stability and hence to be effective for stiff equations.

11.4.2 Backward Differentiation Formulas (BDF Methods)

The first use of BDF methods appears to be by Curtiss and Hirschfelder [29] in 1952. Their development as practical numerical methods, along with computational software, was largely pioneered by Bill Gear in the 1960s and 1970s [44].

Definition. An mth-order m-step method of the form

$$\sum_{\ell=0}^{m} a_\ell y_{k+\ell} = hb_m f(t_{k+m}, y_{k+m}), \tag{11.43}$$

is a **backward differentiation formula** or **BDF method.**

Note that all BDF methods are implicit.

We can use theorem 11.3.1 to derive the BDF methods for different values of m. For $m = 1$, the requirement that the method be first order means that $a_0 + a_1 = 0$ and $0 \cdot a_0 + 1 \cdot a_1 = b_1$. By convention we take $a_1 = 1$, so that $a_0 = -1$ and $b_1 = 1$, giving the method

$$y_{k+1} = y_k + hf(t_{k+1}, y_{k+1}).$$

This is the backward Euler method, which we have seen before.

For $m = 2$, the requirements in (11.33) for second-order accuracy become

$$a_0 + a_1 + a_2 = 0, \quad 0 \cdot a_0 + 1 \cdot a_1 + 2 \cdot a_2 = b_2, \quad 0^2 \cdot a_0 + 1^2 \cdot a_1 + 2^2 \cdot a_2 = 2(2 \cdot b_2).$$

Setting $a_2 = 1$, these equations become $a_0 + a_1 = -1$, $a_1 = b_2 - 2$, and $a_1 = 4b_2 - 4$, and solving for a_0, a_1, and b_2, we find $b_2 = 2/3$, $a_1 = -4/3$, and $a_0 = 1/3$. Thus the second-order BDF method is

$$y_{k+2} - \frac{4}{3}y_{k+1} + \frac{1}{3}y_k = \frac{2}{3}hf(t_{k+2}, y_{k+2}).$$

Note that the characteristic polynomial of this multistep method is

$$\chi(z) = z^2 - \frac{4}{3}z + \frac{1}{3} = (z-1)(z-\frac{1}{3}),$$

which has roots $z = 1$ and $z = 1/3$. The method is zero-stable since the roots are both less than or equal to 1 in absolute value and the root $z = 1$ is simple. Therefore by the Dahlquist equivalence theorem (theorem 11.3.2) the method is convergent and the global error is of the same order as the local error, $O(h^2)$.

One can continue in this way to derive higher-order BDF methods. By looking at their characteristic polynomials it can be shown that the BDF methods are zero-stable only for $1 \leq m \leq 6$, but this is a high enough order for practical purposes.

The BDF method for $m = 3$ is

$$y_{k+3} - \frac{18}{11}y_{k+2} + \frac{9}{11}y_{k+1} - \frac{2}{11}y_k = \frac{6}{11}hf(t_{k+3}, y_{k+3}). \tag{11.44}$$

According to theorem 11.4.1, its region of absolute stability cannot contain the entire open left half plane, but it contains most of this area, as illustrated in figure 11.12.

11.4.3 Implicit Runge–Kutta (IRK) Methods

Another class of methods that are useful for stiff equations are the **implicit Runge–Kutta (IRK) methods**. These have the form

$$\xi_j = y_k + h\sum_{i=1}^{\nu} a_{ji} f(t_k + c_i h, \xi_i), \quad j = 1, \ldots, \nu, \tag{11.45}$$

$$y_{k+1} = y_k + h\sum_{j=1}^{\nu} b_j f(t_k + c_j h, \xi_j). \tag{11.46}$$

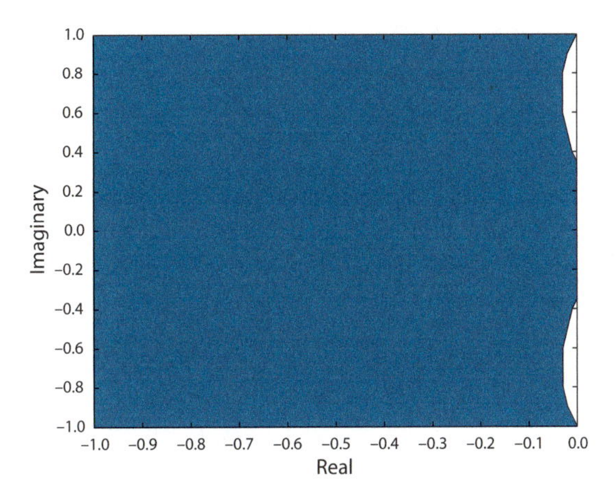

Figure 11.12. Region of absolute stability for BDF with $m = 3$.

The values a_{ji}, b_j, and c_j can be chosen arbitrarily, but for consistency it is required that

$$\sum_{i=1}^{\nu} a_{ji} = c_j, \quad j = 1, \ldots, \nu.$$

It can be shown that for every $\nu \geq 1$, there is a unique IRK method of order 2ν, and it is *A-stable*. The IRK method of order 2ν is obtained by taking the values $c_1, \ldots, c_\nu$ to be the zeros of the νth orthogonal polynomial on $[0, 1]$. The IRK method then corresponds to a Gauss quadrature formula of the type discussed in section 10.3.

The ($\nu = 1$) IRK method of order 2 is the trapezoidal method (11.18). The ($\nu = 2$) IRK method of order 4 has $c_1 = 1/2 - \sqrt{3}/6$ and $c_2 = 1/2 + \sqrt{3}/6$, and the other parameters are then derived to enforce fourth-order accuracy.

11.5 SOLVING SYSTEMS OF NONLINEAR EQUATIONS IN IMPLICIT METHODS

We have seen the importance of implicit methods in solving stiff systems of ODEs. But we have not discussed how to solve the nonlinear equations that must be solved at each time step in order to implement such a method. Chapter 4 describes methods for solving a single nonlinear equation in one unknown, and some of these easily generalize to systems of n nonlinear equations in n unknowns.

The nonlinear equations that arise from solving systems of ODEs have a special form. Suppose, for example, that we use a multistep method (11.31) with $b_m \neq 0$. The system of equations that we must solve for the *vector* $\mathbf{y}_{k+m}$ can be written in the form

$$\mathbf{y}_{k+m} = h b_m \mathbf{f}(t_{k+m}, \mathbf{y}_{k+m}) + \boldsymbol{\gamma}, \tag{11.47}$$

where

$$\gamma = h \sum_{\ell=0}^{m-1} b_\ell \mathbf{f}(t_{k+\ell}, \mathbf{y}_{k+\ell}) - \sum_{\ell=0}^{m-1} a_\ell \mathbf{y}_{k+\ell}$$

is known.

The IRK method (11.45–11.46) can be written in the form

$$\begin{pmatrix} \xi_1 \\ \vdots \\ \xi_\nu \\ \mathbf{y}_{k+1} \end{pmatrix} = h \begin{pmatrix} \sum_{i=1}^{\nu} a_{1i}\mathbf{f}(t_k + c_i h, \xi_i) \\ \vdots \\ \sum_{i=1}^{\nu} a_{\nu i}\mathbf{f}(t_k + c_i h, \xi_i) \\ \sum_{j=1}^{\nu} b_j \mathbf{f}(t_k + c_j h, \xi_j) \end{pmatrix} + \begin{pmatrix} \mathbf{y}_k \\ \vdots \\ \mathbf{y}_k \\ \mathbf{y}_k \end{pmatrix}. \tag{11.48}$$

Both types of method, (11.47) and (11.48), are of the general form

$$\mathbf{w} = h\mathbf{g}(\mathbf{w}) + \gamma, \tag{11.49}$$

where γ is a vector of known values and $\mathbf{w}$ is the vector of values that we seek.

11.5.1 Fixed Point Iteration

The *fixed point iteration methods* described in section 4.5 can be applied to systems of equations of the form (11.49):

$$\mathbf{w}^{(j+1)} = h\mathbf{g}(\mathbf{w}^{(j)}) + \gamma, \tag{11.50}$$

where $\mathbf{w}^{(0)}$ is the initial guess and superscripts denote the iteration number. (Remember that we are solving at a fixed time step k.) The following theorem gives a sufficient condition for convergence.

Theorem 11.5.1. Let $\mathbf{w}_*$ be a solution of (11.49) and suppose that there exists a neighborhood $S_r(\mathbf{w}_*) = \{\mathbf{v} : \|\mathbf{v} - \mathbf{w}_*\| < r\}$ on which $h\mathbf{g}$ is contractive; that is, there exists $\alpha < 1$ such that for all $\mathbf{u}, \mathbf{v} \in S_r(\mathbf{w}_*)$,

$$h\|\mathbf{g}(\mathbf{u}) - \mathbf{g}(\mathbf{v})\| \le \alpha\|\mathbf{u} - \mathbf{v}\|. \tag{11.51}$$

Then if $\mathbf{w}^{(0)} \in S_r(\mathbf{w}_*)$ then $\mathbf{w}^{(j)} \in S_r(\mathbf{w}_*)$ for all j and

$$\|\mathbf{w}^{(j)} - \mathbf{w}_*\| \le \alpha\|\mathbf{w}^{(j-1)} - \mathbf{w}_*\| \le \ldots \le \alpha^j\|\mathbf{w}^{(0)} - \mathbf{w}_*\|. \tag{11.52}$$

Thus $\mathbf{w}^{(j)}$ converges to $\mathbf{w}_*$ at least linearly, with convergence factor α.

Proof. The result follows immediately from the contractivity of $h\mathbf{g}$:

$$\begin{aligned} \|\mathbf{w}^{(j)} - \mathbf{w}_*\| &= \|h\mathbf{g}(\mathbf{w}^{(j-1)}) + \gamma - (h\mathbf{g}(\mathbf{w}_*) + \gamma)\| \\ &= h\|\mathbf{g}(\mathbf{w}^{(j-1)}) - \mathbf{g}(\mathbf{w}_*)\| \\ &\le \alpha\|\mathbf{w}^{(j-1)} - \mathbf{w}_*\|. \end{aligned}$$

From this it follows by induction that each $\mathbf{w}^{(j)} \in S_r(\mathbf{w}_*)$ if $\mathbf{w}^{(0)} \in S_r(\mathbf{w}_*)$ and that (11.52) holds. □

Now, assuming that (11.49) is a stable method for solving the initial value problem, the function $\mathbf{g}$ will satisfy a Lipschitz condition; that is, there will exist a constant L such that for all vectors $\mathbf{u}$ and $\mathbf{v}$:

$$\|\mathbf{g}(\mathbf{u}) - \mathbf{g}(\mathbf{v})\| \leq L\|\mathbf{u} - \mathbf{v}\|.$$

Hence (11.51) will be satisfied if $h < 1/L$. But this is similar to the restriction on h that is necessary for an *explicit* method to be absolutely stable. For example, when Euler's method was applied to the test problem $y' = \lambda y$, we found that a necessary condition to avoid growth in the approximate solution when it should be decaying was $|1 + h\lambda| < 1$; for $\lambda = -L$, this means that we need $h < 2/L$. Thus, if the simple iteration (11.50) is used to solve the nonlinear system in an implicit method, it is not clear that we have gained anything over an explicit method. Therefore a standard rule of thumb is that while the fixed point iteration (11.50) may be acceptable for nonstiff problems, it is not appropriate for stiff systems.

Example 11.5.1. Consider the trapezoidal method with fixed point iteration for the (scalar) test equation $y' = \lambda y$. The trapezoidal method is

$$y_{k+1} = y_k + \frac{h}{2}\lambda\,(y_k + y_{k+1}),$$

and it is A-stable. To solve for y_{k+1}, however, we use the iteration

$$y_{k+1}^{(j+1)} = y_k + \frac{h}{2}\lambda\left(y_k + y_{k+1}^{(j)}\right).$$

Subtracting these two equations, we find

$$y_{k+1}^{(j+1)} - y_{k+1} = \frac{h}{2}\lambda\left(y_{k+1}^{(j)} - y_{k+1}\right),$$

and this difference converges to 0 as $j \to \infty$ if and only if $|h\lambda| < 2$. Thus we lose the benefits of A-stability!

11.5.2 Newton's Method

Suppose we wish to solve a system of n nonlinear equations in n unknowns:

$$\mathbf{q}(\mathbf{w}) = 0, \quad \text{or,} \quad \begin{pmatrix} q_1(w_1, \ldots, w_n) \\ \vdots \\ q_n(w_1, \ldots, w_n) \end{pmatrix} = \begin{pmatrix} 0 \\ \vdots \\ 0 \end{pmatrix}.$$

Given an initial guess $\mathbf{w}^{(0)} = (w_1^{(0)}, \ldots, w_n^{(0)})^T$, we can expand $\mathbf{q}(\mathbf{w})$ in a Taylor series about $\mathbf{w}^{(0)}$:

$$\begin{pmatrix} q_1(w_1, \ldots, w_n) \\ \vdots \\ q_n(w_1, \ldots, w_n) \end{pmatrix} = \begin{pmatrix} q_1(w_1^{(0)}, \ldots, w_n^{(0)}) \\ \vdots \\ q_n(w_1^{(0)}, \ldots, w_n^{(0)}) \end{pmatrix} + \begin{pmatrix} \sum_{i=1}^n \frac{\partial q_1}{\partial w_i}(\mathbf{w}^{(0)})\,(w_i - w_i^{(0)}) \\ \vdots \\ \sum_{i=1}^n \frac{\partial q_n}{\partial w_i}(\mathbf{w}^{(0)})\,(w_i - w_i^{(0)}) \end{pmatrix}$$

$$+ \begin{pmatrix} O(\|\mathbf{w} - \mathbf{w}^{(0)}\|^2) \\ \vdots \\ O(\|\mathbf{w} - \mathbf{w}^{(0)}\|^2) \end{pmatrix},$$

or, equivalently,

$$\mathbf{q}(\mathbf{w}) = \mathbf{q}(\mathbf{w}^{(0)}) + J_{\mathbf{q}}(\mathbf{w}^{(0)})(\mathbf{w} - \mathbf{w}^{(0)}) + O(\|\mathbf{w} - \mathbf{w}^{(0)}\|^2), \tag{11.53}$$

where $J_{\mathbf{q}}(\mathbf{w}^{(0)})$ is the **Jacobian** of $\mathbf{q}$ evaluated at $\mathbf{w}^{(0)}$:

$$J_{\mathbf{q}}(\mathbf{w}^{(0)}) \equiv \left. \begin{pmatrix} \frac{\partial q_1}{\partial w_1} & \cdots & \frac{\partial q_1}{\partial w_n} \\ \vdots & & \vdots \\ \frac{\partial q_n}{\partial w_1} & \cdots & \frac{\partial q_n}{\partial w_n} \end{pmatrix} \right|_{\mathbf{w}^{(0)}} .$$

Dropping the $O(\|\mathbf{w} - \mathbf{w}^{(0)}\|^2)$ term in (11.53) and setting the resulting expression to 0, we obtain *Newton's method* for n equations in n unknowns:

$$0 = \mathbf{q}(\mathbf{w}^{(0)}) + J_{\mathbf{q}}(\mathbf{w}^{(0)})(\mathbf{w}^{(1)} - \mathbf{w}^{(0)}) \implies \mathbf{w}^{(1)} = \mathbf{w}^{(0)} - \left[J_{\mathbf{q}}(\mathbf{w}^{(0)})\right]^{-1} \mathbf{q}(\mathbf{w}^{(0)}),$$

and, in general, for $j = 0, 1, \ldots,$

$$\mathbf{w}^{(j+1)} = \mathbf{w}^{(j)} - \left[J_{\mathbf{q}}(\mathbf{w}^{(j)})\right]^{-1} \mathbf{q}(\mathbf{w}^{(j)}). \tag{11.54}$$

Note that although we have written $[J_{\mathbf{q}}(\mathbf{w}^{(j)})]^{-1}$ in formula (11.54), it is not necessary to actually compute the inverse; to compute $[J_{\mathbf{q}}(\mathbf{w}^{(j)})]^{-1}\mathbf{q}(\mathbf{w}^{(j)})$, one simply solves the linear system $[J_{\mathbf{q}}(\mathbf{w}^{(j)})]\mathbf{v} = \mathbf{q}(\mathbf{w}^{(j)})$ for $\mathbf{v}$, which is then subtracted from $\mathbf{w}^{(j)}$ to obtain $\mathbf{w}^{(j+1)}$.

As in the one-dimensional case, it can be shown that Newton's method in n dimensions converges quadratically to a root $\mathbf{w}_*$ of $\mathbf{q}$, *provided* $J_{\mathbf{q}}(\mathbf{w}_*)$ is nonsingular and the initial guess $\mathbf{w}^{(0)}$ is sufficiently close to $\mathbf{w}_*$.

Suppose Newton's method is applied to the system (11.49), where this equation is written in the form

$$\mathbf{q}(\mathbf{w}) \equiv \mathbf{w} - h\mathbf{g}(\mathbf{w}) - \boldsymbol{\gamma} = 0.$$

The resulting equations are

$$\mathbf{w}^{(j+1)} = \mathbf{w}^{(j)} - \left[I - hJ_{\mathbf{g}}(\mathbf{w}^{(j)})\right]^{-1} \left(\mathbf{w}^{(j)} - h\mathbf{g}(\mathbf{w}^{(j)}) - \boldsymbol{\gamma}\right),$$

where $J_{\mathbf{g}}(\mathbf{w}^{(j)})$ is the Jacobian of $\mathbf{g}$ evaluated at $\mathbf{w}^{(j)}$. Now, for h sufficiently small, the matrix $I - hJ_{\mathbf{g}}(\mathbf{w}_*)$ is close to the identity and hence nonsingular. If the initial guess $\mathbf{w}^{(0)}$ is obtained by an explicit method (called a **predictor step**), then it should differ from $\mathbf{w}_*$ by at most $O(h)$. Thus, for h sufficiently small the conditions for convergence of Newton's method can be guaranteed to hold.

The restrictions on h in order for Newton's method to converge usually turn out to be much less restrictive than those needed for fixed point iteration to converge. Therefore, Newton's method is usually the method of choice for solving the nonlinear systems arising from implicit methods for stiff ODEs. The drawbacks are the need to compute the Jacobian at every step and the need to solve a linear system with the Jacobian as coefficient matrix at every step. Each of these can be quite expensive. For this reason, quasi-Newton methods are sometimes used. The analog of the constant slope method described in chapter 4 is to compute the Jacobian at the initial guess $\mathbf{w}^{(0)}$, factor it in the form LU, and then use this same Jacobian matrix at every iteration. Recall that the time to backsolve with the L and U factors is only $O(n^2)$ as opposed to the $O(n^3)$ operations needed to compute L and U. Other methods have been developed using approximate Jacobians, but they usually require somewhat more iterations than Newton's method to converge and they may require a smaller time step h in order to converge at all.

11.6 CHAPTER 11 EXERCISES

1. Solve the following initial value problems analytically:
   (a) $y' = t^3$, $y(0) = 0$.
   (b) $y' = 2y$, $y(1) = 3$.
   (c) $y' = ay + b$, where a and b are given scalars, $y(0) = y_0$. [Hint: Multiply by the integrating factor e^{-ta} and integrate from 0 to T.]
2. Use theorem 11.1.3 to prove that each of the initial value problems in the previous exercise has a unique solution for all t.
3. Verify that the function $y(t) = t^{3/2}$ solves the initial value problem
$$y' = \frac{3}{2} y^{1/3}, \quad y(0) = 0.$$
Apply Euler's method to this problem and explain why the numerical approximation differs from the solution $t^{3/2}$.
4. Write down the result of applying one step of Euler's method to the initial value problem $y' = (t+1)e^{-y}$, $y(0) = 0$, using step size $h = 0.1$. Do the same for the midpoint method and for Heun's method.
5. Write down the result of applying one step of Euler's method to the predator–prey equations
$$R' = (2 - F)R,$$
$$F' = (R - 2)F,$$
starting with $R_0 = 2$ and $F_0 = 1$ and using step size $h = 0.1$. Do the same for the midpoint method and for Heun's method.
6. Show that when the classical fourth-order Runge–Kutta method is applied to the problem $y' = \lambda y$, the formula for advancing the solution is
$$y_{k+1} = \left[1 + h\lambda + \frac{1}{2}h^2\lambda^2 + \frac{1}{6}h^3\lambda^3 + \frac{1}{24}h^4\lambda^4\right] y_k,$$
and the local truncation error is $O(h^4)$.
7. Show that the classical fourth-order Runge–Kutta method is *stable* by showing that when it is written in the form (11.24), the function ψ satisfies a Lipschitz condition.
8. Continuing the love saga from section 11.2.6, Juliet's emotional swings lead to many sleepless nights, which consequently dampens her emotions. Mathematically, the pair's love can now be expressed as
$$\frac{dx}{dt} = -0.2y,$$
$$\frac{dy}{dt} = 0.8x - 0.1y.$$

Suppose this state of the romance begins again when Romeo is smitten with Juliet ($x(0) = 2$) and Juliet is indifferent ($y(0) = 0$).

(a) Explain how the change in (11.23) that produces the equations above reflects Juliet's dampened emotions.
(b) As in section 11.2.6, use **ode45** to produce 3 graphs, like those in figure 11.7, showing Romeo and Juliet's love for $0 \le t \le 60$.
(c) From your graphs, describe how the change in Juliet described in this exercise will affect the relationship and its eventual outcome.

9. Show that the local truncation error in the multistep method (11.34) is $O(h^2)$.
10. Show that the characteristic polynomial $\det(A - \lambda I)$ of the matrix A in (11.38) is

$$\chi(\lambda) = (-1)^m \left[\lambda^m + \sum_{\ell=0}^{m-1} \alpha_\ell \lambda^\ell \right].$$

[Hint: Expand the determinant using the last row.]

11. Which of the following multistep methods are *convergent*? Justify your answers.

(a) $y_k - y_{k-2} = h(f_k - 3f_{k-1} + 4f_{k-2})$.
(b) $y_k - 2y_{k-1} + y_{k-2} = h(f_k - f_{k-1})$.
(c) $y_k - y_{k-1} - y_{k-2} = h(f_k - f_{k-1})$.

12. Use theorem 11.3.1 to show that the local truncation error in the ($m = 3$)-BDF method (11.44) is $O(h^3)$. Show that the method is zero-stable and hence convergent.
13. Write down the system of nonlinear equations that must be solved in order to apply the trapezoidal method to the predator–prey equations

$$R' = (2 - F)R,$$
$$F' = (R - 2)F,$$

starting with $R_0 = 2$ and $F_0 = 1$ and using step size $h = 0.1$. Using Euler's method to obtain an initial guess $R_1^{(0)}$ and $F_1^{(0)}$ for R_1 and F_1, write down the first Newton step that would be taken to solve for R_1 and F_1.

14. This question concerns the motion of a satellite around the sun: the satellite could be a planet, a comet or an unpowered spacecraft, whose mass is assumed to be negligible compared with the mass of the sun. The motion of such a satellite takes place in a plane, so we can describe its coordinates by just two space dimensions x and y. Assume that the sun is at the origin $(0, 0)$. The system of differential equations describing the satellite's motion is Newton's law of gravitation, discovered by Newton in the late 17th century, namely

$$x'' = \frac{-x}{(\sqrt{x^2 + y^2})^3}, \quad y'' = \frac{-y}{(\sqrt{x^2 + y^2})^3}.$$

(a) This system of two second-order differential equations can be simplified to a system of four first-order differential equations by introducing two new variables $z = x'(t)$ (the speed of the satellite in the x space direction) and $w = y'(t)$ (the speed of the satellite in the y space direction). Write down this system of equations.

(b) Solve the system you wrote down in (a) using first Euler's method and then the fourth-order Runge–Kutta method. Start with $x(0) = 4$, $y(0) = 0$, $z(0) = 0$, and $w(0) = 0.5$. For Euler's method take $h = 0.0025$ and for the Runge-Kutta method take $h = 0.25$ and run out to $tmax = 50$. Plot the position of the satellite at each time step. You should see an almost circular orbit. If you do not, then there is a bug in your code. [Note: If you plot things in MATLAB, be sure to say `axis('equal')`, or the axes will be different lengths and your circle will look like an ellipse!]

Keeping $x(0)$, $y(0)$, and $z(0)$ fixed, try varying $w(0)$, now using only the Runge–Kutta code. You will have to adjust h for accuracy and $tmax$ so that you see a complete orbit. Some recommended values are as follows.

$w(0)$	h	$tmax$
0.5	0.25	50
0.6	0.5	150
0.8	0.5	200
0.4	0.25	35
0.2	0.25	30
0.2	0.05	30

For initial velocities slightly larger than 0.5, you should see an elliptical orbit. The fact that planets and comets have elliptical orbits was observed by Kepler in the early 17th century, but it was not explained until Newton developed his law of gravitation 60 years later. The eccentricity of the ellipse depends on the size of the initial velocity $w(0)$. The orbits of planets, such as the Earth, are typically not very eccentric. They look almost like circles. But the orbits of comets, such as Halley's comet and Hale–Bopp, can be very eccentric. That is why Halley's comet can be seen from Earth only once every 76 years and Hale–Bopp about once every 10,000 years: the rest of the time they are very far from the sun traveling around their highly eccentric orbits.

If you make the initial velocity large enough, the orbit becomes a hyperbola, with the satellite approaching the direction of a straight line heading right out of the solar system; in this case the satellite escapes the gravitational attraction of the sun. Many comets pass through the solar system on hyperbolic orbits: they enter the solar system, pass by the sun once and then leave the solar system again. Some spacecraft launched from Earth, such as Voyager II which was launched in 1977 and flew past Pluto in 1989, will soon leave the solar system on a hyperbolic path, never to return. The borderline case between elliptical and hyperbolic orbits is when $w(0)$ is exactly equal to the *escape velocity*: the value just large enough for the satellite to escape from the solar system. In this case the orbit is actually a parabola, which is the borderline case between an ellipse and a hyperbola.

For initial velocities less than 0.5 but greater than 0, the orbit is again an ellipse. However, if you make $w(0)$ too small, you may find that your plots show the satellite spiraling into the sun and then swinging off out of the solar system. This is a numerical effect; it is not physical. With a sufficiently small step size, you should be able to see the elliptical orbit for any $w(0) > 0$, but the computation might take too long.

Turn in plots showing the types of orbit described, labeled with the value of $w(0)$ which gives rise to that path.

(c) Try using the MATLAB routine `ode45` to see if it can obtain accurate solutions for small values of $w(0)$ in a reasonable amount of time. Comment on the advantages or disadvantages of this routine over your own Runge–Kutta code.

15. Consider the system of equations

$$\begin{pmatrix} x \\ y \end{pmatrix}' = \begin{pmatrix} -1000 & 1 \\ 0 & -1/10 \end{pmatrix} \begin{pmatrix} x \\ y \end{pmatrix},$$

$$x(0) = 1, \quad y(0) = 2,$$

whose exact solution is

$$x(t) = e^{-1000t}(9979/9999) + e^{-t/10}(20/9999), \quad y(t) = 2e^{-t/10}.$$

Use the fourth-order Runge–Kutta method to solve this system of equations, integrating out to $t = 1$. What size time step is necessary to achieve a reasonably accurate approximate solution? Turn in a plot of $x(t)$ and $y(t)$ that shows what happens if you choose the time step too large, and also turn in a plot of $x(t)$ and $y(t)$ once you have found a good size time step.

Now try solving this system of ODEs using MATLAB's `ode23s` routine (which uses a second-order implicit method). How many time steps does it require? Explain why a second-order implicit method can solve this problem accurately using fewer time steps than the fourth-order Runge–Kutta method.

16. The following simple model describes the switching behavior for a muscle that controls a valve in the heart. Let $x(t)$ denote the position of the muscle at time t and let $\alpha(t)$ denote the concentration at time t of a chemical stimulus. Suppose that the dynamics of x and α are controlled by the system of differential equations

$$\frac{dx}{dt} = -\frac{x^3}{3} + x + \alpha,$$

$$\frac{d\alpha}{dt} = -\epsilon x.$$

Here $\epsilon > 0$ is a parameter; its inverse estimates roughly the time that x spends near one of its rest positions.

(a) Taking $\epsilon = 1/100$, $x(0) = 2$, and $\alpha(0) = 2/3$, solve this system of differential equations using an explicit method of your choice. Integrate out to, say, $t = 400$, and turn in plots of $x(t)$ and $\alpha(t)$. Explain why you chose the method that you used and approximately how accurate you think your computed solution is and why. Comment on whether you seemed to need restrictions on the step size for stability or whether accuracy was the only consideration in choosing your step size.

(b) Solve the same problem using the backward Euler method

$$\mathbf{y}_{k+1} = \mathbf{y}_k + h\mathbf{f}(t_{k+1}, \mathbf{y}_{k+1})$$

and solving the nonlinear equations at each step via Newton's method. Write down the Jacobian matrix for the system and explain what initial guess you will use. Were you able to take a larger time step using the backward Euler method than you were with the explicit method used in part (a)?

12 MORE NUMERICAL LINEAR ALGEBRA: EIGENVALUES AND ITERATIVE METHODS FOR SOLVING LINEAR SYSTEMS

Two related problems in numerical linear algebra are how to solve eigenvalue/eigenvector problems and how to solve very large systems of linear equations, when the matrix is simply too large to be stored on the computer and the work involved in Gaussian elimination is prohibitive. Both problems call for *iterative methods*, where one supplies an initial guess for the solution (which could be just a random vector, for example), and one successively refines this estimate until an acceptable level of accuracy is achieved. In neither problem do we necessarily expect to find the *exact* solution (even assuming that we use exact arithmetic): Most eigenvalue problems *cannot* be solved exactly, and while systems of linear equations can be solved exactly using Gaussian elimination (with exact arithmetic), a good approximate solution may be perfectly acceptable if it can be computed in a reasonable amount of time.

Interestingly, both of these problems employ similar types of algorithms. Iterative linear system solvers typically yield some information about the eigenvalues of the coefficient matrix, and some of the iteration techniques used for eigenvalue problems can also generate approximate solutions to a linear system whose right-hand side is the initial vector.

12.1 EIGENVALUE PROBLEMS

Let A be an n by n matrix. We wish to find a real or complex number λ, called an **eigenvalue**, and a nonzero vector $\mathbf{v}$, called an **eigenvector**, such that $A\mathbf{v} = \lambda\mathbf{v}$. Eigenvalue problems arise in many areas, such as the study of the *resonance* of musical instruments, or the stability of fluid flows. We saw in the previous chapter how eigenvalue analysis can be used to study the behavior of a system of linear ODEs, $\mathbf{y}' = A\mathbf{y}$. If λ is an eigenvalue of A with corresponding eigenvector $\mathbf{v}$, then if $\mathbf{y}(0) = \mathbf{v}$, then $\mathbf{y}(t) = e^{\lambda t}\mathbf{v}$. More generally, if $\lambda_1, \ldots, \lambda_n$ are the eigenvalues of A with corresponding eigenvectors $\mathbf{v}_1, \ldots, \mathbf{v}_n$, then if $\mathbf{y}(0) = \sum_{j=1}^{n} c_j \mathbf{v}_j$, then $\mathbf{y}(t) = \sum_{j=1}^{n} c_j e^{\lambda_j t}\mathbf{v}_j$.

The eigenvalues of A are the roots of its **characteristic polynomial:**

$$\det(A - \lambda I) = 0.$$

To find the eigenvalues of

$$A = \begin{pmatrix} 1 & 2 \\ 4 & 3 \end{pmatrix}$$

we can write

$$\det(A - \lambda I) = (1 - \lambda)(3 - \lambda) - 8 = \lambda^2 - 4\lambda - 5 = 0,$$

and solve the quadratic equation to find

$$\lambda = -1 \quad \text{or} \quad \lambda = 5.$$

Even real matrices can have *complex* eigenvalues, as the following example shows:

$$A = \begin{pmatrix} 1 & 2 \\ -4 & 3 \end{pmatrix}, \quad \det(A - \lambda I) = (1 - \lambda)(3 - \lambda) + 8 = \lambda^2 - 4\lambda + 11,$$

$$\lambda = \frac{4 \pm \sqrt{-28}}{2} = 2 \pm i\sqrt{7}.$$

Throughout this section, unless otherwise stated, we will assume that the matrix A is real, but complex numbers will come into play when we consider eigenvalues and eigenvectors.

In most cases, there is no analytic formula for the eigenvalues of a matrix, since the eigenvalues are the roots of the characteristic polynomial, and Abel proved in 1824 that there can be no formula (involving rational numbers, addition, subtraction, multiplication, division, and kth roots) for the roots of a general polynomial of degree 5 or higher. Hence the best one can hope to do is to approximate the eigenvalues numerically. One possible approach would be to determine the coefficients of the characteristic polynomial and then use a root-finding routine, as described in chapter 4, to approximate its roots. Unfortunately, this strategy is a bad one because the polynomial root-finding problem may be ill conditioned, even when the underlying eigenvalue problem is not. A tiny change in the coefficients of the polynomial can make a large change in its roots. We will find other methods for approximating eigenvalues.

Associated with each eigenvalue λ is a collection of nonzero vectors $\mathbf{v}$, called eigenvectors, satisfying $A\mathbf{v} = \lambda\mathbf{v}$. The eigenvectors $\mathbf{v}$ associated with an eigenvalue λ, together with the zero vector, form a vector space, sometimes called the **eigenspace** of λ. In many cases, this space is one-dimensional and $\mathbf{v}$ is determined up to nonzero scalar multiples; clearly if $\mathbf{v}$ is an eigenvector with eigenvalue λ, then so is $c\mathbf{v}$ for any nonzero scalar c, since $A\mathbf{v} = \lambda\mathbf{v} \Rightarrow A(c\mathbf{v}) = \lambda(c\mathbf{v})$. Sometimes there are two or more linearly independent eigenvectors, say $\mathbf{v}_1, \ldots, \mathbf{v}_m$, associated with an eigenvalue λ, and in this case, for any scalars

$c_1, \ldots, c_m$ not all zero, the vector $\sum_{j=1}^{m} c_j \mathbf{v}_j$ is also an eigenvector for λ since $A(\sum_{j=1}^{m} c_j \mathbf{v}_j) = \sum_{j=1}^{m} c_j A\mathbf{v}_j = \sum_{j=1}^{m} c_j \lambda \mathbf{v}_j = \lambda(\sum_{j=1}^{m} c_j \mathbf{v}_j)$. The dimension of the eigenspace is called the **geometric multiplicity** of λ. The eigenspace of λ is the same as the null space of $A - \lambda I$, and to find the eigenvectors associated with λ we can look for nonzero vectors $\mathbf{v}$ that satisfy $(A - \lambda I)\mathbf{v} = \mathbf{0}$.

In the first example above, for instance, to find the eigenvector(s) associated with the eigenvalue $\lambda = -1$, we look for nonzero vectors $\mathbf{v} = (v_1, v_2)^T$ satisfying

$$\begin{pmatrix} 1 & 2 \\ 4 & 3 \end{pmatrix} \begin{pmatrix} v_1 \\ v_2 \end{pmatrix} = - \begin{pmatrix} v_1 \\ v_2 \end{pmatrix},$$

or, equivalently,

$$\begin{pmatrix} 2 & 2 \\ 4 & 4 \end{pmatrix} \begin{pmatrix} v_1 \\ v_2 \end{pmatrix} = \begin{pmatrix} 0 \\ 0 \end{pmatrix}.$$

If we begin to solve this set of equations by Gaussian elimination, subtracting 2 times the first row from the second, then we find:

$$\left(\begin{array}{cc|c} 2 & 2 & 0 \\ 4 & 4 & 0 \end{array}\right) \longrightarrow \left(\begin{array}{cc|c} 2 & 2 & 0 \\ 0 & 0 & 0 \end{array}\right),$$

and the solution set consists of all vectors $\mathbf{v}$ such that $2v_1 + 2v_2 = 0$; that is, such that $v_2 = -v_1$. We can take v_1 to have any nonzero value, say, $v_1 = 1$, and then the eigenvector is $(1, -1)^T$.

Following is a 3 by 3 example:

$$A = \begin{pmatrix} 1 & 2 & 3 \\ 4 & 5 & 6 \\ 7 & 8 & 9 \end{pmatrix}.$$

To find the eigenvalues of A, we form the characteristic polynomial, $\det(A - \lambda I)$:

$$\begin{aligned} \det(A - \lambda I) &= \det \begin{pmatrix} 1-\lambda & 2 & 3 \\ 4 & 5-\lambda & 6 \\ 7 & 8 & 9-\lambda \end{pmatrix} \\ &= (1-\lambda) \det \begin{pmatrix} 5-\lambda & 6 \\ 8 & 9-\lambda \end{pmatrix} - 2 \det \begin{pmatrix} 4 & 6 \\ 7 & 9-\lambda \end{pmatrix} \\ &\quad +3 \det \begin{pmatrix} 4 & 5-\lambda \\ 7 & 8 \end{pmatrix} \\ &= (1-\lambda)[(5-\lambda)(9-\lambda) - 6 \cdot 8] - 2[4(9-\lambda) - 6 \cdot 7] \\ &\quad +3[4 \cdot 8 - (5-\lambda) \cdot 7] \\ &= -\lambda(\lambda^2 - 15\lambda - 18). \end{aligned}$$

The roots of this polynomial are

$$\lambda = 0 \quad \text{and} \quad \lambda = \frac{15 \pm 3\sqrt{33}}{2}.$$

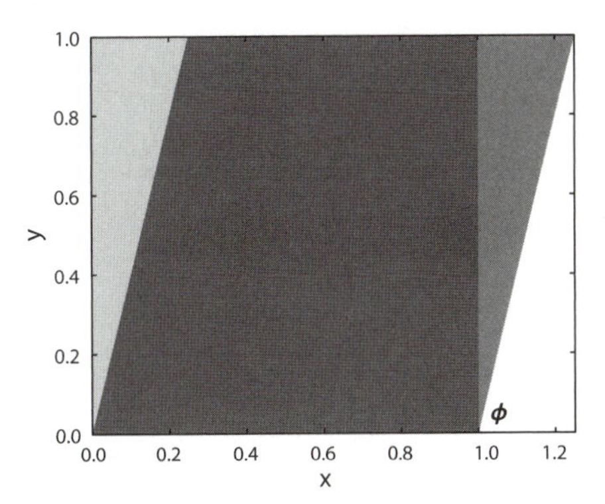

Figure 12.1. A horizontal shear of a unit square.

Let us find the eigenvector(s) associated with the eigenvalue 0. To do this we must find all nonzero vectors $\mathbf{v} = (v_1, v_2, v_3)^T$ satisfying

$$\begin{pmatrix} 1 & 2 & 3 \\ 4 & 5 & 6 \\ 7 & 8 & 9 \end{pmatrix} \begin{pmatrix} v_1 \\ v_2 \\ v_3 \end{pmatrix} = \begin{pmatrix} 0 \\ 0 \\ 0 \end{pmatrix}.$$

Proceeding by Gaussian elimination we find

$$\left(\begin{array}{ccc|c} 1 & 2 & 3 & 0 \\ 4 & 5 & 6 & 0 \\ 7 & 8 & 9 & 0 \end{array}\right) \longrightarrow \left(\begin{array}{ccc|c} 1 & 2 & 3 & 0 \\ 0 & -3 & -6 & 0 \\ 0 & -6 & -12 & 0 \end{array}\right) \longrightarrow \left(\begin{array}{ccc|c} 1 & 2 & 3 & 0 \\ 0 & -3 & -6 & 0 \\ 0 & 0 & 0 & 0 \end{array}\right).$$

The last equation $(0 \cdot v_1 + 0 \cdot v_2 + 0 \cdot v_3 = 0)$ holds for any $\mathbf{v}$, the second equation $(-3v_2 - 6v_3 = 0)$ holds provided $v_2 = -2v_3$, and the first equation $(v_1 + 2v_2 + 3v_3 = 0)$ holds as well provided $v_1 = -2v_2 - 3v_3 = v_3$. Setting v_3 arbitrarily to 1 gives the eigenvector $\mathbf{v} = (1, -2, 1)^T$. You should check that $A\mathbf{v}$ is indeed equal to $0 \cdot \mathbf{v}$. The *eigenspace* associated with the eigenvalue 0 consists of all scalar multiples of the vector $(1, -2, 1)^T$.

Example 12.1.1. Eigenvectors of linear transformations are vectors that are either left unchanged or multiplied by a scale factor when the transformation is performed. For instance, a *shear mapping* leaves points along one axis unchanged while shifting other points in a direction parallel to that axis by an amount that is proportional to their perpendicular distance from the axis. A horizontal shear in the plane is performed by multiplying vectors by the matrix

$$\begin{pmatrix} 1 & k \\ 0 & 1 \end{pmatrix}.$$

Using this matrix, a coordinate $(x, y)^T$ is mapped to $(x + ky, y)^T$. It can be seen in figure 12.1, that $k = \cot \phi$, where ϕ is the angle of a sheared square to the x-axis. Note that the characteristic polynomial of this matrix is $(1 - \lambda)^2$, which has the repeated root $\lambda = 1$. Any nonzero scalar multiple of the vector $(1, 0)^T$ is a corresponding eigenvector and these are the only eigenvectors, which means

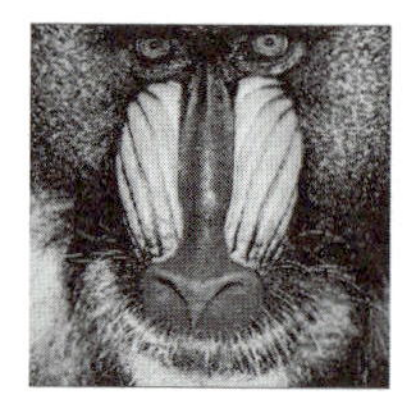

Figure 12.2. Applying a horizontal shear once and then again to an image.

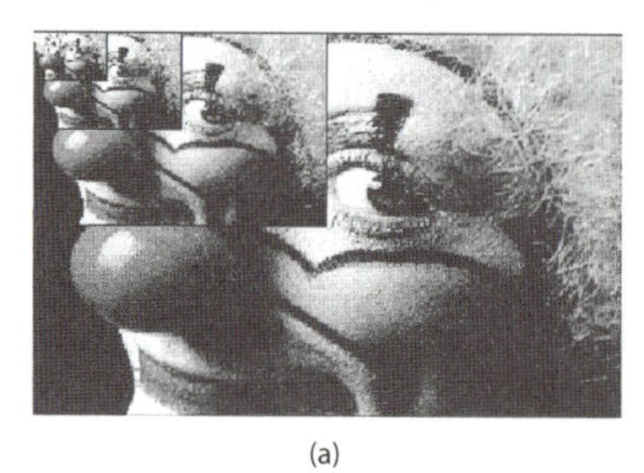
(a)

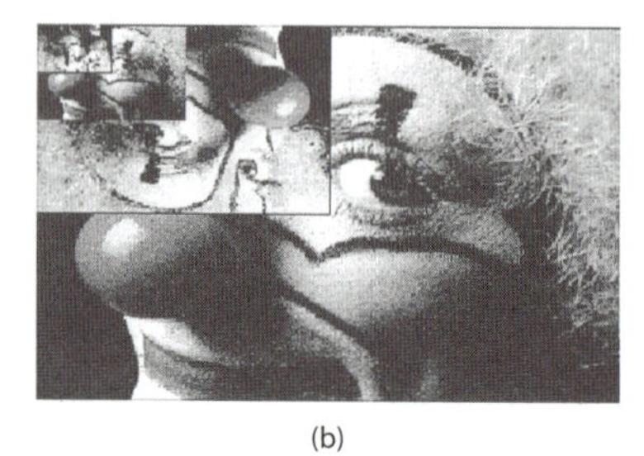
(b)

Figure 12.3. An original image is repeatedly scaled via a linear transformation.

that points along the x-axis are left unchanged by the transformation but all other points are shifted.

In figure 12.2 we see the effect of applying a horizontal shear once and then twice to a picture. Repeatedly applying a shear transformation changes the direction of all vectors in the plane closer to the direction of the eigenvector as can be seen in the picture.

Example 12.1.2. As another example, consider a rubber sheet that is stretched equally in all directions. This means that all vectors $(x, y)^T$ in the plane are multiplied by the same scalar λ; that is, vectors are multiplied by the *diagonal* matrix

$$\begin{pmatrix} \lambda & 0 \\ 0 & \lambda \end{pmatrix}.$$

This matrix has the single eigenvalue λ, and all nonzero 2-vectors are corresponding eigenvectors. Figure 12.3 shows the effect of repeatedly applying this transformation with $\lambda = 1/2$ in (a) and with $\lambda = -1/2$ in (b). In both figures, we have placed the original image and the images produced by three applications of this transformation on top of each other. Knowing the values of λ, can you determine which of the four images in (a) and in (b) is the starting image?

Example 12.1.3. A third example of the use of eigenvalues is in describing the sound of musical instruments. A guitar string, for example, fluctuates at certain natural frequencies. The frequencies are the imaginary parts of the eigenvalues of a linear operator and the decay rates are the negatives of the real parts. Thus, one can give an approximate idea of the sound of a musical instrument by plotting the eigenvalues [54].

The eigenvalues of an actual minor third $A_4\#$ carillon were measured in [90] and reported in [85]. These values are plotted in figure 12.4, where the

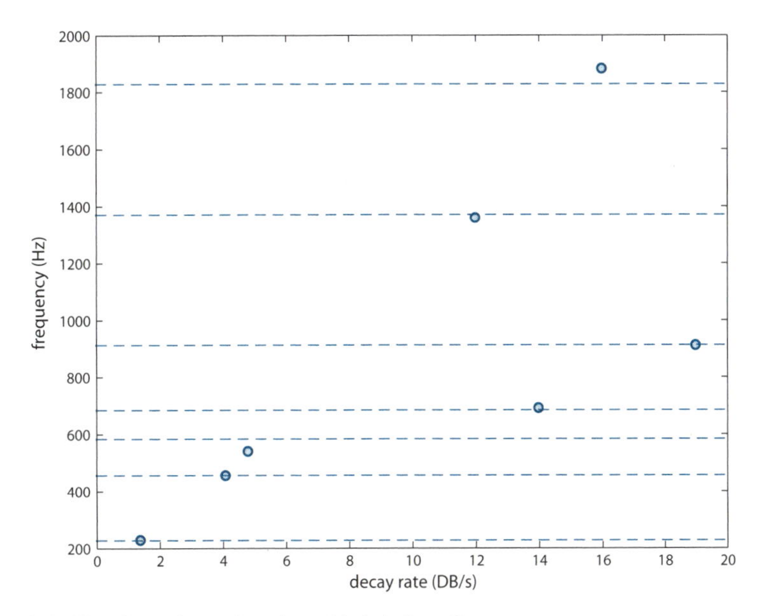

Figure 12.4. The eigenvalues of a minor third $A_4\#$ carillon.

grid lines show the frequencies corresponding to a minor third chord at 456.8 Hz and up to two octaves above and one below. Note the close correspondence between these ideal values and the measured eigenvalues of the bell.

An n by n matrix A that has n linearly independent eigenvectors is said to be *diagonalizable*. If $\mathbf{v}_1, \ldots, \mathbf{v}_n$ are a set of linearly independent eigenvectors with corresponding eigenvalues $\lambda_1, \ldots, \lambda_n$, then we can write

$$A(\mathbf{v}_1, \ldots, \mathbf{v}_n) = (\lambda_1 \mathbf{v}_1, \ldots, \lambda_n \mathbf{v}_n) = (\mathbf{v}_1, \ldots, \mathbf{v}_n) \begin{pmatrix} \lambda_1 & & \\ & \ddots & \\ & & \lambda_n \end{pmatrix}.$$

If $V \equiv (\mathbf{v}_1, \ldots, \mathbf{v}_n)$ is the matrix whose columns are the eigenvectors $\mathbf{v}_1, \ldots, \mathbf{v}_n$, and Λ is the diagonal matrix of eigenvalues, then this becomes

$$AV = V\Lambda, \quad \text{or } A = V\Lambda V^{-1}. \tag{12.1}$$

The rows of V^{-1} (or their complex-conjugate transposes; i.e., the column vectors whose entries are the complex conjugates of those of each row of V^{-1}) are known as **left eigenvectors** of A. If we multiply equation (12.1) by V^{-1} on the left, then we see that $V^{-1}A = \Lambda V^{-1}$, so that if $\mathbf{y}^*$ denotes the ith row of V^{-1}, then $\mathbf{y}^* A = \lambda_i \mathbf{y}^*$. Taking complex-conjugate transposes, we see that $A^* \mathbf{y} = \bar{\lambda}_i \mathbf{y}$, so that $\mathbf{y}$ is a **right eigenvector** of A^* corresponding to the eigenvalue $\bar{\lambda}_i$. Taking ordinary transposes gives $A^T \bar{\mathbf{y}} = \lambda_i \bar{\mathbf{y}}$, so that $\bar{\mathbf{y}}$ is a right eigenvector of A^T corresponding to the eigenvalue λ_i. Unless otherwise stated, the term *eigenvector* will refer to a right eigenvector.

A **similarity transformation** $A \to W^{-1}AW$, where W is any nonsingular matrix, preserves eigenvalues since if $A\mathbf{v} = \lambda \mathbf{v}$, then $W^{-1}AW(W^{-1}\mathbf{v}) = W^{-1}A\mathbf{v} = \lambda(W^{-1}\mathbf{v})$, but the eigenvectors of $W^{-1}AW$ are W^{-1} times the eigenvectors of A. A similarity transformation corresponds to a change of variables. For example, consider the linear system $A\mathbf{x} = \mathbf{b}$, where $A = V\Lambda V^{-1}$.

Making the change of variables $\mathbf{y} = V^{-1}\mathbf{x}$ and $\mathbf{c} = V^{-1}\mathbf{b}$, this becomes the diagonal system $V^{-1}AV\mathbf{y} \equiv \Lambda\mathbf{y} = \mathbf{c}$. Following are two important theorems about classes of matrices that are diagonalizable.

Theorem 12.1.1. If A is an n by n matrix with n distinct eigenvalues then A is diagonalizable.

Proof. Let $\lambda_1, \ldots, \lambda_n$ be the distinct eigenvalues of A, and let $\mathbf{v}_1, \ldots, \mathbf{v}_n$ be corresponding eigenvectors. If $\mathbf{v}_1, \ldots, \mathbf{v}_n$ were *not* linearly independent, then there would be a linear combination of them with nonzero coefficients that was equal to the zero vector. Let $\sum_{j=1}^{m} c_j \mathbf{v}_j$ be the *shortest* such linear combination (i.e., the one involving the fewest $\mathbf{v}_j$'s). Then we have the two equations $\sum_{j=1}^{m} c_j \mathbf{v}_j = \mathbf{0}$ and $A\sum_{j=1}^{m} c_j \mathbf{v}_j = \sum_{j=1}^{m} c_j \lambda_j \mathbf{v}_j = \mathbf{0}$. Subtracting λ_1 times the first equation from the second gives $\sum_{j=2}^{m} c_j(\lambda_j - \lambda_1)\mathbf{v}_j = \mathbf{0}$, but this is a contradiction since this is a shorter linear combination with nonzero coefficients (since $c_j \neq 0$ and $\lambda_j - \lambda_1 \neq 0$, $j = 2, \ldots, m$) that is equal to $\mathbf{0}$. Therefore there must be no such linear combination and $\mathbf{v}_1, \ldots, \mathbf{v}_n$ must be linearly independent. □

Theorem 12.1.2. If A is real and symmetric $(A = A^T)$ then the eigenvalues of A are real and A is diagonalizable via an orthogonal similarity transformation; that is, $A = Q\Lambda Q^T$, where $Q^T = Q^{-1}$ and Λ is the diagonal matrix of eigenvalues.

Not all n by n matrices have n linearly independent eigenvectors. We have already seen an example in the matrix corresponding to a shear transformation. Consider, for instance, the following 2 by 2 matrix:

$$A = \begin{pmatrix} 1 & 1 \\ 0 & 1 \end{pmatrix}.$$

The characteristic polynomial is $\det(A - \lambda I) = (1 - \lambda)^2$, which has a double root at $\lambda = 1$. The eigenspace associated with $\lambda = 1$ is the set of solutions to

$$(A - I)\mathbf{v} = \begin{pmatrix} 0 & 1 \\ 0 & 0 \end{pmatrix} \begin{pmatrix} v_1 \\ v_2 \end{pmatrix} = \begin{pmatrix} 0 \\ 0 \end{pmatrix},$$

which consists of vectors $\mathbf{v}$ for which $v_2 = 0$; that is, the one-dimensional space of scalar multiples of $(1, 0)^T$.

While not all square matrices are diagonalizable, it can be shown that every square matrix is similar to a matrix in *Jordan form*.

Theorem 12.1.3 (Jordan canonical form). Every n by n matrix A is similar to one of the form

$$J = \begin{pmatrix} J_1 & & \\ & \ddots & \\ & & J_m \end{pmatrix},$$

where each block J_i has the form

$$J_i = \begin{pmatrix} \lambda_i & 1 & & \\ & \ddots & \ddots & \\ & & \ddots & 1 \\ & & & \lambda_i \end{pmatrix}.$$

ISSAI SCHUR

Issai Schur (1875–1941) built a famous school at Berlin University where he spent most of his career and served as chair until he was dismissed by the Nazis in 1935. The school that Schur built in Berlin was an active place for collaboration, and discussions there influenced a group of students who extended Schur's impressive results. Schur's charisma played an important role in the school, as did his teaching. Mathematician, Walter Ledermann, who worked under Schur, gave the following account:

> *Schur was a superb lecturer... I remember attending his algebra course which was held in a lecture theatre filled with about 400 students. Sometimes, when I had to be content with a seat at the back of the lecture theatre, I used a pair of opera glasses to get at least a glimpse of the speaker.*

(Photo courtesy of Mathematisches Forschungsinstitut Oberwolfach.)

The number of linearly independent eigenvectors is the number of blocks m. The matrix is diagonalizable if and only if $m = n$. The geometric multiplicity of an eigenvalue λ_i is the number of Jordan blocks with eigenvalue λ_i. The algebraic multiplicity of λ_i (i.e., its degree as a root of the characteristic polynomial) is the sum of the orders of all Jordan blocks with eigenvalue λ_i.

While the Jordan form is an important theoretical tool, it is less useful in computations. A tiny change in a matrix can make a huge change in its Jordan form. For example, suppose we have a single n by n Jordan block with eigenvalue 1. Changing the diagonal entries by arbitrarily small but distinct amounts ϵ_i, we obtain a matrix with distinct eigenvalues $1 + \epsilon_i$, $i = 1, \ldots, n$, which, by theorem 12.1.1, is diagonalizable; that is, has a Jordan form consisting of n one by one blocks instead of one n by n block.

Often the most useful similarity transformations are those that can be carried out with orthogonal or, more generally, unitary matrices. Recall that an orthogonal matrix is a real matrix Q such that $Q^T = Q^{-1}$. The complex analog is a *unitary matrix*: a complex matrix for which the complex-conjugate transpose (the matrix Q^* whose (i, j)-entry is equal to the complex conjugate of Q_{ji}) is equal to Q^{-1}. We state one more important theorem, called *Schur's theorem*, saying that every square matrix is unitarily similar to an upper triangular matrix.

Theorem 12.1.4 (Schur form). Every square matrix A can be written in the form $A = QTQ^*$ where Q is a unitary matrix and T is upper triangular.

For certain types of matrices, the eigenvalues and eigenvectors are easy to find. For example, the eigenvalues of a *diagonal* matrix

$$A = \begin{pmatrix} d_1 & & \\ & \ddots & \\ & & d_n \end{pmatrix}$$

are just the diagonal entries $d_1, \ldots, d_n$, and the corresponding eigenvectors are the unit vectors, $\mathbf{e}_1, \ldots, \mathbf{e}_n$, where $\mathbf{e}_j$ has a 1 in position j and 0s everywhere else; for we have $A\mathbf{e}_j = d_j\mathbf{e}_j$. If A is *triangular*, either upper or lower triangular, then its eigenvalues are again equal to its diagonal entries, and the eigenvectors can be determined by back substitution. For example, if

$$A = \begin{pmatrix} 1 & 2 & 3 \\ 0 & 4 & 5 \\ 0 & 0 & 6 \end{pmatrix}$$

then its eigenvalues are 1, 4, and 6. The eigenvector associated with 1 is $\mathbf{e}_1 = (1, 0, 0)^T$. The eigenvector associated with 4 satisfies

$$\begin{pmatrix} -3 & 2 & 3 \\ 0 & 0 & 5 \\ 0 & 0 & 2 \end{pmatrix} \begin{pmatrix} v_1 \\ v_2 \\ v_3 \end{pmatrix} = \begin{pmatrix} 0 \\ 0 \\ 0 \end{pmatrix},$$

and so must have $v_3 = 0$ and $-3v_1 + 2v_2 = 0$, say, $\mathbf{v} = (2/3, 1, 0)^T$. The eigenvector associated with 6 satisfies

$$\begin{pmatrix} -5 & 2 & 3 \\ 0 & -2 & 5 \\ 0 & 0 & 0 \end{pmatrix} \begin{pmatrix} v_1 \\ v_2 \\ v_3 \end{pmatrix} = \begin{pmatrix} 0 \\ 0 \\ 0 \end{pmatrix},$$

and so v_3 can be any nonzero number, while $v_2 = (5/2)v_3$ and $v_1 = (2/5)v_2 + (3/5)v_3 = (8/5)v_3$, giving, for example, $\mathbf{v} = (8/5, 5/2, 1)^T$.

A very useful theorem for obtaining bounds on eigenvalues, without actually doing much computation, is *Gerschgorin's theorem*. We will prove the first part of this theorem.

Theorem 12.1.5 (Gerschgorin). Let A be an n by n matrix with entries a_{ij} and let r_i denote the sum of the absolute values of the off-diagonal entries in row i: $r_i = \sum_{\substack{j=1 \\ j \neq i}}^{n} |a_{ij}|$. Let D_i denote the disk in the complex plane centered at a_{ii} and of radius r_i:

$$D_i = \{z \in \mathbf{C} : |z - a_{ii}| \leq r_i\}.$$

Then all eigenvalues of A lie in the union $\cup_{i=1}^{n} D_i$ of the Gerschgorin disks. If m of these disks are connected and disjoint from the others, then exactly m eigenvalues of A lie in this connected component.

Proof of first part. Let λ be an eigenvalue of A with corresponding eigenvector $\mathbf{v}$. Then for each $i = 1, \ldots, n$, we have $\sum_{j=1}^{n} a_{ij}v_j = \lambda v_i$, or equivalently, $(\lambda - a_{ii})v_i = \sum_{j \neq i} a_{ij}v_j$. Let v_k be a component of $\mathbf{v}$ of largest absolute value, $|v_k| \geq |v_j|$ for all $j = 1, \ldots, n$. Then using this formula with $i = k$ and dividing

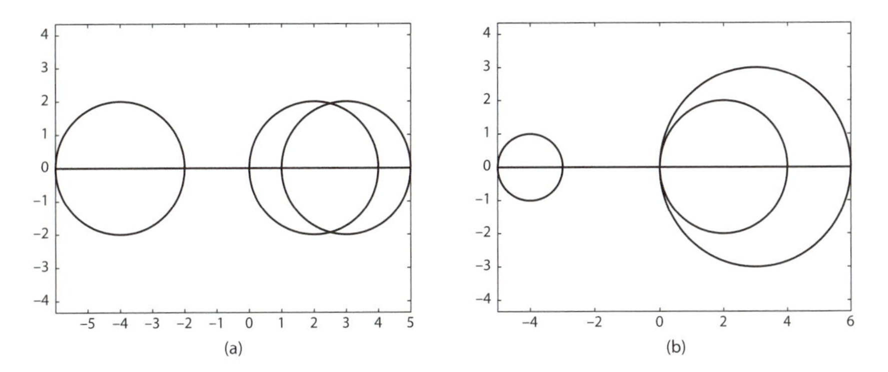

Figure 12.5. Gerschgorin (a) row and (b) column disks.

each side by v_k we find

$$\lambda - a_{kk} = \sum_{\substack{j=1\\ j\neq k}}^{n} a_{ij}(v_j/v_k),$$

and hence, taking absolute values on each side,

$$|\lambda - a_{kk}| \le \sum_{\substack{j=1\\ j\neq k}}^{n} |a_{ij}| \cdot |v_j/v_k| \le \sum_{\substack{j=1\\ j\neq k}}^{n} |a_{ij}|.$$

□

The theorem is stated using the *Gerschgorin row disks*. Since the eigenvalues of A^T are the same as those of A, however, (because if $A = SJS^{-1}$ where J is in Jordan form then $A^T = (S^T)^{-1}J^TS^T$ and the eigenvalues of J^T are the same as those of J), one can state a similar result using the *Gerschgorin column disks*:
Let c_j denote the sum of the absolute values of the off-diagonal entries in column j: $c_j = \sum_{\substack{i=1\\ i\neq j}}^{n} |a_{ij}|$. Let E_j denote the disk in the complex plane centered at a_{jj} and of radius c_j:

$$E_j = \{z \in \mathbf{C} : \ |z - a_{jj}| \le c_j\}.$$

Then all eigenvalues of A lie in the union $\cup_{j=1}^{n} E_j$ of the Gerschgorin column disks. If m of these disks are connected and disjoint from the others, then exactly m eigenvalues of A lie in this connected component.

Example 12.1.4. Consider the matrix

$$A = \begin{pmatrix} 3 & 1 & 1 \\ 2 & 2 & 0 \\ 1 & -1 & -4 \end{pmatrix}.$$

The Gerschgorin row disks, $D_1 = \{z \in \mathbf{C} : \ |z - 3| \le 2\}$, $D_2 = \{z \in \mathbf{C} : \ |z - 2| \le 2\}$, and $D_3 = \{z \in \mathbf{C} : \ |z + 4| \le 2\}$, are pictured in figure 12.5(a). According to the theorem, one eigenvalue lies in the disk centered at -4, while two eigenvalues lie in the union of the disks centered at 2 and 3.

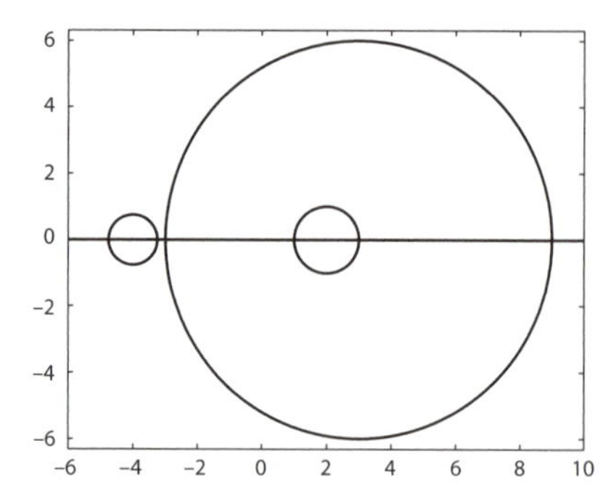

Figure 12.6. Gerschgorin row disks for $D^{-1}AD$.

The Gerschgorin column disks, $E_1 = \{z \in \mathbf{C} : \; |z-3| \leq 3\}$, $E_2 = \{z \in \mathbf{C} : \; |z-2| \leq 2\}$, and $E_3 = \{z \in \mathbf{C} : \; |z+4| \leq 1\}$, are pictured in figure 12.5(b). Combining information from each of the two figures, we obtain the stronger result that one eigenvalue lies in E_3 while two lie in $D_1 \cup D_2$.

Sometimes a simple similarity transformation can produce a matrix, say, $D^{-1}AD$, whose Gerschgorin disks tell us even more about all or some of the eigenvalues of A. In the above example, suppose we take $D = \text{diag}(1, 2, 4)$. Then

$$D^{-1}AD = \begin{pmatrix} 3 & 2 & 4 \\ 1 & 2 & 0 \\ 1/4 & -1/2 & -4 \end{pmatrix}.$$

Looking at the Gerschgorin row disks for $D^{-1}AD$ ($D_1 = \{z \in \mathbf{C} : |z-3| \leq 6\}$, $D_2 = \{z \in \mathbf{C} : |z-2| \leq 1\}$, and $D_3 = \{z \in \mathbf{C} : |z+4| \leq 3/4\}$) in figure 12.6 we see that while $D_1 \cup D_2$ covers a larger region than the corresponding disks associated with A and thus gives less information about the location of the two eigenvalues in that region, the disk D_3 about $z = -4$ is smaller than that in the original matrix and thus tells us more: there is an eigenvalue within 3/4 of −4.

12.1.1 The Power Method for Computing the Largest Eigenpair

Suppose one starts with a vector $\mathbf{w}$ and applies the matrix A to it many, many times, obtaining $A^k\mathbf{w}$, $k = 1, 2, \ldots$. Let us assume that A is **diagonalizable,** meaning that it has n linearly independent eigenvectors $\mathbf{v}_1, \ldots, \mathbf{v}_n$ forming a basis for $\mathbf{C}^n$. The vector $\mathbf{w}$ can be expressed as a linear combination of these eigenvectors: $\mathbf{w} = \sum_{j=1}^n c_j \mathbf{v}_j$, for some scalars $c_1, \ldots, c_n$. Then $A\mathbf{w} = \sum_{j=1}^n c_j A\mathbf{v}_j = \sum_{j=1}^n c_j \lambda_j \mathbf{v}_j$, where λ_j is the eigenvalue corresponding to $\mathbf{v}_j$. Multiplying by A again, we find that $A^2\mathbf{w} = \sum_{j=1}^n c_j \lambda_j A\mathbf{v}_j = \sum_{j=1}^n c_j \lambda_j^2 \mathbf{v}_j$, and continuing in this way we see that for each $k = 1, 2, \ldots$,

$$A^k\mathbf{w} = \sum_{j=1}^n c_j \lambda_j^k \mathbf{v}_j. \tag{12.2}$$

Now suppose that the eigenvalue of *largest absolute value* is strictly greater in absolute value than all of the others: $|\lambda_1| > |\lambda_2| \geq \ldots \geq |\lambda_n|$. Then, assuming that $c_1 \neq 0$, the term in the sum (12.2) that dominates the others when k is very large is the first term: $c_1 \lambda_1^k \mathbf{v}_1$. Hence as k increases, $A^k \mathbf{w}$ looks more and more like a multiple of the first eigenvector $\mathbf{v}_1$. Put another way, if we divide each side of (12.2) by λ_1^k, then we can write

$$\lambda_1^{-k} A^k \mathbf{w} = c_1 \mathbf{v}_1 + \sum_{j=2}^{n} c_j (\lambda_j/\lambda_1)^k \mathbf{v}_j, \tag{12.3}$$

and each of the terms $c_j(\lambda_j/\lambda_1)^k \mathbf{v}_j$, $j = 2, \ldots, n$, approaches 0 as $k \to \infty$, since $|\lambda_j/\lambda_1| < 1$. Hence $\lim_{k\to\infty} \lambda_1^{-k} A^k \mathbf{w} = c_1 \mathbf{v}_1$.

The **power method** for computing the eigenvector of A corresponding to the eigenvalue of largest absolute value is usually implemented as follows:

> Given a nonzero vector $\mathbf{w}$, set $\mathbf{y}^{(0)} = \mathbf{w}/\|\mathbf{w}\|$.
> For $k = 1, 2, \ldots,$
>   Set $\mathbf{y}^{(k)} = A\mathbf{y}^{(k-1)}/\|A\mathbf{y}^{(k-1)}\|$.

By normalizing the vector $\mathbf{y}^{(k)}$, one avoids possible problems with overflow or underflow, without affecting the convergence of the method. Since $\mathbf{y}^{(k)}$ converges to the eigenvector $\hat{\mathbf{v}}_1 \equiv \pm \mathbf{v}_1/\|\mathbf{v}_1\|$, the corresponding eigenvalue λ_1 can be approximated in several different ways. If $\mathbf{y}^{(k)}$ were the true eigenvector, then the ratio of any component of $A\mathbf{y}^{(k)}$ to the corresponding component of $\mathbf{y}^{(k)}$ would be equal to λ_1. Hence one way to approximate λ_1 would be to take the ratio $(A\mathbf{y}^{(k)})_i/\mathbf{y}_i^{(k)}$, for any nonzero component $i = 1, \ldots, n$. Another way is to take the inner product $\langle A\mathbf{y}^{(k)}, \mathbf{y}^{(k)} \rangle \equiv \mathbf{y}^{(k)T} A\mathbf{y}^{(k)}$, which would be equal to $\langle \lambda_1 \hat{\mathbf{v}}_1, \hat{\mathbf{v}}_1 \rangle = \lambda_1 \langle \hat{\mathbf{v}}_1, \hat{\mathbf{v}}_1 \rangle = \lambda_1$, if $\mathbf{y}^{(k)}$ were the true normalized eigenvector $\hat{\mathbf{v}}_1$. The latter method is usually preferred, and the power method becomes:

> **Power Method** for computing the eigenvalue of largest absolute value and the corresponding normalized eigenvector.
>
> Given a nonzero vector $\mathbf{w}$, set $\mathbf{y}^{(0)} = \mathbf{w}/\|\mathbf{w}\|$.
> For $k = 1, 2, \ldots,$
>   Compute $\tilde{\mathbf{y}}^{(k)} = A\mathbf{y}^{(k-1)}$.
>   Set $\lambda^{(k)} = \langle \tilde{\mathbf{y}}^{(k)}, \mathbf{y}^{(k-1)} \rangle$. [Note that $\lambda^{(k)} = \langle A\mathbf{y}^{(k-1)}, \mathbf{y}^{(k-1)} \rangle$.]
>   Form $\mathbf{y}^{(k)} = \tilde{\mathbf{y}}^{(k)}/\|\tilde{\mathbf{y}}^{(k)}\|$.

Note that each iteration requires one matrix–vector multiplication plus some additional work with vectors only.

We have argued that the power method converges to the eigenvector corresponding to the eigenvalue of largest absolute value, *provided* the absolute value of this eigenvalue is strictly greater than that of the others. It can be seen from (12.3) that the *rate* of convergence depends on the ratio $|\lambda_2/\lambda_1|$, where λ_2 is the eigenvalue (or eigenvalues) with second largest absolute value;

Figure 12.7. The eigenvalues and shifted eigenvalues of a matrix.

that is,

$$\|\lambda_1^{-k} A^k \mathbf{w} - c_1 \mathbf{v}_1\| = O(|\lambda_2/\lambda_1|^k). \tag{12.4}$$

If one could decrease the ratio $|\lambda_2/\lambda_1|$, then one could achieve faster convergence.

Suppose, for example, that A has eigenvalues $10, 9, 8, 7, 6, 5$. Then the power method applied to A will converge to the eigenvector corresponding to 10, with a convergence factor of 0.9. Consider the shifted matrix $A - 7I$, whose eigen*vectors* are the same as those of A and whose eigen*values* are $3, 2, 1, 0, -1, -2$, as pictured in figure 12.7.

If the power method is applied to this shifted matrix, then it will converge to the same eigenvector (now corresponding to eigenvalue 3 in the shifted matrix), but the rate of convergence will be governed by the ratio $2/3 \approx 0.667$. We can recover the eigenvalues of the original matrix from those of the shifted matrix just by adding 7.

This leads to the idea of the *power method with a shift*.

Power Method with Shift s for computing the eigenvalue farthest from s and the corresponding normalized eigenvector.

Given a shift s and a nonzero vector $\mathbf{w}$, set $\mathbf{y}^{(0)} = \mathbf{w}/\|\mathbf{w}\|$.
For $k = 1, 2, \ldots,$
  Compute $\tilde{\mathbf{y}}^{(k)} = (A - sI)\mathbf{y}^{(k-1)}$.
  Set $\lambda^{(k)} = \langle \tilde{\mathbf{y}}^{(k)}, \mathbf{y}^{(k-1)} \rangle + s$. [Note that $\lambda^{(k)} = \langle A\mathbf{y}^{(k-1)}, \mathbf{y}^{(k-1)} \rangle$.]
  Form $\mathbf{y}^{(k)} = \tilde{\mathbf{y}}^{(k)}/\|\tilde{\mathbf{y}}^{(k)}\|$.

This iteration converges to the eigenvector corresponding to the eigenvalue of $A - sI$ with largest absolute value. Note that in the previous example, if we shift by more than 7.5, say, we shift by 8, giving eigenvalues $2, 1, 0, -1, -2, -3$, then the eigenvalue of largest absolute value is now the algebraically *smallest* eigenvalue, -3. In this case, the shifted power method converges to the smallest eigenvalue, 5, and corresponding eigenvector of the original matrix. Thus, by appropriately choosing shifts, the power method can be made to converge to the largest or smallest eigenvalue of a matrix with real eigenvalues. It cannot find the other eigenvalues, however, because the shift can never make one of the interior eigenvalues lie farthest from the origin. To find these eigenvalues one can use a method called *inverse iteration*. Another way to accomplish this when A is symmetric is through *deflation*.

Deflation

Assume that A is *symmetric* with eigenvalues $\lambda_1, \ldots, \lambda_n$ satisfying $|\lambda_1| > |\lambda_2| > |\lambda_3| \geq \ldots \geq |\lambda_n|$, and suppose that we have used the power method to compute the eigenvalue λ_1 of largest absolute value and the corresponding normalized eigenvector $\mathbf{v}_1$. Suppose we have used an initial vector $\mathbf{w} = \sum_{j=1}^{n} c_j \mathbf{v}_j$. Knowing $\mathbf{v}_1$, we can construct an initial vector $\hat{\mathbf{w}}$ that is a linear combination of only $\mathbf{v}_2, \ldots, \mathbf{v}_n$:

$$\hat{\mathbf{w}} = \mathbf{w} - \langle \mathbf{w}, \mathbf{v}_1 \rangle \mathbf{v}_1 = \sum_{j=2}^{n} c_j \mathbf{v}_j.$$

This works because the eigenvectors of the symmetric matrix A are orthogonal; orthogonalizing any vector $\mathbf{w}$ against one set of eigenvectors produces a vector that is a linear combination of only the other eigenvectors. This process of modifying the initial vector or other vectors generated by the algorithm so that they are orthogonal to the already computed eigenvector(s) is called **deflation**.

If the power method is run with initial vector $\hat{\mathbf{w}}$, then instead of converging to $\mathbf{v}_1$, it will converge to $\mathbf{v}_2$ (assuming $c_2 \neq 0$), since

$$A^k \hat{\mathbf{w}} = \sum_{j=2}^{n} c_j A^k \mathbf{v}_j = \sum_{j=2}^{n} c_j \lambda_j^k \mathbf{v}_j = \lambda_2^k \left[c_2 \mathbf{v}_2 + \sum_{j=3}^{n} c_j (\lambda_j / \lambda_2)^k \mathbf{v}_j \right],$$

and each of the coefficients $(\lambda_j/\lambda_2)^k$, $j \geq 3$, converges to 0 as $k \to \infty$.

Now, in practice, we do not know $\mathbf{v}_1$ exactly, so the vector $\hat{\mathbf{w}}$ constructed above will likely have at least a small component in the direction of $\mathbf{v}_1$. Even if it did not, rounding errors in the computation of powers of A times $\hat{\mathbf{w}}$ would perturb subsequent vectors so that they had components in the direction of $\mathbf{v}_1$, and the method would probably end up converging to $\mathbf{v}_1$ again. To prevent this, it is necessary to periodically orthogonalize the vectors in the power method against $\mathbf{v}_1$. That is, if $\tilde{\mathbf{y}}^{(k)}$ is the vector produced at the kth step of the power method, then we replace $\tilde{\mathbf{y}}^{(k)}$ by

$$\tilde{\mathbf{y}}^{(k)} - \langle \tilde{\mathbf{y}}^{(k)}, \mathbf{v}_1 \rangle \mathbf{v}_1.$$

This need not be done at every step, but only occasionally, to prevent the reappearance of the eigenvector $\mathbf{v}_1$.

12.1.2 Inverse Iteration

Suppose the matrix A is invertible and the power method is applied to A^{-1} or, more generally, suppose s is a given shift, not equal to an exact eigenvalue of A, and the power method is applied to the matrix $(A - sI)^{-1}$. Again let $\lambda_1, \ldots, \lambda_n$ denote the eigenvalues of A and $\mathbf{v}_1, \ldots, \mathbf{v}_n$ the corresponding eigenvectors. Then the eigen*vectors* of $(A - sI)^{-1}$ are the same as those of A, and the corresponding eigen*values* are $(\lambda_1 - s)^{-1}, \ldots, (\lambda_n - s)^{-1}$, since

$$A\mathbf{v}_j = \lambda_j \mathbf{v}_j \iff (A - sI)\mathbf{v}_j = (\lambda_j - s)\mathbf{v}_j \iff \frac{1}{\lambda_j - s}\mathbf{v}_j = (A - sI)^{-1}\mathbf{v}_j.$$

The eigenvalue of $(A - sI)^{-1}$ that is *farthest* from the origin is the one for which $1/|\lambda_j - s|$ is largest; that is, the one for which $|\lambda_j - s|$ is *smallest*. Thus, for example, if A has eigenvalues $1, 2, \ldots, 10$, then A^{-1} has eigenvalues

$1, 1/2, \ldots, 1/10$, and the power method applied to A^{-1} will converge to 1. Moreover, its rate of convergence will be governed by the ratio of the second largest eigenvalue of A^{-1} to the largest, which is $1/2$. Note that this is a faster rate of convergence than we would have obtained by using the power method on A with a shift; had we shifted by, say, 6 so that $A - 6I$ had eigenvalues $-5, -4, \ldots, 4$, then the power method applied to the shifted matrix would have converged to the eigenpair corresponding to 1 in the original matrix, but the rate would have been governed by the ratio $4/5$. Thus the method of *inverse iteration* may provide faster convergence to the smallest eigenvalue of A.

Moreover, with a properly chosen shift, the inverse iteration method can be made to converge to *any* of the eigenvalues of A. For instance, suppose we choose $s = 3.2$. Then the eigenvalues of $A - sI$ are $-2.2, -1.2, -0.2, 0.8, \ldots, 6.8$, so the eigenvalues of $(A - sI)^{-1}$ are approximately $-0.45, -0.83, -5, 1.25, \ldots, 0.15$. The one that is farthest from the origin is -5, and this corresponds to the interior eigenvalue 3 of A. Thus the inverse iteration method can be made to converge to interior eigenpairs of A. Additionally, if we already have a fairly good estimate of the eigenvalue that we are looking for, then the rate of convergence of the inverse iteration method can be greatly enhanced. Since the shift $s = 3.2$ is already fairly close to the eigenvalue 3, when we shift A by s, the shifted eigenvalue -0.2 will be much closer to the origin than the next closest one, and when we take the inverse of $A - sI$, the inverse of the shifted eigenvalue $1/(-0.2)$ will be much farther from the origin than the next farthest one. In this case, the next farthest eigenvalue of $(A - sI)^{-1}$ from the origin is 1.25, and so the rate of convergence of inverse iteration is governed by the ratio $1.25/5 = 0.15$. Had we chosen an even closer shift, say, $s = 3.01$, then convergence would have been even faster. The eigenvalues of $(A - 3.01I)^{-1}$ are $-1/2.01, -1/1.01, -100, 1/0.99, \ldots, 1/6.99$, so the convergence rate of the power method applied to this matrix is given by the ratio $(1/0.99)/100 \approx 0.01$.

The inverse iteration algorithm can be written as follows:

Inverse Iteration with Shift s for computing the eigenvalue of A that is closest to s and the corresponding normalized eigenvector.

Given a shift s and a nonzero vector $\mathbf{w}$, set $\mathbf{y}^{(0)} = \mathbf{w}/\|\mathbf{w}\|$.
For $k = 1, 2, \ldots,$
  Solve $(A - sI)\tilde{\mathbf{y}}^{(k)} = \mathbf{y}^{(k-1)}$ for $\tilde{\mathbf{y}}^{(k)}$.
  Set $\lambda^{(k)} = 1/\langle \tilde{\mathbf{y}}^{(k)}, \mathbf{y}^{(k-1)} \rangle + s$. [Note that $1/(\lambda^{(k)} - s) = \langle (A - sI)^{-1}\mathbf{y}^{(k-1)}, \mathbf{y}^{(k-1)} \rangle$.]
  Form $\mathbf{y}^{(k)} = \tilde{\mathbf{y}}^{(k)}/\|\tilde{\mathbf{y}}^{(k)}\|$.

Note that we do not actually compute the inverse of $A - sI$ but we solve the linear system $(A - sI)\tilde{\mathbf{y}}^{(k)} = \mathbf{y}^{(k-1)}$ to obtain $\tilde{\mathbf{y}}^{(k)} = (A - sI)^{-1}\mathbf{y}^{(k-1)}$. Recall from chapter 7 that this requires less work and is more stable than computing the inverse.

Note also the eigenvalue approximation that is used: $\lambda^{(k)} = 1/\langle \tilde{\mathbf{y}}^{(k)}, \mathbf{y}^{(k-1)} \rangle + s$. Since $\mathbf{y}^{(k-1)}$ is an approximate normalized eigenvector of $(A - sI)^{-1}$, the

JOHN WILLIAM STRUTT, LORD RAYLEIGH

John William Strutt (1842–1919) was an English physicist who earned a Nobel Prize for Physics in 1904 for discovering the element argon with William Ramsay. He also discovered the phenomenon now called Rayleigh scattering, which provided the first correct explanation of why the sky is blue. Rayleigh, among his many contributions, also predicted the existence of the surface waves now known as Rayleigh waves.

quantity $\langle (A - sI)^{-1}\mathbf{y}^{(k-1)}, \mathbf{y}^{(k-1)} \rangle$ is an approximate eigenvalue of $(A - sI)^{-1}$. The eigenvalues of $(A - sI)^{-1}$ are the inverses of the eigenvalues of $A - sI$; that is, they are of the form $1/(\lambda - s)$, where λ is an eigenvalue of A. Hence λ is of the form $1/(\text{eigenvalue of } (A - sI)^{-1}) + s$. Now, since the eigen*vectors* of $(A - sI)^{-1}$ are the same as those of A, the vector $\mathbf{y}^{(k-1)}$ is also an approximate eigenvector of A, and one might approximate the corresponding eigenvalue with the inner product: $\langle A\mathbf{y}^{(k-1)}, \mathbf{y}^{(k-1)} \rangle$, as was done in the power method. The disadvantage of using this approximation here is that it requires an extra matrix–vector multiplication. Nevertheless, this approximation is often used in *Rayleigh quotient iteration*, which will be described shortly.

Computing Eigenvectors with Inverse Iteration

Often inverse iteration is used to compute eigenvectors, once the eigenvalues have been computed by other means, such as the *QR algorithm* to be described later. If the shift s is a very good approximation to an eigenvalue, then inverse iteration will converge to the corresponding eigenvector very quickly, as explained above. Note, however, that if s were an *exact* eigenvalue, then the matrix $A - sI$ would be singular, and if s is very close to an eigenvalue then one might expect rounding errors to prevent one from obtaining an accurate solution to the linear system: $(A - sI)\tilde{\mathbf{y}}^{(k)} = \mathbf{y}^{(k-1)}$. It turns out that, although this is indeed the case, the error in solving this linear system lies mostly in the direction of the eigenvector that we are approximating. Hence this error does not detract from the performance of inverse iteration.

12.1.3 Rayleigh Quotient Iteration

Suppose one wishes to use inverse iteration to approximate an eigenpair, but one does not know a good shift s that is close to the desired eigenvalue. Since the inverse iteration method generates vectors that approximate an eigenvector of A, one can approximate the corresponding eigenvalue in the same way that this is done in the power method: $\lambda^{(k)} = \langle A\mathbf{y}^{(k-1)}, \mathbf{y}^{(k-1)} \rangle$. These values can then be used as successive shifts in the inverse iteration algorithm. This is called *Rayleigh quotient iteration*:

Rayleigh Quotient Iteration for computing an eigenvalue and corresponding eigenvector of A.

Given a nonzero vector $\mathbf{w}$, set $\mathbf{y}^{(0)} = \mathbf{w}/\|\mathbf{w}\|$ and $\lambda^{(0)} = \langle A\mathbf{y}^{(0)}, \mathbf{y}^{(0)}\rangle$.
For $k = 1, 2, \ldots,$
  Solve $(A - \lambda^{(k-1)}I)\tilde{\mathbf{y}}^{(k)} = \mathbf{y}^{(k-1)}$ for $\tilde{\mathbf{y}}^{(k)}$.
  Form $\mathbf{y}^{(k)} = \tilde{\mathbf{y}}^{(k)}/\|\tilde{\mathbf{y}}^{(k)}\|$.
  Set $\lambda^{(k)} = \langle A\mathbf{y}^{(k)}, \mathbf{y}^{(k)}\rangle$.

Note that each iteration in Rayleigh quotient iteration requires one linear system solution ($(A - \lambda^{(k-1)}I)\tilde{\mathbf{y}}^{(k)} = \mathbf{y}^{(k-1)}$) and one matrix–vector multiplication ($A\mathbf{y}^{(k)}$).

How good an approximation to an eigenvalue is the Rayleigh quotient $\langle A\mathbf{v}, \mathbf{v}\rangle/\langle \mathbf{v}, \mathbf{v}\rangle$, when $\mathbf{v}$ is an approximate eigenvector? Suppose $\mathbf{v} = \gamma\mathbf{v}_j + \mathbf{u}$, where $\mathbf{v}_j$ is a normalized eigenvector of A corresponding to eigenvalue λ_j and $\mathbf{u}$ is orthogonal to $\mathbf{v}_j$. Then

$$\frac{\langle A\mathbf{v}, \mathbf{v}\rangle}{\langle \mathbf{v}, \mathbf{v}\rangle} = \frac{\langle \gamma\lambda_j\mathbf{v}_j + A\mathbf{u}, \gamma\mathbf{v}_j + \mathbf{u}\rangle}{\langle \gamma\mathbf{v}_j + \mathbf{u}, \gamma\mathbf{v}_j + \mathbf{u}\rangle} = \frac{|\gamma|^2\lambda_j + \bar{\gamma}\langle A\mathbf{u}, \mathbf{v}_j\rangle + \langle A\mathbf{u}, \mathbf{u}\rangle}{|\gamma|^2 + \langle \mathbf{u}, \mathbf{u}\rangle}.$$

If A is *symmetric*, then $\langle A\mathbf{u}, \mathbf{v}_j\rangle = \langle \mathbf{u}, A\mathbf{v}_j\rangle = \langle \mathbf{u}, \lambda_j\mathbf{v}_j\rangle = 0$, and so this becomes

$$\frac{\langle A\mathbf{v}, \mathbf{v}\rangle}{\langle \mathbf{v}, \mathbf{v}\rangle} = \frac{|\gamma|^2\lambda_j + \langle A\mathbf{u}, \mathbf{u}\rangle}{|\gamma|^2 + \langle \mathbf{u}, \mathbf{u}\rangle} = \lambda_j + \frac{\langle A\mathbf{u}, \mathbf{u}\rangle - \lambda_j\langle \mathbf{u}, \mathbf{u}\rangle}{|\gamma|^2 + \langle \mathbf{u}, \mathbf{u}\rangle}$$
$$= \lambda_j + \frac{\langle A\mathbf{u}, \mathbf{u}\rangle - \lambda_j\langle \mathbf{u}, \mathbf{u}\rangle}{\|\mathbf{v}\|^2},$$

and the difference between the Rayleigh quotient and the eigenvalue λ_j satisfies

$$\left|\frac{\langle A\mathbf{v}, \mathbf{v}\rangle}{\langle \mathbf{v}, \mathbf{v}\rangle} - \lambda_j\right| \leq (\|A\| + |\lambda_j|)\left(\frac{\|\mathbf{u}\|}{\|\mathbf{v}\|}\right)^2 \leq 2\|A\|\left(\frac{\|\mathbf{u}\|}{\|\mathbf{v}\|}\right)^2.$$

The error in the Rayleigh quotient approximation to λ_j is proportional to the *square* of the norm of the portion of $\mathbf{v}$ that is orthogonal to $\mathbf{v}_j$. Thus if $\mathbf{v}$ is a fairly good approximation to the eigen*vector* $\mathbf{v}_j$, say, $\|\mathbf{v}\| = 1$ and $\|\mathbf{u}\| = 0.1$, then the Rayleigh quotient gives a much better approximation to the eigen*value* λ_j, with error about $2\|A\|$ times 0.01. For a general nonsymmetric matrix, the term $\langle A\mathbf{u}, \mathbf{v}_j\rangle$ does not vanish and the error in the Rayleigh quotient approximation to λ_j is on the same order, $\|\mathbf{u}\|/\|\mathbf{v}\|$, as the error in the eigenvector approximation. For this reason, Rayleigh quotient iteration is used mainly for symmetric eigenproblems.

12.1.4 The QR Algorithm

Thus far we have looked at methods for approximating selected eigenvalues and corresponding eigenvectors. Can one find all of the eigenvalues/vectors at once? The *QR algorithm* applies a sequence of orthogonal similarity transformations to A until it approaches an upper triangular matrix. Then the approximate eigenvalues are taken to be the diagonal entries in this (nearly) upper triangular matrix.

Recall the QR decomposition of a matrix, described in section 7.6.2: Every n by n matrix A can be written in the form $A = QR$, where Q is an *orthogonal matrix* (its columns are orthonormal) and R is upper triangular. This decomposition was carried out using the *Gram–Schmidt algorithm*. In practice, it is usually carried using *Householder reflections*, which will not be described here but see, for example, [96]. Consider now the following procedure: First factor $A \equiv A_0$ in the form $A_0 = QR$; then form the matrix $A_1 = RQ$. Note that $A_1 = (Q^T Q)RQ = Q^T(QR)Q = Q^T A_0 Q$, so that A_1 is similar to A_0 since $Q^T = Q^{-1}$. We can repeat this process of doing a QR factorization and then multiplying the factors in reverse order to obtain a sequence of matrices that are all orthogonally similar to A. This is called the *QR algorithm*:

> Let $A_0 = A$. For $k = 0, 1, \ldots,$
> Compute the QR factorization of A_k: $A_k = Q_k R_k$.
> Form the product $A_{k+1} = R_k Q_k$.

The following theorem, which we state without proof, gives the basic convergence result for the QR algorithm.

Theorem 12.1.6. Suppose the eigenvalues $\lambda_1, \ldots, \lambda_n$ of A satisfy $|\lambda_1| > |\lambda_2| > \ldots > |\lambda_n|$. Then the matrices A_k produced by the QR algorithm converge to an upper triangular matrix whose diagonal entries are the eigenvalues of A. If, in addition, $A = V \Lambda V^{-1}$, where V has an LU decomposition (i.e., pivoting is not required for Gaussian elimination on a matrix V of eigenvectors), then the rate of convergence to 0 of the elements $a_{ij}^{(k)}$, $i > j$, in the strict lower triangle of A_k is given by

$$\left| a_{ij}^{(k)} \right| = O\left(\left| \frac{\lambda_i}{\lambda_j} \right|^k \right).$$

The technical condition that V have an LU decomposition ensures that the eigenvalues appear on the diagonal in descending order of magnitude. This is the usual case. The proof of this theorem is based on a comparison of the QR algorithm with a block version of the power method described previously.

Hessenberg Form

The QR algorithm described so far is too expensive to be practical for a general matrix A. The first problem is that each step requires $O(n^3)$ operations. To reduce the number of operations per step, one can first apply a similarity transformation to put A into a form where the QR factorization is not so expensive. A general matrix A can first be reduced to **upper Hessenberg** form:

$$\begin{pmatrix} * & \cdots & \cdots & * \\ * & \ddots & & \vdots \\ & \ddots & \ddots & \vdots \\ & & * & * \end{pmatrix},$$

which has only one nonzero diagonal below the main diagonal. For a description of the details of this transformation using *Householder reflections*, see, for example, [96]. If the matrix A is *symmetric*, then this symmetry is maintained and hence the upper Hessenberg form is actually *tridiagonal*; that is, only the three center diagonals have nonzero entries.

A single step of the QR algorithm applied to an upper Hessenberg matrix can be carried out using only $O(n^2)$ operations, although the details of how to do so are subtle; see, for example, [32]. The idea is to avoid a direct computation of the QR factorization of the upper Hessenberg matrix and simply compute $A_{k+1} = R_k Q_k = Q_k^T A_k Q_k$ without explicitly forming Q_k. Moreover, the orthogonal matrix Q_k is still upper Hessenberg and the product $R_k Q_k$ maintains this upper Hessenberg form. You can convince yourself of this by considering the Gram–Schmidt algorithm applied to the columns of an upper Hessenberg matrix and noting that orthogonalizing column j against each of the previous columns does not introduce nonzeros below row $j+1$ because all of the previous columns have zeros there. It is easy to see then that the product of an upper triangular matrix and an upper Hessenberg matrix is still upper Hessenberg.

In the symmetric case, the savings are even greater since the upper Hessenberg form is actually tridiagonal. The QR factorization of a tridiagonal matrix can be computed using only $O(n)$ operations. To see this, note that when the Gram–Schmidt algorithm is applied to the columns of a tridiagonal matrix, each column need only be orthogonalized against the previous two (since it is already orthogonal to earlier ones), and the work to do this orthogonalization is $O(1)$. Moreover, as was already noted, the orthogonal matrix Q_k and the product $R_k Q_k$ are both upper Hessenberg. But since $R_k Q_k = Q_k^T (Q_k R_k) Q_k$ and $Q_k R_k$ is symmetric, $R_k Q_k$ is also symmetric and therefore tridiagonal. Thus the tridiagonal form is maintained throughout the QR algorithm and each step requires only $O(n)$ operations when the matrix A is symmetric.

Shifts

An initial transformation of the matrix A to upper Hessenberg form reduces the *work per iteration* of the QR algorithm from $O(n^3)$ to $O(n^2)$ in the general case and to $O(n)$ in the symmetric case. But the *number of iterations* required for the algorithm described so far is still too large; it is *slow* to converge. To improve convergence, one typically uses approximate eigenvalues as *shifts*, as was done in the power method, inverse iteration, and Rayleigh quotient iteration; that is, instead of factoring A_k as $Q_k R_k$, one chooses a shift s_k and factors $A_k - s_k I = Q_k R_k$ and then sets $A_{k+1} = R_k Q_k + s_k I$. Note that A_{k+1} is still similar to A_k: $A_{k+1} = Q_k^T Q_k (R_k Q_k + s_k I) = Q_k^T (Q_k R_k + s_k I) Q_k = Q_k^T A_k Q_k$. Thus all matrices A_k have the same eigenvalues as A.

There are many strategies for choosing shifts, one of the most widely used being the **Wilkinson shift** [108]. Here s_k is taken to be an eigenvalue of the bottom right 2 by 2 block of A_k; the one that is closer to the (n, n)-element of A_k is chosen, or, in case of a tie, either eigenvalue can be used as the shift.

Deflation

One final strategy is important in making the QR algorithm efficient; that is *deflation*. If a subdiagonal entry in the upper Hessenberg (or tridiagonal) matrix A_k is 0, then the problem can be split into two separate problems. This is because the eigenvalues of a matrix of the form

$$\left(\begin{array}{ccc|ccc} x & x & x & x & x & x \\ x & x & x & x & x & x \\ 0 & x & x & x & x & x \\ \hline 0 & 0 & 0 & x & x & x \\ 0 & 0 & 0 & x & x & x \\ 0 & 0 & 0 & 0 & x & x \end{array} \right)$$

are the same as those of its two diagonal blocks.

If a subdiagonal entry in A_k is zero or sufficiently close to zero (say, on the order of the machine precision), then it can be set to zero and the QR algorithm continued on the two diagonal blocks of A_k separately. Note that if the work per iteration for the QR algorithm is Cn^2 for some constant C, then if the n by n matrix A_k can be split into two blocks, say, each of size $n/2$, then the work per iteration to apply the QR algorithm to each of these blocks separately is just $C(n/2)^2 + C(n/2)^2 = \frac{1}{2}Cn^2$. Additionally, one would expect to find better shifts to use with each block separately and so to improve the convergence rate of the algorithm.

Practical QR Algorithm

A version of the QR algorithm incorporating all of the suggested improvements is given below:

Let $A_0 = A$. For $k = 0, 1, \ldots,$
- Choose a shift s_k by finding the eigenvalues of the lower right 2 by 2 block of A_k and taking the one that is closer to the bottom right element of A_k.
- Compute the QR factorization of $A_k - s_k I$: $A_k - s_k I = Q_k R_k$.
- Form the product $A_{k+1} = R_k Q_k + s_k I$.
- If any subdiagonal entry in A_{k+1} is sufficiently close to zero, set it to zero to obtain a block upper triangular matrix $\tilde{A}_{k+1}$, with diagonal blocks B and C. Apply the QR algorithm to B and C separately.

This is essentially the algorithm used by MATLAB when you type `eig(A)`.

12.1.5 Google's PageRank

We now look at an important application involving eigenvectors and the power method for computing the eigenvector corresponding to the eigenvalue of largest magnitude. This is the PageRank computation in Google. In section 1.5, we introduced the PageRank model for ranking web pages. In section 3.4, we saw how to compute PageRank via Monte Carlo simulation of randomly surfing the Internet. We now study the problem from the point of view of Markov chains. For more information, see, for example, [21].

A popular and often effective form of information acquisition is submitting queries to a search engine. In fact, in the past minute an average[1] of just over 2.9 million searches were conducted worldwide by users age 15 or older [25].

Suppose that we submit the query mathematics to Google. In the summer of 2010, Google returned the following pages (plus thousands more) in order:

- Mathematics - Wikipedia, the free encyclopedia, http://en.wikipedia.org/wiki/Mathematics
- Wolfram MathWorld: The Web's Most Extensive Mathematics Resource, mathworld.wolfram.com/
- Mathematics in the Yahoo! Directory, http://dir.yahoo.com/science/mathematics/
- Math tutorials, resources, help, resources and Math Worksheets, http://math.about.com

The mathematics category in the Wikipedia directory is deemed the "best" page related to the query. Being at the top of the list is an enviable position, since being listed 40th or 50th can essentially bury a page into obscurity, as it relates to the query.

Google searches billions of pages and solves a huge linear algebra problem that yields an "importance" ranking of pages. We will see that eigenvectors are deeply integrated into Google and the results that are returned from a query.

Lining Up the Web by Rank

A component of Google's success is the PageRank algorithm developed by Google's founders, Larry Page and Sergey Brin, who were graduate students at Stanford University when they developed the foundational ideas of Google [81]. The results of the algorithm are determined by the link structure of the WWW and involve no content of any page. As a historical note, Jon Kleinberg first suggested link analysis and spectral graph theory [58] as a means to improve web searches, although the development and analysis differed from what Page and Brin would later develop.[2]

How does the PageRank algorithm work? A page A is considered more "important" if more pages link to it. However, Google also considers the

[1] Using data from December 2009.

[2] Note, the citation for the reference to Kleinberg's work is later than that of Brin and Page's seminal paper. Brin and Page, however, cite Kleinberg's work as it appeared in 1998 in the *Proceedings of the ACM–SIAM Symposium on Discrete Algorithms.*

importance of the pages that link to page A; links from "important" pages are given more weight. In the end, pages with high weight sums are given a high PageRank, which Google uses to order results of a query. PageRank is combined with text-matching algorithms to find pages that are both "important" and relevant to the query. Keeping the rankings updated presents its own issues since the structure of the Web changes continually as links are altered and pages are added and deleted.

How does the PageRank algorithm determine the weights of links? Try the following exercise:

> Start at one of your favorite pages. Randomly select a link and click it. On the next page, randomly select a link. Continue this process of following random links.

There are two possible outcomes:

- You visit a page with no outgoing links—a **dangling node**, a dead end.
- Otherwise, you eventually revisit a page. You cannot visit new pages indefinitely, since there are finitely many pages.

Surfing with Markov

Your random walk in the exercise above can be modeled mathematically with a Markov chain. A Markov chain modeling a discrete system requires determining the states of the system, and for each ordered pair of states i and j, the probability m_{ij} of moving from state i to state j. The Markov transition matrix is $M = (m_{ij})$. [Note: Sometimes the transition matrix is defined as the *transpose* of that defined here, so that the (i, j)-entry is the probability of moving from state j to state i. This definition is more common in the numerical analysis literature, since then advancing from one stage to the next means multiplying the state vector (a *column* vector) on the left by the transition matrix. In the probability and statistics literature, however, it is standard to define M as we have done here, and then advancing from one stage to the next will correspond to multiplying the state vector (a *row* vector) on the right by M.]

For surfing the Web, the states are the indexed pages. Markov processes model the behavior of a random system whose probability distribution of possible next states depends only on the current state. The random walk ("surf") that we performed had this attribute. The pages that we visit depend in part on the links available on the current page. For Google's purposes, it is undesirable for the Markov chain to get stuck at a dangling node, so we also include the possibility that the surfer jumps to another page, not linked from the current one, choosing such a page uniformly at random. Once we specify the probability of this random jump, we have a complete conceptual description of the Markov chain used by Google. But determining the transition matrix—the *Google matrix*[3], with all the probabilities of moving from page i to page j, seems a monumental undertaking!

[3] Enter Google matrix as a query in Google and see what you find!

A Googled Matrix

Let W be the set of web pages contained in Google's index. Let n be the cardinality of W (the number of web pages in W). Note that n varies with time. As of November 2004, n was about 8 billion [28] and growing fast. Google's indexing of billions of web pages is a nontrivial task. We assume that such a set W has been constructed.

Consider a directed graph of the WWW where a vertex represents a web page, and an edge from vertex i to vertex j corresponds to a link from page i to page j. Let G be the n by n adjacency matrix of this graph; that is, $g_{ij} = 1$ if there is a hyperlink from page i to page j and 0 otherwise. The number of nonzeros in row i is the number of hyperlinks on page i, out of the very large number n of possible links. Hence, most entries in every row are zeros, and the matrix is sparse. Recall the size of n. The sparsity of G will be an asset in manipulating it on a computer. In particular, only the row and column locations of the nonzero entries need be stored. Because the average outdegree of pages on the Web is about seven [59], this saves a factor on the order of a billion in storage space and since n is growing over time while the average number of links on each page appears to remain about constant, the savings will only increase over time.

The information captured in G leads to the transition matrix. Let c_j and r_i be the column and row sums of G, respectively; that is,

$$c_j = \sum_{1 \le i \le n} g_{ij}, \qquad r_i = \sum_{1 \le j \le n} g_{ij}. \tag{12.5}$$

Then c_k and r_k are the indegree and outdegree of the state (page) k.

Let p be the fraction of time that a random walk follows one of the links available on a page, so that $(1 - p)$ is the fraction of time that a surfer jumps to a random web page chosen uniformly from W. Google typically sets $p = 0.85$.

The transition matrix M has elements

$$m_{ij} = p\left(\frac{g_{ij}}{r_i}\right) + \frac{1-p}{n}, \tag{12.6}$$

where m_{ij} is the probability that we visit page j in the next step, given that we are currently at page i.

Google's Eigenvectors

A nonnegative left eigenvector that satisfies $\mathbf{v}M = \mathbf{v}$ is called a **steady-state vector** of the Markov process (where $\mathbf{v}$ is normalized so that $\sum \mathbf{v}_i = 1$, which results in a vector of probabilities). (Recall that right-multiplication of row vectors $\mathbf{v}$ by M is standard notation in the literature of Markov chains, but we also could write $M^T\mathbf{w} = \mathbf{w}$, where $\mathbf{w} = \mathbf{v}^T$ is a right eigenvector of M^T.) The theory of Markov chains shows that such a vector exists and that, from any starting state, the limiting probability of visiting page i at time t as $t \to \infty$ is $\mathbf{v}_i$.

Google defines the PageRank of page i to be $\mathbf{v}_i$. Therefore, the largest element of $\mathbf{v}$ corresponds to the page with the highest PageRank, the second

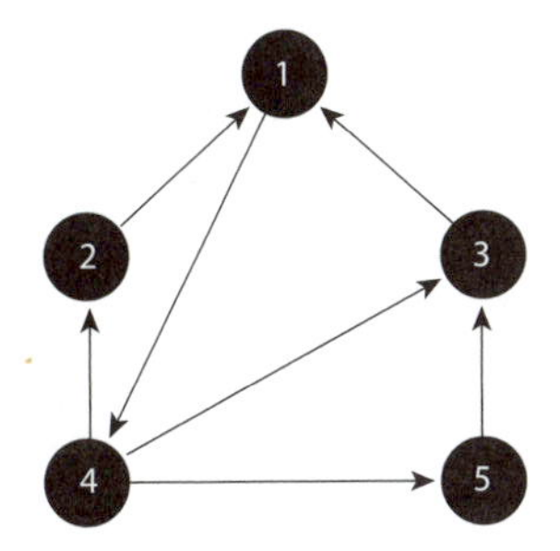

Figure 12.8. A small network of web pages.

largest to the page with the second highest PageRank, and so on. The limiting frequency at which an infinitely dedicated random surfer visits any particular page is that page's PageRank.

The following theorem guarantees the uniqueness of the steady-state vector and that it will have positive entries.

Theorem 12.1.7 (Perron). Every real square matrix P whose entries are all positive has a unique (up to multiplication by a positive scalar) eigenvector with all positive entries. Its corresponding eigenvalue has multiplicity one, and it is the dominant eigenvalue, in that every other eigenvalue has a strictly smaller magnitude.

Recall that the rows of M sum to 1. Therefore, $M\mathbf{1} = \mathbf{1}$, where $\mathbf{1}$ is the *column* vector of all ones. That is, $\mathbf{1}$ is a *right eigenvector* of M associated with the eigenvalue 1, and its entries are all positive. Since $1 - p > 0$, the entries in M are all positive, and therefore Perron's theorem ensures that $\mathbf{1}$ is the *unique* (up to multiplication by a positive scalar) right eigenvector with all positive entries and that its eigenvalue 1 must be the dominant one. The right and left eigenvalues of a matrix are the same, so 1 is the dominant left eigenvalue as well. Thus, there exists a unique steady-state vector $\mathbf{v}$ that satisfies $\mathbf{v}M = \mathbf{v}$ and is normalized so that $\sum \mathbf{v}_i = 1$.

With the theory behind the PageRank algorithm in place, we apply it to the small network in figure 12.8.

The adjacency matrix G of this network is

$$G = \begin{pmatrix} 0 & 0 & 0 & 1 & 0 \\ 1 & 0 & 0 & 0 & 0 \\ 1 & 0 & 0 & 0 & 0 \\ 0 & 1 & 1 & 0 & 1 \\ 0 & 0 & 1 & 0 & 0 \end{pmatrix}.$$

Recall that r_k is the outdegree of page k. We see from G that each page except page 4 has outdegree 1, while the outdegree of page 4 is 3.

From G, we form the transition matrix M. Note from (12.6) that if $g_{ij} = 0$, then $m_{ij} = (1 - p)/n = 0.15/5 = 0.03$. Let us consider the entry m_{14}. From (12.6), we have $m_{14} = (0.85)\left(\frac{1}{1}\right) + 0.03 = 0.88$. Similarly,

TABLE 12.1
PageRanks for the pages in the web of figure 12.8.

Page	PageRank
1	0.2959
4	0.2815
3	0.2031
2	0.1098
5	0.1098

$m_{42} = (0.85)\left(\frac{1}{3}\right) + 0.03 \approx 0.3133$. Evaluating each entry in this way, we find

$$M = \begin{pmatrix} 0.0300 & 0.0300 & 0.0300 & 0.8800 & 0.0300 \\ 0.8800 & 0.0300 & 0.0300 & 0.0300 & 0.0300 \\ 0.8800 & 0.0300 & 0.0300 & 0.0300 & 0.0300 \\ 0.0300 & 0.3133 & 0.3133 & 0.0300 & 0.3133 \\ 0.0300 & 0.0300 & 0.8800 & 0.0300 & 0.0300 \end{pmatrix}.$$

Each row of M sums to 1; equation (12.6) guarantees this. Since m_{1j} is the probability of moving from page 1 to page j, if a surfer is currently on page 1, there is an 88% chance that page 4 will be visited next and only a 3% chance of jumping from page 1 to a specific page other than page 4 (including page 1, which the surfer might jump back to).

Solving $\mathbf{v}M = \mathbf{v}$ yields

$$\mathbf{v} = (0.2959, 0.1098, 0.2031, 0.2815, 0.1098).$$

Therefore, a random surfer will visit page 1 approximately 30% of the time and page 2 about 11% of the time. The PageRanks (the elements of $\mathbf{v}$) for this small web are given in table 12.1.

Compare these PageRanks to the network in figure 12.8. Pages 1 and 3 have the same indegree and outdegree, yet page 1 has higher PageRank because it has links from pages that are either more important (have higher PageRank) than those that link to page 3 or they have lower outdegree making it more likely that one would move from these pages to page 1. Page 4 receives a high PageRank because it is the only page linked from page 1. If a surfer lands on page 1 (which occurs about 30% of the time), then 85% of the time the surfer will follow the link to page 4.

Getting Practical

In this description of PageRank, we have simplified many of the issues of information retrieval. We now introduce some of the complexities inherent to a problem of this magnitude.

Fortunately, the PageRank algorithm requires finding only the *dominant* eigenvector; to try to compute all of the eigenvalues and eigenvectors of a matrix of this size would be computationally infeasible (by orders of magnitude, in terms of both work and storage). To compute just the dominant eigenvector, however, the power method can be used.

Will the power method converge? The answer is yes, since the dominant eigenvalue, 1, is larger in magnitude than the others. This would not necessarily be the case if $p = 1$ were allowed since then the entries in M would be nonnegative but could be 0, so Perron's theorem would not apply. In this case, a generalization of Perron's theorem to nonnegative matrices, known as the Perron–Frobenius theorem, would be relevant, but it requires an additional assumption, namely, that the matrix be *irreducible*. This occurs if and only if every web page is reachable from every other page via a path of hyperlinks. This is *not* the case if there are dangling nodes. With this additional assumption, it can be shown that 1 is a simple eigenvalue of the nonnegative matrix M and that all other eigenvalues have magnitude less than or equal to 1. But still another assumption is needed to show that all other eigenvalues have magnitude strictly less than 1, namely, the assumption that M is *primitive*. This property has to do with the lengths of possible paths starting and ending at a given page P_i, and it could not be expected to hold in general. Hence allowing the surfer a small probability of jumping to a random page, not only results in a more realistic model, but also simplifies the mathematics considerably!

Assume, then, that $|\lambda_1| > |\lambda_2| \geq \ldots \geq |\lambda_n|$. Then, from (12.4), we know that the rate of convergence of the power method is $|\lambda_2|/|\lambda_1|$. From Perron's theorem, we know that $\lambda_1 = 1$ and that $|\lambda_i| < 1$ for $i > 1$, but it gives no other information about $|\lambda_2|$. In [50], it is shown that $|\lambda_2|$ is bounded above by $p = 0.85$. This means that the power method will converge at the same rate (require about the same number of iterations) even as the size of the problem increases.

Even more powerful is the fact that the entire matrix M need not be stored for the power method. In fact, it is enough to know p, n, the nonzeros of G, and the outdegree of each page. For more information on this, see [63]: this paper contains an extensive survey of research on all facets of PageRank, which involves many complexities beyond the scope of this book.

Remarks before Logging Out

To determine the PageRanks of the pages that it indexes, Google solves a linear algebra problem of massive proportions—finding an eigenvector of a matrix of order 8 billion.

Again, keep in mind that PageRank alone does not determine the order of the list returned by Google for a query. For example, http://mathworld.wolfram.com/ has a lower PageRank than http://www.google.com. However, our query mathematics does not produce http://www.google.com in the first 100 pages returned from Google. Why? Using text-matching algorithms, Google determines the relevance of a page to a query and combines this with its PageRank to determine the final ranking for a query.

To reflect the subtlety of language and the challenges of text matching, suppose that we submit the query show boat. The page that tops the list returned is http://www.imdb.com/title/tt0044030/, which gives details on the 1951 film

Google Bombs

Much power lies in the hands of those who can use knowledge of the underlying mathematics and computer science of information retrieval to their advantage. For example, in October 2003, George Johnston initiated a *Google bomb* called the "miserable failure" bomb, which targeted President George W. Bush and detonated by late November of that same year. The result was Google returning the official White House Biography of the President as the highest ranked result to the query miserable failure. Putting together a Google bomb was relatively easy. It involved several web pages linking to a common web page with an agreed upon anchor text (the text that one clicks to go to the hyperlink). In the case of the "miserable failure" project, of the over 800 links that pointed to the Bush biography, only 32 used the phrase "miserable failure" in the anchor text [64]. This ignited a sort of political sparring match among the web savvy, and by January 2004, a miserable failure query returned results for Michael Moore, President Bush, Jimmy Carter and Hillary Clinton in the top four positions. As expected, Google is fully aware of such tactics and works to outsmart these and other initiatives that can dilute the effectiveness of its results. For more information, search the Internet on Google bombs or link farms or see the recent text on information retrieval by Langville and Meyer [64].

Show Boat based on the musical by Jerome Kern and Oscar Hammerstein II. For the query boat show, topping the list is http://www.showmanagement.com/, containing information on the Show Management company that produces boat shows throughout the year.

By inverting the order of the words in our query, we changed the list of pages returned. This is what we want. If we are looking for information on boat shows, we probably do not want information on the musical *Show Boat*. Carefully designed text-matching algorithms play the role in differentiating between pages that have the words "boat" and "show" in their content.

How do text-matching algorithms work? How is relevance to a string quantified? These questions step into thought-provoking areas beyond the scope of this textbook. The answers, however, may simply be a query on Google away.

12.2 ITERATIVE METHODS FOR SOLVING LINEAR SYSTEMS

12.2.1 Basic Iterative Methods for Solving Linear Systems

We saw in chapter 7 how Gaussian elimination can be used to solve a nonsingular n by n system of linear equations $A\mathbf{x} = \mathbf{b}$. For moderate size n, say, n on the order of a few hundred or perhaps a few thousand, this is a reasonable approach. Sometimes, however, n is *very* large, on the order of hundreds of thousands or even millions or billions. We saw earlier how the Google matrix, depicting connections between all web pages on the World Wide Web, has dimensions on the order of billions.

Many large linear systems arise from differencing partial differential equations, as was illustrated in section 7.2.4, with the matrix arising from Poisson's equation. That matrix was relatively small, coming from a small two-dimensional problem, but much larger matrices can arise from differencing partial differential equations in three dimensions. In chapter 14 we will see precisely how such linear systems arise, but to give an indication here, consider Poisson's equation on the unit cube in three dimensions:

$$\frac{\partial^2 u}{\partial x^2} + \frac{\partial^2 u}{\partial y^2} + \frac{\partial^2 u}{\partial z^2} = f(x, y, z), \quad 0 \le x, y, z \le 1.$$

To approximate the solution $u(x, y, z)$, we divide the cube into many small subcubes and approximate the partial derivatives using finite difference quotients (described in chapter 9). These involve the values u_{ijk} of the unknown solution u at each corner (x_i, y_j, z_k) of a subcube. These values become the unknowns in a system of linear equations. If there are m subcubes in each direction, then the number of unknowns is approximately m^3. Thus, with 100 subcubes in each direction, we end up with a one million by one million linear system.

For such problems, both the work and the storage required by Gaussian elimination are prohibitive. Recall that Gaussian elimination requires about $(2/3)n^3$ operations (additions, subtractions, multiplications, and divisions) to solve a general n by n linear system. For $n = 10^6$, this is $(2/3) \cdot 10^{18}$ operations, and on a computer that performs, say, 10^9 operations per second, this would require $(2/3) \cdot 10^9$ seconds, or, about 21 years! Some advantage can be taken of the band structure of such a matrix, as discussed in section 7.2.4. The half bandwidth of this matrix, using what is known as the *natural* ordering of unknowns (see chapter 14), is about m^2, and using this fact the work can be reduced to about $2(m^2)^2 n = 2 \cdot 10^{14}$ operations, which, at 10^9 operations per second, could be performed in about 2.3 days. Still, work is not the only prohibiting factor. Storing the matrix is another problem. To store a dense n by n matrix, where $n = 10^6$, requires 10^{12} words of storage. This is far more storage than is typically available on a single processor, and the matrix elements would have to be distributed among many processors. Again, some advantage can be taken of the band structure; since zeros outside of the band remain zero during Gaussian elimination, they need not be stored. Still, everything within the band (including zeros since they fill in with nonzeros during Gaussian elimination) must be stored, and this requires about $m^2 n = 10^{10}$ words of storage, which is still far more than is generally available on a single processor.

To circumvent both of these difficulties, iterative methods can be used to solve $A\mathbf{x} = \mathbf{b}$. These methods require only matrix–vector multiplication and, perhaps, solution of a preconditioning system (to be described later). To compute the product of A with a given vector $\mathbf{v}$, one need store only the nonzeros of A along with their row and column numbers. The matrix arising from Poisson's equation above with a standard differencing scheme has only about 7 nonzeros per row, resulting in a storage requirement of just 7 million words. This property of **sparseness**—having most entries equal to 0—is typical of the matrices arising from partial differential equations and from many other mathematical formulas as well. Moreover, if the matrix entries are sufficiently easy to compute, then they need not be stored at all; knowing the nonzero entries and their locations one can write a routine to compute the product of A with a given vector $\mathbf{v}$ without using any additional storage for the matrix. For a dense n by n matrix, the cost of a matrix–vector multiplication is about $2n^2$ operations, but if A is sparse then this cost will be significantly less. For a matrix with an average of r nonzeros per row, the cost is about $2rn$. Thus, if we think of r as fixed and the matrix size n as increasing, then the number of operations required for a matrix–vector multiplication is $O(n)$. If an acceptably good approximate solution to the linear system $A\mathbf{x} = \mathbf{b}$ can be found by using a moderate number of matrix–vector multiplications and, say, $O(n)$ additional work, then an iterative method that does this will be *much* faster than Gaussian elimination.

It may turn out that the linear system $A\mathbf{x} = \mathbf{b}$ is not well suited to solution by the means described above; that is, if an iteration consists of simply computing the product of A with an appropriate vector and doing $O(n)$ additional work, then it may turn out that many, many iterations are required to obtain a good approximate solution to $A\mathbf{x} = \mathbf{b}$. In this case, the original linear system can be modified through the use of a *preconditioner* or *matrix splitting*. All of the iterative methods that we will discuss find the solution immediately if A is the identity matrix. Hence one might look for a matrix M with the property that linear systems with coefficient matrix M are easy to solve and yet $M^{-1}A$ is, in some sense, close to the identity. One might then replace the original linear system $A\mathbf{x} = \mathbf{b}$ by the preconditioned system $M^{-1}A\mathbf{x} = M^{-1}\mathbf{b}$ and apply an iterative method to this new linear system. One need not form the matrix $M^{-1}A$; one need only be able to compute the product of $M^{-1}A$ with a given vector $\mathbf{v}$. This can be done by computing the product $A\mathbf{v}$ and then solving the linear system $M\mathbf{w} = A\mathbf{v}$ for $\mathbf{w} = M^{-1}A\mathbf{v}$. Examples of preconditioners M that are sometimes used include the diagonal of A and the lower (or upper) triangle of A, since linear systems with diagonal or triangular coefficient matrices are relatively easy to solve.

12.2.2 Simple Iteration

Suppose one has an initial guess $\mathbf{x}^{(0)}$ for the solution to $A\mathbf{x} = \mathbf{b}$. If one could compute the *error* $\mathbf{e}^{(0)} \equiv A^{-1}\mathbf{b} - \mathbf{x}^{(0)}$, then one could find the solution, $\mathbf{x} = \mathbf{x}^{(0)} + \mathbf{e}^{(0)}$. Unfortunately, one cannot compute the error in $\mathbf{x}^{(0)}$ (without

David M. Young

David M. Young (1923–2008) received his doctorate from Harvard University in 1950. Young's research focused on developing relaxation methods suitable for automatically solving linear systems on digital computers. This was not the first work on relaxation methods. For instance, Sir Richard Southwell had developed such methods, designed for hand implementation, which involved, as Young wrote in his dissertation, "scanning of the residuals, a process which is easy for a human computer but which cannot be done efficiently by any large automatic computing machine in existence or being built." Not surprisingly, Young studied Southwell's work. Reflecting on the difficulty of Young's dissertation topic, Southwell said on a visit with Garrett Birkhoff who was Young's advisor, "Any attempt to mechanize relaxation methods would be a waste of time!" [41, 109, 110]

solving a linear system that is as difficult as the original one), but one can compute the *residual*, $\mathbf{r}^{(0)} \equiv \mathbf{b} - A\mathbf{x}^{(0)}$. Note that $\mathbf{r}^{(0)} = A\mathbf{e}^{(0)}$, so that if M is a matrix with the property that $M^{-1}A$ approximates the identity, yet linear systems with coefficient matrix M are easy to solve, then one could approximate the error by $M^{-1}\mathbf{r}^{(0)}$. Thus, one might solve the linear system $M\mathbf{z}^{(0)} = \mathbf{r}^{(0)}$ for $\mathbf{z}^{(0)}$ and then replace the approximation $\mathbf{x}^{(0)}$ by the, hopefully, improved approximation $\mathbf{x}^{(1)} \equiv \mathbf{x}^{(0)} + \mathbf{z}^{(0)}$. This process could then be repeated with $\mathbf{x}^{(1)}$; that is, compute the residual $\mathbf{r}^{(1)} \equiv \mathbf{b} - A\mathbf{x}^{(1)}$, solve $M\mathbf{z}^{(1)} = \mathbf{r}^{(1)}$ to obtain an approximation $\mathbf{z}^{(1)}$ to the error in $\mathbf{x}^{(1)}$, then add $\mathbf{z}^{(1)}$ to $\mathbf{x}^{(1)}$ to form what is, hopefully, an improved approximation. This is the simplest type of iterative algorithm.

Simple Iteration.

Given an initial guess $\mathbf{x}^{(0)}$, compute $\mathbf{r}^{(0)} = \mathbf{b} - A\mathbf{x}^{(0)}$, and solve $M\mathbf{z}^{(0)} = \mathbf{r}^{(0)}$ for $\mathbf{z}^{(0)}$.
For $k = 1, 2, \ldots,$

Set $\mathbf{x}^{(k)} = \mathbf{x}^{(k-1)} + \mathbf{z}^{(k-1)}$.

Compute $\mathbf{r}^{(k)} = \mathbf{b} - A\mathbf{x}^{(k)}$.

Solve $M\mathbf{z}^{(k)} = \mathbf{r}^{(k)}$ for $\mathbf{z}^{(k)}$.

This algorithm goes by different names, according to the preconditioner M that is used. For M equal to the diagonal of A, it is called **Jacobi** iteration;

for M equal to the lower triangle of A, it is the **Gauss–Seidel** method; for $M = \omega^{-1}D - L$, where D is the diagonal of A, $-L$ is the strict lower triangle of A, and ω is a parameter chosen to accelerate convergence, it is the **successive overrelaxation (SOR)** method. The SOR method was devised simultaneously by David M. Young and by H. Frankel. Young presented the method in his Ph.D. dissertation from Harvard University [109] and Frankel in his paper "Convergence rates of iterative treatments of partial differential equations" [41], in which he called the process the "extrapolated Liebmann" method. Both publications appeared in 1950.

Another way to derive the method is as follows. Write the matrix A in the form $A = M - N$. This is called a **matrix splitting.** Now, the true solution $\mathbf{x}$ satisfies

$$M\mathbf{x} = N\mathbf{x} + \mathbf{b}, \qquad \text{or} \qquad \mathbf{x} = M^{-1}N\mathbf{x} + M^{-1}\mathbf{b}.$$

Recall section 4.5 on fixed point iteration, where we looked for a solution to the scalar equation $x = \varphi(x)$. Here we are looking for a solution to the vector equation $\mathbf{x} = \boldsymbol{\varphi}(\mathbf{x})$, where $\boldsymbol{\varphi}(\mathbf{x}) = M^{-1}N\mathbf{x} + M^{-1}\mathbf{b}$, using the same fixed point iteration:

$$M\mathbf{x}^{(k)} = N\mathbf{x}^{(k-1)} + \mathbf{b}, \qquad \text{or} \qquad \mathbf{x}^{(k)} = M^{-1}N\mathbf{x}^{(k-1)} + M^{-1}\mathbf{b}. \tag{12.7}$$

In section 4.5 we showed that fixed point iteration converged to the unique solution of $x = \varphi(x)$ if φ was a contraction. We will see in the next subsection that essentially the same result holds for the vector equation $\mathbf{x} = \boldsymbol{\varphi}(\mathbf{x})$. Now, however, we must consider different norms since $\boldsymbol{\varphi}$ may be a contraction in one norm but not in another. The result in the next subsection shows that this iteration converges to the unique solution $\mathbf{x}$ if $\boldsymbol{\varphi}$ is a contraction in *some* norm.

To see that (12.7) is the same as the simple iteration algorithm described previously, note that $M^{-1}N = I - M^{-1}A$, so that (12.7) is equivalent to

$$\mathbf{x}^{(k)} = (I - M^{-1}A)\mathbf{x}^{(k-1)} + M^{-1}\mathbf{b} = \mathbf{x}^{(k-1)} + M^{-1}(\mathbf{b} - A\mathbf{x}^{(k-1)}) = \mathbf{x}^{(k-1)} + \mathbf{z}^{(k-1)}. \tag{12.8}$$

Example 12.2.1. Write down the approximate solutions generated in two steps of the Jacobi and Gauss–Seidel methods for solving $A\mathbf{x} = \mathbf{b}$, where

$$A = \begin{pmatrix} 2 & -1 \\ -1 & 2 \end{pmatrix}, \qquad \mathbf{b} = \begin{pmatrix} 1 \\ 1 \end{pmatrix}.$$

Start with the initial guess $\mathbf{x}^{(0)} = (0, 0)^T$.

For Jacobi's method M is the diagonal of A, and so we find after one step that

$$\begin{pmatrix} x_1 \\ x_2 \end{pmatrix}^{(1)} = \begin{pmatrix} 0 \\ 0 \end{pmatrix} + \begin{pmatrix} 2 & 0 \\ 0 & 2 \end{pmatrix}^{-1} \begin{pmatrix} 1 \\ 1 \end{pmatrix} = \begin{pmatrix} 1/2 \\ 1/2 \end{pmatrix}.$$

The residual is

$$\begin{pmatrix} r_1 \\ r_2 \end{pmatrix}^{(1)} = \begin{pmatrix} 1 \\ 1 \end{pmatrix} - \begin{pmatrix} 2 & -1 \\ -1 & 2 \end{pmatrix} \begin{pmatrix} 1/2 \\ 1/2 \end{pmatrix} = \begin{pmatrix} 1/2 \\ 1/2 \end{pmatrix},$$

and the vector $\mathbf{z}^{(1)} \equiv M^{-1}\mathbf{r}^{(1)}$ is

$$\begin{pmatrix} z_1 \\ z_2 \end{pmatrix}^{(1)} = \begin{pmatrix} 1/4 \\ 1/4 \end{pmatrix}.$$

The second iterate, $\mathbf{x}^{(2)} = \mathbf{x}^{(1)} + \mathbf{z}^{(1)}$, is

$$\begin{pmatrix} x_1 \\ x_2 \end{pmatrix}^{(2)} = \begin{pmatrix} 3/4 \\ 3/4 \end{pmatrix}.$$

Note that for this problem, each successive iterate $\mathbf{x}^{(0)}$, $\mathbf{x}^{(1)}$, $\mathbf{x}^{(2)}$ moves closer to the true solution $\mathbf{x} = (1, 1)^T$.

For the Gauss–Seidel method M is the lower triangle of A, and so we find after one step that

$$\begin{pmatrix} x_1 \\ x_2 \end{pmatrix}^{(1)} = \begin{pmatrix} 0 \\ 0 \end{pmatrix} + \begin{pmatrix} 2 & 0 \\ -1 & 2 \end{pmatrix}^{-1} \begin{pmatrix} 1 \\ 1 \end{pmatrix} = \begin{pmatrix} 1/2 \\ 3/4 \end{pmatrix}.$$

The residual is

$$\begin{pmatrix} r_1 \\ r_2 \end{pmatrix}^{(1)} = \begin{pmatrix} 1 \\ 1 \end{pmatrix} - \begin{pmatrix} 2 & -1 \\ -1 & 2 \end{pmatrix} \begin{pmatrix} 1/2 \\ 3/4 \end{pmatrix} = \begin{pmatrix} 3/4 \\ 0 \end{pmatrix},$$

and the vector $\mathbf{z}^{(1)} \equiv M^{-1}\mathbf{r}^{(1)}$ is

$$\begin{pmatrix} z_1 \\ z_2 \end{pmatrix}^{(1)} = \begin{pmatrix} 3/8 \\ 3/16 \end{pmatrix}.$$

The second iterate, $\mathbf{x}^{(2)} = \mathbf{x}^{(1)} + \mathbf{z}^{(1)}$, is

$$\begin{pmatrix} x_1 \\ x_2 \end{pmatrix}^{(2)} = \begin{pmatrix} 7/8 \\ 15/16 \end{pmatrix}.$$

Note that the Gauss–Seidel iterates are closer to the true solution $\mathbf{x} = (1, 1)^T$ than the corresponding Jacobi iterates.

Example 12.2.2. The plot in figure 12.9 shows the convergence of the Jacobi, Gauss–Seidel, and SOR iterative methods for a linear system arising from Poisson's equation on a square in two dimensions. The number of grid points in each direction was 50, and the dimension of the matrix was $n = 50^2 = 2500$. For this model problem, the optimal value of the SOR parameter ω is known, and that value was used to obtain the results in figure 12.9.

Note the slow convergence of the Jacobi and Gauss–Seidel methods. After 500 iterations, the residual norm is still on the order of 10^{-2} or 10^{-3}. This convergence rate only gets worse as the problem size increases. The SOR method, with a well-chosen parameter ω, is considerably faster to converge, but, except for special model problems like this one, the optimal value of ω is not known. Software implementing the SOR method must either ask the user to supply a good value for ω, or it must try to estimate it dynamically. In later sections, we will discuss iterative methods that achieve SOR-type convergence rates or *better* and do not require estimation of any parameters.

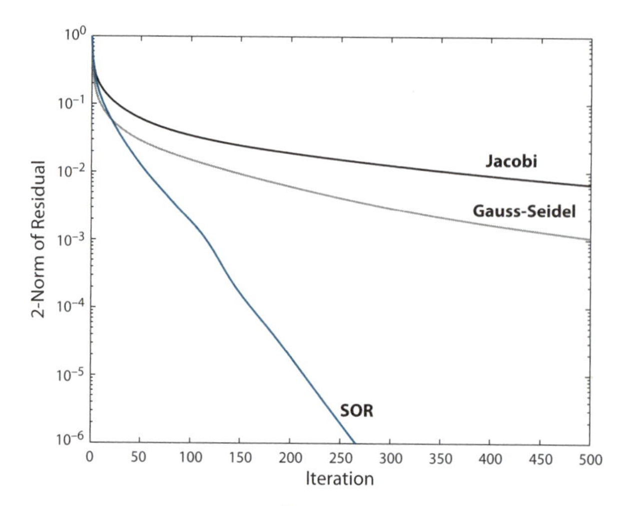

Figure 12.9. Convergence of the Jacobi, Gauss–Seidel, and SOR iterative methods.

12.2.3 Analysis of Convergence

The questions that we will be concerned with, in studying this and other iterative methods for solving linear systems, are: (1) Does the method converge to the solution $A^{-1}\mathbf{b}$ of the linear system, and (2) If so, how rapidly does it converge (i.e., how many iterations are required to compute an approximate solution that differs from the true solution by no more than some given tolerance)? We must also keep in mind the work per iteration; if one method requires more work per iteration than another, then it must converge in correspondingly fewer iterations in order to be competitive.

To answer these questions for the simple iteration methods described here, we need to look at an equation for the error $\mathbf{e}^{(k)} \equiv A^{-1}\mathbf{b} - \mathbf{x}^{(k)}$ at step k. Subtracting each side of (12.8) from $A^{-1}\mathbf{b}$, we obtain the equation

$$\mathbf{e}^{(k)} = \mathbf{e}^{(k-1)} - M^{-1}\mathbf{r}^{(k-1)} = (I - M^{-1}A)\mathbf{e}^{(k-1)}.$$

An analogous formula holds for $\mathbf{e}^{(k-1)}$, and substituting that formula into the one above and continuing, we find

$$\mathbf{e}^{(k)} = (I - M^{-1}A)\mathbf{e}^{(k-1)} = (I - M^{-1}A)^2\mathbf{e}^{(k-2)} = \ldots = (I - M^{-1}A)^k\mathbf{e}^{(0)}.$$

Taking norms on each side, we find

$$\|\mathbf{e}^{(k)}\| \leq \|(I - M^{-1}A)^k\| \cdot \|\mathbf{e}^{(0)}\|, \tag{12.9}$$

where $\|\cdot\|$ can be any vector norm and we take the matrix norm to be the one *induced* by the vector norm: $\|B\| \equiv \max_{\|\mathbf{v}\|=1} \|B\mathbf{v}\|$. In this case, the bound in (12.9) is *sharp*; that is, for each k there is an initial error $\mathbf{e}^{(0)}$ for which equality holds (although the initial error for which equality holds at step k may be different from the one for which equality holds at some other step j).

Theorem 12.2.1. The norm of the error in iteration (12.8) converges to 0 and $\mathbf{x}^{(k)}$ converges to $A^{-1}\mathbf{b}$ as $k \to \infty$, for *every* initial error $\mathbf{e}^{(0)}$, if and only if

$$\lim_{k\to\infty} \|(I - M^{-1}A)^k\| = 0.$$

Proof. Taking limits on each side in (12.9) shows that if $\lim_{k\to\infty} \|(I - M^{-1}A)^k\| = 0$, then $\lim_{k\to\infty} \|\mathbf{e}^{(k)}\| = 0$.

The "only if" part requires a bit more machinery to prove. Suppose, conversely, that $\lim_{k\to\infty} \|(I - M^{-1}A)^k\| \neq 0$. Then there is a positive number α such that $\|(I - M^{-1}A)^k\| \geq \alpha$ for infinitely many values of k. The vectors $\mathbf{e}^{(0,k)}$, $k = 1, 2, \ldots$, with norm 1 for which equality holds in (12.9) at step k form a bounded infinite set in $\mathbf{C}^n$, and so, by the Bolzano–Weierstrass theorem, they contain a convergent subsequence. Let $\mathbf{e}^{(0)}$ be the limit of such a subsequence. Then for any $\epsilon > 0$, for k sufficiently large in this subsequence, we will have $\|\mathbf{e}^{(0)} - \mathbf{e}^{(0,k)}\| \leq \epsilon$, and hence

$$\begin{aligned}\|(I - M^{-1}A)^k\mathbf{e}^{(0)}\| &\geq \|(I - M^{-1}A)^k\mathbf{e}^{(0,k)}\| - \|(I - M^{-1}A)^k(\mathbf{e}^{(0,k)} - \mathbf{e}^{(0)})\| \\ &\geq \|(I - M^{-1}A)^k\|(1-\epsilon) \geq \alpha(1-\epsilon).\end{aligned}$$

For $\epsilon < 1$, the right-hand side is positive, and since this inequality holds for infinitely many values of k, it follows that $\lim_{k\to\infty} \|(I - M^{-1}A)^k\mathbf{e}^{(0)}\|$ cannot be 0; either this limit does not exist or it is greater than 0. □

The *spectral radius*, or largest absolute value of an eigenvalue of a matrix G, will be used to further explain the result of theorem 12.2.1.

Definition. The **spectral radius** of an n by n matrix G is

$$\rho(G) = \max\{|\lambda| : \lambda \text{ is an eigenvalue of } G\}.$$

The following theorem relates the spectral radius to matrix norms. Its proof relies on the Schur form introduced in section 12.1.

Theorem 12.2.2. For every matrix norm $\|\cdot\|$ and every n by n matrix G, $\rho(G) \leq \|G\|$. Given an n by n matrix G and any $\epsilon > 0$, there is a matrix norm $|||\cdot|||$, induced by a certain vector norm, such that $|||G||| \leq \rho(G) + \epsilon$.

Proof. To see that $\rho(G) \leq \|G\|$, note there is at least one eigenvalue λ of G for which $|\lambda| = \rho$. Let $\mathbf{v}$ be an eigenvector associated with such an eigenvalue λ, and let V be an n by n matrix each of whose columns is $\mathbf{v}$. Then $GV = \lambda V$, and it follows that for any matrix norm $\|\cdot\|$, $|\lambda|\cdot\|V\| = \|\lambda V\| = \|GV\| \leq \|G\|\cdot\|V\|$. Therefore $\rho(G) = |\lambda| \leq \|G\|$.

Now suppose we are given an n by n matrix G and a number $\epsilon > 0$. We wish to determine a norm $|||\cdot|||$ for which $|||G||| \leq \rho(G) + \epsilon$. By the Schur triangularization theorem 12.1.4, there is a unitary matrix Q and an upper triangular matrix T such that $G = QTQ^*$. Let $\delta > 0$ be a parameter to be determined later and let $D = \text{diag}(1, \delta, \delta^2, \ldots, \delta^{n-1})$. Then

$$D^{-1}TD = \begin{pmatrix} t_{11} & \delta t_{12} & \delta^2 t_{13} & \ldots & \delta^{n-1} t_{1n} \\ & t_{22} & \delta t_{23} & \ldots & \delta^{n-2} t_{2n} \\ & & \ddots & \ddots & \vdots \\ & & & \ddots & \delta t_{n-1,n} \\ & & & & t_{nn} \end{pmatrix}.$$

For $\delta > 0$ sufficiently small, we can make the entries in the strict upper triangle of this matrix as small as we like. In particular, we can make the sum of absolute values of the entries in any column of the strict upper triangle less than ϵ. Since the diagonal entries are the eigenvalues and the absolute value of each eigenvalue is less than or equal to $\rho(G)$, it follows that $\|D^{-1}TD\|_1 \leq \rho(G) + \epsilon$. (Recall that $\|\cdot\|_1$ is the maximum absolute column sum.) Define a matrix norm $|||\cdot|||$ for any n by n matrix C by

$$|||C||| \equiv \|D^{-1}Q^*CQD\|_1.$$

Then $|||G||| = \|D^{-1}TD\|_1 \leq \rho(G) + \epsilon$. To see that $|||\cdot|||$ is indeed a matrix norm, induced by a certain vector norm, note that if, for every n-vector $\mathbf{y}$ we define $|||\mathbf{y}||| \equiv \|D^{-1}Q^*\mathbf{y}\|_1$, then

$$\begin{aligned} \max_{|||\mathbf{y}|||=1} |||C\mathbf{y}||| &= \max_{\|D^{-1}Q^*\mathbf{y}\|_1=1} \|D^{-1}Q^*C\mathbf{y}\|_1 \\ &= \max_{\|D^{-1}Q^*\mathbf{y}\|_1=1} \|D^{-1}Q^*CQD(D^{-1}Q^*\mathbf{y})\|_1 \\ &= \max_{\|\mathbf{w}\|_1=1} \|D^{-1}Q^*CQD\mathbf{w}\|_1 \\ &= \|D^{-1}Q^*CQD\|_1 \equiv |||C|||. \end{aligned}$$

Thus if $|||\mathbf{y}||| \equiv \|D^{-1}Q^*\mathbf{y}\|_1$ defines a vector norm, then the matrix norm $|||\cdot|||$ is induced by this norm. It is left as an exercise to check that the requirements of a vector norm are satisfied by $|||\cdot|||$. □

The following is a consequence of theorem 12.2.2.

Theorem 12.2.3. Let G be an n by n matrix. Then $\lim_{k\to\infty} G^k = 0$ if and only if $\rho(G) < 1$.

Proof. First suppose $\lim_{k\to\infty} G^k = 0$. Let λ be an eigenvalue of G with eigenvector $\mathbf{v}$. Since $G^k\mathbf{v} = \lambda^k\mathbf{v} \to 0$ as $k \to \infty$, this implies that $|\lambda| < 1$.

Conversely, if $\rho(G) < 1$, then by theorem 12.2.2 there is a matrix norm $|||\cdot|||$ such that $|||G||| < 1$. It follows that $|||G^k||| \leq |||G|||^k \to 0$ as $k \to \infty$. Since all norms on the set of n by n matrices are equivalent (i.e., for any two norms $\|\cdot\|$ and $|||\cdot|||$, there exist constants c and C such that for all n by n matrices G, $c\|G\| \leq |||G||| \leq C\|G\|$), it follows that all norms of G^k approach 0 as $k \to \infty$. Taking the norm to be, say, the 1-norm (maximum absolute column sum) or the ∞-norm (maximum absolute row sum), it is clear that this implies that all entries in G^k approach 0 as $k \to \infty$; that is, $\lim_{k\to\infty} G^k = 0$. □

With this result, theorem 12.2.1 can be restated as follows:

Theorem 12.2.4. The norm of the error in iteration (12.8) converges to 0 and $\mathbf{x}^{(k)}$ converges to $A^{-1}\mathbf{b}$ as $k \to \infty$, for *every* initial error $\mathbf{e}^{(0)}$, if and only if

$$\rho(I - M^{-1}A) < 1.$$

While theorem 12.2.4 gives a necessary and sufficient condition for convergence of the iterative method (12.8), it may not be easy to check. In general, one does not know the spectral radius of the iteration matrix $I - M^{-1}A$ or whether or not it is less than 1. There are some circumstances, however, in which this condition can be verified. There is a vast literature on properties of A and M that are sufficient to guarantee $\rho(I - M^{-1}A) < 1$. Here we prove just two such results.

Recall that an n by n matrix A is said to be *strictly diagonally dominant* (or *strictly row diagonally dominant*) if for each $i = 1, \ldots, n$, $|a_{ii}| > \sum_{j \neq i} |a_{ij}|$; in words, each diagonal entry is greater in magnitude than the sum of the magnitudes of the off-diagonal entries in its row.

Theorem 12.2.5. If A is strictly diagonally dominant, then the Jacobi iteration converges to the unique solution of the linear system $A\mathbf{x} = \mathbf{b}$, for any initial vector $\mathbf{x}^{(0)}$.

Proof. Since $M = \mathrm{diag}(A)$, the matrix $I - M^{-1}A$ has 0s on its main diagonal, and its (i, j)-entry, for $i \neq j$, is $-a_{ij}/a_{ii}$. Consider the ∞-norm of this matrix:

$$\|I - M^{-1}A\|_\infty = \max_{i=1,\ldots,n} \sum_{j \neq i} \left| \frac{a_{ij}}{a_{ii}} \right| = \max_{i=1,\ldots,n} \frac{1}{|a_{ii}|} \sum_{j \neq i} |a_{ij}|. \tag{12.10}$$

Since A is strictly diagonally dominant, $|a_{ii}| > \sum_{j \neq i} |a_{ij}|$ for all i, and it follows from expression (12.10) that $\|I - M^{-1}A\|_\infty < 1$. It then follows from theorem 12.2.1 that the Jacobi iteration converges to $A^{-1}\mathbf{b}$. (The strict diagonal dominance of A implies also that it is nonsingular, by Gerschgorin's theorem, and hence that the solution to $A\mathbf{x} = \mathbf{b}$ is unique.) □

If A is strictly diagonally dominant, it also can be shown that the Gauss–Seidel method converges. Instead of proving this, however, we will prove convergence of the Gauss–Seidel method when A is symmetric and positive definite. For convenience, let us assume that A has been prescaled by its diagonal; that is, the real symmetric positive definite matrix A has the form $A = I + L + L^T$, where I (the identity) is the diagonal of A and L is the strict lower triangle of A; the strict upper triangle is L^T since A is symmetric. We noted earlier that a real symmetric matrix is *positive definite* if all of its eigenvalues are positive. An equivalent criterion is: A is positive definite if, for all nonzero vectors $\mathbf{v}$ (real or complex),

$$\mathbf{v}^* A\mathbf{v} > 0. \tag{12.11}$$

Here $\mathbf{v}^*$ denotes the *Hermitian transpose*. If $\mathbf{v}$ is real, then it is just the ordinary transpose, but if $\mathbf{v}$ is complex then it is the complex-conjugate transpose: $\mathbf{v}^* \equiv (\bar{v}_1, \ldots, \bar{v}_n)$. Thus, $\mathbf{v}^* A\mathbf{v} = \sum_{i,j=1}^{n} a_{ij}\bar{v}_i v_j$.

Theorem 12.2.6. If A is real symmetric and positive definite and has the form $A = I + L + L^T$, where I is the diagonal of A and L is the strict lower triangle of A, then the Gauss–Seidel iteration converges to the unique solution of the linear system $A\mathbf{x} = \mathbf{b}$, for any initial vector $\mathbf{x}^{(0)}$.

Proof. First note that since A has positive eigenvalues, it is nonsingular and hence the linear system $A\mathbf{x} = \mathbf{b}$ has a unique solution. The Gauss–Seidel iteration matrix is $I - M^{-1}A$, where $M = I + L$; hence

$$I - M^{-1}A = I - (I + L)^{-1}(I + L + L^T) = -(I + L)^{-1}L^T.$$

Based on theorem 12.2.4, our task is to show that all eigenvalues of this matrix have absolute value less than 1. Let λ be an eigenvalue and $\mathbf{v}$ a corresponding normalized eigenvector ($\|\mathbf{v}\|_2 = 1$). Then $-(I + L)^{-1}L^T\mathbf{v} = \lambda\mathbf{v}$, or, multiplying by $(I + L)$ on each side,

$$-L^T\mathbf{v} = \lambda(I + L)\mathbf{v}.$$

Multiply each side of this equation on the left by $\mathbf{v}^*$ to find

$$-\mathbf{v}^*L^T\mathbf{v} = \lambda(1 + \mathbf{v}^*L\mathbf{v}). \tag{12.12}$$

Let the complex number $\mathbf{v}^*L\mathbf{v}$ be denoted as $\alpha + i\beta$, where $i = \sqrt{-1}$. Then, since this quantity is just a number, its Hermitian transpose $\mathbf{v}^*L^T\mathbf{v}$, is just its complex conjugate, $\alpha - i\beta$. Hence equation (12.12) can be written as

$$-\alpha + i\beta = \lambda(1 + \alpha + i\beta).$$

Taking absolute values on each side, we find that

$$\sqrt{\alpha^2 + \beta^2} = |\lambda|\sqrt{1 + 2\alpha + \alpha^2 + \beta^2},$$

and it will follow that $|\lambda| < 1$ if $1+2\alpha > 0$. But since $A = I+L+L^T$ is positive definite, it follows from (12.11) that $\mathbf{v}^*A\mathbf{v} > 0$, and $\mathbf{v}^*A\mathbf{v} = 1+\mathbf{v}^*L\mathbf{v}+\mathbf{v}^*L^T\mathbf{v} = 1 + (\alpha + i\beta) + (\alpha - i\beta) = 1 + 2\alpha$. □

For any nonsingular matrix A, it can be shown that the matrix A^TA is symmetric and positive definite. Hence, if the linear system $A\mathbf{x} = \mathbf{b}$ is replaced by the symmetric positive definite system $A^TA\mathbf{x} = A^T\mathbf{b}$, then the Gauss–Seidel iteration is guaranteed to converge. The problem is that the *rate* of convergence may be extremely slow. In general, replacing $A\mathbf{x} = \mathbf{b}$ by this symmetric positive definite system, known as the **normal equations**, is a *bad* idea. We have already seen in section 7.6 that the condition number of A^TA is the square of that of A, leading to possible loss of accuracy in the direct solution of a linear system or least squares problem.

12.2.4 The Conjugate Gradient Algorithm

In this subsection, we restrict our attention to matrices A that are *symmetric and positive definite*. If a preconditioner M is used then we also require that M be symmetric and positive definite.

Suppose, for the moment, that M is the identity and that one wishes to solve $A\mathbf{x} = \mathbf{b}$ using only matrix–vector multiplication and perhaps some additional operations with vectors. As before, one starts with an initial guess $\mathbf{x}^{(0)}$ for the solution and computes the initial residual $\mathbf{r}^{(0)} = \mathbf{b} - A\mathbf{x}^{(0)}$. Since $\mathbf{r}^{(0)}$ is the only available vector that is related to the error in $\mathbf{x}^{(0)}$, it seems reasonable to replace $\mathbf{x}^{(0)}$ by $\mathbf{x}^{(0)}$ plus some scalar multiple of $\mathbf{r}^{(0)}$ in hopes of obtaining an improved approximation $\mathbf{x}^{(1)} = \mathbf{x}^{(0)} + c_0\mathbf{r}^{(0)}$. (In the simple iteration method, one takes $c_0 = 1$.) The new residual $\mathbf{r}^{(1)} \equiv \mathbf{b} - A\mathbf{x}^{(1)}$ then satisfies $\mathbf{r}^{(1)} = \mathbf{b} - A\mathbf{x}^{(0)} - c_0 A\mathbf{r}^{(0)} = (I - c_0 A)\mathbf{r}^{(0)}$. Having computed the new residual $\mathbf{r}^{(1)}$, it seems reasonable to replace $\mathbf{x}^{(1)}$ by $\mathbf{x}^{(1)}$ plus some scalar multiple of $\mathbf{r}^{(1)}$ in hopes of obtaining a still better approximate solution. Alternatively, the new approximate solution $\mathbf{x}^{(2)}$ could be set equal to $\mathbf{x}^{(0)}$ plus some linear combination of $\mathbf{r}^{(0)}$ and $\mathbf{r}^{(1)}$, thus allowing even more flexibility in the choice of $\mathbf{x}^{(2)}$. Since $\mathbf{r}^{(1)} = \mathbf{r}^{(0)} - c_0 A\mathbf{r}^{(0)}$, this is equivalent to taking $\mathbf{x}^{(2)}$ to be $\mathbf{x}^{(0)}$ plus a linear combination of $\mathbf{r}^{(0)}$ and $A\mathbf{r}^{(0)}$. When one computes the new residual $\mathbf{r}^{(2)} \equiv \mathbf{b} - A\mathbf{x}^{(2)}$, it is easily seen that the vectors $\mathbf{r}^{(0)}$, $\mathbf{r}^{(1)}$, and $\mathbf{r}^{(2)}$ span the same space as $\mathbf{r}^{(0)}$, $A\mathbf{r}^{(0)}$, and $A^2\mathbf{r}^{(0)}$. Hence the next approximation $\mathbf{x}^{(3)}$ might be taken to be $\mathbf{x}^{(0)}$ plus some linear combination of these vectors. In general, at step k, one might take $\mathbf{x}^{(k)}$ to be equal to $\mathbf{x}^{(0)}$ plus some linear combination of

$$\{\mathbf{r}^{(0)}, A\mathbf{r}^{(0)}, A^2\mathbf{r}^{(0)}, \ldots, A^{k-1}\mathbf{r}^{(0)}\}. \tag{12.13}$$

The space spanned by the vectors in (12.13) is called a **Krylov space**. The simple iteration method, and almost every other iterative method with $M = I$, generates an approximation $\mathbf{x}^{(k)}$ for which $\mathbf{x}^{(k)} - \mathbf{x}^{(0)}$ lies in this Krylov space. Methods differ according to which linear combination of these vectors is chosen as the approximate solution.

If a preconditioner M is used, then the original linear system $A\mathbf{x} = \mathbf{b}$ is replaced by the modified problem $M^{-1}A\mathbf{x} = M^{-1}\mathbf{b}$, and for $k = 1, 2, \ldots$, the approximation $\mathbf{x}^{(k)}$ is set to $\mathbf{x}^{(0)}$ plus some linear combination of

$$\{\mathbf{z}^{(0)}, (M^{-1}A)\mathbf{z}^{(0)}, (M^{-1}A)^2\mathbf{z}^{(0)}, \ldots, (M^{-1}A)^{k-1}\mathbf{z}^{(0)}\}, \tag{12.14}$$

where $\mathbf{z}^{(0)} = M^{-1}(\mathbf{b} - A\mathbf{x}^{(0)})$.

Ideally, one would like to choose the vector $\mathbf{x}^{(k)}$ from the affine space, $\mathbf{x}^{(0)}$ plus the span of the vectors in (12.13) or (12.14), in such a way as to *minimize* the error $\mathbf{e}^{(k)} \equiv A^{-1}\mathbf{b} - \mathbf{x}^{(k)}$ in some desired norm. Typically, one might be interested in minimizing, say, the 2-norm of the error, but this is difficult to do. It turns out that if A and M are symmetric and positive definite, and if one instead aims to minimize the A-norm of the error, $\|\mathbf{e}^{(k)}\|_A \equiv \langle \mathbf{e}^{(k)}, A\mathbf{e}^{(k)}\rangle^{1/2}$, then there is a simple algorithm that will do this. It is called the **conjugate gradient algorithm**. It was invented by M. Hestenes and E. Stiefel [51] in 1952, but in the same year (and in the same volume of *Jour. of the Nat. Bur. Standards*), Cornelius Lanczos [62] pointed out its equivalence to an algorithm that he had described in 1950 for computing eigenvalues [61].

The conjugate gradient algorithm can be derived in a number of different ways. Since least squares problems were discussed in section 7.6, we will derive it in a way that is analogous to what was done there, except that now we are minimizing the A-norm of the error so that we will need an A-orthogonal basis for the space over which we are minimizing. To simplify notation, let us again assume that the preconditioner is just the identity. Suppose that

$\{\mathbf{p}^{(0)}, \ldots, \mathbf{p}^{(k-1)}\}$ form an A-orthogonal basis for the Krylov space

$$K_k(A, \mathbf{r}^{(0)}) \equiv \text{span}\{\mathbf{r}^{(0)}, A\mathbf{r}^{(0)}, \ldots, A^{k-1}\mathbf{r}^{(0)}\}; \tag{12.15}$$

that is, the vectors $\{\mathbf{p}^{(0)}, \ldots, \mathbf{p}^{(k-1)}\}$ span the same space as $\{\mathbf{r}^{(0)}, A\mathbf{r}^{(0)}, \ldots, A^{k-1}\mathbf{r}^{(0)}\}$ and $\langle \mathbf{p}^{(i)}, A\mathbf{p}^{(j)} \rangle = 0$ for $i \neq j$. The approximate solution $\mathbf{x}^{(k)}$ is of the form

$$\mathbf{x}^{(k)} = \mathbf{x}^{(0)} + \sum_{j=0}^{k-1} c_j \mathbf{p}^{(j)}, \tag{12.16}$$

for certain coefficients c_j, so that the error $\mathbf{e}^{(k)}$ is of the form

$$\mathbf{e}^{(k)} = \mathbf{e}^{(0)} - \sum_{j=0}^{k-1} c_j \mathbf{p}^{(j)}.$$

To minimize the A-norm of $\mathbf{e}^{(k)}$ one must choose the coefficients to make $\mathbf{e}^{(k)}$ A-orthogonal to $\mathbf{p}^{(0)}, \ldots, \mathbf{p}^{(k-1)}$; that is,

$$c_j = \frac{\langle \mathbf{e}^{(0)}, A\mathbf{p}^{(j)} \rangle}{\langle \mathbf{p}^{(j)}, A\mathbf{p}^{(j)} \rangle} = \frac{\langle \mathbf{r}^{(0)}, \mathbf{p}^{(j)} \rangle}{\langle \mathbf{p}^{(j)}, A\mathbf{p}^{(j)} \rangle}. \tag{12.17}$$

Here we have used the fact that A is symmetric (so that $\langle \mathbf{e}^{(0)}, A\mathbf{p}^{(j)} \rangle = \langle A\mathbf{e}^{(0)}, \mathbf{p}^{(j)} \rangle$) and that $A\mathbf{e}^{(0)} = \mathbf{r}^{(0)}$ to derive an expression for the coefficients that can be computed without knowing the error vector $\mathbf{e}^{(0)}$. [Note that if we had worked with a set of orthogonal (in the usual Euclidean sense) basis vectors and attempted to minimize the 2-norm of $\mathbf{e}^{(k)}$ instead of the A-norm, the required coefficient formulas $\frac{\langle \mathbf{e}^{(0)}, \mathbf{p}^{(j)} \rangle}{\langle \mathbf{p}^{(j)}, \mathbf{p}^{(j)} \rangle}$ would not have been computable, without knowing the initial error $\mathbf{e}^{(0)}$ and hence the solution $\mathbf{x}^{(0)} + \mathbf{e}^{(0)}$.]

While equation (12.16) with coefficients given by (12.17) provides a formula for the optimal approximation from each successive Krylov space, its implementation in this form is not practical. Let us see what savings can be made, both in constructing the A-orthogonal basis vectors $\mathbf{p}^{(0)}, \mathbf{p}^{(1)}, \ldots,$ and in forming the approximate solutions $\mathbf{x}^{(k)}$. First note that if $\mathbf{x}^{(k-1)}$ has been constructed so that $\mathbf{e}^{(k-1)}$ is A-orthogonal to $\mathbf{p}^{(0)}, \ldots, \mathbf{p}^{(k-2)}$, then since $\mathbf{p}^{(k-1)}$ is also A-orthogonal to these vectors, if we take

$$\mathbf{x}^{(k)} = \mathbf{x}^{(k-1)} + a_{k-1}\mathbf{p}^{(k-1)},$$

where

$$a_{k-1} = \frac{\langle \mathbf{e}^{(k-1)}, A\mathbf{p}^{(k-1)} \rangle}{\langle \mathbf{p}^{(k-1)}, A\mathbf{p}^{(k-1)} \rangle} = \frac{\langle \mathbf{r}^{(k-1)}, \mathbf{p}^{(k-1)} \rangle}{\langle \mathbf{p}^{(k-1)}, A\mathbf{p}^{(k-1)} \rangle},$$

then $\mathbf{e}^{(k)} = \mathbf{e}^{(k-1)} - a_{k-1}\mathbf{p}^{(k-1)}$ will be A-orthogonal to $\mathbf{p}^{(k-1)}$ as well as to $\mathbf{p}^{(0)}, \ldots, \mathbf{p}^{(k-2)}$.

To construct the A-orthogonal basis vectors $\mathbf{p}^{(0)}, \mathbf{p}^{(1)}, \ldots,$ we start by setting $\mathbf{p}^{(0)} = \mathbf{r}^{(0)}$. Suppose that $\mathbf{p}^{(0)}, \ldots, \mathbf{p}^{(k-1)}$ have been constructed with the desired orthogonality properties, and from these we have computed $\mathbf{x}^{(1)}, \ldots, \mathbf{x}^{(k)}$ and $\mathbf{r}^{(1)}, \ldots, \mathbf{r}^{(k)}$. To construct the next A-orthogonal basis vector $\mathbf{p}^{(k)}$, we first note that $\mathbf{r}^{(k)} = \mathbf{r}^{(k-1)} - a_{k-1}A\mathbf{p}^{(k-1)}$ lies in the $(k+1)$-dimensional Krylov space $K_{k+1}(A, \mathbf{r}^{(0)})$ and since $\mathbf{e}^{(k)}$ is A-orthogonal to span$\{\mathbf{p}^{(0)}, \ldots, \mathbf{p}^{(k-1)}\} = K_k(A, \mathbf{r}^{(0)})$, it follows that $\mathbf{r}^{(k)} = A\mathbf{e}^{(k)}$ is A-orthogonal to all but the last of these vectors; that is, $\langle \mathbf{r}^{(k)}, A\mathbf{p}^{(j)} \rangle = 0$ for $j = 0, \ldots, k-2$. Hence if we set

$$\mathbf{p}^{(k)} = \mathbf{r}^{(k)} + b_{k-1}\mathbf{p}^{(k-1)},$$

where

$$b_{k-1} = -\frac{\langle \mathbf{r}^{(k)}, A\mathbf{p}^{(k-1)} \rangle}{\langle \mathbf{p}^{(k-1)}, A\mathbf{p}^{(k-1)} \rangle},$$

then $\mathbf{p}^{(k)}$ will be A-orthogonal to $\mathbf{p}^{(k-1)}$ as well as to $\mathbf{p}^{(0)}, \ldots, \mathbf{p}^{(k-2)}$. This is how the conjugate gradient algorithm is implemented.

In practice, slightly different but equivalent coefficient formulas are often used:

Conjugate Gradient (CG) Algorithm for Symmetric Positive Definite Problems.

Given an initial guess $\mathbf{x}^{(0)}$, compute $\mathbf{r}^{(0)} = \mathbf{b} - A\mathbf{x}^{(0)}$, and set $\mathbf{p}^{(0)} = \mathbf{r}^{(0)}$.
For $k = 1, 2, \ldots,$

Compute $A\mathbf{p}^{(k-1)}$.
Set $\mathbf{x}^{(k)} = \mathbf{x}^{(k-1)} + a_{k-1}\mathbf{p}^{(k-1)}$, where $a_{k-1} = \frac{\langle \mathbf{r}^{(k-1)}, \mathbf{r}^{(k-1)} \rangle}{\langle \mathbf{p}^{(k-1)}, A\mathbf{p}^{(k-1)} \rangle}$.

Compute $\mathbf{r}^{(k)} = \mathbf{r}^{(k-1)} - a_{k-1} A\mathbf{p}^{(k-1)}$.

Set $\mathbf{p}^{(k)} = \mathbf{r}^{(k)} + b_{k-1}\mathbf{p}^{(k-1)}$, where $b_{k-1} = \frac{\langle \mathbf{r}^{(k)}, \mathbf{r}^{(k)} \rangle}{\langle \mathbf{r}^{(k-1)}, \mathbf{r}^{(k-1)} \rangle}$.

Note that each iteration requires one matrix–vector multiplication (to compute $A\mathbf{p}^{(k-1)}$) and two inner products (to compute $\langle \mathbf{p}^{(k-1)}, A\mathbf{p}^{(k-1)} \rangle$ and $\langle \mathbf{r}^{(k)}, \mathbf{r}^{(k)} \rangle$; the inner product $\langle \mathbf{r}^{(k-1)}, \mathbf{r}^{(k-1)} \rangle$ can be saved from the previous iteration), as well as three additions of a vector and a scalar multiple of another vector. In most cases, the matrix–vector multiplication is the most expensive part of the iteration, and the reward for doing the few extra vector operations is that one obtains the *optimal* (in A-norm) approximation from the Krylov space. Thus, in terms of the number of iterations, the conjugate gradient algorithm is faster than *any* other (unpreconditioned) iterative method to reduce the A-norm of the error below a given threshold.

Note also that since the error at each step k is of the form $\mathbf{e}^{(0)}$ plus a linear combination of $\{\mathbf{r}^{(0)}, A\mathbf{r}^{(0)}, \ldots, A^{k-1}\mathbf{r}^{(0)}\}$, and of all vectors of this form it is the one with minimal A-norm, if, at some step k, we find that the set of vectors $\{\mathbf{e}^{(0)}, \mathbf{r}^{(0)} = A\mathbf{e}^{(0)}, \ldots, A^{k-1}\mathbf{r}^{(0)} = A^k\mathbf{e}^{(0)}\}$ is linearly dependent, then $\mathbf{e}^{(k)}$ will be 0 and the algorithm will have found the exact solution; for, in this case, there is a linear combination of these vectors $\sum_{j=0}^{k} c_j A^j \mathbf{e}^{(0)}$ that is equal to 0 but not all c_j's are 0. If c_J is the first nonzero coefficient in this linear combination, then it follows that $c_J A^J \mathbf{e}^{(0)} = -\sum_{j=J+1}^{k} c_j A^j \mathbf{e}^{(0)}$, or, $\mathbf{e}^{(0)} = -\sum_{j=1}^{k-J} \frac{c_{j+J}}{c_J} A^j \mathbf{e}^{(0)}$, so, in fact, $\mathbf{e}^{(k-J)}$ would be 0. Since there can be at most n linearly independent

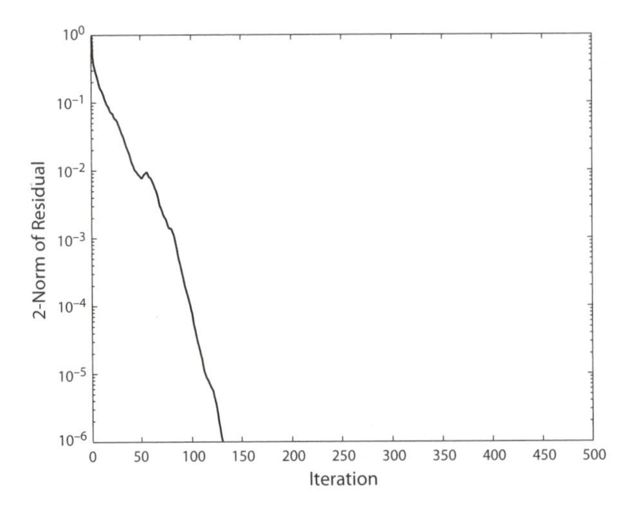

Figure 12.10. Convergence of the unpreconditioned CG algorithm.

n-vectors, it follows that the conjugate gradient algorithm always finds the exact solution within n steps. Thus, the algorithm could actually be viewed as a direct method for solving linear systems. Since the hope is that it will generate a good approximate solution long before step n, it is more useful to view it as an iterative technique and ask questions about how many steps are required to obtain a good approximate solution.

The convergence analysis of the conjugate gradient algorithm, especially in finite-precision arithmetic, is a fascinating subject that, unfortunately, we will barely be able to touch upon here. For more information, see, for example, [48]. We do state the following theorem giving an upper bound on the A-norm of the error at step k of the CG algorithm in terms of the 2-norm *condition number* (i.e., the ratio of largest to smallest eigenvalue) of the symmetric positive definite matrix A.

Theorem 12.2.7. Let $\mathbf{e}^{(k)}$ be the error at step k of the CG algorithm applied to the real symmetric positive definite linear system $A\mathbf{x} = \mathbf{b}$. Then

$$\frac{\|\mathbf{e}^{(k)}\|_A}{\|\mathbf{e}^{(0)}\|_A} \leq 2\left(\frac{\sqrt{\kappa}-1}{\sqrt{\kappa}+1}\right)^k, \tag{12.18}$$

where $\kappa = \lambda_{max}/\lambda_{min}$ is the ratio of the largest to smallest eigenvalue of A.

Example 12.2.3. Figure 12.10 shows the convergence of the unpreconditioned CG algorithm for the same linear system that was solved with the Jacobi, Gauss–Seidel, and SOR iterative methods in the previous subsection, namely, the linear system arising from Poisson's equation on a square with 50 grid points in each direction.

For this problem, the SOR method, even with the optimal parameter ω, required more than 250 iterations to reduce the 2-norm of the residual below 10^{-6}, while unpreconditioned CG required fewer than 150 iterations. Still, there is plenty of room for improvement!

Although the CG method finds the optimal (in A-norm) approximation of the form (12.16), for certain matrices A one may have to run for many steps k before there *is* a good approximate solution of this form. One may be able to reduce the number of iterations by using a preconditioner. Assume that the preconditioner M is symmetric and positive definite. Then it was shown in section 7.2.4 that M can be factored in the form $M = LL^T$, where L is a lower triangular matrix with positive diagonal entries. Imagine applying the CG algorithm to the symmetrically preconditioned linear system $L^{-1}AL^{-T}\mathbf{y} = L^{-1}\mathbf{b}$, whose solution $\mathbf{y}$ is L^T times the solution of the original linear system. Then, starting with an initial guess $\mathbf{y}^{(0)}$ for the solution and setting $\hat{\mathbf{p}}^{(0)} = \hat{\mathbf{r}}^{(0)} = L^{-1}\mathbf{b} - L^{-1}AL^{-T}\mathbf{y}^{(0)}$, the CG iteration would take the form

$$\mathbf{y}^{(k)} = \mathbf{y}^{(k-1)} + a_{k-1}\hat{\mathbf{p}}^{(k-1)}, \quad a_{k-1} = \frac{\langle \hat{\mathbf{r}}^{(k-1)}, \hat{\mathbf{r}}^{(k-1)} \rangle}{\langle \hat{\mathbf{p}}^{(k-1)}, L^{-1}AL^{-T}\hat{\mathbf{p}}^{(k-1)} \rangle},$$

$$\hat{\mathbf{r}}^{(k)} = \hat{\mathbf{r}}^{(k-1)} - a_{k-1}L^{-1}AL^{-T}\hat{\mathbf{p}}^{(k-1)},$$

$$\hat{\mathbf{p}}^{(k)} = \hat{\mathbf{r}}^{(k)} + b_{k-1}\hat{\mathbf{p}}^{(k-1)}, \quad b_{k-1} = \frac{\langle \hat{\mathbf{r}}^{(k)}, \hat{\mathbf{r}}^{(k)} \rangle}{\langle \hat{\mathbf{r}}^{(k-1)}, \hat{\mathbf{r}}^{(k-1)} \rangle}.$$

Defining

$$\mathbf{x}^{(k)} \equiv L^{-T}\mathbf{y}^{(k)}, \quad \mathbf{r}^{(k)} \equiv L\hat{\mathbf{r}}^{(k)}, \quad \mathbf{p}^{(k)} \equiv L^{-T}\hat{\mathbf{p}}^{(k)},$$

the above formulas can be written in terms of $\mathbf{x}^{(k)}$, $\mathbf{r}^{(k)}$, and $\mathbf{p}^{(k)}$, and doing so leads to the following preconditioned CG algorithm for $A\mathbf{x} = \mathbf{b}$.

Preconditioned Conjugate Gradient (PCG) Algorithm for Symmetric Positive Definite Problems with a symmetric positive definite preconditioner.

Given an initial guess $\mathbf{x}^{(0)}$, compute $\mathbf{r}^{(0)} = \mathbf{b} - A\mathbf{x}^{(0)}$, and solve $M\mathbf{z}^{(0)} = \mathbf{r}^{(0)}$. Set $\mathbf{p}^{(0)} = \mathbf{z}^{(0)}$.
For $k = 1, 2, \ldots,$

Compute $A\mathbf{p}^{(k-1)}$.
Set $\mathbf{x}^{(k)} = \mathbf{x}^{(k-1)} + a_{k-1}\mathbf{p}^{(k-1)}$, where $a_{k-1} = \frac{\langle \mathbf{r}^{(k-1)}, \mathbf{z}^{(k-1)} \rangle}{\langle \mathbf{p}^{(k-1)}, A\mathbf{p}^{(k-1)} \rangle}$.

Compute $\mathbf{r}^{(k)} = \mathbf{r}^{(k-1)} - a_{k-1}A\mathbf{p}^{(k-1)}$.

Solve $M\mathbf{z}^{(k)} = \mathbf{r}^{(k)}$.
Set $\mathbf{p}^{(k)} = \mathbf{z}^{(k)} + b_{k-1}\mathbf{p}^{(k-1)}$, where $b_{k-1} = \frac{\langle \mathbf{r}^{(k)}, \mathbf{z}^{(k)} \rangle}{\langle \mathbf{r}^{(k-1)}, \mathbf{z}^{(k-1)} \rangle}$.

Again, this algorithm minimizes the $L^{-1}AL^{-T}$-norm of the error in the approximate solution $\mathbf{y}^{(k)}$ of the preconditioned system, but this is the same as the A-norm of the error in $\mathbf{x}^{(k)}$, since

$$\begin{aligned}&\langle (L^{-1}AL^{-T})^{-1}L^{-1}\mathbf{b} - \mathbf{y}^{(k)}, L^{-1}\mathbf{b} - (L^{-1}AL^{-T})\mathbf{y}^{(k)}\rangle \\ &= \langle L^T(A^{-1}\mathbf{b} - L^{-T}\mathbf{y}^{(k)}), L^{-1}(\mathbf{b} - AL^{-T}\mathbf{y}^{(k)})\rangle \\ &= \langle A^{-1}\mathbf{b} - \mathbf{x}^{(k)}, \mathbf{b} - A\mathbf{x}^{(k)}\rangle = \langle \mathbf{e}^{(k)}, A\mathbf{e}^{(k)}\rangle.\end{aligned}$$

Incomplete Cholesky Decomposition

The subject of preconditioners is a vast one, on which much current research is focused. We have already mentioned a few possible choices for preconditioners. Taking $M = \text{diag}(A)$ and using simple iteration leads to the Jacobi method. If A is symmetric and positive definite then so is diag(A), so this preconditioner can also be used with the conjugate gradient algorithm. The other two preconditioners that were mentioned, when combined with simple iteration, gave rise to the Gauss–Seidel and SOR methods. If $A = D - L - U$, where D is diagonal, L is strictly lower triangular, and U is strictly upper triangular, then the SOR preconditioner has the form $M = \omega^{-1}D - L$ for some parameter ω, where $\omega = 1$ corresponds to the Gauss–Seidel preconditioner. Preconditioners of this form cannot be used with the conjugate gradient algorithm because, even if A is symmetric so that $U = L^T$, the preconditioner $M = \omega^{-1}D - L$ is nonsymmetric. A symmetrized form of the SOR preconditioner is

$$M = \frac{\omega}{2-\omega}(\omega^{-1}D - L)D^{-1}(\omega^{-1}D - U),$$

and if this preconditioner is used with simple iteration, the resulting method is called the **symmetric successive overrelaxation (SSOR)** method. When $U = L^T$, it is easy to check that this preconditioner is symmetric, and if A is positive definite then so is the preconditioner. Hence the SSOR preconditioner can be used with the conjugate gradient algorithm. Whether used with simple iteration or with the conjugate gradient algorithm, one must solve a linear system with coefficient matrix M at each iteration. This is easy to do since M is the product of known lower and upper triangular factors. In general, if $M = \mathcal{L}\mathcal{U}$, where $\mathcal{L}$ is lower triangular and $\mathcal{U}$ is upper triangular, then to solve $M\mathbf{z} = \mathbf{r}$ for $\mathbf{z}$, one first solves the lower triangular system $\mathcal{L}\mathbf{y} = \mathbf{r}$ for $\mathbf{y}$ and one then solves the upper triangular system $\mathcal{U}\mathbf{z} = \mathbf{y}$ for $\mathbf{z}$.

When A is symmetric and positive definite, we saw in section 7.2.4 that it can be factored in the form LL^T, where L is lower triangular. In most cases, however, L is much less sparse than the lower triangle of A; if A is banded ($a_{ij} = 0$ for $|i - j|$ greater than the half bandwidth), then L is also banded, but any zeros within the band of A usually fill in with nonzeros in L. One might obtain an approximate factorization by forcing certain entries in L to be 0 and choosing the nonzero entries so that LL^T matches A as closely as possible. While any sparsity pattern can be chosen for L, one frequently chooses L to have the same sparsity pattern as the lower triangle of A; that is, if $a_{ij} = 0$, for some $i > j$, then $l_{ij} = 0$. It turns out that if this choice is made, then (under one additional assumption about the matrix A), it can be shown that the nonzero entries in L can be chosen so that LL^T matches A in places where

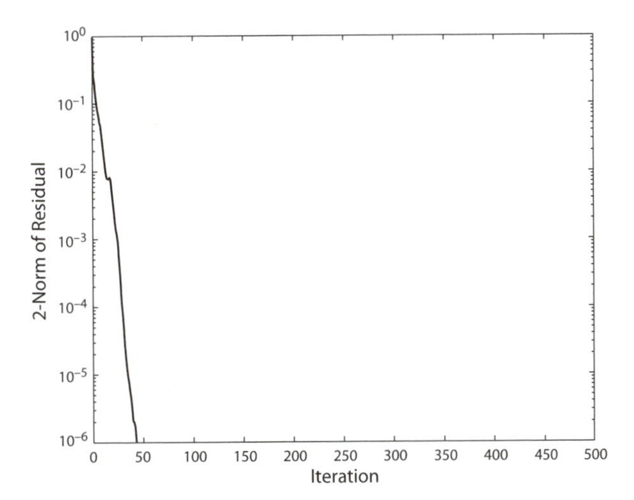

Figure 12.11. Convergence of the CG algorithm with the incomplete Cholesky preconditioner.

A has nonzeros, but LL^T has some nonzero entries in places where A has zeros. More generally, Meijerink and van der Vorst [73] proved the following result about the existence of an incomplete LU factorization.

Theorem 12.2.8. If $A = [a_{ij}]$ is an n by n M-matrix (i.e., the diagonal entries in A are positive, the off-diagonal entries are nonpositive, A is nonsingular, and the entries in A^{-1} are all greater than or equal to 0), then for every subset P of off-diagonal indices there exists a lower triangular matrix $L = [l_{ij}]$ with unit diagonal and an upper triangular matrix $U = [u_{ij}]$ such that $A = LU - R$, where

$$l_{ij} = 0 \text{ if } (i, j) \in P, \quad u_{ij} = 0 \text{ if } (i, j) \in P, \quad \text{and} \quad r_{ij} = 0 \text{ if } (i, j) \notin P.$$

In other words, the theorem states that no matter what sparsity pattern is chosen for L and U, one can find factors with that sparsity pattern such that the product LU matches A in positions where L and U are allowed to have nonzeros. If A is a symmetric M-matrix and L and U^T are chosen to have the same sparsity pattern (i.e., l_{ij}, $i > j$, is allowed to be nonzero if and only if u_{ji} is allowed to be nonzero), then, by properly choosing the diagonal entries, one can force $U = L^T$. The approximate factorization described in the theorem is called an **incomplete LU** decomposition, and when $U = L^T$ it is known as an **incomplete Cholesky** decomposition.

The proof of this theorem proceeds by construction. At the kth stage of Gaussian elimination, one first replaces the entries in the current coefficient matrix with indices $(k, j) \in P$ and $(i, k) \in P$ by 0. One then performs a Gaussian elimination step in the usual way: Eliminate entries in rows $k + 1$ through n of column k by subtracting appropriate multiples of row k from rows $k + 1$ through n. This is how the incomplete LU or incomplete Cholesky factorization is carried out.

Example 12.2.4. Figure 12.11 shows the convergence of the CG algorithm with the incomplete Cholesky preconditioner (known as ICCG) for the same linear system that was used to illustrate the Jacobi, Gauss–Seidel, SOR, and unpreconditioned CG methods; that is, the linear system arising from Poisson's equation on a square with 50 grid points in each direction. The sparsity pattern of L was chosen to match that of A; that is, l_{ij} was allowed to be nonzero only for $j = i,\ i-1,\ i-50$.

The number of iterations required to reduce the residual norm to 10^{-6} was reduced by about a factor of 3 over the unpreconditioned CG algorithm. The cost for this reduction in number of iterations was about $7n$ operations to compute the incomplete Cholesky factorization and then an additional $11n$ operations at each iteration to solve the preconditioning system $LL^T\mathbf{z} = \mathbf{r}$.

12.2.5 Methods for Nonsymmetric Linear Systems

The conjugate gradient algorithm for symmetric positive definite linear systems has the nice property of generating the *optimal* (in A-norm) approximation from successive Krylov spaces while requiring just a few more vector operations per iteration than, say, the simple iteration method. Unfortunately, no such method exists for nonsymmetric linear systems. One must either perform more and more work per iteration to find the optimal approximation from successive Krylov spaces, or one must give up on optimality. The (full) GMRES (generalized minimal residual) algorithm [87] finds the approximation for which the 2-norm of the residual is minimal but requires extra work and storage to do so. Other methods such as restarted GMRES, QMR (quasi-minimal residual method) [42], BiCGSTAB [98], and IDR [91] generate nonoptimal approximations but require less work per iteration than full GMRES and so may be cost-effective if they require only slightly more iterations. In this section we will describe only the GMRES algorithm. We will again do so from a least squares point of view, although this is not the most common implementation.

As in the symmetric case, we wish to solve $A\mathbf{x} = \mathbf{b}$ by choosing an approximate solution $\mathbf{x}^{(k)}$ at each step $k = 1, 2, \ldots,$ that is of the form $\mathbf{x}^{(0)}$ plus some linear combination of $\{\mathbf{r}^{(0}, A\mathbf{r}^{(0)}, \ldots, A^{k-1}\mathbf{r}^{(0)}\}$. Hence the residual $\mathbf{r}^{(k)} \equiv \mathbf{b} - A\mathbf{x}^{(k)}$ is of the form

$$\mathbf{r}^{(k)} = \mathbf{r}^{(0)} - \sum_{j=1}^{k} c_j A^j \mathbf{r}^{(0)},$$

where the coefficients $c_1, \ldots c_k$ are chosen to minimize the 2-norm of $\mathbf{r}^{(k)}$. We know from section 7.6 that if we have an *orthonormal basis* $\hat{\mathbf{q}}^{(1)}, \ldots, \hat{\mathbf{q}}^{(k)}$ for the space span$\{A\mathbf{r}^{(0)}, \ldots, A^k\mathbf{r}^{(0)}\}$, then $\|\mathbf{r}^{(k)}\|_2$ is minimized by taking $\mathbf{r}^{(k)} = \mathbf{r}^{(0)} - \sum_{j=1}^{k} \langle \mathbf{r}^{(0)}, \hat{\mathbf{q}}^{(j)} \rangle \hat{\mathbf{q}}^{(j)}$. In this case, the approximate solution $\mathbf{x}^{(k)}$ is $\mathbf{x}^{(k)} = \mathbf{x}^{(0)} + \sum_{j=1}^{k} \langle \mathbf{r}^{(0)}, \hat{\mathbf{q}}^{(j)} \rangle A^{-1}\hat{\mathbf{q}}^{(j)}$, so that the vectors $A^{-1}\hat{\mathbf{q}}^{(j)}$ must be available as well in order to implement the GMRES algorithm in this form.

To construct the vectors $\hat{\mathbf{q}}^{(j)}$ and $\mathbf{q}^{(j)} \equiv A^{-1}\hat{\mathbf{q}}^{(j)}$, one can use the Gram–Schmidt algorithm discussed in chapter 7, or a slightly modified version of this:

Compute $\beta_0 = \|A\mathbf{r}^{(0)}\|$ and set $\hat{\mathbf{q}}^{(1)} = A\mathbf{r}^{(0)}/\beta_0$, $\mathbf{q}^{(1)} = \mathbf{r}^{(0)}/\beta_0$. [Note that $\hat{\mathbf{q}}^{(1)}$ has norm 1 and that $\mathbf{q}^{(1)} = A^{-1}\hat{\mathbf{q}}^{(1)}$.]

For $j = 1, 2, \ldots,$
  Start with $\hat{\mathbf{q}}^{(j+1)} = A\hat{\mathbf{q}}^{(j)}$ and $\mathbf{q}^{(j+1)} = \hat{\mathbf{q}}^{(j)}$. For $i = 1, \ldots, j$,
    Compute $\gamma_{ij} = \langle \hat{\mathbf{q}}^{(j+1)}, \hat{\mathbf{q}}^{(i)} \rangle$.
    Replace $\hat{\mathbf{q}}^{(j+1)}$ by $\hat{\mathbf{q}}^{(j+1)} - \gamma_{ij}\hat{\mathbf{q}}^{(i)}$ and $\mathbf{q}^{(j+1)}$ by $\mathbf{q}^{(j+1)} - \gamma_{ij}\mathbf{q}^{(i)}$. [Note that $\hat{\mathbf{q}}^{(j+1)}$ is now orthogonal to $\hat{\mathbf{q}}^{(1)}, \ldots, \hat{\mathbf{q}}^{(i)}$ and that $\mathbf{q}^{(j+1)}$ is still equal to $A^{-1}\hat{\mathbf{q}}^{(j+1)}$.]
  Normalize by computing $\beta_j = \|\hat{\mathbf{q}}^{(j+1)}\|$ and replacing $\hat{\mathbf{q}}^{(j+1)}$ by $\hat{\mathbf{q}}^{(j+1)}/\beta_j$. Also replace $\mathbf{q}^{(j+1)}$ by $\mathbf{q}^{(j+1)}/\beta_j$.

One can update the approximate solution $\mathbf{x}^{(k)}$ and the residual vector $\mathbf{r}^{(k)} = \mathbf{b} - A\mathbf{x}^{(k)}$ at each iteration by computing $a_k = \langle \mathbf{r}^{(k-1)}, \hat{\mathbf{q}}^{(k)} \rangle$ and setting

$$\mathbf{x}^{(k)} = \mathbf{x}^{(k-1)} + a_k\mathbf{q}^{(k)}, \quad \mathbf{r}^{(k)} = \mathbf{r}^{(k-1)} - a_k\hat{\mathbf{q}}^{(k)}.$$

This is one version of the GMRES algorithm.

Note that, unlike the symmetric case, each new basis vector $\hat{\mathbf{q}}^{(j+1)}$ must be orthogonalized against *all* of the previous vectors $\hat{\mathbf{q}}^{(1)}, \ldots, \hat{\mathbf{q}}^{(j)}$. This requires $O(nj)$ work and storage at iteration j, so that each iteration takes longer than the previous one. Often a restarted version of the GMRES algorithm is used. One runs for a certain maximum number of iterations K (based on available storage and/or time), then restarts the algorithm using the latest iterate $\mathbf{x}^{(K)}$ as the initial guess.

As with other iterative methods, a preconditioner can be used with the GMRES algorithm to improve convergence. Here the preconditioner M can be applied in several different ways. One can precondition from the *left* by applying the GMRES algorithm to the linear system $M^{-1}A\mathbf{x} = M^{-1}\mathbf{b}$, or one can precondition from the *right* by applying the GMRES algorithm to the linear system $AM^{-1}\mathbf{y} = \mathbf{b}$ and setting $\mathbf{x} = M^{-1}\mathbf{y}$. If M is a product of known factors M_1 and M_2, then the original linear system $A\mathbf{x} = \mathbf{b}$ can be replaced by $M_1^{-1}AM_2^{-1}\mathbf{y} = M_1^{-1}\mathbf{b}$ and the GMRES algorithm can be applied to this linear system, with the solution to the original system being given by $\mathbf{x} = M_2^{-1}\mathbf{y}$. In any of these cases, one need not compute M^{-1} or M_1 and M_2; it is necessary only to be able to solve linear systems with coefficient matrix M. Approximations to the solution of the original linear system can be computed directly by exploiting their relationship to the approximate solutions of the modified systems that would be generated by the GMRES algorithm.

12.3 CHAPTER 12 EXERCISES

1. Show that the normalized vectors $\mathbf{y}^{(k)}$ generated by the power method are scalar multiples of the vectors $A^k\mathbf{w}$ that would be generated without normalization; that is, $\mathbf{y}^{(k)} = (A^k\mathbf{w})/\|A^k\mathbf{w}\|$. Show that $\lambda^{(k+1)} \equiv \langle A\mathbf{y}^{(k)}, \mathbf{y}^{(k)} \rangle$ in the power method is equal to $\langle A^{k+1}\mathbf{w}, A^k\mathbf{w} \rangle \;/\; \langle A^k\mathbf{w}, A^k\mathbf{w} \rangle$.
2. Suppose the matrix A is real but has complex eigenvalues. If the initial vector $\mathbf{w}$ in the power method is real, then all other quantities, including approximate eigenvalues, generated by the algorithm will be real. Hence if the eigenvalue of A of largest absolute value has nonzero imaginary part, then the power method will not converge to it. Can you explain this

apparent paradox? [Hint: Can a real matrix have a complex eigenvalue λ_1 with nonzero imaginary part that satisfies $|\lambda_1| > |\lambda_2| \geq \ldots \geq |\lambda_n|$? Why or why not?]

3. Determine a shift that can be used in the power method in order to compute λ_1, when the eigenvalues of A satisfy $\lambda_1 = -\lambda_2 > |\lambda_3| \geq \ldots \geq |\lambda_n|$.
4. Let the eigenvalues of A satisfy $\lambda_1 > \lambda_2 > \ldots > \lambda_n$ (all real but not necessarily positive). What shift should be used in the power method in order to make it converge most rapidly to λ_1?
5. Consider the matrix

$$A = \begin{pmatrix} 2 & -1 & & & \\ -1 & 2 & \ddots & & \\ & \ddots & \ddots & \ddots & \\ & & \ddots & \ddots & -1 \\ & & & -1 & 2 \end{pmatrix}.$$

Taking A to be a 10 by 10 matrix, try the following:

(a) What information does Gerschgorin's theorem give you about the eigenvalues of this matrix?

(b) Implement the power method to compute an approximation to the eigenvalue of largest absolute value and its corresponding eigenvector. [Note: Use a random initial vector. If you choose something special, like the vector of all 1s, it may turn out to be orthogonal to the eigenvector you are looking for.] Turn in a listing of your code together with the eigenvalue/eigenvector pair that you computed. Once you have a good approximate eigenvalue, look at the error in previous approximations and comment on the rate of convergence of the power method.

(c) Implement the QR algorithm to take A to diagonal form. You may use MATLAB routine `[Q,R] = qr(A);` to perform the necessary QR decompositions. Comment on the rate at which the off-diagonal entries in A are reduced.

(d) Take one of the eigenvalues computed in your QR algorithm and, using it as a shift in inverse iteration, compute the corresponding eigenvector.

6. Generate some random diagonally dominant matrices in MATLAB and try solving linear systems with these coefficient matrices using the Jacobi and Gauss–Seidel methods. You might generate the matrices in the following way:

```
n = input('Enter dimension of matrix: ');
dd = input('Enter diagonal dominance factor dd > 1: ');

A = randn(n,n);      % Start with a random matrix.
for i=1:n,           % Make it diagonally dominant.
```

```
    A(i,i) = 0;
    A(i,i) = sum(abs(A(i,:)))*dd;
end;
```

Set a random right-hand side vector `b = randn(n,1);`, and use a zero initial guess `x0 = zeros(n,1)`. At each step, compute the residual norm, `norm(b-A*xk)`, and make a plot of residual norm versus iteration number for each method. (Use the `semilogy` plotting command to get a log scale for the residual norm.) You can run for a fixed number of iterations or stop when the residual norm drops below a certain level. Experiment with different matrix sizes `n` and different diagonal dominance factors `dd`. (You might even try some values of `dd` that are less than one, in which case the Jacobi and/or Gauss–Seidel methods might or might not converge.) Turn in your plots, labeled with the matrix size and diagonal dominance factor for each.

7. Show that the residual $\mathbf{r}^{(k)} \equiv \mathbf{b} - A\mathbf{x}^{(k)}$ in iteration (12.8) satisfies

$$\mathbf{r}^{(k)} = (I - AM^{-1})\mathbf{r}^{(k-1)},$$

and that the vector $\mathbf{z}^{(k)} \equiv M^{-1}\mathbf{r}^{(k)}$ (sometimes called the *preconditioned* residual) satisfies

$$\mathbf{z}^{(k)} = (I - M^{-1}A)\mathbf{z}^{(k-1)}.$$

8. The simple iteration algorithm (12.8) for solving linear systems bears some resemblance to the *power method* for computing eigenvalues. It follows from the previous exercise that $\mathbf{r}^{(k)} = (I - AM^{-1})^k\mathbf{r}^{(0)}$ and $\mathbf{z}^{(k)} = (I - M^{-1}A)^k\mathbf{z}^{(0)}$. Explain why, for large k, one might expect $\mathbf{r}^{(k)}$ to approximate an eigenvector of $I - AM^{-1}$ and $\mathbf{z}^{(k)}$ to approximate an eigenvector of $I - M^{-1}A$. Show that the matrices $I - AM^{-1}$ and $I - M^{-1}A$ have the same eigenvalues, and explain how $\mathbf{r}^{(k)}$ or $\mathbf{z}^{(k)}$ might be used to estimate the eigenvalue of largest absolute value.

9. Complete the proof of theorem 12.2.2 by showing that if C is any nonsingular n by n matrix and we define $|||\mathbf{y}||| \equiv \|C\mathbf{y}\|_1$ for n-vectors $\mathbf{y}$, then $|||\cdot|||$ is a vector norm. More generally, show that if C is any nonsingular matrix and $\|\cdot\|$ is *any* vector norm, then $|||\mathbf{y}||| \equiv \|C\mathbf{y}\|$ is a vector norm.

10. Implement the conjugate gradient algorithm for the matrix A in Exercise 5; that is,

$$A = \begin{pmatrix} 2 & -1 & & & \\ -1 & 2 & \ddots & & \\ & \ddots & \ddots & \ddots & \\ & & \ddots & \ddots & -1 \\ & & & -1 & 2 \end{pmatrix}.$$

Try some different matrix sizes, say, $n = 10, 20, 40, 80$, and base your convergence criterion on the 2-norm of the residual divided by the

2-norm of the right-hand side vector: $\|\mathbf{b} - A\mathbf{x}^{(k)}\|/\|\mathbf{b}\|$. Comment on how the number of iterations required to reduce this quantity below a given level, say, 10^{-6}, appears to be related to the matrix size. For a right-hand side vector, take $\mathbf{b} = (b_1, \ldots, b_n)^T$, where $b_i = (i/(n+1))^2$, $i = 1, \ldots, n$, and take the initial guess $\mathbf{x}^{(0)}$ to be the zero vector. (We will see in chapter 13 that this corresponds to a finite difference method for the two-point boundary value problem: $u''(\chi) = \chi^2$, $0 < \chi < 1$, $u(0) = u(1) = 0$, whose solution is $u(\chi) = \frac{1}{12}(\chi^4 - \chi)$. The entries in the solution vector $\mathbf{x}$ of the linear system are approximations to u at the points $i/(n+1)$, $i = 1, \ldots, n$.)

11. Implement the GMRES algorithm for the same linear systems used in Exercise 6. Compare its performance with that of the Jacobi and Gauss–Seidel methods.
12. Suppose we wish to solve $A\mathbf{x} = \mathbf{b}$. Writing A in the form

$$A = D - L - U,$$

where D is diagonal, L is strictly lower triangular, and U is strictly upper triangular, the Jacobi method can be written as

$$\mathbf{x}^{(k+1)} = D^{-1}(L + U)\mathbf{x}^{(k)} + D^{-1}\mathbf{b}.$$

The file gsMorph.m on the book's web page solves a linear system with this formulation of the Jacobi method. The matrix A is constructed as the 5-point finite difference Laplacian on a square using the delsq command in MATLAB. In this problem, the vector $\mathbf{b}$ is the product $A\mathbf{x}$ where $\mathbf{x}$ consists of the grayscale values of an image of David M. Young. As an initial guess, we let $\mathbf{x}^{(0)}$ be the vector of grayscale values of an image of Carl Gustav Jacobi. We then visualize $\mathbf{x}^{(k)}$ as a vector of grayscale values after each iteration of the Jacobi method to create a morph from the image of Jacobi to the image of Young.

Jacobi

Young

(a) In the code, you will find the value of the variable diagonalIncrement to be 0. Run the code with diagonalIncrement = 0.5. This increments every diagonal entry in A by 0.5. Then run the code again with this variable set to 1 and finally to 2. Describe the effect on the morph from these different values. How does incrementing this value affect the rate of convergence of the iterative method? What effect must incrementing the diagonal entries in A have on the spectral radius of the resulting iteration matrix?

(b) Set diagonalIncrement to be 0.5. The Gauss–Seidel method can be written as

$$\mathbf{x}^{(k+1)} = (D - L)^{-1}U\mathbf{x}^{(k)} + (D - L)^{-1}\mathbf{b}.$$

Modify the code to implement the Gauss–Seidel method instead of the Jacobi method. Also, change the morph to begin with an image of Gauss rather than Jacobi. You will find an image of Gauss along with the code on the book's web page.

13

NUMERICAL SOLUTION OF TWO-POINT BOUNDARY VALUE PROBLEMS

In chapter 11 we discussed initial value problems: $y' = f(t, y(t))$, where $y(t_0)$ is given. For a single first-order equation like this, the value of y at any one point determines its value for all time. If y is a vector, however, so that we have a system of first-order equations, or if we have a higher-order equation such as

$$u''(x) = F(x, u(x), u'(x)), \quad 0 \le x \le 1,$$

then more conditions are needed. If the values of u and u' are given at some point $x_0 \in [0, 1]$, then this fits the paradigm of an initial value problem, but if instead, say, the values of u are given at $x = 0$ and $x = 1$, then this becomes a *two-point boundary value problem*. The restriction to the interval $[0, 1]$ is for convenience only; the equation could be defined over any interval $[a, b]$ by making the change of variable $x \to a + x(b - a)$.

If F is nonlinear in either $u(x)$ or $u'(x)$, the boundary value problem (BVP) is said to be *nonlinear*. A second-order *linear* two-point BVP takes the form

$$-\frac{d}{dx}\left(p(x)\frac{du}{dx}\right) + q(x)\frac{du}{dx} + r(x)u = f(x), \quad 0 \le x \le 1, \tag{13.1}$$

$$u(0) = \alpha, \quad u(1) = \beta. \tag{13.2}$$

If $p(x)$ is differentiable, then the first term can be differentiated to obtain $-(pu'' + p'u')$, but there may be good reasons to leave it in the form (13.1).

13.1 AN APPLICATION: STEADY-STATE TEMPERATURE DISTRIBUTION

Consider a thin rod, as pictured in figure 13.1.

Suppose a heat source is applied to this rod, and let $u(x, t)$ denote the temperature at position x, time t. By Newton's law of cooling, the amount

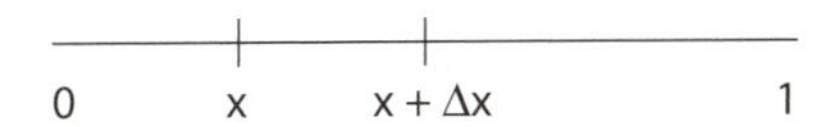

Figure 13.1. A thin rod.

of heat flowing from left to right across x in time Δt is $-\kappa(x)\frac{\partial u}{\partial x}(x,t)\Delta t$, where $\kappa(x)$ is the heat conductivity at position x in the rod; thus heat flow is proportional to the *gradient* of temperature. The *net* heat flowing into the part of the rod between x and $x+\Delta x$ in time interval t to $t+\Delta t$ is therefore

$$\kappa(x+\Delta x)\frac{\partial u}{\partial x}(x+\Delta x,t)\Delta t - \kappa(x)\frac{\partial u}{\partial x}(x,t)\Delta t.$$

The net heat flowing into this part of the rod in this time interval can also be expressed as

$$(\rho\Delta x)\cdot c\cdot\left(\frac{\partial u}{\partial t}\Delta t\right),$$

where ρ is the density and $\rho\Delta x$ the mass of the section of rod, c is the specific heat (the amount of heat needed to raise one unit of mass by one unit of temperature), and $(\partial u/\partial t)\Delta t$ is the change in temperature over the given time period. Equating these two expressions, we find

$$\frac{(\kappa\frac{\partial u}{\partial x})(x+\Delta x,t)-(\kappa\frac{\partial u}{\partial x})(x,t)}{\Delta x} = \rho c\frac{\partial u}{\partial t},$$

and taking the limit as $\Delta x \to 0$ gives

$$\frac{\partial}{\partial x}\left(\kappa\frac{\partial u}{\partial x}\right) = \rho c\frac{\partial u}{\partial t}.$$

When the system reaches a *steady-state* temperature distribution with $\partial u/\partial t = 0$, this equation becomes

$$\frac{\partial}{\partial x}\left(\kappa\frac{\partial u}{\partial x}\right) = 0.$$

This has the same form as equation (13.1) with $p(x)=\kappa$ and $q(x)\equiv r(x)\equiv f(x)\equiv 0$, where u, which is now a function of x only, represents the steady-state temperature. If the ends of the rod are held at fixed temperatures $u(0)=\alpha$ and $u(1)=\beta$, then we obtain the boundary conditions in (13.2). These are called **Dirichlet** boundary conditions. If, instead, one or both endpoints are insulated so that no heat flows in or out at $x=0$ and/or $x=1$, then the Dirichlet boundary conditions are replaced by homogeneous **Neumann** conditions: $u'(0)=0$ and/or $u'(1)=0$. Sometimes one has mixed or **Robin** boundary conditions: $c_{00}u(0)+c_{01}u'(0)=g_0$, $c_{10}u(1)+c_{11}u'(1)=g_1$, where c_{00}, c_{01}, c_{10}, c_{11}, g_0 , and g_1 are given.

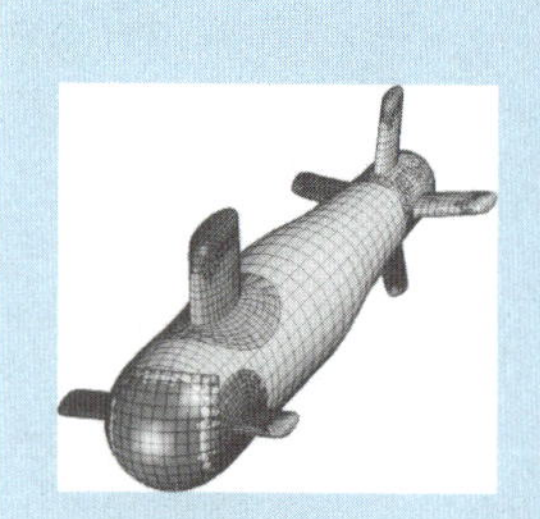

Creating Grids for Finite Differences

Finite difference methods are used in scientific simulations often involving complex, moving geometries. To deal with such geometries, sophisticated software packages have been developed. Sometimes, overlapping grids are used, as seen in the grid to the left [80]. Note that simulation of fluid flow past the submarine would involve solving partial differential equations, which will be covered in chapter 14. (Image courtesy of Bill Henshaw.)

Figure 13.2. Dividing the interval $[0, 1]$ into small subintervals of width $h = 1/n$.

13.2 FINITE DIFFERENCE METHODS

Let us consider the special case where $p(x) \equiv 1$ and $q(x) \equiv 0$ in (13.1):

$$-u''(x) + r(x)u(x) = f(x), \quad 0 \le x \le 1, \tag{13.3}$$

$$u(0) = \alpha, \quad u(1) = \beta. \tag{13.4}$$

Then it can be shown that if $r(x) \ge 0$, then the problem (13.3–13.4) has a unique solution. In general, proving the existence and uniqueness of solutions to boundary value problems is more difficult than proving such results for initial value problems. Given that the problem has a unique solution, however, one can then set about approximating this solution numerically.

The first step is to divide the interval $[0, 1]$ into small subintervals of width $h = 1/n$, as pictured in figure 13.2.
The *mesh points* or *nodes* are labeled $x_0 = 0, x_1 = h, \ldots, x_i = ih, \ldots, x_n = nh = 1$.

Next one approximates derivatives using finite difference quotients, as described in chapter 9. For example, to approximate the second derivative $u''(x_i)$, one might first approximate $u'(x_i \pm h/2)$ using centered-difference quotients:

$$u'(x_i + h/2) \approx \frac{u(x_{i+1}) - u(x_i)}{h}, \quad u'(x_i - h/2) \approx \frac{u(x_i) - u(x_{i-1})}{h},$$

and then combine these two approximations to obtain

$$u''(x_i) \approx \frac{u'(x_i + h/2) - u'(x_i - h/2)}{h} \approx \frac{u(x_{i+1}) - 2u(x_i) + u(x_{i-1})}{h^2}.$$

Substituting this expression into (13.3) and letting u_i denote the value of the approximate solution at x_i, we obtain the system of linear equations

$$-\frac{1}{h^2}[u_{i+1} - 2u_i + u_{i-1}] + r(x_i)u_i = f(x_i), \quad i = 1, \ldots, n-1. \tag{13.5}$$

Combining this with the boundary conditions, $u_0 = \alpha$, $u_n = \beta$, gives a system of $n-1$ equations in the $n-1$ unknown values $u_1, \ldots, u_{n-1}$. Moving all the known terms to the right-hand side, these equations can be written in matrix form as

$$\frac{1}{h^2}\begin{pmatrix} 2+h^2r(x_1) & -1 & & & \\ -1 & 2+h^2r(x_2) & -1 & & \\ & \ddots & \ddots & \ddots & \\ & & -1 & 2+h^2r(x_{n-2}) & -1 \\ & & & -1 & 2+h^2r(x_{n-1}) \end{pmatrix}\begin{pmatrix} u_1 \\ u_2 \\ \vdots \\ u_{n-2} \\ u_{n-1} \end{pmatrix}$$

$$= \begin{pmatrix} f(x_1) + \alpha/h^2 \\ f(x_2) \\ \vdots \\ f(x_{n-2}) \\ f(x_{n-1}) + \beta/h^2 \end{pmatrix}. \tag{13.6}$$

Note that this system of linear equations is *tridiagonal* and can be solved with Gaussian elimination using only $O(n)$ operations (see section 7.2.4).

Example 13.2.1. Let $r(x) = e^x$. We can construct a test problem for which we know the solution by deciding what solution $u(x)$ we would like to have and then determining what $f(x)$ must be. Suppose we wish to have $u(x) = x^2 + \sin(\pi x)$. Then $u'(x) = 2x + \pi \cos(\pi x)$ and $u''(x) = 2 - \pi^2 \sin(\pi x)$. Hence we must set $f(x) = -2 + (\pi^2 + e^x)\sin(\pi x) + e^x x^2$. Note that u satisfies the boundary conditions $u(0) = 0$ and $u(1) = 1$, so we will set $\alpha = 0$ and $\beta = 1$ in (13.6). Following is a MATLAB code that sets up and solves this problem.

```
%  Define r and f via inline functions that can work on
%  vectors of x values.
r = inline('exp(x)');
f = inline('-2 + (pi^2 + exp(x)).*sin(pi*x) + (exp(x).*x.^2)');

%  Set boundary values.
alpha = 0; beta = 1;

%  Ask user for number of subintervals.
n = input('Enter no. of subintervals n: ');

%  Set up the grid.
h = 1/n;
x = [h:h:(n-1)*h]';
```

```
% Form the matrix A.  Store it as a sparse matrix.
A = sparse(diag(2 + h^2*r(x),0) + diag(-ones(n-2,1),1)
    + diag(-ones(n-2,1),-1));
A = (1/h^2)*A;

% Form right-hand side vector b.
b = f(x);  b(n-1) = b(n-1) + beta/h^2;

% Compute approximate solution.
u_approx = A\b;

% Compare approximate solution with true solution.
u_true = x.^2 + sin(pi*x);
err = max(abs(u_true - u_approx))
```

Using this code with $n = 10,\ 20,\ 40,$ and 80 subintervals, we found that the maximum absolute difference between the true and approximate solutions at the nodes was 0.0071, 0.0018, 4.4×10^{-4}, and 1.1×10^{-4}, respectively. When the number of subintervals was doubled, the error went down by about a factor of 4. We will see why this is so in the next subsection.

13.2.1 Accuracy

As always, we will be concerned with the accuracy of the approximate solution generated by such a finite difference method, and especially in what happens in the limit as $h \to 0$. The **global error** (or the ∞-norm of the global error) is defined as

$$\max_{i=0,\dots,n} |u(x_i) - u_i|,$$

where $u(x_i)$ is the true solution and u_i the approximate solution at node i. The **local discretization error** or **truncation error** is the amount by which the true solution fails to satisfy the difference equations:

$$\tau(x,h) \equiv -\frac{1}{h^2}[u(x+h) - 2u(x) + u(x-h)] + r(x)u(x) - f(x), \tag{13.7}$$

where $u(x)$ is an exact solution of the differential equation. It follows from (13.3) that

$$\tau(x,h) = -\frac{1}{h^2}[u(x+h) - 2u(x) + u(x-h)] + u''(x). \tag{13.8}$$

The local truncation error can be analyzed using Taylor's theorem. Assume that $u \in C^4[0,1]$, and expand $u(x+h)$ and $u(x-h)$ in Taylor series about x:

$$u(x+h) = u(x) + hu'(x) + \frac{h^2}{2!}u''(x) + \frac{h^3}{3!}u'''(x) + \frac{h^4}{4!}u''''(\xi), \quad \xi \in [x, x+h],$$

$$u(x-h) = u(x) - hu'(x) + \frac{h^2}{2!}u''(x) - \frac{h^3}{3!}u'''(x) + \frac{h^4}{4!}u''''(\eta), \quad \eta \in [x-h, x].$$

Substituting these expressions into (13.8) gives

$$\tau(x, h) = -\frac{1}{h^2}\left[\frac{h^4}{4!}u''''(\xi) + \frac{h^4}{4!}u''''(\eta)\right] = O(h^2).$$

Next we would like to relate the local truncation error to the global error. This is usually the more difficult part of the analysis. We have from (13.5) and (13.7):

$$-\frac{1}{h^2}[u_{i+1} - 2u_i + u_{i-1}] + r(x_i)u_i - f(x_i) = 0,$$

$$-\frac{1}{h^2}[u(x_{i+1}) - 2u(x_i) + u(x_{i-1})] + r(x_i)u(x_i) - f(x_i) = \tau(x_i, h).$$

Subtracting these two equations and letting $e_i \equiv u(x_i) - u_i$ denote the error at node i, we find

$$-\frac{1}{h^2}[e_{i+1} - 2e_i + e_{i-1}] + r(x_i)e_i = \tau(x_i, h), \quad i = 1, \ldots, n-1. \tag{13.9}$$

Letting $\mathbf{e} \equiv (e_1, \ldots, e_{n-1})^T$ denote the vector of error values, these equations can be written in matrix form as

$$A\mathbf{e} = \boldsymbol{\tau},$$

where $\boldsymbol{\tau} \equiv (\tau(x_1, h), \ldots, \tau(x_{n-1}, h))^T$ is the vector of truncation errors and

$$A = \frac{1}{h^2}\begin{pmatrix} 2 + h^2 r(x_1) & -1 & & \\ -1 & \ddots & \ddots & \\ & \ddots & \ddots & -1 \\ & & -1 & 2 + h^2 r(x_{n-1}) \end{pmatrix} \tag{13.10}$$

is the same matrix as in (13.6). Assuming that A^{-1} exists, the error is then given by

$$\mathbf{e} = A^{-1}\boldsymbol{\tau}, \tag{13.11}$$

and we must study the behavior of A^{-1} as $h \to 0$.

Assume first that $r(x) \geq \gamma > 0$ for all $x \in [0, 1]$. Multiplying (13.9) by h^2 on each side, we find that

$$(2 + r(x_i)h^2)e_i = e_{i+1} + e_{i-1} + h^2\tau(x_i, h),$$

and hence that

$$(2 + \gamma h^2)|e_i| \leq |e_{i+1}| + |e_{i-1}| + h^2|\tau(x_i, h)| \leq 2\|\mathbf{e}\|_\infty + h^2\|\boldsymbol{\tau}\|_\infty.$$

Since this holds for all $i = 1, \ldots, n-1$, it follows that

$$(2 + \gamma h^2)\|\mathbf{e}\|_\infty \leq 2\|\mathbf{e}\|_\infty + h^2\|\boldsymbol{\tau}\|_\infty,$$

or,

$$\|\mathbf{e}\|_\infty \leq \|\boldsymbol{\tau}\|_\infty/\gamma = O(h^2).$$

Thus the global error is of the same order as the local truncation error. This same result holds when $r(x) = 0$, but the proof is more difficult.

Let us consider the case where $r(x) \equiv 0$. There are various norms in which one can measure the error, and now we will use something like the L_2-norm:

$$\left(\int_0^1 (u(x) - \hat{u}(x))^2 \, dx \right)^{1/2},$$

where $u(x)$ is the true solution and $\hat{u}(x)$ our approximate solution. Unfortunately, the approximate solution is defined only at mesh points x_i, but if we think of $\hat{u}$ as being a function that is equal to u_i at each x_i, $i = 1, \ldots, n-1$ and equal to α at $x = 0$ and to β at $x = 1$ and defined in some reasonable way throughout the rest of the interval $[0, 1]$, then this integral can be approximated by

$$\|\mathbf{e}\|_{L_2} \equiv \left(h \sum_{i=1}^{n-1} e_i^2 \right)^{1/2} = \frac{1}{\sqrt{n}} \|\mathbf{e}\|_2,$$

where $\|\cdot\|_2$ denotes the usual Euclidean norm for vectors. This is the composite trapezoid rule approximation to the above integral: $\frac{h}{2} e_0^2 + h \sum_{i=1}^{n-1} e_i^2 + \frac{h}{2} e_n^2 = h \sum_{i=1}^{n-1} e_i^2$, since $e_0 = e_n = 0$ (see section 10.2). Note the factor $\frac{1}{\sqrt{n}}$. As we refine the mesh, the number of mesh points n increases, and so if we measured the standard Euclidean norm of the error vector we would be summing the errors at more and more points. This weights the squared error values by the size of the subinterval over which they occur.

When $r(x) \equiv 0$, the matrix A in (13.10) becomes

$$A = \frac{1}{h^2} \begin{pmatrix} 2 & -1 & & \\ -1 & \ddots & \ddots & \\ & \ddots & \ddots & -1 \\ & & -1 & 2 \end{pmatrix}, \tag{13.12}$$

which is a **tridiagonal symmetric Toeplitz** (TST) matrix. It is tridiagonal because only the three center diagonals have nonzero entries, it is symmetric because $A = A^T$, and it is Toeplitz because it is constant along any diagonal. (The main diagonal consists of $2/h^2$'s, the first upper and lower diagonals of $-1/h^2$'s, and all other diagonals of 0s.) The eigenvalues and eigenvectors of such matrices are known.

Theorem 13.2.1. The eigenvalues of the m by m TST matrix

$$G = \begin{pmatrix} a & b & & \\ b & \ddots & \ddots & \\ & \ddots & \ddots & b \\ & & b & a \end{pmatrix}$$

are

$$\lambda_k = a + 2b\cos\left(\frac{k\pi}{m+1}\right), \quad k = 1, \ldots, m. \tag{13.13}$$

The corresponding orthonormal eigenvectors are

$$v_\ell^{(k)} = \sqrt{\frac{2}{m+1}}\sin\left(\frac{k\ell\pi}{m+1}\right), \quad k, \ell = 1, \ldots, m. \tag{13.14}$$

Proof. This theorem can be proved simply by checking that $G\mathbf{v}^{(k)} = \lambda_k \mathbf{v}^{(k)}$, $k = 1, \ldots, m$, and that $\|\mathbf{v}^{(k)}\| = 1$. Since G is symmetric and has distinct eigenvalues, its eigenvectors are orthogonal (see section 12.1).

To *derive* the formulas, however, suppose that λ is an eigenvalue of G and $\mathbf{v}$ a corresponding eigenvector. Letting $v_0 = v_{m+1} = 0$, the equation $A\mathbf{v} = \lambda\mathbf{v}$ can be written in the form

$$bv_{\ell-1} + (a - \lambda)v_\ell + bv_{\ell+1} = 0, \quad \ell = 1, \ldots, m. \tag{13.15}$$

This is a homogeneous linear difference equation of the sort discussed in section 11.3.3. To find its solutions, we consider the characteristic polynomial

$$\chi(z) = b + (a - \lambda)z + bz^2.$$

If ζ is a root of χ, then $v_\ell = \zeta^\ell$ is a solution of (13.15). If the roots of χ are denoted ζ_1 and ζ_2, then the general solution of (13.15) is $v_\ell = c_1\zeta_1^\ell + c_2\zeta_2^\ell$, if ζ_1 and ζ_2 are distinct, or $v_\ell = c_1\zeta^\ell + c_2\ell\zeta^\ell$ if $\zeta_1 = \zeta_2 = \zeta$, where c_1 and c_2 are determined by the conditions $v_0 = v_{m+1} = 0$.

The roots of χ are

$$\zeta_1 = \frac{\lambda - a + \sqrt{(\lambda - a)^2 - 4b^2}}{2b}, \quad \zeta_2 = \frac{\lambda - a - \sqrt{(\lambda - a)^2 - 4b^2}}{2b}. \tag{13.16}$$

If these two roots are equal (which would be the case if $(\lambda - a)^2 - 4b^2 = 0$; i.e., if $\lambda = a \pm 2b$), then the condition $v_0 = 0$ implies that $c_1 = 0$, and the condition $v_{m+1} = 0$ implies that $c_2(m+1)\zeta^{m+1} = 0$, which implies that either $c_2 = 0$ or $\zeta = 0$, and in either case $v_\ell = 0$ for all ℓ. Hence values of λ for which $\zeta_1 = \zeta_2$ are *not* eigenvalues since there are no corresponding nonzero eigenvectors.

Assuming, then, that ζ_1 and ζ_2 are distinct, the condition $v_0 = 0$ implies that $c_1 + c_2 = 0$, or, $c_2 = -c_1$. Together with this, the condition $v_{m+1} = 0$ implies that $c_1\zeta_1^{m+1} - c_1\zeta_2^{m+1} = 0$, which implies that either $c_1 = 0$ (in which case $c_2 = 0$ and $v_\ell = 0$ for all ℓ) or $\zeta_1^{m+1} = \zeta_2^{m+1}$. The latter condition is equivalent to

$$\zeta_2 = \zeta_1 \exp\left(\frac{2\pi i k}{m+1}\right) \tag{13.17}$$

for some $k = 0, 1, \ldots, m$, where $i = \sqrt{-1}$. We can discard the case $k = 0$, since it corresponds to $\zeta_2 = \zeta_1$.

Substituting expressions (13.16) for ζ_1 and ζ_2 into relation (13.17) and multiplying each side by $\exp(-ik\pi/(m+1))$, we obtain the equation

$$\left[\lambda - a - \sqrt{(\lambda-a)^2 - 4b^2}\right] \exp\left(\frac{-ik\pi}{m+1}\right)$$
$$= \left[\lambda - a + \sqrt{(\lambda-a)^2 - 4b^2}\right] \exp\left(\frac{ik\pi}{m+1}\right).$$

Replacing $\exp(\pm ik\pi/(m+1))$ by $\cos(k\pi/(m+1)) \pm i\sin(k\pi/(m+1))$, this becomes

$$\sqrt{(\lambda-a)^2 - 4b^2}\cos\left(\frac{k\pi}{m+1}\right) = -(\lambda-a)i\sin\left(\frac{k\pi}{m+1}\right).$$

Squaring both sides gives

$$[(\lambda-a)^2 - 4b^2]\cos^2\left(\frac{k\pi}{m+1}\right) = -(\lambda-a)^2\sin^2\left(\frac{k\pi}{m+1}\right),$$

or,

$$(\lambda-a)^2 = 4b^2\cos^2\left(\frac{k\pi}{m+1}\right).$$

Finally, taking square roots gives

$$\lambda = a \pm 2b\cos\left(\frac{k\pi}{m+1}\right), \quad k = 1, \dots, m.$$

We need only take the plus sign since the minus sign gives the same set of values, and this establishes (13.13).

Substituting (13.13) into (13.16), we readily find that

$$\zeta_1 = \cos\left(\frac{k\pi}{m+1}\right) + i\sin\left(\frac{k\pi}{m+1}\right) = \exp\left(\frac{ik\pi}{m+1}\right),$$

$$\zeta_2 = \cos\left(\frac{k\pi}{m+1}\right) - i\sin\left(\frac{k\pi}{m+1}\right) = \exp\left(\frac{-ik\pi}{m+1}\right),$$

and therefore

$$v_\ell^{(k)} = c_1(\zeta_1^\ell - \zeta_2^\ell) = 2c_1 i\sin\left(\frac{k\ell\pi}{m+1}\right).$$

Since this is a multiple of the vector $\mathbf{v}^{(k)}$ in (13.14), it remains only to show that by choosing c_1 so that $\mathbf{v}^{(k)}$ satisfies (13.14), we obtain a vector with norm 1. It is left as an exercise to show that

$$\sum_{\ell=1}^{m}\sin^2\left(\frac{k\ell\pi}{m+1}\right) = \frac{m+1}{2},$$

and hence the vectors in (13.14) have unit norm. □

Using theorem 13.2.1, we can write down the eigenvalues of the matrix A in (13.12):

$$\lambda_k = \frac{1}{h^2}\left[2 - 2\cos\left(\frac{k\pi}{n}\right)\right], \quad k = 1, \dots, n-1. \tag{13.18}$$

Note that the eigenvalues of A are all positive. A symmetric matrix with all positive eigenvalues is said to be **positive definite.**

Note also that the eigenvectors $\mathbf{v}^{(k)}$ defined in (13.14) are related to the *eigenfunctions* $v^{(k)}(x) = \sin(k\pi x)$, $k = 1, 2, \dots$ of the differential operator $-\frac{d^2}{dx^2}$. These functions are called eigenfunctions because they satisfy

$$-\frac{d^2}{dx^2} v^{(k)}(x) = k^2\pi^2 v^{(k)}(x),$$

and have value 0 at $x = 0$ and $x = 1$. The *infinite* set of values $k^2\pi^2$, $k = 1, 2, \dots$ are eigenvalues of the differential operator. The eigen*vectors* $\mathbf{v}^{(k)}$, $k = 1, \dots, n-1$ of the discretized system (i.e., of the matrix A) are the values of the first $n-1$ eigenfunctions at the nodes. The eigen*values* of A are not exactly the same as the first $n-1$ eigenvalues of the differential operator, but for small values of k they are close; expanding $\cos(k\pi/n)$ in (13.18) in a Taylor series about 0, we find

$$\lambda_k = \frac{1}{h^2}\left[2 - 2\left(1 - \frac{1}{2}\left(\frac{k\pi}{n}\right)^2 + O(h^4)\right)\right] = k^2\pi^2 + O(h^2),$$

as $h \to 0$ for fixed k.

In order to estimate $\|\mathbf{e}\|_{L_2}$ using (13.11), we must estimate

$$\|A^{-1}\|_{L_2} \equiv \max_{\|\mathbf{w}\|_{L_2}=1} \|A^{-1}\mathbf{w}\|_{L_2} = \max_{\|(1/\sqrt{n})\mathbf{w}\|_2=1} \|A^{-1}(1/\sqrt{n})\mathbf{w}\|_2 \equiv \|A^{-1}\|_2.$$

Recall that the 2-norm of a symmetric matrix is the largest absolute value of its eigenvalues and that the eigenvalues of A^{-1} are the inverses of the eigenvalues of A. Hence $\|A^{-1}\|_{L_2}$ is one over the smallest eigenvalue of A.

It follows from (13.18) that the smallest eigenvalue of A is

$$\lambda_1 = \frac{1}{h^2}\left[2 - 2\cos\left(\frac{\pi}{n}\right)\right].$$

Expanding $\cos(\pi/n) = \cos(\pi h)$ in a Taylor series about 0, we find

$$\cos(\pi h) = 1 - \frac{(\pi h)^2}{2} + O(h^4),$$

and substituting this into the expression for λ_1 gives

$$\lambda_1 = \frac{1}{h^2}[\pi^2 h^2 + O(h^4)] = \pi^2 + O(h^2).$$

Hence $\|A^{-1}\|_{L_2} = \pi^{-2} + O(h^2) = O(1)$, and it follows from (13.11) that

$$\|\mathbf{e}\|_{L_2} \le \|A^{-1}\|_{L_2} \cdot \|\boldsymbol{\tau}\|_{L_2} = O(1) \cdot O(h^2) = O(h^2).$$

We have thus established, in the case where $r(x) \equiv 0$, that the L_2-norm of the global error is of the same order as that of the local truncation error; that is, $O(h^2)$.

Note also from (13.18) that the largest eigenvalue of A is

$$\lambda_{n-1} = \frac{1}{h^2}\left[2 - 2\cos\left(\frac{(n-1)\pi}{n}\right)\right].$$

Expanding $\cos((n-1)\pi/n) = \cos(\pi - \pi h)$ in a Taylor series about π gives

$$\cos(\pi - \pi h) = -1 + \frac{(\pi h)^2}{2} + O(h^4),$$

and it follows that

$$\lambda_{n-1} = \frac{1}{h^2}[4 - \pi^2 h^2 + O(h^4)] = O(h^{-2}).$$

Recall that the 2-norm **condition number** of a symmetric matrix is the ratio of its largest to smallest eigenvalue in absolute value. Hence the condition number of A is $O(h^{-2})$. The condition number plays an important role in determining the number of iterations required by the conjugate gradient algorithm to solve the linear system (13.6) (see theorem 12.2.7).

13.2.2 More General Equations and Boundary Conditions

Suppose $p(x)$ in (13.1) is not identically 1 but is a function of x. If $p(x) > 0$ for all $x \in [0, 1]$ and $r(x) \geq 0$ for all $x \in [0, 1]$, then it can again be shown that the boundary value problem

$$-\frac{d}{dx}\left(p(x)\frac{du}{dx}\right) + r(x)u = f(x), \quad 0 \leq x \leq 1, \tag{13.19}$$

$$u(0) = \alpha, \quad u(1) = \beta, \tag{13.20}$$

has a unique solution.

We can use the same approach to difference this more general equation. Approximate $-(d/dx)\,(p(x)du/dx)$ by the centered-difference formula

$$\begin{aligned}
-\frac{d}{dx}\left(p(x)\frac{du}{dx}\right) &\approx -\frac{(p\,du/dx)(x+h/2) - (p\,du/dx)(x-h/2)}{h} \\
&\approx -\frac{p(x+h/2)(u(x+h)-u(x))/h - p(x-h/2)(u(x)-u(x-h))/h}{h} \\
&= \frac{1}{h^2}\big[(p(x+h/2) + p(x-h/2))u(x) - p(x+h/2)u(x+h) \\
&\qquad - p(x-h/2)u(x-h)\big],
\end{aligned}$$

which is also second-order accurate. The difference equation at node i then becomes

$$\begin{aligned}
\frac{1}{h^2}[(p(x_i+h/2) + p(x_i-h/2) + h^2 r(x_i))u_i - p(x_i+h/2)u_{i+1} \\
- p(x_i-h/2)u_{i-1}] = f(x_i),
\end{aligned}$$

and writing the difference equations at nodes $i = 1, \ldots, n-1$ in matrix form we obtain the linear system

$$\frac{1}{h^2}\begin{pmatrix} a_1 & b_1 & & & \\ b_1 & a_2 & b_2 & & \\ & \ddots & \ddots & \ddots & \\ & & b_{n-3} & a_{n-2} & b_{n-2} \\ & & & b_{n-2} & a_{n-1} \end{pmatrix} \begin{pmatrix} u_1 \\ u_2 \\ \vdots \\ u_{n-2} \\ u_{n-1} \end{pmatrix} = \begin{pmatrix} f(x_1) - \alpha b_0/h^2 \\ f(x_2) \\ \vdots \\ f(x_{n-2}) \\ f(x_{n-1}) - \beta b_{n-1}/h^2 \end{pmatrix},$$

where

$$a_i = p(x_i + h/2) + p(x_i - h/2) + h^2 r(x_i), \quad i = 1, \ldots, n-1,$$

$$b_i = -p(x_i + h/2), \quad i = 0, 1, \ldots, n-1.$$

Note that the coefficient matrix for this problem is again symmetric and tridiagonal, but it is no longer Toeplitz. Still it can be shown that this matrix is also positive definite.

If $p(x)$ is differentiable, then one could replace the term $-(d/dx)(p(x)\, du/dx)$ in (13.19) by $-(pu'' + p'u')$, using the rule for differentiating a product. Usually this is not a good idea, however, because, as we will see next, when one differences a *first* derivative the resulting matrix is nonsymmetric. Since the original differential operator $-(d/dx)(p(x)du/dx)$ is *self-adjoint* (which is the operator equivalent of *symmetric* and will be defined in the next section), the matrix should share this property.

Suppose now that $q(x)$ in (13.1) is not identically 0. For simplicity, let us return to the case where $p(x) \equiv 1$, so that the differential equation becomes

$$-u'' + q(x)u' + r(x)u = f(x), \quad 0 \le x \le 1.$$

If we use a centered-difference approximation to $u'(x)$:

$$u'(x) \approx \frac{u(x+h) - u(x-h)}{2h}, \tag{13.21}$$

which is accurate to order h^2, then the difference equations become

$$\frac{1}{h^2}[2u_i - u_{i+1} - u_{i-1}] + q(x_i)\frac{u_{i+1} - u_{i-1}}{2h} + r(x_i)u_i = f(x_i), \quad i = 1, \ldots, n-1.$$

This is the *nonsymmetric* tridiagonal system

$$\begin{pmatrix} a_1 & b_1 & & & \\ c_2 & a_2 & b_2 & & \\ & \ddots & \ddots & \ddots & \\ & & c_{n-2} & a_{n-2} & b_{n-2} \\ & & & c_{n-1} & a_{n-1} \end{pmatrix} \begin{pmatrix} u_1 \\ u_2 \\ \vdots \\ u_{n-2} \\ u_{n-1} \end{pmatrix} = \begin{pmatrix} f(x_1) - c_1\alpha \\ f(x_2) \\ \vdots \\ f(x_{n-2}) \\ f(x_{n-1}) - b_{n-1}\beta \end{pmatrix}, \tag{13.22}$$

where

$$a_i = \frac{2}{h^2} + r(x_i), \quad b_i = -\frac{1}{h^2} + \frac{q(x_i)}{2h}, \quad c_i = -\frac{1}{h^2} - \frac{q(x_i)}{2h}, \quad i = 1, \ldots, n-1. \tag{13.23}$$

For h sufficiently small, the sub- and super-diagonal entries, b_i and c_i, are negative and satisfy

$$|b_i| + |c_i| = \frac{2}{h^2}.$$

If $r(x_i) > 0$ for all $i = 1, \ldots, n-1$, then the diagonal entries a_i satisfy

$$|a_i| = \frac{2}{h^2} + r(x_i) > |b_i| + |c_i|.$$

For this reason, the matrix is said to be *strictly row diagonally dominant*. We have seen this definition before in section 12.2.3: Each diagonal entry in the matrix is strictly greater in magnitude than the sum of the absolute values of the off-diagonal entries in its row.

The importance of strict diagonal dominance is that it can be used to show that a matrix is nonsingular. Clearly, if the difference equations (13.22) are to have a unique solution, then the matrix there must be nonsingular. It follows from Gerschgorin's theorem [Section 12.1] that a matrix that is strictly row or column diagonally dominant cannot have 0 as an eigenvalue and so must be nonsingular. Hence for h sufficiently small the matrix in (13.22) is nonsingular.

Sometimes the centered-difference approximation (13.21) to $u'(x)$ is replaced by a one-sided difference:

$$u'(x_i) \approx \begin{cases} \frac{1}{h}(u_{i+1} - u_i) & \text{if } q(x_i) < 0, \\ \frac{1}{h}(u_i - u_{i-1}) & \text{if } q(x_i) \geq 0. \end{cases}$$

This is called **upwind** differencing. It is only first-order accurate, but it may have other advantages. With this differencing, (13.22) holds with (13.23) replaced by

$$a_i = \frac{2}{h^2} + r(x_i) + \frac{|q(x_i)|}{h},$$

$$b_i = \begin{cases} -\frac{1}{h^2} + \frac{q(x_i)}{h} & \text{if } q(x_i) < 0, \\ -\frac{1}{h^2} & \text{if } q(x_i) \geq 0, \end{cases} \qquad c_i = \begin{cases} -\frac{1}{h^2} & \text{if } q(x_i) < 0, \\ -\frac{1}{h^2} - \frac{q(x_i)}{h} & \text{if } q(x_i) \geq 0, \end{cases}$$

$i = 1, \ldots, n-1.$

Now the matrix is strictly row diagonally dominant when $r(x_i) > 0$ for all i, *independent* of h. This is because $|b_i| + |c_i| = (2/h^2) + (|q(x_i)|/h)$, while $|a_i| = (2/h^2) + (|q(x_i)|/h) + r(x_i)$.

As noted previously, sometimes different boundary conditions are specified. Let us return to the problem (13.3–13.4). Suppose the second Dirichlet boundary condition, $u(1) = \beta$, is replaced by a Neumann condition $u'(1) = \beta$. Now the value u_n is no longer known from the boundary condition at $x = 1$ but instead will be an additional unknown. We will add an additional equation to approximate the boundary condition.

There are several ways to do this. One idea is to approximate $u''(1)$ using a point outside the interval:

$$u''(1) \approx \frac{1}{h^2}[u_{n-1} - 2u_n + u_{n+1}].$$

Then, approximating the Neumann boundary condition by

$$\beta = u'(1) \approx \frac{1}{2h}[u_{n+1} - u_{n-1}],$$

we can solve for u_{n+1}:

$$u_{n+1} = u_{n-1} + 2h\beta.$$

Substituting this into the approximate expression for $u''(1)$ and requiring that the differential equation (13.3) hold approximately at $x = 1$, we obtain the equation

$$\frac{1}{h^2}[2u_n - 2u_{n-1} - 2h\beta] + r(x_n)u_n = f(x_n).$$

This adds one more equation and one more unknown to the matrix equation in (13.6):

$$\frac{1}{h^2}\begin{pmatrix} 2+h^2r(x_1) & -1 & & & \\ -1 & 2+h^2r(x_2) & -1 & & \\ & \ddots & \ddots & \ddots & \\ & & -1 & 2+h^2r(x_{n-1}) & -1 \\ & & & -2 & 2+h^2r(x_n) \end{pmatrix}\begin{pmatrix} u_1 \\ u_2 \\ \vdots \\ u_{n-1} \\ u_n \end{pmatrix}$$

$$= \begin{pmatrix} f(x_1) + \alpha/h^2 \\ f(x_2) \\ \vdots \\ f(x_{n-1}) \\ f(x_n) + 2\beta/h \end{pmatrix}. \tag{13.24}$$

The matrix is no longer symmetric, but it can be symmetrized easily (see Exercise 4 at the end of this chapter).

If, in addition, the first Dirichlet condition $u(0) = \alpha$ is replaced by the Neumann condition $u'(0) = \alpha$, then u_0 becomes an unknown and a similar argument leads to the equation

$$\frac{1}{h^2}[2u_0 - 2u_1 + 2h\alpha] + r(x_0)u_0 = f(x_0).$$

Adding this equation and unknown to the system in (13.24), we have

$$\frac{1}{h^2}\begin{pmatrix} 2+h^2r(x_0) & -2 & & & \\ -1 & 2+h^2r(x_1) & -1 & & \\ & \ddots & \ddots & \ddots & \\ & & -1 & 2+h^2r(x_{n-1}) & -1 \\ & & & -2 & 2+h^2r(x_n) \end{pmatrix}\begin{pmatrix} u_0 \\ u_1 \\ \vdots \\ u_{n-1} \\ u_n \end{pmatrix}$$

$$= \begin{pmatrix} f(x_0) - 2\alpha/h \\ f(x_1) \\ \vdots \\ f(x_{n-1}) \\ f(x_n) + 2\beta/h \end{pmatrix}. \tag{13.25}$$

If $r(x_i) > 0$ for all $i = 0, 1, \ldots, n-1, n$, then this matrix is strictly row diagonally dominant and hence nonsingular. On the other hand, if $r(x_i) = 0$ for all i, then this matrix is singular, because its row sums are all 0; this means that if one applies the matrix to the vector of all 1s, one obtains the zero vector. This makes sense because the solution of the boundary value problem

$$-u''(x) = f(x), \quad u'(0) = \alpha, \quad u'(1) = \beta,$$

is either nonexistent (for example, if $f(x) = 0$ but $\alpha \neq \beta$, since in this case $f(x) = 0 \Rightarrow u(x) = c_0 + c_1 x \Rightarrow u'(x) = c_1$ for some constant c_1) or is nonunique; if $u(x)$ is any one solution, then $u(x) + \gamma$ is another solution, for any constant γ.

Example 13.2.2. We can modify the example code given at the beginning of this section to handle a Neumann boundary condition at either or both ends of the interval [0, 1]. Following is a modified code that implements the Neumann boundary condition $u'(1) = 2 - \pi$. The true solution is the same as in the previous example, $u(x) = x^2 + \sin(\pi x)$.

```
% Define r and f via inline functions that can work on
% vectors of x values.
r = inline('exp(x)');
f = inline('-2 + (pi^2 + exp(x)).*sin(pi*x) + (exp(x).*x.^2)');

% Set boundary values.  Beta is now the value of u'(1).
alpha = 0; beta = 2 - pi;

% Ask user for number of subintervals.
n = input('Enter no. of subintervals n: ');

% Set up the grid.  There are now n grid points at which to
% find the soln.
```

```
h = 1/n;
x = [h:h:n*h]';

% Form the matrix A.  Store it as a sparse matrix.
% It is now n by n.
% It is nonsymmetric but can easily be symmetrized.
A = sparse(diag(2 + h^2*r(x),0) + diag(-ones(n-1,1),1)
    + diag(-ones(n-1,1),-1));
A(n,n-1) = -2;
A = (1/h^2)*A;

% Form right-hand side vector b.  It now has length n.
b = f(x);  b(n) = b(n) + 2*beta/h;

% Compute approximate solution.
u_approx = A\b;

% Compare approximate solution with true solution.
u_true = x.^2 + sin(pi*x);
err = max(abs(u_true - u_approx))
```

Running this code with $n = 10, 20, 40,$ and 80 subintervals, we found that the ∞-norm of the error was 0.0181, 0.0045, 0.0011, and 2.8×10^{-4}, respectively. While the absolute error is somewhat larger than with the Dirichlet boundary condition, the method is still second-order accurate.

13.3 FINITE ELEMENT METHODS

One can take a different approach to solving problems of the form (13.1–13.2). Instead of approximating derivatives by finite difference quotients, one might attempt to approximate the solution $u(x)$ by a simple function such as a polynomial or piecewise polynomial. While such a function might not be able to satisfy (13.1–13.2) exactly, one could choose the coefficients so that these equations were approximately satisfied. This is one of the ideas behind *finite element methods*.

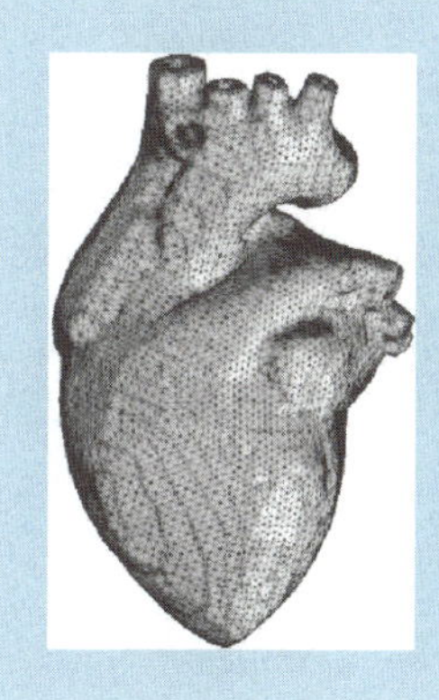

BIOMEDICAL SIMULATIONS

Like finite difference methods, finite element methods are also utilized in scientific simulation. To the left, we see a tetrahedral mesh of a heart model on which a finite element method is used for simulating activities of the heart. Such models play a role in the development of cardiovascular surgical procedures and the design of artificial heart valves. (Created by and courtesy of the Computational Visualization Center at the University of Texas at Austin).

Again let us restrict our attention for now to the BVP (13.3–13.4), and let us assume that $\alpha = \beta = 0$. This problem can be written in the very general form:

$$\mathcal{L}u \equiv -u'' + r(x)u = f(x), \quad 0 \le x \le 1, \tag{13.26}$$

$$u(0) = u(1) = 0. \tag{13.27}$$

In what sense should equation (13.26) be approximately satisfied? One idea is to force it to hold at certain points $x_1, \ldots, x_{n-1}$. This leads to what are called *collocation methods*. Another idea is to choose the approximate solution, say, $\hat{u}(x)$, from the selected class of functions S in order to minimize the L_2-norm of $\mathcal{L}\hat{u} - f$:

$$\min_{\hat{u} \in S} \left(\int_0^1 |\mathcal{L}\hat{u} - f|^2 \, dx \right)^{1/2}.$$

This leads to a *least squares approximation.*

Another idea is to note that if u satisfies (13.26), then for any function φ (sufficiently well behaved so that all of the integrals make sense), the integral of the product of $\mathcal{L}u$ with φ will be equal to the integral of the product of f with φ. Let us define an **inner product** among functions defined on $[0, 1]$ by

$$\langle v, w \rangle \equiv \int_0^1 v(x)w(x)\, dx.$$

Then if u satisfies (13.26), it also satisfies

$$\langle \mathcal{L}u, \varphi \rangle \equiv \int_0^1 (-u''(x) + r(x)u(x))\varphi(x)\, dx = \int_0^1 f(x)\varphi(x)\, dx \equiv \langle f, \varphi \rangle, \tag{13.28}$$

for all functions φ. Equation (13.28) is sometimes called the *weak form* of the differential equation.

One thing to note about the weak form (13.28) is that we can integrate by parts in the integral on the left:

$$\int_0^1 -u''(x)\varphi(x)\, dx = -u'(x)\varphi(x)\big|_0^1 + \int_0^1 u'(x)\varphi'(x)\, dx.$$

Hence (13.28) can be replaced by

$$-u'(x)\varphi(x)\big|_0^1 + \int_0^1 u'(x)\varphi'(x)\, dx + \int_0^1 r(x)u(x)\varphi(x)\, dx = \int_0^1 f(x)\varphi(x)\, dx. \tag{13.29}$$

With this formulation, we can consider "approximate" solutions $\hat{u}$ that are differentiable only once instead of twice.

If $\hat{u}$ is restricted to come from a certain set S, called the *trial space*, then we probably will not be able to find a function $\hat{u} \in S$ that satisfies (13.28) or (13.29) for *all* functions φ; but perhaps we can find a function $\hat{u} \in S$ that satisfies (13.29) for all functions φ in some set T, called the *test space*. The most common choice is $T = S$. Then taking S to be a space of piecewise polynomials, this gives rise to a *Galerkin finite element* approximation.

BORIS GALERKIN

Boris Galerkin (1871–1945) was a Russian mathematician and engineer. His first technical paper, on the buckling of bars, was published while he was imprisoned in 1906 by the Tzar in pre-Revolution Russia. Russian texts often refer to the Galerkin finite element method as the Bubnov–Galerkin method since I. G. Bubnov published a paper using this idea in 1915. The European Community on Computational Methods in Applied Sciences gives the Ritz–Galerkin Medal in honor of Walter Ritz and Boris Galerkin and the synergy between mathematics, numerical analysis, and engineering [23].

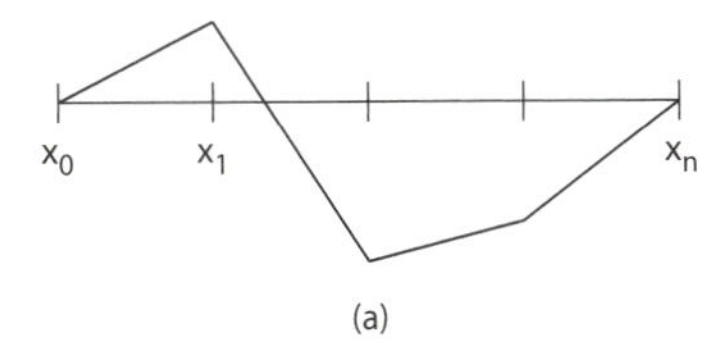

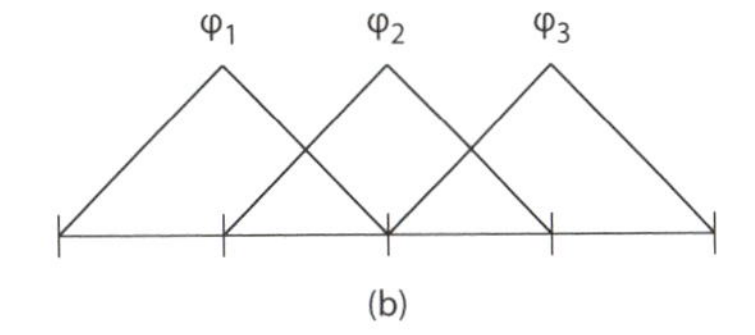

Figure 13.3. (a) A continuous piecewise linear function and (b) a basis of "hat functions" for the finite element method.

More specifically, divide the interval [0,1] into subintervals of width $h = 1/n$, with endpoints $x_0 = 0$, $x_1 = h, \ldots, x_n = nh = 1$, and let S be the space of *continuous piecewise linear functions* that satisfy the boundary conditions (13.27). A function from S might look like that pictured in figure 13.3(a).

We wish to find the function $\hat{u} \in S$ that satisfies (13.29) for all functions $\varphi \in S$. To do this, it will be helpful to have a *basis* for the space S: a set of linearly independent functions in S with the property that every function in S can be written as a linear combination of these. The set of "hat functions" pictured in figure 13.3(b) forms a basis for S.

More precisely, the set of continuous piecewise linear functions φ_i, $i = 1, \ldots n-1$, where φ_i is equal to one at node i and zero at all other nodes; that is,

$$\varphi_i(x) = \begin{cases} \dfrac{x - x_{i-1}}{x_i - x_{i-1}}, & x \in [x_{i-1}, x_i], \\ \dfrac{x_{i+1} - x}{x_{i+1} - x_i}, & x \in [x_i, x_{i+1}], \\ 0 & \text{otherwise}, \end{cases} \qquad i = 1, \ldots, n-1, \tag{13.30}$$

forms a basis for S. To write an arbitrary continuous piecewise linear function φ in S as a linear combination of these basis functions, $\varphi(x) = \sum_{i=1}^{n-1} c_i \varphi_i(x)$, we simply take the coefficients c_i to be the values of the function φ at each node x_i.

Then the linear combination $\sum_{i=1}^{n-1} c_i \varphi_i$ will have the same values as φ at each of the nodes $x_1, \ldots, x_{n-1}$ (as well as at the endpoints x_0 and x_n where they are each zero), and since a piecewise linear function on this grid is completely determined by its values at the nodes, they must be the same function.

Example 13.3.1. Suppose the interval $[0, 1]$ is divided into four equal subintervals, and suppose $\varphi(x)$ is the continuous piecewise linear function defined by

$$\varphi(x) = \begin{cases} x & \text{if } 0 \le x \le 1/4, \\ 2x - 1/4 & \text{if } 1/4 \le x \le 1/2, \\ 7/4 - 2x & \text{if } 1/2 \le x \le 3/4, \\ 1 - x & \text{if } 3/4 \le x \le 1. \end{cases}$$

The basis functions φ_1, φ_2, and φ_3, for the space S of continuous piecewise linear functions that are zero at the endpoints of the interval, have the value one at $x = 1/4$, $x = 1/2$, and $x = 3/4$, respectively, and they each take the value zero at each of the other nodes. Since $\varphi(1/4) = 1/4$, $\varphi(1/2) = 3/4$, and $\varphi(3/4) = 1/4$, it follows that $\varphi(x) = (1/4)\varphi_1(x) + (3/4)\varphi_2(x) + (1/4)\varphi_3(x)$. To see that this is so, note that

$$\varphi_1(x) = \begin{cases} 4x & \text{if } 0 \le x \le 1/4, \\ 2 - 4x & \text{if } 1/4 \le x \le 1/2, \\ 0 & \text{otherwise}, \end{cases} \qquad \varphi_2(x) = \begin{cases} 4x - 1 & \text{if } 1/4 \le x \le 1/2, \\ 3 - 4x & \text{if } 1/2 \le x \le 3/4, \\ 0 & \text{otherwise}, \end{cases}$$

$$\varphi_3(x) = \begin{cases} 4x - 2 & \text{if } 1/2 \le x \le 3/4, \\ 4 - 4x & \text{if } 3/4 \le x \le 1, \\ 0 & \text{otherwise}. \end{cases}$$

Hence

$$\frac{1}{4}\varphi_1(x) + \frac{3}{4}\varphi_2(x) + \frac{1}{4}\varphi_3(x)$$
$$= \begin{cases} \frac{1}{4}4x = x & \text{if } 0 \le x \le 1/4, \\ \frac{1}{4}(2 - 4x) + \frac{3}{4}(4x - 1) = 2x - \frac{1}{4} & \text{if } 1/4 \le x \le 1/2, \\ \frac{3}{4}(3 - 4x) + \frac{1}{4}(4x - 2) = \frac{7}{4} - 2x & \text{if } 1/2 \le x \le 3/4, \\ \frac{1}{4}(4 - 4x) = 1 - x & \text{if } 3/4 \le x \le 1. \end{cases}$$

Let us write our approximate solution $\hat{u}(x)$ as $\hat{u}(x) = \sum_{j=1}^{n-1} c_j \varphi_j(x)$. The coefficients $c_1, \ldots, c_{n-1}$ will be chosen so that $\langle \mathcal{L}\hat{u}, \varphi_i \rangle = \langle f, \varphi_i \rangle$ for all $i = 1, \ldots, n-1$. (The expression $\langle \mathcal{L}\hat{u}, \varphi_i \rangle$ should be understood to mean the integral *after* integrating by parts, as in (13.29); otherwise, we would have trouble defining $\mathcal{L}\hat{u}$, since $\hat{u}'$ is discontinuous; the second derivative of $\hat{u}$ is zero in the interior of each subinterval and infinite at the nodes.) Since any function φ in S can be written as a linear combination of the φ_i's, say, $\varphi(x) = \sum_{i=1}^{n-1} d_i \varphi_i(x)$, and since the inner product is linear in each of its arguments, this ensures that $\langle \mathcal{L}\hat{u} - f, \varphi \rangle = \langle \mathcal{L}\hat{u} - f, \sum_{i=1}^{n-1} d_i \varphi_i \rangle = \sum_{i=1}^{n-1} d_i \langle \mathcal{L}\hat{u} - f, \varphi_i \rangle = 0$, or, equivalently, $\langle \mathcal{L}\hat{u}, \varphi \rangle = \langle f, \varphi \rangle$.

More specifically, using (13.29), we will choose $c_1, \ldots, c_{n-1}$ so that

$$-\hat{u}'(x)\varphi_i(x)\Big|_0^1 + \int_0^1 \left(\sum_{j=1}^{n-1} c_j \varphi_j'(x) \right) \varphi_i'(x)\, dx + \int_0^1 r(x) \left(\sum_{j=1}^{n-1} c_j \varphi_j(x) \right) \varphi_i(x)\, dx$$

$$= \int_0^1 f(x)\varphi_i(x)\, dx, \quad i = 1, \ldots, n-1.$$

This is a system of $n-1$ linear equations in the $n-1$ unknowns $c_1, \ldots, c_{n-1}$. Note first that the boundary term, $-\hat{u}'(x)\varphi(x)\Big|_0^1$ is zero since φ in S implies $\varphi(0) = \varphi(1) = 0$. Note also that the summations can be pulled outside the integrals:

$$\sum_{j=1}^{n-1} c_j \left[\int_0^1 \varphi_j'(x)\varphi_i'(x)\, dx + \int_0^1 r(x)\varphi_j(x)\varphi_i(x)\, dx \right]$$
$$= \int_0^1 f(x)\varphi_i(x)\, dx, \quad i = 1, \ldots, n-1. \tag{13.31}$$

The system (13.31) can be written in matrix form as

$$A\mathbf{c} = \mathbf{f}, \tag{13.32}$$

where $\mathbf{c} \equiv (c_1, \ldots, c_{n-1})^T$ is the vector of unknown coefficients, $\mathbf{f}$ is the vector whose ith entry is $\int_0^1 f(x)\varphi_i(x)\, dx$, $i = 1, \ldots, n-1$, and A is the $(n-1)$ by $(n-1)$ matrix whose (i, j)-entry is

$$a_{ij} \equiv \int_0^1 \varphi_j'(x)\varphi_i'(x)\, dx + \int_0^1 r(x)\varphi_j(x)\varphi_i(x)\, dx. \tag{13.33}$$

In other words, the ith entry in $\mathbf{f}$ is $\langle f, \varphi_i \rangle$ and the (i, j)-entry in A is $\langle \mathcal{L}\varphi_j, \varphi_i \rangle$, after integrating by parts.

Returning to expression (13.30) for $\varphi_i(x)$, it can be seen that φ_i' satisfies

$$\varphi_i'(x) = \begin{cases} \dfrac{1}{x_i - x_{i-1}}, & x \in [x_{i-1}, x_i], \\ \dfrac{-1}{x_{i+1} - x_i}, & x \in [x_i, x_{i+1}], \\ 0 & \text{otherwise.} \end{cases} \tag{13.34}$$

It follows that a_{ij} is nonzero only for $j = i$, $j = i+1$, or $j = i-1$. Hence the matrix A is *tridiagonal*. It follows from (13.33) that A is *symmetric*: $a_{ij} = a_{ji}$ for all i, j.

If $r(x)$ in (13.33) is a simple enough function, then the integral there can be computed exactly. Otherwise it can be approximated using a quadrature formula, as described in chapter 10. The same holds for the integrals in (13.31) involving f.

Let us consider the simplest case in which $r(x) \equiv 0$. Assume that the nodes are equally spaced: $x_i - x_{i-1} \equiv h$ for all i. Then

$$a_{ii} = \int_0^1 (\varphi_i'(x))^2 \, dx = \int_{x_{i-1}}^{x_i} \left(\frac{1}{h}\right)^2 dx + \int_{x_i}^{x_{i+1}} \left(-\frac{1}{h}\right)^2 = \frac{2}{h}, \tag{13.35}$$

$$a_{i,i+1} = a_{i+1,i} = \int_0^1 \varphi_i'(x)\varphi_{i+1}'(x) \, dx = \int_{x_i}^{x_{i+1}} \left(-\frac{1}{h}\right)\left(\frac{1}{h}\right) dx = -\frac{1}{h}. \tag{13.36}$$

If we divide each side of (13.32) by h, the linear system becomes

$$\frac{1}{h^2}\begin{pmatrix} 2 & -1 & & \\ -1 & \ddots & \ddots & \\ & \ddots & \ddots & -1 \\ & & -1 & 2 \end{pmatrix}\begin{pmatrix} c_1 \\ c_2 \\ \vdots \\ c_{n-2} \\ c_{n-1} \end{pmatrix} = \frac{1}{h}\begin{pmatrix} \langle f, \varphi_1 \rangle \\ \langle f, \varphi_2 \rangle \\ \vdots \\ \langle f, \varphi_{n-2} \rangle \\ \langle f, \varphi_{n-1} \rangle \end{pmatrix}.$$

Note that the matrix on the left is the same one that arose in the centered finite difference scheme in (13.12). Finite element methods often turn out to be almost the same as finite difference formulas. Sometimes, however, what appear to be minor differences in the way boundary conditions are handled or in the way the right-hand side vector is formed turn out to be important for accuracy.

Having derived the equations for the continuous piecewise linear finite element approximation for the problem (13.26–13.27), it is easy to generalize to equation (13.1). Simply define the differential operator $\mathcal{L}$ appropriately and integrate by parts as before. For equation (13.1), the (i, j)-entry in the coefficient matrix in (13.32) would be

$$\begin{aligned} a_{ij} &= \langle \mathcal{L}\varphi_j, \varphi_i \rangle \\ &= \int_0^1 -\frac{d}{dx}\left(p(x)\frac{d\varphi_j}{dx}\right)\varphi_i(x)\, dx + \int_0^1 q(x)\varphi_j'(x)\varphi_i(x)\, dx \\ &\quad + \int_0^1 r(x)\varphi_j(x)\varphi_i(x)\, dx \\ &= -\varphi_i(x)p(x)\frac{d\varphi_j}{dx}\bigg|_0^1 + \int_0^1 p(x)\varphi_j'(x)\varphi_i'(x)\, dx + \int_0^1 q(x)\varphi_j'(x)\varphi_i(x)\, dx \\ &\quad + \int_0^1 r(x)\varphi_j(x)\varphi_i(x)\, dx \\ &= \int_0^1 p(x)\varphi_j'(x)\varphi_i'(x)\, dx + \int_0^1 q(x)\varphi_j'(x)\varphi_i(x)\, dx + \int_0^1 r(x)\varphi_j(x)\varphi_i(x)\, dx. \end{aligned}$$

This matrix is still tridiagonal. It is symmetric if $q(x) \equiv 0$.

Example 13.3.2. Suppose we wish to write a finite element code using continuous piecewise linear functions to solve the problem in Example 13.2.1: $-u''(x) + r(x)u(x) = f(x)$ on $[0, 1]$, where $r(x) = e^x$. Since we have discussed only homogeneous Dirichlet boundary conditions, let us modify the solution slightly from that used in the previous section, so that it satisfies $u(0) = u(1) = 0$, say, $u(x) = x^2 - x + \sin(\pi x)$. Then $u'(x) = 2x - 1 + \pi\cos(\pi x)$ and $u''(x) = 2 - \pi^2 \sin(\pi x)$, so the right-hand side function f is $f(x) = -2 + (\pi^2 + e^x)\sin(\pi x) + e^x(x^2 - x)$. The finite element matrix A will be tridiagonal and its nonzero entries will be like those in (13.35) and (13.36) but with the additional term

$$\int_0^1 e^x \varphi_j(x)\varphi_i(x)\,dx. \tag{13.37}$$

This integral might be difficult to compute analytically, but we can approximate it with a quadrature formula. The quadrature formula should have the same or higher-order accuracy than the finite element approximation if we are to expect that order of accuracy in the solution. Suppose, for example, that we use the second-order-accurate *midpoint rule* as the quadrature formula for each subinterval. If $x_{i\pm1/2}$ denotes the midpoint $(x_i + x_{i\pm1})/2$, then

$$\int_0^1 e^x \varphi_i^2(x)\,dx = \int_{x_{i-1}}^{x_i} e^x \varphi_i^2(x)\,dx + \int_{x_i}^{x_i+1} e^x \varphi_i^2(x)\,dx \approx \frac{h}{4}e^{x_{i-1/2}} + \frac{h}{4}e^{x_{i+1/2}},$$

$$\int_0^1 e^x \varphi_i(x)\varphi_{i-1}(x)\,dx = \int_{x_{i-1}}^{x_i} e^x \varphi_i(x)\varphi_{i-1}(x)\,dx \approx \frac{h}{4}e^{x_{i-1/2}},$$

$$\int_0^1 e^x \varphi_i(x)\varphi_{i+1}(x)\,dx = \int_{x_i}^{x_{i+1}} e^x \varphi_i(x)\varphi_{i+1}(x)\,dx \approx \frac{h}{4}e^{x_{i+1/2}}.$$

Likewise, we might use the midpoint rule to approximate the integrals involving the right-hand side function f.

$$\int_0^1 f(x)\varphi_i(x)\,dx = \int_{x_{i-1}}^{x_i} f(x)\varphi_i(x)\,dx + \int_{x_i}^{x_{i+1}} f(x)\varphi_i(x)\,dx$$
$$\approx \frac{h}{2} f(x_{i-1/2}) + \frac{h}{2} f(x_{i+1/2}).$$

Following is a finite element code that approximates all integrals over subintervals involving $r(x)$ and $f(x)$ using the midpoint rule.

```
% Define r and f via inline functions that can work on
% vectors of x values.
r = inline('exp(x)');
f = inline('-2 + (pi^2 + exp(x)).*sin(pi*x)+(exp(x).*(x.^2-x))');

% Boundary conditions are homogeneous Dirichlet: u(0)=u(1) = 0.

% Ask user for number of subintervals.
n = input('Enter no. of subintervals n: ');

% Set up the grid.
h = 1/n;
x = [h:h:(n-1)*h]';
```

```
% Form the matrix A.  Store it as a sparse matrix.
% First form the matrix corres to u'' then add in
% the terms for r(x)u.
A = sparse(diag(2*ones(n-1,1),0)+diag(-ones(n-2,1),1)
   +diag(-ones(n-2,1),-1));
A = (1/h)*A;
xmid = [x(1)/2; (x(1:n-2)+x(2:n-1))/2; (x(n-1)+1)/2];
% Midpoints of subints
A = A + (h/4)*diag(exp(xmid(1:n-1))+exp(xmid(2:n)),0);
A = A + (h/4)*diag(exp(xmid(2:n-1)),-1)
   + (h/4) *diag(exp(xmid(2:n-1)),1);

% Form right-hand side vector b.
fmid = f(xmid);
b = (h/2)*fmid(1:n-1) + (h/2)*fmid(2:n);

% Compute approximate solution.
u_approx = A\b;

% Compare approximate solution with true solution.
u_true = x.^2 - x + sin(pi*x);
err = max(abs(u_true - u_approx))
```

Using this code with $n = 10, 20, 40$, and 80, we obtained the values 0.0022, 5.5×10^{-4}, 1.4×10^{-4}, and 3.4×10^{-5}, respectively, for the ∞-norm of the error. These values would probably be somewhat smaller if we had used a higher-order-accurate approximation to the integrals, such as Simpson's rule. Still, the order of accuracy is $O(h^2)$.

13.3.1 Accuracy

One of the great advantages of finite element methods is that they sometimes permit easy proofs of the order of accuracy; this is because the finite element approximation is *optimal* in a certain sense among all possible approximations from the trial space S.

When the operator $\mathcal{L}$ is *self-adjoint* (i.e., $\langle \mathcal{L}v, w\rangle = \langle v, \mathcal{L}w\rangle$ for all v and w) and *positive definite* (i.e., $\langle \mathcal{L}v, v\rangle > 0$ for all $v \neq 0$), one might choose an approximate solution $\tilde{u} \in S$ to minimize the *energy norm* of the error:

$$\langle \tilde{u} - \mathcal{L}^{-1} f, \mathcal{L}(\tilde{u} - \mathcal{L}^{-1} f)\rangle = \langle \mathcal{L}\tilde{u}, \tilde{u}\rangle - 2\langle f, \tilde{u}\rangle + \text{constant}. \tag{13.38}$$

Because $\mathcal{L}$ is self-adjoint and positive definite, the expression in (13.38) is a norm; it often represents the energy in the system, and sometimes the original problem is stated in this form. Minimizing this expression leads to the *Ritz* finite element approximation.

Note that while we cannot compute $\mathcal{L}^{-1} f$ in (13.38) (since this is the solution that we are seeking), we can compute the expression involving $\tilde{u}$ on the right-hand side, and this expression differs only by a constant (independent of $\tilde{u}$) from the quantity on the left. Hence it is equivalent to choose $\tilde{u} \in S$ to minimize $\langle \mathcal{L}\tilde{u}, \tilde{u} \rangle - 2\langle f, \tilde{u} \rangle$. For the differential operator in (13.19), this is

$$\int_0^1 \left(-\frac{d}{dx}(p(x)\tilde{u}'(x)) + r(x)\tilde{u}(x) \right) \tilde{u}(x)\, dx - 2 \int_0^1 f(x)\tilde{u}(x)\, dx,$$

and after integrating by parts and using the fact that $\tilde{u}(0) = \tilde{u}(1) = 0$, this becomes

$$\int_0^1 \left(p(x)(\tilde{u}'(x))^2 + r(x)(\tilde{u}(x))^2 \right) dx - 2 \int_0^1 f(x)\tilde{u}(x)\, dx.$$

Expressing $\tilde{u}$ as $\sum_{j=1}^{n-1} d_j \varphi_j$, we obtain a quadratic function in $d_1, \ldots, d_{n-1}$:

$$\min_{d_1,\ldots,d_{n-1}} \int_0^1 \left[p(x) \left(\sum_{j=1}^{n-1} d_j \varphi_j'(x) \right)^2 + r(x) \left(\sum_{j=1}^{n-1} d_j \varphi_j(x) \right)^2 \right] dx$$

$$-2 \int_0^1 f(x) \left(\sum_{j=1}^{n-1} d_j \varphi_j(x) \right) dx.$$

The minimum of this function occurs where its derivative with respect to each of the variables d_i is zero. Differentiating with respect to each d_i and setting these derivatives to zero, we obtain the system of linear equations:

$$2 \int_0^1 p(x) \left(\sum_{j=1}^{n-1} d_j \varphi_j'(x) \right) \varphi_i'(x)\, dx + 2 \int_0^1 r(x) \left(\sum_{j=1}^{n-1} d_j \varphi_j(x) \right) \varphi_i(x)\, dx$$

$$-2 \int_0^1 f(x)\varphi_i(x)\, dx = 0, \quad i = 1, \ldots, n-1,$$

or, equivalently,

$$\sum_{j=1}^{n-1} d_j \langle \mathcal{L}\varphi_j, \varphi_i \rangle = \langle f, \varphi_i \rangle, \quad i = 1, \ldots, n-1.$$

These are exactly the same equations that define the *Galerkin* finite element approximation! This is no accident; when $\mathcal{L}$ is self-adjoint and positive definite, the Ritz and Galerkin approximations are the same.

Theorem 13.3.1. If $\mathcal{L}$ is self-adjoint and positive definite, then the Galerkin approximation is the same as the Ritz approximation for the problem $\mathcal{L}u = f$.

Proof. Suppose $\tilde{u}$ minimizes

$$\mathcal{I}(v) \equiv \langle v - \mathcal{L}^{-1} f, \mathcal{L}(v - \mathcal{L}^{-1} f) \rangle$$

over all functions $v \in S$. For any function v in this space and any number ϵ, we can write

$$\mathcal{I}(\tilde{u}+\epsilon v) = \langle \tilde{u} - \mathcal{L}^{-1} f + \epsilon v, \mathcal{L}(\tilde{u} - \mathcal{L}^{-1} f + \epsilon v)\rangle$$
$$= \mathcal{I}(\tilde{u}) + 2\epsilon\langle \mathcal{L}\tilde{u} - f, v\rangle + \epsilon^2\langle \mathcal{L}v, v\rangle. \tag{13.39}$$

The only way that this can be greater than or equal to $\mathcal{I}(\tilde{u})$ for *all* ϵ (i.e., for ϵ small enough in absolute value so that the ϵ^2 term is negligible and for ϵ either positive or negative) is for the coefficient of ϵ to be zero; that is, $\langle \mathcal{L}\tilde{u} - f, v\rangle = 0$ for all $v \in S$.

Conversely, if $\langle \mathcal{L}\tilde{u} - f, v\rangle = 0$ for all $v \in S$, then expression (13.39) is always greater than or equal to $\mathcal{I}(\tilde{u})$, so $\tilde{u}$ minimizes $\mathcal{I}(v)$. □

This theorem is important because it means that of all functions in the trial space S, the finite element approximation is *best* as far as minimizing the energy norm of the error. Now one can derive results about the global error (in the energy norm) just by determining how well $u(x)$ can be approximated by a function from S; for example, by a continuous piecewise linear function, say, its piecewise linear *interpolant*. Other trial spaces are possible too, such as continuous piecewise quadratics or Hermite cubics or cubic splines. Chapter 8 dealt with the errors in approximating a function by its piecewise polynomial interpolant, so results from that chapter can be used to bound the error in the finite element approximation from a piecewise polynomial trial space.

The following theorem, for example, bounds the L_2-norm of the error in the piecewise polynomial interpolant of a function. While the finite element approximation is not optimal in the L_2-norm, it can be shown that its *order* of accuracy in the L_2-norm is the same as that of the optimal approximation from the space; hence its order of accuracy is at least as good as that of the interpolant of u.

Theorem 13.3.2. Let u_I be the $(k-1)$st-degree piecewise polynomial interpolant of u. (That is, u_I matches u at each of the nodes $x_0, x_1, \ldots, x_{n-1}, x_n$, and, if $k > 2$, u_I also matches u at $k-2$ points equally spaced within each subinterval.) Then

$$\|u - u_I\|_{L_2} \le Ch^k \|u^{(k)}\|_{L_2},$$

where $h \equiv max_i |x_{i+1} - x_i|$ and $\|v\|_{L_2} \equiv [\int_0^1 (v(x))^2\, dx]^{1/2}$.

For piecewise linear functions, $k = 2$ in the theorem, and the theorem tells us that the L_2-norm of the error in the piecewise linear interpolant is $O(h^2)$. This is the order of the L_2-norm of the error in the piecewise linear finite element approximation as well.

13.4 SPECTRAL METHODS

Spectral methods have much in common with finite element methods. The difference is that instead of approximating the solution by a *piecewise polynomial*, one approximates the solution by a polynomial of degree equal to the number of nodes minus one, or, by a linear combination of orthogonal basis functions for this space, such as the Chebyshev polynomials defined in section 8.5. The advantage is that, under suitable smoothness conditions, these methods converge to the solution faster than any power of n^{-1}, where n is the number of nodes or basis functions used in the representation. See, for example, theorem 8.5.1 in section 8.5.

As with finite element methods, the approximate solution can be chosen in several different ways. Here we consider a *spectral collocation method*, sometimes called a *pseudospectral method*, in which the approximate solution $u(x)$ is chosen so that $\mathcal{L}u(x_j) = f(x_j)$ at the Chebyshev points x_j defined in (8.15) but scaled and shifted to the interval on which the problem is defined.

Again consider the general linear two-point boundary value problem (13.1–13.2). Suppose that we seek an approximate solution $\hat{u}$ of the form

$$\hat{u}(x) = \sum_{j=0}^{n} c_j T_j(x),$$

where $c_0, \ldots, c_n$ are unknown coefficients and the functions T_j are Chebyshev polynomials but now defined on the interval $[0, 1]$: $T_j(x) = \cos(j \arccos t)$, where $t = 2x - 1$. If we require that the boundary conditions (13.2) hold,

$$\sum_{j=0}^{n} c_j T_j(0) = \alpha, \quad \sum_{j=0}^{n} c_j T_j(1) = \beta,$$

and that the differential equation (13.1) holds at $n - 1$ points,

$$-\sum_{j=0}^{n} c_j \left. \frac{d}{dx} \left(p(x) T_j'(x)\right) \right|_{x_i} + q(x_i) \sum_{j=0}^{n} c_j T_j'(x_i) + r(x_i) \sum_{j=0}^{n} c_j T_j(x_i) = f(x_i),$$
$$i = 1, \ldots, n-1,$$

then we obtain a system of $n + 1$ linear equations for the $n + 1$ unknowns $c_0, \ldots, c_n$. It is no longer *sparse*, as it was for finite difference and finite element approximations; in general the coefficient matrix A will have *all* nonzero entries. Still, it turns out that there are relatively efficient ways to handle such problems, and these are implemented in the MATLAB package chebfun that was introduced in section 8.5.

Example 13.4.1. The following code solves the same example problem that was presented in section 13.3, using a spectral collocation method in the chebfun package.

```
[d,x] = domain(0,1);              % Specify domain as [0,1].
L = -diff(d,2) + diag(exp(x)); % Create differential operator -u''+exp(x)u.
L.lbc = 0; L.rbc = 0;             % Specify homogeneous Dirichlet bcs.
f = chebfun('-2 + (pi^2 + exp(x)).*sin(pi*x) + exp(x).*(x.^2 - x)',d)
                                  % Create right-hand side function.
u = L\f                           % Solve the differential equation using
                                  % same notation as for solving a linear system

u_true = chebfun('x.^2 - x + sin(pi*x)',d)
err = max(abs(u_true - u))
```

First the domain of the problem is specified as [0, 1]. Then the differential operator `L` is formed; the first term is the negative of the second derivative operator and the second represents multiplication by e^x. Homogeneous Dirichlet boundary conditions are specified on the left and right ends of the domain. Then the right-hand side function `f` is defined. The package informs us that the length of `f` is 17, meaning that it has used a polynomial of degree 16 to represent $f(x)$ by interpolating this function at 17 Chebyshev points. To solve the boundary value problem, one uses the same notation as for solving a system of linear algebraic equations `L\f`. Finally, comparing the true and approximate solutions, we find that the maximum absolute difference is 9.3×10^{-15}. This is far more accurate than the finite difference or finite element solutions that we obtained. It requires working with a dense matrix, but for a smooth problem like this the matrix is only 17 by 17, and the solution is returned very quickly. The spectral accuracy also relies on the smoothness of the solution. Had the right-hand side function or other functions in the differential equation (13.1) been discontinuous, then a modified procedure would have been required.

13.5 CHAPTER 13 EXERCISES

1. Consider the two-point boundary value problem

$$u'' + 2xu' - x^2 u = x^2, \quad u(0) = 1, \quad u(1) = 0.$$

   (a) Let $h = 1/4$ and explicitly write out the difference equations, using centered differences for all derivatives.

   (b) Repeat part (a) using the one-sided approximation

$$u'(x_i) \approx \frac{u(x_i) - u(x_{i-1})}{h}.$$

   (c) Repeat part (a) for the boundary conditions $u'(0) + u(0) = 1$, $u'(1) + \frac{1}{2}u(1) = 0$.

2. A rod of length 1 meter has a heat source applied to it and it eventually reaches a steady state where the temperature is not changing. The conductivity of the rod is a function of position x and is given by $c(x) = 1 + x^2$. The left end of the rod is held at a constant temperature of 1 degree. The right end of the rod is insulated so that no heat flows in or out from that end of the rod. This problem is described by the boundary value problem:

$$\frac{d}{dx}\left((1 + x^2)\frac{du}{dx}\right) = f(x), \quad 0 \le x \le 1,$$

$$u(0) = 1, \quad u'(1) = 0.$$

   (a) Write down a set of difference equations for this problem. Be sure to show how you do the differencing at the endpoints.

(b) Write a MATLAB code to solve the difference equations. You can test your code on a problem where you know the solution by choosing a function $u(x)$ that satisfies the boundary conditions and determining what $f(x)$ must be in order for $u(x)$ to solve the problem. Try $u(x) = (1-x)^2$. Then $f(x) = 2(3x^2 - 2x + 1)$.

(c) Try several different values for the mesh size h. Based on your results, what would you say is the order of accuracy of your method?

3. Show that the vectors $\mathbf{v}^{(k)}$ in (13.14) are orthonormal; that is, show that

$$\sum_{\ell=1}^{m} v_\ell^{(k)} v_\ell^{(j)} = \begin{cases} 1 & \text{if } j = k, \\ 0 & \text{otherwise.} \end{cases}$$

4. Show that the matrix in (13.24) is similar to a symmetric matrix. [Hint: Let $D = \text{diag}(1, \ldots, 1, \sqrt{2})$. What is $D^{-1}AD$, if A is the matrix in (13.24)?]
5. Use the Galerkin finite element method with continuous piecewise linear basis functions to solve the problem of Exercise 2, but with homogeneous Dirichlet boundary conditions:

$$\frac{d}{dx}\left((1+x^2)\frac{du}{dx}\right) = f(x), \quad 0 \le x \le 1,$$

$$u(0) = 0, \quad u(1) = 0.$$

(a) Derive the matrix equation that you will need to solve for this problem.

(b) Write a MATLAB code to solve this set of equations. You can test your code on a problem where you know the solution by choosing a function $u(x)$ that satisfies the boundary conditions and determining what $f(x)$ must be in order for $u(x)$ to solve the problem. Try $u(x) = x(1-x)$. Then $f(x) = -2(3x^2 - x + 1)$.

(c) Try several different values for the mesh size h. Based on your results, what would you say is the order of accuracy of the Galerkin method with continuous piecewise linear basis functions?

6. As noted in the text, *collocation methods* choose an approximate solution from a space S so that the given differential equation holds at certain points in the domain. Consider the two-point boundary value problem

$$u''(x) = u(x) + x^2, \quad 0 \le x \le 1,$$

$$u(0) = u(1) = 0.$$

(a) Consider the basis functions $\phi_j(x) = \sin(j\pi x)$, $j = 1, 2, 3$, and the collocation points $x_i = i/4$, $i = 1, 2, 3$. Suppose $u(x)$ is to be approximated by a linear combination of these basis functions: $u(x) \approx \sum_{j=1}^{3} c_j \phi_j(x)$, and suppose the coefficients are to be chosen so that the differential equation holds at the collocation points. Write down the

system of linear equations that you would need to solve to determine c_1, c_2, and c_3.
There is a problem with this method. Do you see what it is? Not all collocation methods can be used with all trial spaces.

(b) Repeat part (a) with $\phi_j(x) = x^j(1-x)$, $j = 1, 2, 3$. Does this have the same problem that you found in part (a)?

(c) Solve this problem using the `chebfun` package. What degree polynomial does it use to approximate $u(x)$ and what collocation points does it use? Check the accuracy of your solution by computing $\max_{x\in[0,1]} |u''(x) - u(x) - x^2|$. Is it close to the machine precision?

14

NUMERICAL SOLUTION PARTIAL DIFFERENTIAL EQUATIONS

Previous chapters have dealt with differential equations in a single independent variable. Chapter 11 dealt with time-dependent problems, while chapter 13 dealt with problems having a single spatial variable. In this chapter we will combine the two and deal with problems whose solutions depend on both time and space and/or on more than one spatial variable. These problems will be described by *partial differential equations*.

A general second-order linear partial differential equation in two independent variables x and y can be written in the form

$$au_{xx} + 2bu_{xy} + cu_{yy} + du_x + eu_y + fu = g, \tag{14.1}$$

where the coefficients a, b, c, d, e, f, and g are given functions of x and y. The equation is classified as one of three types, based on the *discriminant*, $D \equiv b^2 - ac$: If $D < 0$, the equation is **elliptic**; if $D = 0$, it is **parabolic**; and if $D > 0$, it is **hyperbolic**. Sometimes problems are of **mixed** type because the discriminant has one sign in one part of the domain and a different sign in another part. Note that the lower-order terms (those involving u_x, u_y, and u) are ignored in this classification.

The prototype of an elliptic equation is **Poisson's equation**:

$$u_{xx} + u_{yy} = g. \tag{14.2}$$

(Here $b = 0$ and $a = c = 1$, so that $D \equiv b^2 - ac = -1$.) In the special case where $g \equiv 0$, this is called **Laplace's equation**. An elliptic equation with constant coefficients can be transformed to Poisson's equation by a linear change of variables. Elliptic equations usually define stationary solutions that minimize energy.

The prototype of a parabolic equation is the *heat equation* (also called the *diffusion equation*):

$$u_t = u_{xx}. \tag{14.3}$$

SIMÉON-DENIS POISSON

Siméon-Denis Poisson (1781–1840) drew the attention of his teachers, Laplace and Lagrange, for his mathematical abilities as a student at the École Polytechnique. Legendre's attention was caught by the 18-year-old's memoir on finite differences. His rigorous studies in Paris started a productive career that led to between 300 and 400 mathematical works. The breadth of his accomplishments is reflected in the many areas in which his name appears: Poisson's integral, Poisson's equation in potential theory, Poisson brackets in differential equations, Poisson's ratio in elasticity, and Poisson's constant in electricity, to name just a few.

(Here we have changed the variable name y to t because it usually represents time. In the form (14.1), writing the equation as $-u_{xx} + u_t = 0$, we have $a = -1$ and $b = c = 0$, so that $D = 0$.) This equation was described in section 13.1, where it was noted that $u(x, t)$ might represent the temperature in a rod at position x and time t. If there are two independent spatial variables x and y, in addition to time t, the heat equation becomes

$$u_t = u_{xx} + u_{yy}.$$

Parabolic equations usually describe smooth, spreading flows, such as the flow of heat through a rod (in one spatial dimension) or through, say, an electric blanket (in two spatial dimensions). A parabolic equation might also describe the diffusion of a liquid through a porous medium.

The prototype of a hyperbolic equation is the *wave equation*:

$$u_{tt} = u_{xx}, \tag{14.4}$$

(We have again used t instead of y, since this variable usually represents time. In the notation of (14.1), writing the equation as $-u_{xx} + u_{tt} = 0$, we have $a = -1$, $c = 1$, and $b = 0$, so that $D = 1$.) Hyperbolic equations usually describe the motion of a disturbance-preserving wave as, for example, in acoustics.

One classical example of a hyperbolic problem is a vibrating string, such as a plucked guitar string. Assume that the string is held fixed at $x = 0$ and $x = 1$. Let $u(x, t)$ denote the displacement of the string at point x and time t, and assume that the displacement is in the vertical direction only. Let $T(x, t)$ denote the tension in the string (i.e., the force that the part of the string to the right of x exerts on the part to the left of x, at time t). Assume that T is tangential to the string and has magnitude proportional to the local stretching factor $\sqrt{1 + u_x^2}$, as pictured in figure 14.1. [Note that if the point $(x, 0)$ on the unplucked string moves to position $(x, u(x, t))$ when the string is plucked, and the point $(x+h, 0)$ on the unplucked string moves to position $(x + h, u(x + h, t))$, then the length

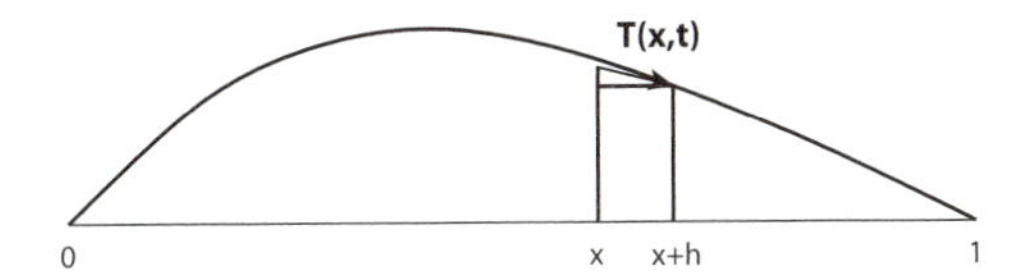

Figure 14.1. A vibrating string and $T(x, t)$, the tension of the string, which is tangential to the string.

of the segment connecting the new points is $\sqrt{h^2 + (u(x+h, t) - u(x, t))^2} = h\sqrt{1 + ((u(x+h, t) - u(x, t))/h)^2}$. The ratio of this length to that of the original segment is $\sqrt{1 + ((u(x+h, t) - u(x, t))/h)^2}$, which approaches $\sqrt{1 + u_x^2}$ as $h \to 0$. This quantity is referred to as the *local stretching factor*.] If τ is the tension in the equilibrium position along the horizontal axis, then the magnitude of $T(x, t)$ is $\tau\sqrt{1 + u_x(x, t)^2}$ and the vertical component of $T(x, t)$ is $\tau u_x(x, t)$. Now consider the part of the string between x and $x + h$. The vertical component of the force applied to this section of the string is $\tau u_x(x+h, t) - \tau u_x(x, t)$, while the mass of this section of string is ρh, where ρ is the density. The acceleration is $u_{tt}(x, t)$. Hence by Newton's second law ($F = ma$, force equals mass times acceleration) applied to this section of string,

$$\tau u_x(x+h, t) - \tau u_x(x, t) = \rho h u_{tt}(x, t).$$

Dividing each side by h and taking the limit as $h \to 0$, gives the equation

$$\tau u_{xx} = \rho u_{tt},$$

and normalizing units so that $\rho = \tau$, we obtain the wave equation (14.4).

14.1 ELLIPTIC EQUATIONS

We start with a study of time-independent elliptic equations such as Poisson's equation in two space dimensions:

$$u_{xx} + u_{yy} = f(x, y), \quad (x, y) \in \Omega, \tag{14.5}$$

where Ω is a region in the plane. To uniquely define a solution, we need boundary values, say,

$$u(x, y) = g(x, y), \quad (x, y) \in \partial\Omega, \tag{14.6}$$

where $\partial\Omega$ denotes the boundary of Ω.

14.1.1 Finite Difference Methods

Assume, for simplicity, that Ω is the unit square $[0, 1] \times [0, 1]$. Divide the interval from $x = 0$ to $x = 1$ into pieces of width $h_x = 1/n_x$, and divide the interval from $y = 0$ to $y = 1$ into pieces of width $h_y = 1/n_y$, as in figure 14.2. Let u_{ij} denote the approximate value of the solution at grid point $(x_i, y_j) \equiv (ih_x, jh_y)$. Approximate the second derivatives u_{xx} and u_{yy} by the

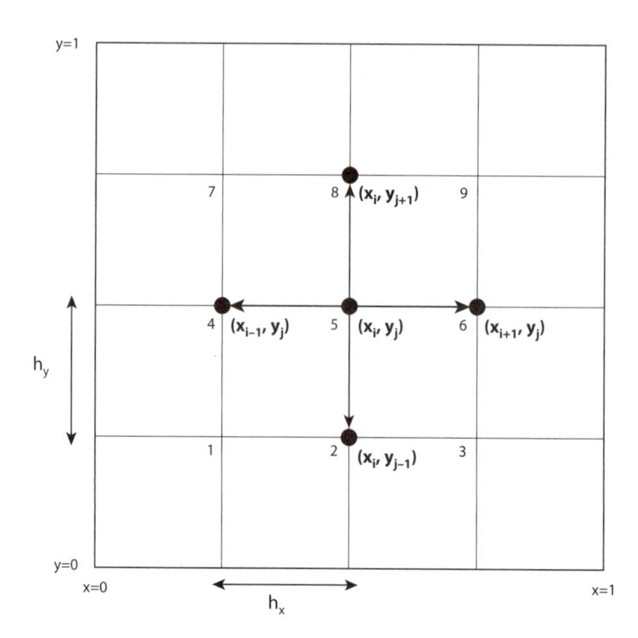

Figure 14.2. Five-point finite difference operator for Poisson's equation, natural ordering of grid points 1 through 9.

finite difference quotients

$$u_{xx}(x_i, y_j) \approx \frac{-2u_{ij} + u_{i+1,j} + u_{i-1,j}}{h_x^2}, \quad u_{yy}(x_i, y_j) \approx \frac{-2u_{ij} + u_{i,j+1} + u_{i,j-1}}{h_y^2}. \tag{14.7}$$

There are $(n_x - 1)(n_y - 1)$ unknown values u_{ij}, $i = 1, \ldots, n_x - 1$, $j = 1, \ldots, n_y - 1$, since the boundary values $u_{0,j} = g(0, y_j)$, $u_{n_x,j} = g(1, y_j)$, $u_{i,0} = g(x_i, 0)$, and $u_{i,n_y} = g(x_i, 1)$ are given. If we substitute expressions (14.7) into equation (14.5) and require that the resulting difference equations hold at the interior nodes, then we obtain $(n_x - 1)(n_y - 1)$ linear equations for the unknown values u_{ij}:

$$\frac{-2u_{ij} + u_{i+1,j} + u_{i-1,j}}{h_x^2} + \frac{-2u_{ij} + u_{i,j+1} + u_{i,j-1}}{h_y^2} = f(x_i, y_j),$$
$$i = 1, \ldots, n_x - 1, \quad j = 1, \ldots, n_y - 1.$$

The finite difference operator on the left is sometimes called a *five-point* operator, because it couples u_{ij} to its four neighbors $u_{i+1,j}$, $u_{i-1,j}$, $u_{i,j+1}$, and $u_{i,j-1}$. It is represented schematically in figure 14.2. In the special case where $n_x = n_y \equiv n$ and hence $h_x = h_y \equiv 1/n$, these equations become

$$\frac{-4u_{ij} + u_{i+1,j} + u_{i-1,j} + u_{i,j+1} + u_{i,j-1}}{h^2} = f(x_i, y_j), \quad i, j = 1, \ldots, n - 1. \tag{14.8}$$

We can assemble these equations into a matrix equation, but the exact form of the matrix will depend on the ordering of equations and unknowns. Using the **natural ordering**, equations and unknowns are ordered from left to right across each row of the grid, with those in the bottom row numbered first, followed by those in the next row above, etc., as indicated in figure 14.2. Thus, point (i, j) in the grid has index $k = i + (j-1)(n_x - 1)$ in the set of equations. Using this ordering, the matrix equation for (14.8) takes the form

$$\frac{1}{h^2}\begin{pmatrix} T & I & & & \\ I & T & I & & \\ & \ddots & \ddots & \ddots & \\ & & I & T & I \\ & & & I & T \end{pmatrix}\begin{pmatrix} \mathbf{u}_1 \\ \mathbf{u}_2 \\ \vdots \\ \mathbf{u}_{n-2} \\ \mathbf{u}_{n-1} \end{pmatrix} = \begin{pmatrix} \mathbf{f}_1 - (1/h^2)\mathbf{u}_0 \\ \mathbf{f}_2 \\ \vdots \\ \mathbf{f}_{n-2} \\ \mathbf{f}_{n-1} - (1/h^2)\mathbf{u}_n \end{pmatrix}, \tag{14.9}$$

where T is the $(n-1)$ by $(n-1)$ tridiagonal matrix

$$T = \begin{pmatrix} -4 & 1 & & \\ 1 & \ddots & \ddots & \\ & \ddots & \ddots & 1 \\ & & 1 & -4 \end{pmatrix},$$

I is the $(n-1)$ by $(n-1)$ identity matrix, and

$$\mathbf{u}_j = \begin{pmatrix} u_{1,j} \\ u_{2,j} \\ \vdots \\ u_{n-2,j} \\ u_{n-1,j} \end{pmatrix}, \quad \mathbf{f}_j = \begin{pmatrix} f(x_1, y_j) - (1/h^2)g(0, y_j) \\ f(x_2, y_j) \\ \vdots \\ f(x_{n-2}, y_j) \\ f(x_{n-1}, y_j) - (1/h^2)g(1, y_j) \end{pmatrix}, \quad j = 1, \ldots, n-1,$$

$$\mathbf{u}_0 = \begin{pmatrix} g(x_1, 0) \\ \vdots \\ g(x_{n-1}, 0) \end{pmatrix}, \quad \mathbf{u}_n = \begin{pmatrix} g(x_1, 1) \\ \vdots \\ g(x_{n-1}, 1) \end{pmatrix}.$$

This matrix is *block tridiagonal* (because only the main diagonal blocks and the first sub- and super-diagonal blocks are nonzero), and it is symmetric. It is also said to be *block Toeplitz* because the blocks along each block diagonal are the same. Note, however, that while the individual blocks are Toeplitz, the matrix as a whole is not, because the first (elementwise) sub- and super-diagonals contain 1s except in positions that are multiples of $(n-1)$, where they contain 0s.

The matrix in (14.9) is *banded* with half bandwidth $n-1$. It was shown in section 7.2.4 that the work required to perform Gaussian elimination without pivoting on a banded matrix of size N with half bandwidth m is about $2m^2 N$ operations. During the process of Gaussian elimination, zeros inside the band fill in with nonzeros which must be stored. Hence the amount of storage required is about mN words. For equation (14.9), $N = (n-1)^2$ and $m = n-1$, so the total work required to solve the linear system using Gaussian elimination is about $2(n-1)^4$ operations, and $(n-1)^3$ words of storage are needed. We saw in section 12.2 that iterative methods could be used to solve such linear systems without storing the matrix or even the part inside the band. Since this matrix is

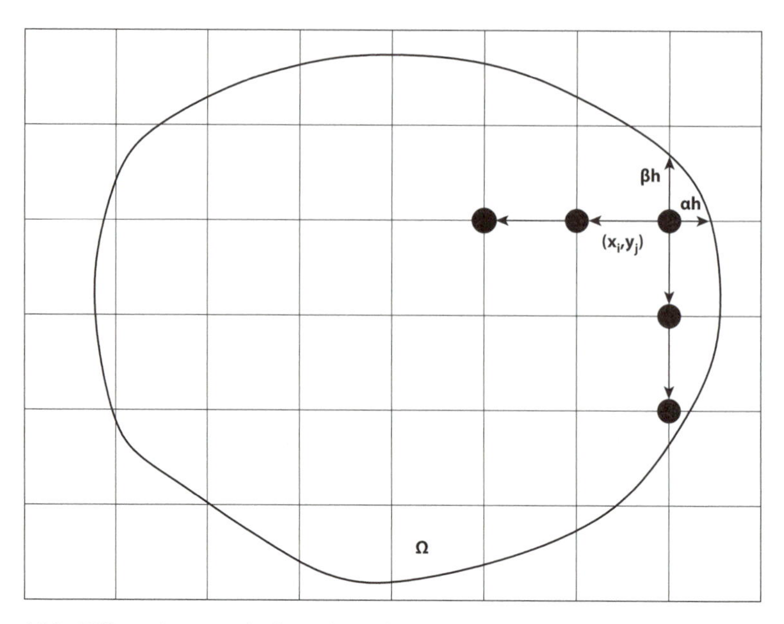

Figure 14.3. Differencing near the boundary of a general region Ω.

symmetric and negative definite (so that multiplying each side of the equation by -1 yields a positive definite matrix), the conjugate gradient method, with an appropriate preconditioner, would probably be the method of choice. We also will see that the particular equations in (14.9) can be solved very rapidly using a *fast Poisson solver*.

As in the case of two-point boundary value problems, one can handle more general equations and more general boundary conditions using finite differences in two spatial dimensions. One also can deal with more general regions Ω, but this may become complicated. Consider, for example, the region Ω pictured in figure 14.3.

Suppose we wish to solve equation (14.5) on this domain, with the same Dirichlet boundary conditions (14.6). Consider the difference equations at the point (x_i, y_j) in the figure. Since (x_{i+1}, y_j) lies outside the domain, we cannot use it in our approximation of u_{xx} at (x_i, y_j). Instead, we would like to use the given boundary value, which lies a distance, say, αh to the right of x_i along the grid line $y = y_j$. Again we will use Taylor series to try and derive a second-order-accurate approximation to u_{xx} at (x_i, y_j). Since all functions in our expansion will be evaluated at $y = y_j$, we will simplify the notation by omitting the second argument. First expanding $u(x_{i-1})$ about the point (x_i, y_j) gives

$$u(x_{i-1}) = u(x_i) - hu_x(x_i) + \frac{h^2}{2!}u_{xx}(x_i) - \frac{h^3}{3!}u_{xxx}(x_i) + O(h^4). \tag{14.10}$$

Now expanding $u(x_i + \alpha h)$ about the point (x_i, y_j) gives

$$u(x_i + \alpha h) = u(x_i) + \alpha h u_x(x_i) + \frac{(\alpha h)^2}{2!}u_{xx}(x_i) + \frac{(\alpha h)^3}{3!}u_{xxx}(x_i) + O(h^4). \tag{14.11}$$

Adding α times equation (14.10) to equation (14.11), we find

$$\alpha u(x_{i-1}) + u(x_i + \alpha h) = (\alpha + 1)u(x_i) + \frac{\alpha(1+\alpha)h^2}{2!}u_{xx}(x_i) - \frac{\alpha(1-\alpha^2)h^3}{3!}u_{xxx}(x_i) + O(h^4),$$

and solving for $u_{xx}(x_i)$ gives

$$u_{xx}(x_i) = 2\frac{\alpha u(x_{i-1}) + u(x_i + \alpha h) - (\alpha + 1)u(x_i)}{\alpha(1+\alpha)h^2} + \frac{(1-\alpha)h}{3}u_{xxx}(x_i) + O(h^2).$$

Thus, while one can approximate u_{xx} at (x_i, y_j) by using the grid values u_{ij} and $u_{i-1,j}$ and the boundary value $u(x_i+\alpha h, y_j) = g(x_i+\alpha h, y_j)$, the approximation is only first-order accurate; the error is $((1-\alpha)h/3)u_{xxx}(x_i)$ plus higher-order terms in h.

In order to maintain a second-order local truncation error, one must use a difference formula that involves an additional point, say, $u_{i-2,j}$. Expanding $u(x_{i-2}, y_j)$ in a Taylor series about (x_i, y_j) and again dropping the second argument from the expression since it is always y_j, we find

$$u(x_{i-2}) = u(x_i) - 2hu_x(x_i) + \frac{(2h)^2}{2!}u_{xx}(x_i) - \frac{(2h)^3}{3!}u_{xxx}(x_i) + O(h^4). \quad (14.12)$$

Let A, B, and C be coefficients to be determined and add A times equation (14.10) and B times equation (14.11) and C times equation (14.12):

$$Au(x_{i-1}) + Bu(x_i + \alpha h) + Cu(x_{i-2}) - (A+B+C)u(x_i) = h(-A+\alpha B-2C)u_x(x_i)$$

$$+ \frac{h^2(A+\alpha^2 B+4C)}{2!}u_{xx}(x_i) + \frac{h^3(-A+\alpha^3 B-8C)}{3!}u_{xxx}(x_i) + O(h^4).$$

In order for the left-hand side, divided by the coefficient of u_{xx} on the right-hand side, to be a second-order-accurate approximation to u_{xx}, we must have

$$-A+\alpha B-2C=0, \quad -A+\alpha^3 B-8C=0.$$

With a little algebra, one finds the following second-order-accurate approximation to u_{xx}:

$$u_{xx}(x_i) \approx \frac{1}{h^2}\left[\frac{\alpha-1}{\alpha+2}u(x_{i-2}) + \frac{2(2-\alpha)}{\alpha+1}u(x_{i-1}) + \frac{\alpha-3}{\alpha}u(x_i) + \frac{6}{\alpha(\alpha+1)(\alpha+2)}u(x_i+\alpha h)\right].$$

A similar procedure must be carried out in order to find a second-order-accurate approximation to u_{yy}, using values of u at (x_i, y_{j-2}), (x_i, y_{j-1}), (x_i, y_j), and $(x_i, y_j + \beta h)$, where βh is the distance from y_j to the boundary along the grid line $x = x_i$. Combining the two approximations, one obtains the difference equation at node (x_i, y_j).

14.1.2 Finite Element Methods

Finite element methods can be applied in two (or more) spatial dimensions as well. Here we just briefly describe the procedure for solving a differential equation $\mathcal{L}u = f$, such as (14.5), with homogeneous Dirichlet boundary conditions, $u(x, y) = 0$ on $\partial\Omega$.

One can approximate $u(x, y)$ by a function $\hat{u}(x, y)$ from a space S, called the trial space. We will choose $\hat{u}$ so that

$$\langle \mathcal{L}\hat{u}, \hat{v}\rangle = \langle f, \hat{v}\rangle$$

for all functions $\hat{v}$ in a space T, called the test space. Taking $T = S$ gives the **Galerkin** finite element approximation. Here the inner product $\langle\cdot,\cdot\rangle$ is defined by

$$\langle v, w\rangle \equiv \iint_\Omega v(x, y)w(x, y)\, dx\, dy.$$

When evaluating $\langle \mathcal{L}\hat{u}, \hat{v}\rangle$, we will use the two-dimensional analog of integration by parts, namely, *Green's theorem*:

$$\iint_\Omega (\hat{u}_{xx} + \hat{u}_{yy})\hat{v}\, dx\, dy = -\iint_\Omega (\hat{u}_x\hat{v}_x + \hat{u}_y\hat{v}_y)\, dx\, dy + \int_{\partial\Omega} \hat{u}_{\mathbf{n}}\hat{v}\, d\gamma.$$

Here $\hat{u}_{\mathbf{n}}$ denotes the derivative of $\hat{u}$ in the direction of the outward normal to the boundary $\partial\Omega$. Since we will be dealing with functions $\hat{v}$ that satisfy the homogeneous Dirichlet boundary condition $\hat{v} = 0$ on $\partial\Omega$, however, the boundary term will vanish.

As in the one-dimensional case, one must choose a trial space S and then find a basis $\varphi_1, \ldots, \varphi_N$ of S. One option is to take S to consist of continuous piecewise *bilinear* functions defined on rectangles. Suppose, for example, that Ω is the unit square $[0, 1] \times [0, 1]$. One can divide the square into rectangles of width $h_x = 1/n_x$ in the x direction and of width $h_y = 1/n_y$ in the y direction. The trial space would consist of functions that are continuous, satisfy the boundary conditions, and in each subrectangle have the form $a+bx+cy+dxy$ for some coefficients a, b, c, and d. A basis for this space can be formed by "pyramid" functions that are 1 at one node and 0 at all the others, much like the "hat" functions in one dimension. There is one such basis function for each interior node (x_i, y_j), and a sample basis function is pictured in figure 14.4.

Each basis function is nonzero in only four rectangles. The four parameters in its bilinear representation (within a single rectangle) are chosen to force it to have the desired values at each of the four corners of the rectangle. For example, the basis function $\varphi_{i+(j-1)(n_x-1)}$ associated with node (x_i, y_j) has the

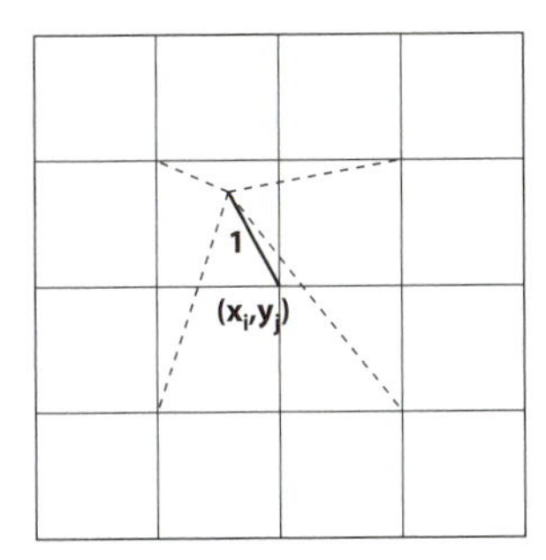

Figure 14.4. A sample bilinear basis function.

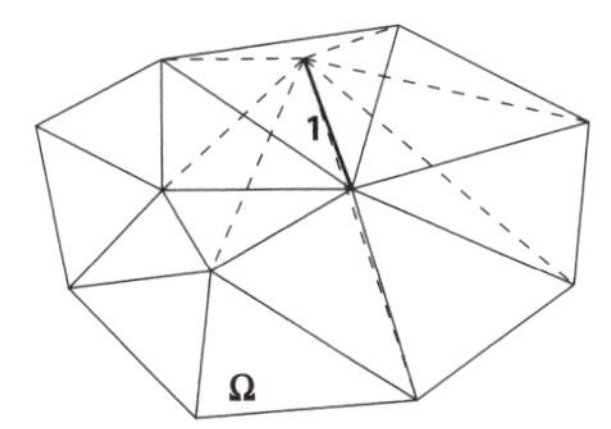

Figure 14.5. Triangle basis functions that are 1 at one node and 0 at all the others.

equation

$$\varphi_{i+(j-1)(n_x-1)}(x, y) = \begin{cases} \dfrac{(x - x_{i-1})(y - y_{j-1})}{(x_i - x_{i-1})(y_j - y_{j-1})} & \text{in } [x_{i-1}, x_i] \times [y_{j-1}, y_j], \\ \dfrac{(x - x_{i+1})(y - y_{j-1})}{(x_i - x_{i+1})(y_j - y_{j-1})} & \text{in } [x_i, x_{i+1}] \times [y_{j-1}, y_j], \\ \dfrac{(x - x_{i-1})(y - y_{j+1})}{(x_i - x_{i-1})(y_j - y_{j+1})} & \text{in } [x_{i-1}, x_i] \times [y_j, y_{j+1}], \\ \dfrac{(x - x_{i+1})(y - y_{j+1})}{(x_i - x_{i+1})(y_j - y_{j+1})} & \text{in } [x_i, x_{i+1}] \times [y_j, y_{j+1}], \\ 0 & \text{elsewhere.} \end{cases}$$

Note that because the bilinear basis function is zero along the entire edges corresponding to $x = x_{i-1}$, $x = x_{i+1}$, $y = y_{j-1}$, and $y = y_{j+1}$, it is continuous.

Another option is to use continuous piecewise *linear* functions defined on triangles. This has the advantage that a general region Ω can usually be better approximated by a union of triangles than by a union of similar sized rectangles. Again one can choose basis functions that are 1 at one node and 0 at all the others, as pictured in figure 14.5.

Within each triangle, a basis function has the form $a + bx + cy$, where the three parameters a, b, and c are chosen to give the correct values at the three nodes of the triangle. A basis function associated with node k is nonzero over all triangles that have node k as a vertex.

As in the one-dimensional case, one writes the approximate solution $\hat{u}$ as a linear combination of the basis functions, $\hat{u}(x, y) = \sum_{k=1}^{N} c_k \varphi_k(x, y)$, and then chooses the coefficients $c_1, \ldots, c_N$ so that

$$\langle \mathcal{L}\hat{u}, \varphi_\ell \rangle = \sum_{k=1}^{N} c_k \langle \mathcal{L}\varphi_k, \varphi_\ell \rangle = \langle f, \varphi_\ell \rangle, \quad \ell = 1, \ldots, N.$$

Redistributing the Earth's Heat

Image (a) shows annually-averaged near-surface air temperatures of the Earth from 1961 to 1990. These range from about −58°F (at the South Pole) to 85°F (along the equator). The plot in (b) is a contour map of the image in (a) so regions of the globe along a contour line in the figure have the same temperature. This data is used as the initial condition $u(x, y, 0)$ for the heat equation, defined not on the three-dimensional earth but on the two-dimensional projection shown below with homogeneous Neumann boundary conditions (meaning that no heat flows in or out but it simply redistributes itself). In (c), we see a contour plot of the solution at $t = 1$ and in (d) a contour plot of the solution at $t = 10$. (Image (a) created by Robert A. Rohde for Global Warming Art.)

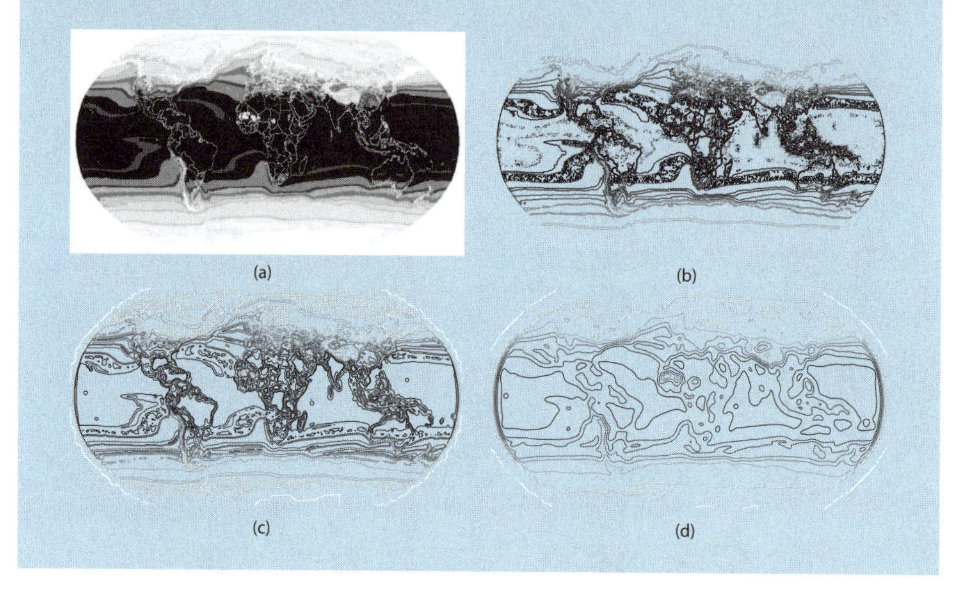

(The expression $\langle \mathcal{L}\varphi_k, \varphi_\ell \rangle$ is expressed using Green's theorem so that, instead of involving second derivatives, it involves only first derivatives $(\varphi_\ell)_x$, $(\varphi_k)_x$, $(\varphi_\ell)_y$ and $(\varphi_k)_y$.) This is a system of N equations (where N is the number of interior nodes in the grid) in the N unknown coefficients (which are actually the values of $\hat{u}$ at the interior nodes). It has the form $A\mathbf{c} = \mathbf{f}$, where $\mathbf{c} = (c_1, \ldots, c_N)^T$, $\mathbf{f} = (\langle f, \varphi_1 \rangle, \ldots, \langle f, \varphi_N \rangle)^T$, and $A_{\ell k} = \langle \mathcal{L}\varphi_k, \varphi_\ell \rangle$.

14.2 PARABOLIC EQUATIONS

We return to the time-dependent heat equation in one spatial dimension:

$$u_t = u_{xx}. \tag{14.13}$$

In order to uniquely specify a solution, one must have *initial conditions* (say, at time $t = 0$), as well as *boundary conditions* (at the endpoints of the interval in x, unless the equation holds in the infinite domain $x \in (-\infty, \infty)$). For example, if the equation holds for $x \in [0, 1]$ and for $t \geq 0$, then one might have the accompanying initial and boundary conditions,

$$u(x, 0) = q(x), \quad 0 \leq x \leq 1, \tag{14.14}$$

$$u(0, t) = \alpha, \quad u(1, t) = \beta, \quad t \geq 0, \tag{14.15}$$

where q is a given function of x and α and β are given constants. Such a problem is called an *initial boundary value problem* (IBVP).

14.2.1 Semidiscretization and the Method of Lines

If we divide the spatial interval $[0, 1]$ into subintervals of width $h = 1/n$ and let $u_1(t), \ldots, u_{n-1}(t)$ denote approximate solution values at each of the nodes $x_1, \ldots, x_{n-1}$ at time t, then we can replace the spatial derivative u_{xx} in (14.13) by a finite difference quotient, just as before:

$$u_{xx}(x_i, t) \approx \frac{-2u_i(t) + u_{i+1}(t) + u_{i-1}(t)}{h^2}.$$

Substituting this approximation into equation (14.13), we end up with a system of *ordinary differential equations* (ODEs) in time, for the approximate solution values $u_1(t), \ldots, u_{n-1}(t)$:

$$\frac{du_i}{dt} = \frac{-2u_i(t) + u_{i+1}(t) + u_{i-1}(t)}{h^2}, \quad i = 1, \ldots, n-1.$$

Letting $\mathbf{u}(t)$ denote the vector $(u_1(t), \ldots, u_{n-1}(t))^T$, this system of ODEs can be written in the form $\mathbf{u}' = A\mathbf{u} + \mathbf{b}$, where

$$A = \frac{1}{h^2} \begin{pmatrix} -2 & 1 & & & \\ 1 & -2 & 1 & & \\ & \ddots & \ddots & \ddots & \\ & & 1 & -2 & 1 \\ & & & 1 & -2 \end{pmatrix}, \quad \mathbf{b} = \frac{1}{h^2} \begin{pmatrix} \alpha \\ 0 \\ \vdots \\ 0 \\ \beta \end{pmatrix}.$$

Any of the methods discussed in chapter 11 for solving the initial value problem for ODEs can then be applied to solve this system of ODEs, with the initial value $\mathbf{u}(0) = (q(x_1), \ldots, q(x_{n-1}))^T$. The analysis of that section can be used to estimate the error in the numerically computed approximation to the solution of the ODEs, while one also must estimate the difference between the true solution of the ODEs and that of the original PDE problem. This approach of discretizing in space and then using an ODE solver in time is sometimes called the **method of lines**.

14.2.2 Discretization in Time

Let us consider some specific time discretizations. Just as the spatial domain was divided into subintervals of width h, let us divide the time domain into intervals of size Δt. We will then approximate u_t using a finite difference in time and let $u_i^{(k)}$ denote the approximate solution at $x_i = ih$ and $t_k = k\Delta t$.

Explicit Methods and Stability

Using a *forward difference* in time, we approximate $u_t(x,t)$ by $(u(x,t+\Delta t) - u(x,t))/\Delta t$, giving rise to the difference equations

$$\frac{u_i^{(k+1)} - u_i^{(k)}}{\Delta t} = \frac{-2u_i^{(k)} + u_{i+1}^{(k)} + u_{i-1}^{(k)}}{h^2}, \quad i = 1, \ldots, n-1, \quad k = 0, 1, \ldots. \tag{14.16}$$

Equivalently, expressing the value at the new time step in terms of the (already computed) values at the old time step, we have

$$u_i^{(k+1)} = u_i^{(k)} + \frac{\Delta t}{h^2}[-2u_i^{(k)} + u_{i+1}^{(k)} + u_{i-1}^{(k)}], \quad i = 1, \ldots, n-1, \tag{14.17}$$

where the boundary conditions (14.15) provide the values

$$u_0^{(k)} = \alpha, \quad u_n^{(k)} = \beta, \quad k = 0, 1, \ldots, \tag{14.18}$$

while the initial conditions (14.14) give the values

$$u_i^{(0)} = q(x_i), \quad i = 1, \ldots, n-1. \tag{14.19}$$

Thus formula (14.17) gives a procedure for advancing from one time step to the next: Starting with the values $u_i^{(0)}$, we use formula (14.17) with $k = 0$ to obtain the values $u_i^{(1)}$; knowing those values we can then compute the values $u_i^{(2)}$ at time step 2, etc.

Using Taylor's theorem, one can show that locally the method is second-order accurate in space and first order in time; that is, the *local truncation error* is $O(h^2) + O(\Delta t)$. To see this, substitute the true solution $u(x,t)$ into the difference equations (14.16) and look at the difference between the left- and right-hand sides:

$$\tau(x,t) \equiv \frac{u(x,t+\Delta t) - u(x,t)}{\Delta t} - \frac{-2u(x,t) + u(x+h,t) + u(x-h,t)}{h^2}.$$

Expanding $u(x, t+\Delta t)$ about (x,t) gives

$$u(x,t+\Delta t) = u(x,t) + (\Delta t)u_t(x,t) + O((\Delta t)^2),$$

and since $u_t = u_{xx}$, this becomes

$$u(x,t+\Delta t) = u(x,t) + (\Delta t)u_{xx}(x,t) + O((\Delta t)^2).$$

We have already seen that

$$u_{xx}(x,t) = \frac{-2u(x,t) + u(x+h,t) + u(x-h,t)}{h^2} + O(h^2),$$

and it follows that

$$u(x,t+\Delta t) = u(x,t) + (\Delta t)\left(\frac{-2u(x,t) + u(x+h,t) + u(x-h,t)}{h^2} + O(h^2)\right)$$
$$+O((\Delta t)^2),$$

or, subtracting $u(x, t)$ from each side and dividing by Δt,

$$\tau(x, t) = \frac{u(x, t + \Delta t) - u(x, t)}{\Delta t} - \frac{-2u(x, t) + u(x + h, t) + u(x - h, t)}{h^2}$$
$$= O(h^2) + O(\Delta t). \tag{14.20}$$

It follows from (14.20) that the local truncation error converges to 0 as h and Δt go to 0, but it does not follow that the *global error*—the difference between the true and approximate solution—converges to 0. To show that the approximate solution converges to the true solution on some time interval $[0, T]$ is more difficult and usually requires additional conditions on the relation between h and Δt. The condition that $\tau(x, t) \to 0$ as $h \to 0$ and $\Delta t \to 0$ is a *necessary* condition for convergence, and if this holds then the difference equations are said to be **consistent**. In order to establish convergence of the approximate solution to the true solution, one also must show that the difference method is *stable*.

There are a number of different notions of *stability*. One definition requires that if the true solution decays to zero over time then the approximate solution must do the same. If we let $\alpha = \beta = 0$ in (14.15), then since $u(x, t)$ represents the temperature in a rod whose ends are held fixed at temperature 0, it follows from physical arguments that $\lim_{t\to\infty} u(x, t) = 0$ for all x. Hence one should expect the same from the approximate solution values at the nodes $x_1, \ldots, x_{n-1}$. To see under what circumstances this holds, we will write equations (14.17) in matrix form and use matrix analysis to determine necessary and sufficient conditions on h and Δt to ensure that $\lim_{k\to\infty} u_i^{(k)} = 0$ for all $i = 1, \ldots, n - 1$.

Defining $\mu \equiv (\Delta t)/h^2$, and still assuming that $\alpha = \beta = 0$, equations (14.17) can be written in the form $\mathbf{u}^{(k+1)} = A\mathbf{u}^{(k)}$, where $\mathbf{u}^{(k)} \equiv (u_1^{(k)}, \ldots, u_{n-1}^{(k)})^T$ and

$$A = \begin{pmatrix} 1-2\mu & \mu & & \\ \mu & \ddots & \ddots & \\ & \ddots & \ddots & \mu \\ & & \mu & 1-2\mu \end{pmatrix}. \tag{14.21}$$

It follows that $\mathbf{u}^{(k)} = A^k\mathbf{u}^{(0)}$ and hence that $\mathbf{u}^{(k)}$ converges to 0 as $k \to \infty$ (for all $\mathbf{u}^{(0)}$) if and only if the matrix powers A^k converge to 0 as $k \to \infty$. This will be the case if and only if the *spectral radius* $\rho(A)$, which is the maximum absolute value of an eigenvalue of A, is less than 1.

Fortunately, we know the eigenvalues of the matrix A in (14.21). According to theorem 13.2.1, they are

$$1 - 2\mu + 2\mu\cos\left(\frac{\ell\pi}{n}\right), \quad \ell = 1, \ldots, n - 1.$$

For stability we need all of these eigenvalues to be less than 1 in magnitude; that is,

$$-1 < 1 - 2\mu\left(1 - \cos\frac{\ell\pi}{n}\right) < 1, \quad \ell = 1, \ldots, n - 1. \tag{14.22}$$

Subtracting 1 from each side and dividing by -2, this inequality becomes

$$1 > \mu\left(1 - \cos\frac{\ell\pi}{n}\right) > 0.$$

This holds for all $\ell = 1, \ldots, n-1$ provided $\mu < 1/2$ (since $1 - \cos(\ell\pi/n)$ is positive and is maximal when $\ell = n-1$ and then its value is close to 2 if n is large). Thus the necessary and sufficient condition for stability of the explicit method (14.17) is

$$\frac{\Delta t}{h^2} < \frac{1}{2}. \tag{14.23}$$

This is a *severe* restriction on the time step. If $h = 0.01$, for example, this says that we must use a time step $\Delta t < 1/20,000$. Thus to solve the problem for t between 0 and 1 would require more than $20,000$ time steps and would be very slow. For this reason, explicit time differencing methods are seldom used for solving the heat equation.

Example 14.2.1. To see the effects of instability, let us take, say, $h = 1/4$ and $\Delta t = 1/4$, so that $\mu \equiv \Delta t/h^2 = 4$. Starting with the initial solution $q(x) = x(1-x)$, we set $u_1^{(0)} = q(1/4) = 3/16$, $u_2^{(0)} = q(1/2) = 1/4$, and $u_3^{(0)} = q(3/4) = 3/16$. Formula (14.17) for advancing the solution in time becomes

$$u_i^{(k+1)} = -7u_i^{(k)} + 4u_{i+1}^{(k)} + 4u_{i-1}^{(k)}, \quad i = 1, 2, 3, \quad k = 0, 1, \ldots,$$

and using this formula we find

$$\mathbf{u}^{(1)} = \begin{pmatrix} -5/16 \\ -1/4 \\ -5/16 \end{pmatrix}, \quad \mathbf{u}^{(2)} = \begin{pmatrix} 19/16 \\ -3/4 \\ 19/16 \end{pmatrix},$$

$$\mathbf{u}^{(3)} = \begin{pmatrix} -181/16 \\ 59/4 \\ -181/16 \end{pmatrix}, \quad \mathbf{u}^{(4)} = \begin{pmatrix} 2211/16 \\ -775/4 \\ 2211/16 \end{pmatrix}.$$

After only 4 time steps, the ∞-norm of $\mathbf{u}$ has grown from an initial value of $1/4$ to $775/4 \approx 194$, and after 10 time steps it is about 10^{+9}. The reason for this is clear: The matrix

$$A = \begin{pmatrix} -7 & 4 & 0 \\ 4 & -7 & 4 \\ 0 & 4 & -7 \end{pmatrix}$$

by which we are multiplying the "approximate solution" vector at each step has spectral radius $|-7 + 8\cos(3\pi/4)| \approx 12.66$, and so as we proceed in time, this vector is being multiplied by approximately this factor at every step. This is despite the fact that the true solution is decaying to 0 over time. A plot of u_1 and u_2 is shown in figure 14.6 for the first 4 time steps.

Another way to think of stability (that leads to the same conclusion) is to imagine that an error is introduced at some step (which can be taken to be step 0) and to require that the error be damped rather than amplified over

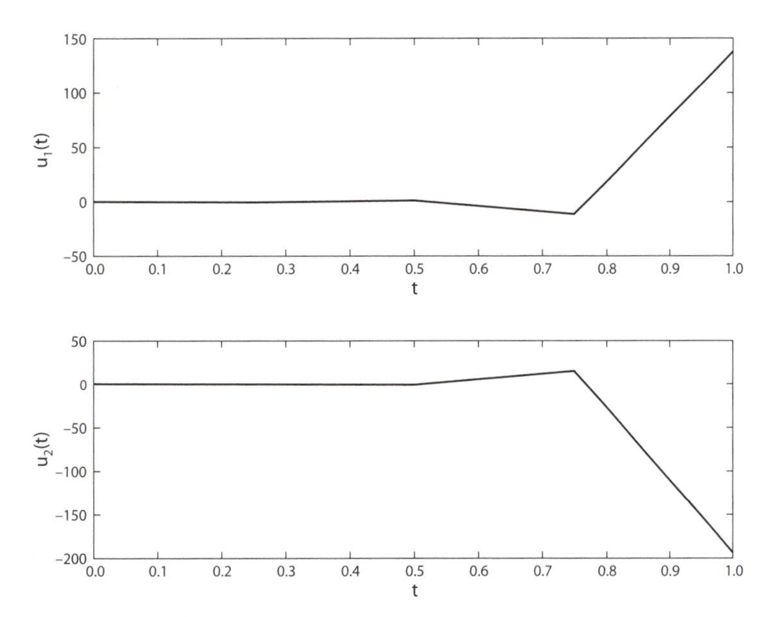

Figure 14.6. Unstable method, $h = 1/4$, $\Delta t = 1/4$.

time. Thus, if instead of starting with the initial vector $\mathbf{u}^{(0)}$ one started with a slightly different vector $\tilde{\mathbf{u}}^{(0)}$, then the solution produced at step k would be $\tilde{\mathbf{u}}^{(k)} = A^k \tilde{\mathbf{u}}^{(0)}$ instead of $\mathbf{u}^{(k)} = A^k \mathbf{u}^{(0)}$. The difference is

$$\mathbf{u}^{(k)} - \tilde{\mathbf{u}}^{(k)} = A^k(\mathbf{u}^{(0)} - \tilde{\mathbf{u}}^{(0)}),$$

and this difference approaches 0 as $k \to \infty$ if and only if $A^k \to 0$ as $k \to \infty$; that is, if and only if $\rho(A) < 1$.

Another approach to analyzing stability is sometimes called **von Neumann stability analysis** or the **Fourier method**. We will now change notation and let the spatial variable be denoted with subscript j since i will denote $\sqrt{-1}$. For this analysis, one often assumes an infinite spatial domain. Assume that the solution $u_j^{(k)}$ of the difference equations (14.17) can be written in the form

$$u_j^{(k)} = g^k e^{ijh\xi} = g^k(\cos(jh\xi) + i \sin(jh\xi)), \tag{14.24}$$

where g represents an amplification factor (since $|u_j^{(k)}| = |g|^k$) and ξ is a frequency in the Fourier transform of the solution. Substituting this expression into (14.17) gives

$$g^{k+1} e^{ijh\xi} = (1 - 2\mu) g^k e^{ijh\xi} + \mu g^k (e^{i(j+1)h\xi} + e^{i(j-1)h\xi}),$$

from which it follows that

$$g = (1 - 2\mu) + \mu(e^{ih\xi} + e^{-ih\xi}) = 1 - 2\mu(1 - \cos(h\xi)).$$

The condition for stability is that $|g|$ be less than 1 for all frequencies ξ, and this is the same as condition (14.22). Hence we are again led to the conclusion that the method is stable if and only if (14.23) holds.

Implicit Methods

The finite difference method (14.17) is called an **explicit** method because it gives an explicit formula for the solution at the new time step in terms of the solution at the old time step. In an **implicit** method one must solve a system of equations to determine the solution at the new time step. This requires significantly more work per time step, but it may be worth the extra expense because of better stability properties, allowing larger time steps.

As an example, suppose that the forward difference in time used to derive formula (14.17) is replaced by a *backward* difference in time: $u_t(x,t) \approx (u(x,t) - u(x, t-\Delta t))/(\Delta t)$. The resulting set of difference equations is

$$\frac{u_i^{(k+1)} - u_i^{(k)}}{\Delta t} = \frac{-2u_i^{(k+1)} + u_{i+1}^{(k+1)} + u_{i-1}^{(k+1)}}{h^2}, \quad i = 1, \ldots, n-1, \quad k = 0, 1, \ldots. \tag{14.25}$$

Now to determine $\mathbf{u}^{(k+1)} \equiv (u_1^{(k+1)}, \ldots, u_{n-1}^{(k+1)})^T$ from $\mathbf{u}^{(k)}$, one must solve a system of linear equations:

$$A\mathbf{u}^{(k+1)} = \mathbf{u}^{(k)} + \mathbf{b}, \tag{14.26}$$

where

$$A = \begin{pmatrix} 1+2\mu & -\mu & & \\ -\mu & \ddots & \ddots & \\ & \ddots & \ddots & -\mu \\ & & -\mu & 1+2\mu \end{pmatrix}, \quad \mathbf{b} = \mu \begin{pmatrix} \alpha \\ 0 \\ \vdots \\ 0 \\ \beta \end{pmatrix}. \tag{14.27}$$

The local truncation error in this method is again $O(h^2) + O(\Delta t)$. To analyze stability, we again assume that $\alpha = \beta = 0$ so that $\mathbf{b} = 0$ in (14.26) and $\mathbf{u}^{(k+1)} = A^{-1}\mathbf{u}^{(k)}$. Then $\mathbf{u}^{(k)} = A^{-k}\mathbf{u}^{(0)}$ and $\mathbf{u}^{(k)}$ will converge to 0 as $k \to \infty$ (for all $\mathbf{u}^{(0)}$) if and only if $\rho(A^{-1}) < 1$.

Once again, theorem 13.2.1 tells us the eigenvalues of A:

$$1 + 2\mu - 2\mu \cos\left(\frac{\ell\pi}{n}\right), \quad \ell = 1, \ldots, n-1,$$

and the eigenvalues of A^{-1} are the inverses of these:

$$\frac{1}{1 + 2\mu\left(1 - \cos\frac{\ell\pi}{n}\right)}, \quad \ell = 1, \ldots, n-1. \tag{14.28}$$

Since $\mu \equiv \Delta t/h^2 > 0$ and $1 - \cos(\ell\pi/n) \in (0, 2)$, the denominator in (14.28) is always greater than 1. Thus the eigenvalues of A^{-1} are positive and less than 1 for every value of μ. The method is therefore said to be *unconditionally stable*, and it can be shown that the approximate solution converges to the true solution at the rate $O(h^2) + O(\Delta t)$, independent of the relation between h and Δt.

Example 14.2.2. When the implicit method (14.25) is used with $h = 1/4$ and $\Delta t = 1/4$, the resulting approximate solution decays over time as it should. Starting with $q(x) = x(1-x)$ so that $\mathbf{u}^{(0)} = (q(1/4), q(1/2), q(3/4))^T = (3/16,$

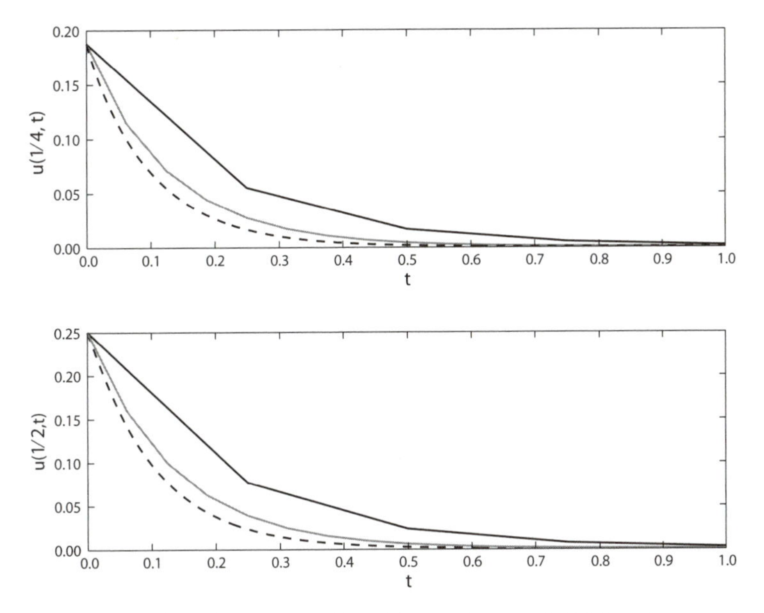

Figure 14.7. Stable method, $h = \Delta t = 1/4$ and $h = 1/8$, $\Delta t = 1/16$. Exact solution is plotted with dashed line.

$1/4, 3/16)^T$, the approximate solution vectors produced at time steps 1 through 4 are

$$\mathbf{u}^{(1)} = \begin{pmatrix} 0.0548 \\ 0.0765 \\ 0.0548 \end{pmatrix}, \quad \mathbf{u}^{(2)} = \begin{pmatrix} 0.0163 \\ 0.0230 \\ 0.0163 \end{pmatrix}, \quad \mathbf{u}^{(3)} = \begin{pmatrix} 0.0049 \\ 0.0069 \\ 0.0049 \end{pmatrix}, \quad \mathbf{u}^{(4)} = \begin{pmatrix} 0.0015 \\ 0.0021 \\ 0.0015 \end{pmatrix}.$$

The first two components are plotted versus time in figure 14.7. Also plotted, with a dashed line, is the exact solution at $x = 1/4$ and $x = 1/2$. While the method is stable, it is not very accurate with $h = 1/4$ and $\Delta t = 1/4$. Also plotted in the figure are the approximate solution values using $h = 1/8$ and $\Delta t = 1/16$. Since the method is second-order accurate in space and first order in time, the error in these values should be about 1/4 of that in the original approximation.

A potentially better implicit method is the **Crank–Nicolson method,** which uses the average of the right-hand sides in (14.16) and (14.25):

$$\frac{u_i^{(k+1)} - u_i^{(k)}}{\Delta t} = \frac{1}{2}\left[\frac{-2u_i^{(k)} + u_{i+1}^{(k)} + u_{i-1}^{(k)}}{h^2} + \frac{-2u_i^{(k+1)} + u_{i+1}^{(k+1)} + u_{i-1}^{(k+1)}}{h^2}\right]. \tag{14.29}$$

The average of these two right-hand sides is an approximation to $u_{xx}(x_i, (t_k + t_{k+1})/2)$ that is second order in both space and time; that is, one can show using Taylor's theorem that the error is $O(h^2) + O((\Delta t)^2)$. The left-hand side of (14.29) is a centered-difference approximation to $u_t(x_i, (t_k + t_{k+1})/2)$ and so it is second-order accurate; that is, the error is $O((\Delta t)^2)$. Hence the local truncation error for the Crank–Nicolson method is $O(h^2) + O((\Delta t)^2)$.

Moving terms involving the solution at the new time step to the left and those involving the solution at the old time step to the right, formula (14.29) can be written as

$$(1+\mu)u_i^{(k+1)} - \frac{1}{2}\mu u_{i+1}^{(k+1)} - \frac{1}{2}\mu u_{i-1}^{(k+1)} = (1-\mu)u_i^{(k)} + \frac{1}{2}\mu u_{i+1}^{(k)} + \frac{1}{2}\mu u_{i-1}^{(k)}. \tag{14.30}$$

Once again we can analyze the stability of the method by using matrix analysis and theorem 11.2.1, or we can use the Fourier approach and substitute a Fourier component of the form (14.24) into (14.30) and determine whether such components grow or decay over time. The two approaches again lead to the same conclusion. Taking the latter approach (and again letting j denote the spatial variable so that i can be used for $\sqrt{-1}$), we find

$$(1+\mu)g^{k+1}e^{ijh\xi} - \frac{1}{2}\mu g^{k+1}[e^{i(j+1)h\xi} + e^{i(j-1)h\xi}] = (1-\mu)g^k e^{ijh\xi} + \frac{1}{2}\mu g^k[e^{i(j+1)h\xi} + e^{i(j-1)h\xi}],$$

or, dividing by $g^k e^{ijh\xi}$ and solving for g,

$$g = \frac{1-\mu(1-\cos(h\xi))}{1+\mu(1-\cos(h\xi))}.$$

Since the term $\mu(1-\cos(h\xi))$ in the numerator and denominator is always positive, the numerator is always smaller in absolute value than the denominator:

$$|1-\mu(1-\cos(h\xi))| < |1| + |\mu(1-\cos(h\xi))| = 1 + \mu(1-\cos(h\xi)).$$

Hence this method also is *unconditionally stable*.

Example 14.2.3. Applying the Crank–Nicolson method with the same values of h and Δt used in the previous examples yields the results in figure 14.8. For $h = \Delta t = 1/4$, the method is stable but not very accurate. For $h = 1/8$, $\Delta t = 1/16$, the solution produced by the Crank–Nicolson method can barely be distinguished from the exact solution, plotted with a dashed line in the figure.

14.3 SEPARATION OF VARIABLES

We have discussed the numerical solution of a number of model problems involving partial differential equations. Actually, some of these problems can be solved analytically, using a technique called *separation of variables*. This technique works only for problems in which the solution can be written as a product of two functions, each of which involves only one of the independent variables. In contrast, the numerical techniques that we have described can be applied to general problems of the form (14.1).

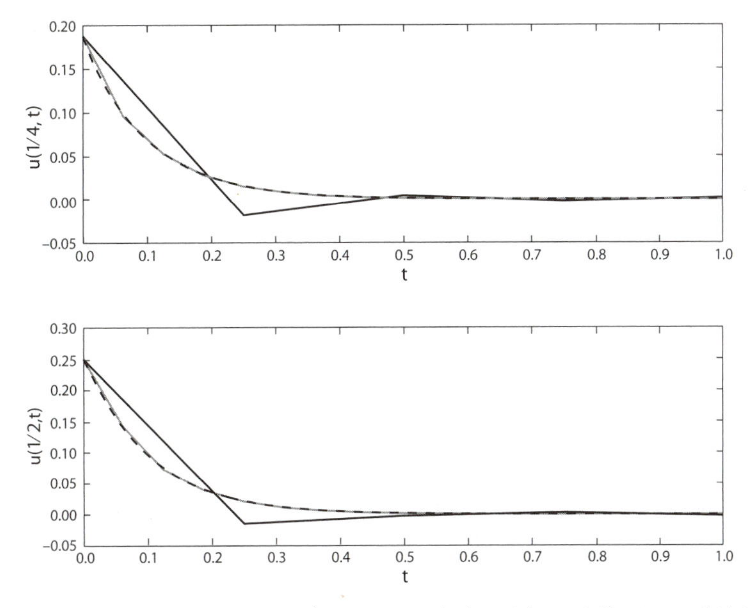

Figure 14.8. Crank–Nicolson method, $h = \Delta t = 1/4$ and $h = 1/8$, $\Delta t = 1/16$. Exact solution is plotted with dashed line.

To solve the heat equation via separation of variables, let us consider a problem with *periodic boundary conditions*:

$$u_t = u_{xx}, \quad 0 \le x \le 2\pi, \quad t > 0.$$

$$u(x, 0) = q(x), \quad 0 \le x \le 2\pi,$$

$$u(0, t) = u(2\pi, t), \quad u_x(0, t) = u_x(2\pi, t), \quad t > 0. \tag{14.31}$$

This was a problem that Fourier studied. It models the temperature in a circular rod of length 2π.

To solve the IBVP (14.31) via separation of variables, assume that the solution $u(x, t)$ can be written in the form

$$u(x, t) = v(x)w(t),$$

for some functions v and w. Then the differential equation becomes

$$v(x)w'(t) = v''(x)w(t).$$

Assuming that $v(x) \neq 0$ and $w(t) \neq 0$, we can divide each side by $v(x)w(t)$ to obtain

$$\frac{w'(t)}{w(t)} = \frac{v''(x)}{v(x)}.$$

Since the left-hand side depends only on t and the right-hand side depends only on x, both sides must be equal to some constant, say, $-\lambda$. Hence

$$w'(t) = -\lambda w(t), \quad t > 0, \tag{14.32}$$

and

$$v''(x) = -\lambda v(x), \quad 0 \leq x \leq 2\pi. \tag{14.33}$$

The general solution of equation (14.32) is

$$w(t) = Ce^{-\lambda t}. \tag{14.34}$$

Equation (14.33) for v is an eigenvalue problem and must be coupled with the boundary conditions

$$v(0) = v(2\pi), \quad v'(0) = v'(2\pi), \tag{14.35}$$

in order for $v(x)w(t)$ to satisfy the boundary conditions in (14.31). Equation (14.33) for v always has $v = 0$ as a solution, but its other solutions depend on the value of λ.

We must consider three cases: Case (i) $\lambda < 0$: It can be shown in this case that the only solution to (14.33) that satisfies the boundary conditions (14.35) is $v = 0$. Case (ii) $\lambda = 0$: In this case, the general solution of (14.33) is $v(x) = c_1 + c_2 x$. The condition $v(0) = v(2\pi)$ implies that $c_1 = c_1 + c_2(2\pi)$, or, $c_2 = 0$. Hence $v(x)$ must be a constant, and in this case the boundary condition $v'(0) = v'(2\pi)$ holds as well since $v' \equiv 0$. Case (iii) $\lambda > 0$: In this case, the general solution of (14.33) is

$$v(x) = c_1 \cos(\sqrt{\lambda} x) + c_2 \sin(\sqrt{\lambda} x).$$

The boundary conditions (14.35) imply that

$$\begin{aligned} c_1 &= c_1 \cos(\sqrt{\lambda}\, 2\pi) + c_2 \sin(\sqrt{\lambda}\, 2\pi), \\ c_2 &= -c_1 \sin(\sqrt{\lambda}\, 2\pi) + c_2 \cos(\sqrt{\lambda}\, 2\pi), \end{aligned}$$

or, equivalently,

$$\begin{pmatrix} 1 - \cos(\sqrt{\lambda}\, 2\pi) & -\sin(\sqrt{\lambda}\, 2\pi) \\ \sin(\sqrt{\lambda}\, 2\pi) & 1 - \cos(\sqrt{\lambda}\, 2\pi) \end{pmatrix} \begin{pmatrix} c_1 \\ c_2 \end{pmatrix} = \begin{pmatrix} 0 \\ 0 \end{pmatrix}.$$

This equation has a nonzero solution for (c_1, c_2) if and only if the matrix on the left-hand side is singular; that is, if and only if its determinant is 0:

$$(1 - \cos(\sqrt{\lambda}\, 2\pi))^2 + \sin^2(\sqrt{\lambda}\, 2\pi) = 0.$$

In other words, it must be the case that $\sin(\sqrt{\lambda}\, 2\pi) = 0$ and $1 - \cos(\sqrt{\lambda}\, 2\pi) = 0$; that is, $\sqrt{\lambda} = 1, 2, \ldots$. Equivalently, $\lambda = m^2$, $m = 1, 2, \ldots$. In this case, $\cos(mx)$ and $\sin(mx)$ are two linearly independent solutions of the differential equation (14.33) with the boundary conditions (14.35). Substituting for λ in (14.34) and considering all of the cases (i)–(iii), we find that the solutions of the differential equation in (14.31) that satisfy the periodic boundary conditions consist of all linear combinations of

$$1, \quad \cos(mx)e^{-m^2 t}, \quad \sin(mx)e^{-m^2 t}, \quad m = 1, 2, \ldots.$$

These are called **fundamental modes** of the differential equation with periodic boundary conditions.

To satisfy the initial condition $u(x, 0) = q(x)$, we must expand $q(x)$ in terms of these fundamental modes; that is, in a *Fourier series*:

$$q(x) = a_0 + \sum_{m=1}^{\infty} (a_m \cos(mx) + b_m \sin(mx)).$$

A large class of 2π-periodic functions q can be represented in this way. (In our problem, q is defined only on the interval $[0, 2\pi]$, but assuming that q satisfies the boundary conditions $q(0) = q(2\pi)$ and $q'(0) = q'(2\pi)$, it can be extended to be periodic throughout the real line.) Then, assuming that the series converges appropriately, the solution of the IBVP (14.31) is

$$u(x, t) = a_0 + \sum_{m=1}^{\infty} (a_m \cos(mx) + b_m \sin(mx))e^{-m^2 t}.$$

One can check that this satisfies the differential equation by differentiating the series term by term (again, assuming that the series converges in such a way that term by term differentiation is justified):

$$u_t = \sum_{m=1}^{\infty} (a_m \cos(mx) + b_m \sin(mx))(-m^2)e^{-m^2 t},$$

$$u_{xx} = \sum_{m=1}^{\infty} (a_m(-m^2) \cos(mx) + b_m(-m^2) \sin(mx))e^{-m^2 t} = u_t.$$

Had the boundary conditions been different, say, homogeneous Dirichlet boundary conditions, instead of those in (14.31), our analysis of problem (14.33) would have been slightly different. For convenience, let us change the interval on which the problem is defined from $[0, 2\pi]$ to $[0, \pi]$. In this case, the boundary conditions (14.35) are replaced by

$$v(0) = 0, \quad v(\pi) = 0. \tag{14.36}$$

One again finds that in case (i), $\lambda < 0$, the only solution of (14.33) that satisfies the boundary conditions (14.36) is $v = 0$. In case (ii), $\lambda = 0$, the general solution of (14.33) is again $v(x) = c_1 + c_2 x$, but now $v(0) = 0 \Rightarrow c_1 = 0$ and hence $v(\pi) = 0 \Rightarrow c_2 = 0$. Thus $v = 0$ is the only solution in this case as well. In case (iii), $\lambda > 0$, we again have the general solution $v(x) = c_1 \cos(\sqrt{\lambda} x) + c_2 \sin(\sqrt{\lambda} x)$, but the boundary condition $v(0) = 0$ implies that $c_1 = 0$, and the boundary condition $v(\pi) = 0$ then implies that $\sin(\sqrt{\lambda}\pi) = 0$, or, $\sqrt{\lambda} = 1, 2, \ldots$. Thus the fundamental modes for this problem are

$$\sin(mx), \quad m = 1, 2, \ldots.$$

Again to satisfy the initial condition, one must expand $q(x)$ in terms of these fundamental modes; that is, in a *Fourier sine series*:

$$q(x) = \sum_{m=1}^{\infty} b_m \sin(mx).$$

This is the appropriate series to use for functions that satisfy the boundary conditions $q(0) = q(\pi) = 0$. Assuming that the series converges in the right way, the solution to the IBVP is then

$$u(x,t) = \sum_{m=1}^{\infty} b_m \sin(mx) e^{-m^2 t}. \qquad (14.37)$$

The method of separation of variables works for a given problem, provided the set of fundamental modes for the differential equation with the given boundary conditions is rich enough that the initial function can be represented as a linear combination of these fundamental modes. Otherwise, the solution $u(x,t)$ cannot be written as a product $v(x)w(t)$, and a different approach must be used to solve or approximately solve the IBVP.

We argued previously based on physical grounds that the solution to the heat equation in a rod with the ends held fixed at temperature 0 would decay to 0 over time. Now this can be seen analytically through expression (14.37), each of whose terms approaches 0 as $t \to \infty$, for all x.

14.3.1 Separation of Variables for Difference Equations

Just as separation of variables and Fourier series were used to solve the IBVP (14.31), they can be used to solve the difference equations (14.17–14.19), assuming that $\alpha = \beta = 0$. Assume that $u_i^{(k)}$ can be written in the form

$$u_i^{(k)} = v_i w_k.$$

Making this substitution in (14.16), we have

$$\frac{v_i w_{k+1} - v_i w_k}{\Delta t} = \frac{-2v_i w_k + v_{i+1} w_k + v_{i-1} w_k}{h^2}, \quad i = 1, \ldots, n-1, \quad k = 0, 1, \ldots.$$

Collecting terms involving k on the left and those involving i on the right, and assuming that $v_i \neq 0$ and $w_k \neq 0$, gives

$$\frac{w_{k+1} - w_k}{\mu w_k} = \frac{-2v_i + v_{i+1} + v_{i-1}}{v_i},$$

where $\mu = (\Delta t)/h^2$. Since the left-hand side is independent of i and the right-hand side is independent of k, both sides must be equal to some constant, say, $-\lambda$; thus

$$w_{k+1} = (1 - \lambda\mu) w_k, \quad k = 0, 1, \ldots, \qquad (14.38)$$

$$2v_i - v_{i+1} - v_{i-1} = \lambda v_i, \quad i = 1, \ldots, n-1, \quad v_0 = v_n = 0. \qquad (14.39)$$

Equation (14.39) represents the eigenvalue problem for the TST matrix with 2s on its main diagonal and -1s on its sub- and super-diagonals. From theorem 13.2.1, these eigenvalues are $\lambda_\ell = 2 - 2\cos(\ell\pi/n)$, $\ell = 1, \ldots, n-1$, and the corresponding orthonormal eigenvectors are

$$\mathbf{v}^{(\ell)} = \sqrt{2/n}\, (\sin(\ell\pi/n), \sin(2\ell\pi/n), \ldots, \sin((n-1)\ell\pi/n))^T. \qquad (14.40)$$

Thus for each $\lambda = \lambda_\ell$, the vector $\mathbf{v}$ with $v_i = \sqrt{2/n}\ \sin(i\ell\pi/n)$, $i = 1, \ldots, n-1$ is a solution of (14.39). The solution of (14.38) is $w_k = (1 - \lambda\mu)^k w_0$, and so the solutions of the difference equation (14.16) with homogeneous Dirichlet boundary conditions are linear combinations of

$$\{(1 - 2\mu + 2\mu\cos(\ell\pi/n))^k \sqrt{2/n}\ \sin(i\ell\pi/n) : \ \ell = 1, \ldots, n-1\}.$$

To satisfy the initial conditions $u_i^{(0)} = q(x_i)$, $i = 1, \ldots, n-1$, one must expand the initial vector in terms of the eigenvectors in (14.40); that is, one must find coefficients a_ℓ, $\ell = 1, \ldots, n-1$, such that

$$\sum_{\ell=1}^{n-1} a_\ell \sqrt{2/n}\ \sin(i\ell\pi/n) = q(x_i), \quad i = 1, \ldots, n-1.$$

Because the eigenvectors span $\mathbf{R}^{n-1}$, this expansion is always possible, and because the eigenvectors are orthonormal, it turns out that

$$a_\ell = \sqrt{2/n} \sum_{j=1}^{n-1} q(x_j) \sin(j\ell\pi/n), \quad \ell = 1, \ldots, n-1.$$

Thus the solution of (14.17–14.19) with $\alpha = \beta = 0$ is

$$u_i^{(k)} = \sum_{\ell=1}^{n-1} (1 - 2\mu + 2\mu\cos(\ell\pi/n))^k (2/n) \sum_{j=1}^{n-1} q(x_j) \sin(j\ell\pi/n) \sin(i\ell\pi/n).$$

We can use this expression to answer questions about stability of the difference method, as was done before. The approximate solution $u_i^{(k)}$ will approach 0 as $k \to \infty$ if and only if $|1 - 2\mu + 2\mu\cos(\ell\pi/n)|$ is less than one for all $\ell = 1, \ldots, n-1$. We have seen previously that a sufficient condition for this is $\mu \le 1/2$.

MATLAB'S LOGO

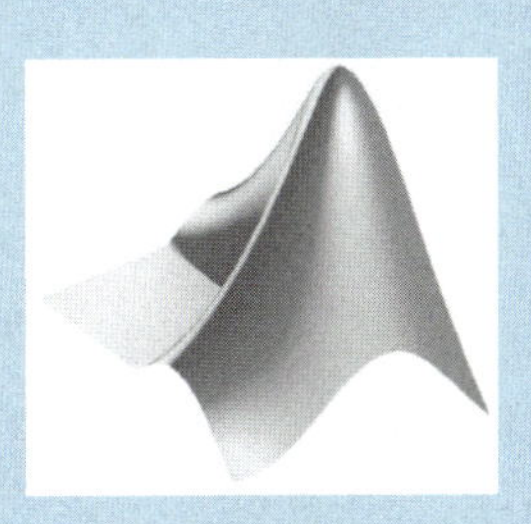

The wave equation on an L-shaped region formed from three unit squares is one of the simplest geometries for which solutions to the wave equation cannot be expressed analytically. Eigenfunctions of the differential operator play an important role, analogous to that of eigenvectors for matrices. Any solution of the wave equation can be expressed as a linear combination of its eigenfunctions. The logo of MathWorks, the company that produces MATLAB, comes from the first eigenfunction of an L-shaped membrane. Given the L-shaped geometry, the MATLAB logo can only be produced with numerical computation.

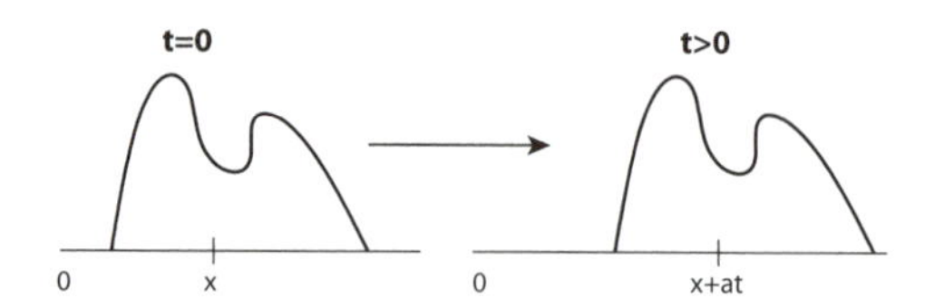

Figure 14.9. A solution to the one-way wave equation is a shifted copy of the original function.

14.4 HYPERBOLIC EQUATIONS

14.4.1 Characteristics

We begin our study of hyperbolic equations by considering the one-way wave equation

$$u_t + au_x = 0, \quad x \in (-\infty, \infty), \quad t > 0, \quad u(x, 0) = q(x). \tag{14.41}$$

Here we consider an infinite spatial domain, so no boundary conditions are present. To uniquely specify the solution, one still needs an initial condition, $u(x, 0) = q(x)$, so this is called an *initial value problem* (IVP). This problem can be solved analytically. The solution is

$$u(x, t) = q(x - at), \tag{14.42}$$

as is easily checked by differentiating: $u_t(x, t) = -aq'(x - at)$ and $u_x(x, t) = q'(x - at)$; hence $u_t + au_x = 0$.

The solution at any time $t > 0$ is a copy of the original function, shifted to the right if a is positive and to the left if a is negative, as pictured in figure 14.9. Thus the solution at (x, t) depends only on the value of q at $x - at$. The lines in the (x, t) plane on which $x - at$ is constant are called **characteristics**. The solution is a wave that propagates with speed a, without change of shape. Note also that while the differential equation seems to make sense only if u is differentiable in x, the "solution" (14.42) is defined even if q is not differentiable. In general, discontinuous solutions are allowed for hyperbolic problems. An example is a *shock wave*, which can occur in nonlinear hyperbolic equations.

Consider the more general equation

$$u_t(x, t) + au_x(x, t) + bu(x, t) = f(x, t), \quad x \in (-\infty, \infty), \quad t > 0. \tag{14.43}$$

Based on the preceding observations, let us change variables from (x, t) to (ξ, τ), where

$$\tau = t, \quad \xi = x - at \quad (x = \xi + a\tau).$$

Define $\tilde{u}(\xi, \tau) = u(x, t)$. Then

$$\frac{\partial \tilde{u}}{\partial \tau} = \frac{\partial u}{\partial t}\frac{\partial t}{\partial \tau} + \frac{\partial u}{\partial x}\frac{\partial x}{\partial \tau} = u_t + au_x,$$

since $\partial t/\partial \tau = 1$ and $\partial x/\partial \tau = a$. Hence equation (14.43) becomes

$$\frac{\partial \tilde{u}}{\partial \tau}(\xi, \tau) + b\tilde{u}(\xi, \tau) = f(\xi + a\tau, \tau).$$

This is an *ordinary differential equation* (ODE) in τ, and the solution is

$$\tilde{u}(\xi,\tau) = q(\xi)e^{-b\tau} + \int_0^{\tau} f(\xi + as, s)e^{-b(\tau - s)}\, ds.$$

Returning to the original variables,

$$u(x,t) = q(x - at)e^{-bt} + \int_0^{t} f(x - a(t - s), s)e^{-b(t-s)}\, ds.$$

Note that $u(x,t)$ depends only on the value of q at $x - at$ and the values of f at points $(\hat{x},\hat{t})$ such that $\hat{x} - a\hat{t} = x - at$; that is, only on values of q and f along the characteristic through (x,t). This method of solving along characteristics can be extended to nonlinear equations of the form $u_t + au_x = f(x,t,u)$.

14.4.2 Systems of Hyperbolic Equations

Definition. A system of the form

$$\mathbf{u}_t + A\mathbf{u}_x + B\mathbf{u} = \mathbf{f}(x,t) \tag{14.44}$$

is **hyperbolic** if A is diagonalizable with real eigenvalues.

Example 14.4.1. As an example of a hyperbolic system, suppose we start with the *wave equation* $u_{tt} = a^2 u_{xx}$, where a is a given scalar, and we introduce the functions $v = au_x$ and $w = u_t$. Then we obtain the system of equations

$$\begin{pmatrix} v \\ w \end{pmatrix}_t = \begin{pmatrix} 0 & a \\ a & 0 \end{pmatrix} \begin{pmatrix} v \\ w \end{pmatrix}_x . \tag{14.45}$$

The eigenvalues of the matrix in (14.45) are the solutions of $\lambda^2 - a^2 = 0$; that is, $\lambda = \pm a$. These are real and the matrix is diagonalizable.

The eigenvalues of A are the characteristic speeds of the system. If $A = S\Lambda S^{-1}$, where $\Lambda = \mathrm{diag}(\lambda_1, \ldots, \lambda_m)$, then under the change of variables $\hat{\mathbf{u}} = S^{-1}\mathbf{u}$, equation (14.44) becomes, upon multiplying by S^{-1} on the left,

$$\hat{\mathbf{u}}_t + \Lambda\hat{\mathbf{u}}_x + (S^{-1}BS)\hat{\mathbf{u}} = S^{-1}\mathbf{f}.$$

When $B = 0$, these equations decouple:

$$(\hat{u}_i)_t + \lambda_i(\hat{u}_i)_x = (S^{-1}\mathbf{f})_i, \ldots, \quad i = 1, \ldots, m.$$

Example 14.4.2. The eigenvectors of the matrix in (14.45) are $(\sqrt{2}/2, \sqrt{2}/2)^T$ corresponding to $\lambda = a$, and $(\sqrt{2}/2, -\sqrt{2}/2)^T$ corresponding to $\lambda = -a$. Hence

$$S = \begin{pmatrix} \sqrt{2}/2 & \sqrt{2}/2 \\ \sqrt{2}/2 & -\sqrt{2}/2 \end{pmatrix}, \quad S^{-1} = \begin{pmatrix} \sqrt{2}/2 & \sqrt{2}/2 \\ \sqrt{2}/2 & -\sqrt{2}/2 \end{pmatrix}.$$

Define $\hat{u}_1 = (\sqrt{2}/2)v + (\sqrt{2}/2)w$ and $\hat{u}_2 = (\sqrt{2}/2)v - (\sqrt{2}/2)w$. Then $(\hat{u}_1)_t = a(\hat{u}_1)_x$ so that $\hat{u}_1(x,t) = \hat{q}_1(x + at)$, and $(\hat{u}_2)_t = -a(\hat{u}_2)_x$ so that $\hat{u}_2(x,t) = \hat{q}_2(x - at)$. Thus the solution is a combination of a wave moving to the left with speed a and another wave moving to the right with speed a.

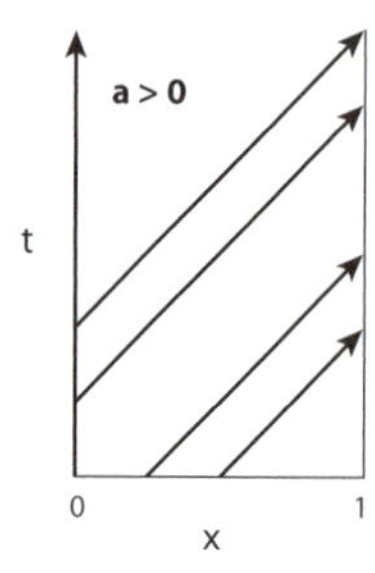

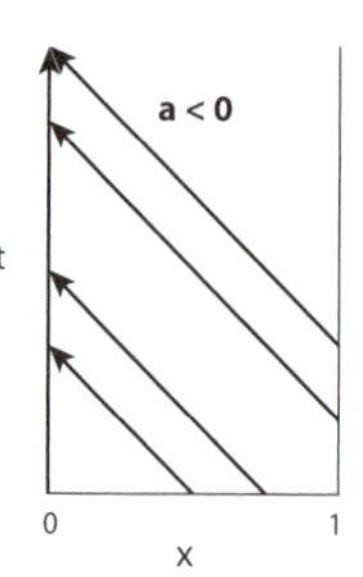

Figure 14.10. The characteristics of the one-way wave equation for $a > 0$ (*left*) and $a < 0$ (*right*).

14.4.3 Boundary Conditions

Hyperbolic PDEs may be defined on a finite interval rather than the entire real line, and then boundary conditions must be specified. This gives rise to an initial boundary value problem, but one must be sure that the correct types of boundary conditions are given.

Consider again the one-way wave equation

$$u_t + au_x = 0, \quad 0 \leq x \leq 1, \quad t > 0, \tag{14.46}$$

defined now on the finite interval $[0, 1]$. If $a > 0$, the characteristics in this region propagate to the right, while if $a < 0$, they propagate to the left, as illustrated in figure 14.10.

It follows that, in addition to the initial data at time $t = 0$, the solution must be given on the left boundary of the region ($x = 0$) when a is positive and on the right boundary ($x = 1$) when a is negative.

If we specify initial data $u(x, 0) = q(x)$ and boundary data $u(0, t) = g(t)$, for $a > 0$, then the solution is

$$u(x, t) = \begin{cases} q(x - at) & \text{if } x - at > 0, \\ g(t - x/a) & \text{if } x - at < 0. \end{cases}$$

For $a < 0$, the roles of the two boundaries are reversed.

14.4.4 Finite Difference Methods

One can imagine any number of ways of differencing the equations (14.46). Letting $u_j^{(k)}$ denote the approximate solution at spatial point $x_j = jh$ and time $t_k = k(\Delta t)$, one could use a forward difference in time and a forward difference in space to obtain the equations

$$\frac{u_j^{(k+1)} - u_j^{(k)}}{\Delta t} + a\frac{u_{j+1}^{(k)} - u_j^{(k)}}{h} = 0, \quad j = 1, \ldots, n-1, \quad k = 0, 1, \ldots. \tag{14.47}$$

Alternatively, one could try a forward difference in time and a backward difference in space:

$$\frac{u_j^{(k+1)} - u_j^{(k)}}{\Delta t} + a\frac{u_j^{(k)} - u_{j-1}^{(k)}}{h} = 0, \quad j = 1, \ldots, n-1, \quad k = 0, 1, \ldots. \tag{14.48}$$

A third alternative is a forward difference in time and a centered difference in space:

$$\frac{u_j^{(k+1)} - u_j^{(k)}}{\Delta t} + a\frac{u_{j+1}^{(k)} - u_{j-1}^{(k)}}{2h} = 0, \quad j = 1, \ldots, n-1, \quad k = 0, 1, \ldots. \tag{14.49}$$

The **leapfrog** scheme uses centered differences in time and space:

$$\frac{u_j^{(k+1)} - u_j^{(k-1)}}{2(\Delta t)} + a\frac{u_{j+1}^{(k)} - u_{j-1}^{(k)}}{2h} = 0, \quad j = 1, \ldots, n-1, \quad k = 0, 1, \ldots. \tag{14.50}$$

The leapfrog method is called a *multistep method* because it uses solutions from two previous time steps (steps k and $k-1$) to compute the solution at the new time step (step $k+1$). In order to start the method, one must use a one-step method to approximate the solution at t_1, given the initial data at time $t_0 = 0$. Once the approximate solution is known at two consecutive time steps, formula (14.50) can then be used to advance to future time steps.

Still another possibility is the **Lax–Friedrichs** scheme:

$$\frac{u_j^{(k+1)} - \frac{1}{2}(u_{j+1}^{(k)} + u_{j-1}^{(k)})}{\Delta t} + a\frac{u_{j+1}^{(k)} - u_{j-1}^{(k)}}{2h} = 0, \quad j = 1, \ldots, n-1, \quad k = 0, 1, \ldots. \tag{14.51}$$

The local truncation error (LTE) for these different methods can be analyzed just as was done for parabolic problems. Using Taylor's theorem to determine the amount by which the true solution fails to satisfy the difference equations, one finds that the LTE for the forward-time, forward-space method (14.47) is $O(\Delta t) + O(h)$, and the same holds for the forward-time, backward-space method (14.48). The LTE for the forward-time, centered-space method (14.49) is $O(\Delta t) + O(h^2)$, since the centered-difference approximation to u_x is second-order accurate.

The leapfrog scheme (14.50) has LTE that is second-order in both space and time: $O((\Delta t)^2) + O(h^2)$. The Lax–Friedrichs scheme (14.51) has LTE $O(\Delta t) + O(h^2) + O(h^2/(\Delta t))$, with the last term arising because $\frac{1}{2}(u(x_{j+1}, t_k) + u(x_{j-1}, t_k))$ is an $O(h^2)$ approximation to $u(x_j, t_k)$, but this error is divided by Δt in formula (14.51).

As we have seen before, however, local truncation error is only part of the story. In order for a difference method to generate a reasonable approximation to the solution of the differential equation, it also must be *stable*.

Stability and the CFL Condition

The **Lax–Richtmyer equivalence theorem** states that if a difference method for solving a well-posed linear initial value problem is consistent (meaning that the local truncation error approaches 0 as h and Δt go to 0), then it is convergent (meaning that the approximate solution converges uniformly to the true solution as h and Δt go to 0) if and only if it is stable. Without giving a formal definition of stability, we note that since the true solution to equation (14.41) does not grow over time, the approximate solution generated by a finite difference scheme should not grow either.

Consider, for example, the forward-time, forward-space difference scheme (14.47). If we define $\lambda \equiv \Delta t/h$, then (14.47) can be written as

$$u_j^{(k+1)} = u_j^{(k)} - a\lambda(u_{j+1}^{(k)} - u_j^{(k)}).$$

Looking at the sum of squares of the components on each side, we find

$$\begin{aligned}\sum_j \left(u_j^{(k+1)}\right)^2 &= \sum_j \left((1+a\lambda)u_j^{(k)} - a\lambda u_{j+1}^{(k)}\right)^2 \\ &= \sum_j \left[(1+a\lambda)^2 \left(u_j^{(k)}\right)^2 - 2(1+a\lambda)a\lambda u_j^{(k)} u_{j+1}^{(k)} + (a\lambda)^2 \left(u_{j+1}^{(k)}\right)^2\right] \\ &\le (|1+a\lambda| + |a\lambda|)^2 \sum_j \left(u_j^{(k)}\right)^2.\end{aligned}$$

A sufficient condition for stability is therefore

$$|1+a\lambda| + |a\lambda| \le 1,$$

or, equivalently, $-1 \le a\lambda \le 0$. It turns out that this is also a necessary condition for stability. Since $\lambda \equiv \Delta t/h$ is positive, the method cannot be stable if $a > 0$. For $a < 0$, the requirement is that $\lambda \le 1/|a|$, or,

$$\Delta t \le \frac{h}{|a|}. \tag{14.52}$$

One can see in another way that condition (14.52) is necessary for the difference method (14.47) to be able to generate a reasonable approximation to the solution of the differential equation (14.41). Consider the approximation $u_j^{(k)}$ to the solution at time $t_k = k(\Delta t)$ and spatial point $x_j = jh$. We know that the true solution is $u(x_j, t_k) = q(x_j - at_k)$. Suppose a is positive. Now $u_j^{(k)}$ is determined from $u_j^{(k-1)}$ and $u_{j+1}^{(k-1)}$. These values are determined from $u_j^{(k-2)}$, $u_{j+1}^{(k-2)}$, and $u_{j+2}^{(k-2)}$. Continuing to trace the values back over time, we obtain a triangular structure of dependence as illustrated in figure 14.11.

Thus $u_j^{(k)}$ depends only on values of the initial solution q between $x = x_j$ and $x = x_j + kh$. Yet if a is positive, then the true solution at (x_j, t_k) is determined by the value of q at a point to the left of x_j, namely, $q(x_j - ak\Delta t)$. Hence, despite the fact that the difference scheme is consistent, it cannot generate a reasonable approximation to the true solution. For a negative, we still need the domain of dependence of $u_j^{(k)}$ to include that of the true solution; that is,

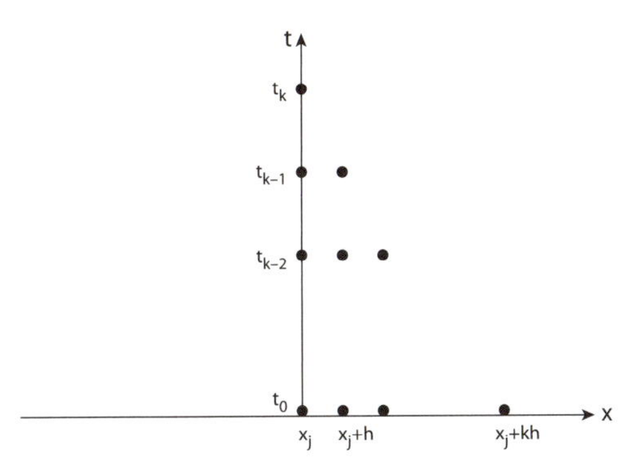

Figure 14.11. A triangular structure of dependence formed in the analysis of stability.

we need $x_j + kh$ to be greater than or equal to $x_j + |a|k(\Delta t)$, which is equivalent to condition (14.52).

The idea that the domain of dependence of the approximate solution must include that of the true solution is called the **CFL (Courant–Friedrichs–Lewy) condition.** More precisely, we can state the following.

Theorem 14.4.1. Consider a consistent difference scheme of the form

$$u_j^{(k+1)} = \alpha u_{j-1}^{(k)} + \beta u_j^{(k)} + \gamma u_{j+1}^{(k)}$$

for equation (14.41). A necessary condition for stability is the CFL condition:

$$\Delta t \leq \frac{h}{|a|}.$$

Proof. The true solution $u(x_j, t_k)$ is $q(x_j - at_k)$. On the other hand, the "approximate solution" $u_j^{(k)}$ is determined by $u_{j-1}^{(k-1)}$, $u_j^{(k-1)}$, and $u_{j+1}^{(k-1)}$; these in turn are determined by $u_{j-2}^{(k-2)}, \ldots, u_{j+2}^{(k-2)}$; etc. Continuing, it follows that $u_j^{(k)}$ depends only on values of q between $x_j - kh$ and $x_j + kh$. Hence in order for the method to be convergent, it is necessary that $x_j - ak\Delta t \in [x_j - kh, x_j + kh]$. This means that $kh \geq |a|k\Delta t$, or, $\Delta t \leq h/|a|$. By the Lax–Richtmyer theorem, a consistent method is convergent if and only if it is stable. Hence the condition $\Delta t \leq h/|a|$ is necessary for stability. □

In addition to using the CFL condition to determine necessary conditions for stability, the stability of difference methods for hyperbolic problems can be analyzed in a way similar to what was done for parabolic problems, namely, by assuming that the approximate solution $u_j^{(k)}$ has the form $g^k e^{ijh\xi}$ for some amplification factor g and some frequency ξ. The condition for stability is $|g| \leq 1$ for all frequencies ξ. Using this approach, let us consider the forward-time, centered-space method (14.49). Substituting $g^k e^{ijh\xi}$ for $u_j^{(k)}$ and dividing each side by $g^k e^{ijh\xi}$ gives

$$\frac{g-1}{\Delta t} + a\frac{e^{ih\xi} - e^{-ih\xi}}{2h} = 0,$$

or, with $\lambda = (\Delta t)/h$,

$$g = 1 - \frac{a}{2}\lambda(e^{ih\xi} - e^{-ih\xi}) = 1 - a\lambda i \sin(h\xi).$$

Since $|g|^2 = 1 + (a\lambda \sin h\xi)^2 > 1$, this method is always unstable!

Let us again consider the two-way wave equation

$$u_{tt} = a^2 u_{xx}, \quad 0 \le x \le 1, \quad t > 0. \tag{14.53}$$

It was shown in Example 14.4.1 that this second-order equation could be reduced to a system of two first-order hyperbolic equations. One way to solve this problem numerically would be to carry out this reduction and then apply an appropriate difference method to the system of first-order equations. Another approach is to difference the equation as it stands. In addition to the differential equation, one needs initial conditions and, if the spatial domain is finite, appropriate boundary conditions, in order to define a unique solution. Since the equation involves a second derivative in time, we will need two initial conditions, such as

$$u(x,0) = q(x), \quad u_t(x,0) = p(x), \quad 0 \le x \le 1.$$

If we take homogeneous Dirichlet boundary conditions,

$$u(0,t) = u(1,t) = 0,$$

then for stability we again require that the approximate solution not grow over time.

Suppose a centered-time, centered-space method is used for equation (14.53):

$$\frac{-2u_j^{(k)} + u_j^{(k+1)} + u_j^{(k-1)}}{(\Delta t)^2} = a^2 \frac{-2u_j^{(k)} + u_{j+1}^{(k)} + u_{j-1}^{(k)}}{h^2}. \tag{14.54}$$

This is called a **multistep** method because the approximate solution at time step $k+1$ depends on that at step k as well as that at step $k-1$. To start the solution process, we need to know $\mathbf{u}^{(0)}$, which can be obtained from the initial condition $u_j^{(0)} = q(x_j)$, but we also need $\mathbf{u}^{(1)}$. This can be obtained from the other initial condition $u_t(x,0) = p(x)$. Approximating $u_t(x_j,0)$ by $(u_j^{(1)} - u_j^{(0)})/(\Delta t)$, we find $u_j^{(1)} = u_j^{(0)} + (\Delta t)p(x_j)$. With the values of $\mathbf{u}$ known at time steps 0 and 1, equation (14.54) can then be used to advance the solution to time step 2, etc.

Since centered differences are used in (14.54) for both u_{tt} and u_{xx}, the local truncation error is $O((\Delta t)^2) + O(h^2)$. One can study the stability of the method by assuming that $u_j^{(k)}$ has the form $g^k e^{ijh\xi}$ and determining conditions under which $|g| \le 1$. Making this substitution in (14.54) and dividing by $g^k e^{ijh\xi}$ gives

$$\frac{-2 + g + g^{-1}}{(\Delta t)^2} = a^2 \frac{-2 + e^{ih\xi} + e^{-ih\xi}}{h^2},$$

or, with $\lambda = (\Delta t)/h$,

$$-2 + g + g^{-1} = a^2\lambda^2(-2 + 2\cos h\xi).$$

After some algebra, it can be shown that $|g| \le 1$ for all ξ if and only if $|a\lambda| \le 1$. Hence we are again led to the stability condition (14.52).

LANDMARK DOODLING

Ideas of the Fast Fourier Transform can be traced back at least as far as Gauss. The method gained notice, however, with an article by James W. Cooley and John W. Tukey (*left*). While in a meeting of President Kennedy's Science Advisory Committee, Richard Garwin of IBM was sitting next to Tukey, who was, in his usual style, doodling on ideas related to improving the discrete Fourier transform. Tukey's formulation would reduce the number of multiplications and additions by an amount that caught Garwin's attention—so much so that Garwin approached Cooley to have it implemented. Needing to hide national security interests in the method, Garwin told Cooley that the algorithm was needed for another application. The work of Tukey and Cooley led to their groundbreaking paper of 1965 which has applications to modern cell phones, computer disk drives, DVDs and JPEGs [24, 26, 43].

14.5 FAST METHODS FOR POISSON'S EQUATION

We have seen that the solution of elliptic equations and parabolic equations often involves solving large systems of linear equations. For Poisson's equation this must be done once, while for the heat equation, if an implicit time-differencing scheme is used, then a linear system must be solved at each time step. If these equations are solved by Gaussian elimination, then the amount of work required is $O(m^2 N)$, where m is the half bandwidth and N is the total number of equations and unknowns. For Poisson's equation on an n by n grid, the half bandwidth is $m = n - 1$ and the total number of equations is $N = (n-1)^2$, for a total of $O(n^4)$ operations. For certain problems, such as Poisson's equation on a rectangular region with Dirichlet boundary conditions, there are fast methods for solving the linear equations that require only $O(N \log N)$ operations. These methods make use of the fast Fourier transform (FFT).

Suppose we have a *block* TST matrix with TST (tridiagonal symmetric Toeplitz) blocks:

$$A = \begin{pmatrix} S & T & & \\ T & \ddots & \ddots & \\ & \ddots & \ddots & T \\ & & T & S \end{pmatrix}, \quad S = \begin{pmatrix} \alpha & \beta & & \\ \beta & \ddots & \ddots & \\ & \ddots & \ddots & \beta \\ & & \beta & \alpha \end{pmatrix}, \quad T = \begin{pmatrix} \gamma & \delta & & \\ \delta & \ddots & \ddots & \\ & \ddots & \ddots & \delta \\ & & \delta & \gamma \end{pmatrix}. \quad (14.55)$$

If A comes from the five-point finite difference approximation (14.7) for Poisson's equation with Dirichlet boundary conditions on an n_x by n_y rectangular grid, then, with the natural ordering of equations and unknowns, S and T are $(n_x - 1)$ by $(n_x - 1)$ matrices given by

$$S = \begin{pmatrix} -2/h_x^2 - 2/h_y^2 & 1/h_x^2 & & \\ 1/h_x^2 & \ddots & \ddots & \\ & \ddots & \ddots & 1/h_x^2 \\ & & 1/h_x^2 & -2/h_x^2 - 2/h_y^2 \end{pmatrix}, \quad T = \frac{1}{h_y^2} I,$$

where $h_x = 1/n_x$ and $h_y = 1/n_y$, and A consists of an $(n_y - 1)$ by $(n_y - 1)$ array of such blocks.

The eigenvalues and eigenvectors of TST matrices are known. It was shown in theorem 11.2.1 that all TST matrices have the same eigenvectors; only their eigenvalues differ. Let A be the block TST matrix in (14.55) and assume that S and T are m_1 by m_1 blocks, while the entire matrix is an m_2 by m_2 array of such blocks. Consider the linear system $A\mathbf{u} = \mathbf{f}$, where the vectors $\mathbf{u}$ and $\mathbf{f}$ can also be divided into m_2 blocks, each of length m_1, corresponding to the lines in a grid of m_1 by m_2 interior nodes: $\mathbf{u} = (\mathbf{u}_1, \ldots, \mathbf{u}_{m_2})^T$, $\mathbf{f} = (\mathbf{f}_1, \ldots, \mathbf{f}_{m_2})^T$. This linear system has the form

$$T\mathbf{u}_{\ell-1} + S\mathbf{u}_\ell + T\mathbf{u}_{\ell+1} = \mathbf{f}_\ell, \quad \ell = 1, \ldots, m_2, \tag{14.56}$$

where $\mathbf{u}_0 = \mathbf{u}_{m_2+1} = \mathbf{0}$. Let Q be the orthogonal matrix whose columns, $\mathbf{q}^1, \ldots, \mathbf{q}^{m_1}$, are the eigenvectors of S and T. Then S and T can be written as $S = Q\Lambda Q^T$, $T = Q\,\Theta\,Q^T$, where Λ and Θ are diagonal matrices of eigenvalues of S and T, respectively. Multiplying equation (14.56) on the left by Q^T gives

$$\Theta(Q^T\mathbf{u}_{\ell-1}) + \Lambda(Q^T\mathbf{u}_\ell) + \Theta(Q^T\mathbf{u}_{\ell+1}) = Q^T\mathbf{f}_\ell.$$

Defining $\mathbf{y}_\ell \equiv Q^T\mathbf{u}_\ell$ and $\mathbf{g}_\ell \equiv Q^T\mathbf{f}_\ell$, this becomes

$$\Theta\mathbf{y}_{\ell-1} + \Lambda\mathbf{y}_\ell + \Theta\mathbf{y}_{\ell+1} = \mathbf{g}_\ell, \quad \ell = 1, \ldots, m_2.$$

Let the entries in block ℓ of $\mathbf{y}$ be denoted $y_{1,\ell}, \ldots, y_{m_1,\ell}$, and similarly for those in each block of $\mathbf{g}$. Look at the equations for the jth entry in each block of $\mathbf{y}$:

$$\theta_j y_{j,\ell-1} + \lambda_j y_{j,\ell} + \theta_j y_{j,\ell+1} = g_{j,\ell}, \quad \ell = 1, \ldots, m_2.$$

Note that these equations decouple from those for all other entries. If, for each $j = 1, \ldots, m_1$, we define $\tilde{\mathbf{y}}_j = (y_{j,1}, \ldots, y_{j,m_2})^T$ and likewise $\tilde{\mathbf{g}}_j = (g_{j,1}, \ldots, g_{j,m_2})^T$, then we have m_1 independent tridiagonal systems, each involving m_2 unknowns:

$$\begin{pmatrix} \lambda_j & \theta_j & & \\ \theta_j & \ddots & \ddots & \\ & \ddots & \ddots & \theta_j \\ & & \theta_j & \lambda_j \end{pmatrix} \tilde{\mathbf{y}}_j = \tilde{\mathbf{g}}_j, \quad j = 1, \ldots, m_1.$$

The work to solve these tridiagonal systems is $O(m_1 m_2)$ operations. This gives us the vectors $\tilde{\mathbf{y}}_1, \ldots, \tilde{\mathbf{y}}_{m_1}$, which can be rearranged to obtain the vectors

$\mathbf{y}_1, \ldots, \mathbf{y}_{m_2}$. These were obtained from $\mathbf{u}_1, \ldots, \mathbf{u}_{m_2}$ by multiplying by Q^T. Thus the only remaining task is to recover the solution $\mathbf{u}$ from $\mathbf{y}$:

$$\mathbf{u}_\ell = Q\mathbf{y}_\ell, \quad \ell = 1, \ldots, m_2. \tag{14.57}$$

This requires m_2 matrix–vector multiplications, where the matrices are m_1 by m_1. Ordinarily, the amount of work required would be $O(m_2 m_1^2)$ operations. A similar set of matrix–vector multiplications was required to compute $\mathbf{g}$ from $\mathbf{f}$: $\mathbf{g}_\ell = Q^T \mathbf{f}_\ell$, $\ell = 1, \ldots, m_2$. It turns out that because of the special form of the matrix Q, these matrix–vector multiplications can each be performed in time $O(m_1 \log_2 m_1)$ using the FFT. Thus the total work for performing these multiplications by Q and Q^T, and the major part of the work in solving the original linear system, is $O(m_2 m_1 \log_2 m_1)$. This is almost the optimal order, $O(m_2 m_1)$.

14.5.1 The Fast Fourier Transform

To compute the entries in $\mathbf{u}_\ell$ in (14.57), one must compute sums of the form

$$u_{k,\ell} = \sum_{j=1}^{m_1} Q_{k,j} y_{j,\ell} = \sqrt{\frac{2}{m_1+1}} \sum_{j=1}^{m_1} \sin\left(\frac{\pi j k}{m_1+1}\right) y_{j,\ell}, \quad k = 1, \ldots, m_1.$$

Such sums are called **discrete sine transforms**, and they are closely related to (and can be computed from) the **discrete Fourier transform** (DFT):

$$F_k = \sum_{j=0}^{N-1} e^{2\pi i j k/N} f_j, \quad k = 0, 1, \ldots, N-1. \tag{14.58}$$

Before discussing how to compute the DFT, let us briefly describe the Fourier transform in general. When one measures a signal f, such as the sound made by pushing a button on a touch-tone telephone, over an interval of time or space, one may ask the question what frequencies comprise the signal and what are their magnitudes. To answer this question, one evaluates the **Fourier transform** of f:

$$F(\omega) = \frac{1}{\sqrt{2\pi}} \int_{-\infty}^{\infty} e^{i\omega t} f(t)\, dt, \quad -\infty < \omega < \infty.$$

A nice feature of the Fourier transform is that the original function f can be recovered from F using a very similar formula for the inverse transform:

$$f(t) = \frac{1}{\sqrt{2\pi}} \int_{-\infty}^{\infty} e^{-i\omega t} F(\omega)\, d\omega, \quad -\infty < t < \infty.$$

The precise definition of the Fourier transform differs throughout the literature. Sometimes the factor $1/\sqrt{2\pi}$ in the definition of F is not present, and then a factor of $1/(2\pi)$ is required in the inversion formula. Sometimes the definition of F involves the factor $e^{-i\omega t}$, and then the inversion formula must have the factor $e^{i\omega t}$. Usually, this difference in notation doesn't matter because one computes the Fourier transform, performs some operation on the transformed data, and then transforms back.

In practice, one cannot sample a signal continuously in time, and so one samples at, say, N points $t_j = j(\Delta t)$, $j = 0, 1, \ldots, N-1$, and obtains a vector of N function values $f_j = f(t_j)$, $j = 0, 1, \ldots, N-1$. Knowing f at only N points, one cannot expect to compute F at all frequencies ω. Instead, one might hope to evaluate F at N frequencies $\omega_k = 2\pi k/(N(\Delta t))$, $k = 0, 1, \ldots, N-1$. The formula for $F(\omega_k)$ involves an integral which might be approximated by a sum involving the measured values f_j:

$$F(\omega_k) = \frac{1}{\sqrt{2\pi}} \int_{-\infty}^{\infty} e^{i\omega_k t} f(t)\, dt \approx \frac{1}{\sqrt{2\pi}} \sum_{j=0}^{N-1} e^{i\omega_k t_j} f_j \cdot (\Delta t) = \frac{\Delta t}{\sqrt{2\pi}} \sum_{j=0}^{N-1} e^{2\pi i jk/N} f_j.$$

With this in mind, the discrete Fourier transform (DFT) is defined as in (14.58). It is an approximation (modulo a factor $\Delta t/\sqrt{2\pi}$) to the continuous Fourier transform at the frequencies ω_k. Like the continuous Fourier transform, the DFT can be inverted using a similar formula:

$$f_j = \frac{1}{N} \sum_{k=0}^{N-1} e^{-2\pi i jk/N} F_k, \quad j = 0, 1, \ldots, N-1. \tag{14.59}$$

Evaluating the DFT (14.58) (or the inverse DFT (14.59)) would appear to require $O(N^2)$ operations: One must compute N values, $F_0, \ldots, F_{N-1}$, and each one requires a sum of N terms. It turns out, however, that this evaluation can be done with only $O(N \log N)$ operations using the FFT algorithm of Cooley and Tukey [27].

The ideas of the FFT had been set forth in earlier papers, dating back at least as far as Gauss. Another significant paper in which the FFT was explained was one by Danielson and Lanczos [30]. It was not until the 1960s however, that the complete FFT algorithm was implemented on a computer.

The key observation in computing the DFT of length N is to note that it can be expressed in terms of two DFTs of length $N/2$, one involving the even terms and one involving the odd terms in (14.58). To this end, assume that N is even, and define $w = e^{2\pi i/N}$. We can write

$$\begin{aligned} F_k &= \sum_{j=0}^{N/2-1} e^{2\pi i (2j)k/N} f_{2j} + \sum_{j=0}^{N/2-1} e^{2\pi i (2j+1)k/N} f_{2j+1} \\ &= \sum_{j=0}^{N/2-1} e^{2\pi i jk/(N/2)} f_{2j} + w^k \sum_{j=0}^{N/2-1} e^{2\pi i jk/(N/2)} f_{2j+1} \\ &= F_k^{(e)} + w^k F_k^{(o)}, \quad k = 0, 1, \ldots, N-1, \end{aligned} \tag{14.60}$$

where $F^{(e)}$ denotes the DFT of the even-numbered data points and $F^{(o)}$ the DFT of the odd-numbered data points. Note also that the DFT is periodic, with period equal to the number of data points. Thus $F^{(e)}$ and $F^{(o)}$ have period $N/2$: $F^{(e,o)}_{N/2+p} = F^{(e,o)}_p$, $p = 0, 1, \ldots, (N/2) - 1$. Therefore once the $N/2$ entries in a period of $F^{(e)}$ and $F^{(o)}$ have been computed, we can obtain the length N DFT by combining the two according to (14.60). If the N coefficients w^k, $k = 0, 1, \ldots, N-1$ have already been computed, then the work to do this is

$2N$ operations; we must multiply the coefficients w^k by the entries $F_k^{(o)}$ and add to the entries $F_k^{(e)}$, $k = 0, 1, \ldots, N-1$.

This process can be repeated! For example, to compute the length $N/2$ DFT $F^{(e)}$ (assuming now that N is a multiple of 4), we can write

$$\begin{aligned} F_k^{(e)} &= \sum_{j=0}^{N/4-1} e^{2\pi\imath(2j)k/(N/2)} f_{4j} + \sum_{j=0}^{N/4-1} e^{2\pi\imath(2j+1)k/(N/2)} f_{4j+2} \\ &= \sum_{j=0}^{N/4-1} e^{2\pi\imath jk/(N/4)} f_{4j} + w^{2k} \sum_{j=0}^{N/4-1} e^{2\pi\imath jk/(N/4)} f_{4j+2} \\ &= F_k^{(ee)} + w^{2k} F_k^{(eo)}, \quad k = 0, 1, \ldots, N/2-1, \end{aligned}$$

where $F^{(ee)}$ denotes the DFT of the "even even" data (f_{4j}) and $F^{(eo)}$ the DFT of the "even odd" data (f_{4j+2}). Since the length $N/4$ DFTs $F^{(ee)}$ and $F^{(eo)}$ are periodic with period $N/4$, once the $N/4$ entries in a period of these two transforms are known, we can combine them according to the above formula to obtain $F^{(e)}$. This requires $2N/2$ operations (if the powers of w have already been computed), and the same number of operations is required to compute $F^{(o)}$, for a total of $2N$ operations.

Assuming that N is a power of 2, this process can be repeated until we reach the length 1 DFTs. The DFT of length 1 is the identity, so it requires no work to compute these. There are $\log_2 N$ stages in this process and each stage requires $2N$ operations to combine the DFTs of length 2^j to obtain those of length 2^{j+1}. This gives a total amount of work that is approximately $2N \log_2 N$.

The only remaining question is how to keep track of which entry in the original vector of input data corresponds to each length 1 transform; for example, $F^{(eooeoe)} = f_?$. This looks like a bookkeeping nightmare! But actually it is not so difficult. Reverse the sequence of e's and o's, assign 0 to e and 1 to o, and you will have the binary representation of the index of the original data point. Do you see why this works? It is because at each stage we separate the data according to the next bit from the right. The even data at stage one consists of those elements that have a 0 in the rightmost position of their binary index, while the odd data consists of those elements that have a 1 in the rightmost bit of their index. The "even even" data points have a 0 in the rightmost two bits of their index, while the "even odd" data points have a 0 in the rightmost bit and a 1 in the next bit from the right, etc. To make the bookkeeping really easy, one can initially order the data using bit reversal of the binary index. For example, if there are 8 data points, then one would order them as

$$\begin{pmatrix} f_{0=000} \\ f_{1=001} \\ f_{2=010} \\ f_{3=011} \\ f_{4=100} \\ f_{5=101} \\ f_{6=110} \\ f_{7=111} \end{pmatrix} \longrightarrow \begin{pmatrix} f_{0=000} \\ f_{4=100} \\ f_{2=010} \\ f_{6=110} \\ f_{1=001} \\ f_{5=101} \\ f_{3=011} \\ f_{7=111} \end{pmatrix}.$$

Then the length 2 transforms are just linear combinations of neighboring entries, the length 4 transforms are linear combinations of neighboring length 2 transforms, etc.

Chebyshev Expansions via the FFT

Recall the `chebfun` package that was described in section 8.5, and further used for numerical differentiation and integration in sections 9.1 and 10.4. The key to efficiency of the algorithms used there lies in the ability to rapidly translate between the values of a function at the Chebyshev points (8.15) and the coefficients in a Chebyshev expansion of its nth-degree polynomial interpolant: $p(x) = \sum_{j=0}^{n} a_j T_j(x)$, where $T_j(x) = \cos(j \arccos x)$ is the jth-degree Chebyshev polynomial.

Knowing the coefficients $a_0, \ldots, a_n$, one can evaluate p at the Chebyshev points $\cos(k\pi/n)$, $k = 0, \ldots, n$, by evaluating the sums

$$p\left(\cos\left(\frac{k\pi}{n}\right)\right) = \sum_{j=0}^{n} a_j \cos\left(\frac{jk\pi}{n}\right), \quad k = 0, \ldots, n.$$

These sums are much like the real part of the sums in (14.58), but the argument of the cosine differs by a factor of two from the value that would make them equal. They can be evaluated with a *fast cosine transform*, or they can be put into a form where the FFT can be used; see, for example, [83, p. 508]. To go in the other direction, and determine the coefficients $a_0, \ldots, a_n$ from the function values $f(\cos(k\pi/n))$, $k = 0, \ldots, n$, one simply uses the inverse cosine transform or the inverse FFT.

14.6 MULTIGRID METHODS

In section 12.2 the idea of preconditioners for use with iterative linear system solvers was discussed. When the linear system comes from differencing a partial differential equation, such as Poisson's equation, a natural idea is to use a coarser grid to obtain a less accurate, but easier to compute, approximate solution to the PDE. Of course, the linear system that arises from differencing on a coarser grid is of smaller dimension than the original, so its solution is a vector of values at a smaller number of grid points. Still, if one thinks of this vector as values of, say, a piecewise bilinear function at the nodes of the coarse grid, then the values of that function at intermediate points (including nodes of the fine grid) can be found using bilinear interpolation.

A coarse grid solver alone does not make a good preconditioner, but it turns out that when it is combined with fine grid relaxation methods such as Jacobi iteration or the Gauss–Seidel method the combination can be extremely effective! The relaxation methods reduce "high frequency" error components rapidly, and the coarse grid solver eliminates "low frequency" error components. This is the idea behind **two-grid** methods. Used with the simple iteration method described in section 12.2.2, the procedure would look as follows:

> Form difference equations $A\mathbf{u} = \mathbf{f}$ for the differential equation problem on the desired grid. Also form the matrix A_C for the problem

using a coarser grid. Given an initial guess $\mathbf{u}^{(0)}$ for the solution to the linear system on the fine grid, compute the initial residual $\mathbf{r}^{(0)} = \mathbf{f} - A\mathbf{u}^{(0)}$.

For $k = 1, 2, \ldots,$

1. Thinking of $\mathbf{r}^{(k-1)}$ as defining, say, a piecewise bilinear function, whose values at the fine grid nodes are the entries in $\mathbf{r}^{(k-1)}$, determine the right-hand side vector $\mathbf{r}_C^{(k-1)}$ for the coarse grid corresponding to this right-hand side function in the differential equation.
2. Solve the coarse grid problem $A_C \mathbf{z}_C^{(k-1)} = \mathbf{r}_C^{(k-1)}$ for $\mathbf{z}_C^{(k-1)}$.
3. Think of $\mathbf{z}_C^{(k-1)}$ as containing the nodal values of a piecewise bilinear function on the coarse grid. Use bilinear interpolation to determine the values of this function at the fine grid nodes and assemble these values into a vector $\mathbf{z}^{(k-1,0)}$.
4. Set $\mathbf{u}^{(k,0)} = \mathbf{u}^{(k-1)} + \mathbf{z}^{(k-1,0)}$, and compute the new residual $\mathbf{r}^{(k,0)} = \mathbf{f} - A\mathbf{u}^{(k,0)}$.
5. Perform one or more relaxation steps on the fine grid. If G is the preconditioner, for $\ell = 0, \ldots, L$, set $\mathbf{u}^{(k,\ell+1)} = \mathbf{u}^{(k,\ell)} + G^{-1}(\mathbf{f} - A\mathbf{u}^{(k,\ell)})$. Set $\mathbf{u}^{(k)} = \mathbf{u}^{(k,L)}$ and $\mathbf{r}^{(k)} = \mathbf{f} - A\mathbf{u}^{(k)}$.

Step 1 is sometimes called a *projection* or *restriction* step since the fine grid residual vector is shortened to coincide with the coarse grid. Step 3 is called an *interpolation* or *prolongation* step since the coarse grid solution vector is lengthened to coincide with the fine grid. Suppose the fine grid has N nodes and the coarse grid has N_C nodes. Then the projection operation (assuming that it is linear) might be thought of as multiplying the fine grid residual on the left by an N_C by N matrix P. The interpolation operation might be thought of as multiplying the coarse grid solution vector on the left by an N by N_C matrix Q. Thus, the result of steps 1 to 3 is the vector $\mathbf{z}^{(k-1,0)} = QA_C^{-1}P\mathbf{r}^{(k-1)}$. The error in $\mathbf{u}^{(k,0)}$ is $\mathbf{e}^{(k,0)} \equiv A^{-1}\mathbf{f} - \mathbf{u}^{(k,0)} = \mathbf{e}^{(k-1)} - \mathbf{z}^{(k-1,0)} = (I - QA_C^{-1}PA)\mathbf{e}^{(k-1)}$. After L relaxation steps the error becomes $\mathbf{e}^{(k)} \equiv A^{-1}\mathbf{f} - \mathbf{u}^{(k)} = (I - G^{-1}A)^L(I - QA_C^{-1}PA)\mathbf{e}^{(k-1)}$. We know from section 12.2.2 that the method converges if the spectral radius $\rho[(I - G^{-1}A)^L(I - QA_C^{-1}PA)]$ is less than 1.

Here we have combined the two-grid approach with simple iteration, but the procedure of starting with $\mathbf{r}^{(k-1)}$ and determining the increment to $\mathbf{u}^{(k-1)}$ in order to obtain $\mathbf{u}^{(k)}$ can be thought of as solving a preconditioning system $M\mathbf{z}^{(k-1)} = \mathbf{r}^{(k-1)}$, and then this preconditioner can be used with any other appropriate iterative method. For example, if A and M are symmetric and positive definite then this preconditioner can be used with the conjugate gradient algorithm.

This idea can be applied recursively. Clearly, in step 2, one would like the coarse grid to be sufficiently fine to represent the differential equation with at least modest accuracy; otherwise the coarse and fine grid solutions may have little to do with each other. On the other hand, for fast solution of the coarse grid problem, one would like a coarse grid with very few nodes; perhaps one for which the N_C by N_C linear system $A_C\mathbf{z}_C^{(k-1)} = \mathbf{r}_C^{(k-1)}$ could be solved directly with only a modest number of arithmetic operations. These two competing goals may be achieved by using a *multigrid* method. Take the "coarse" grid to be only moderately coarser than the fine grid (typically, the coarse grid might have spacing $2h$ if the fine grid has spacing h), but solve the system in step 2 by applying the two-grid method with a still coarser grid. Use as many levels as necessary until you reach a grid with few enough nodes that the linear system on that grid can be solved directly.

In practice, it has been found that one need not solve the linear system $A_C \mathbf{z}_C^{(k-1)} = \mathbf{r}_C^{(k-1)}$ very accurately. Instead, problem solutions at different levels of refinement might be approximated by applying one or a few Gauss–Seidel iterations, for example. When the coarsest grid level is reached, then the problem is solved directly. Let grid levels $0, 1, \ldots, J$ be defined with maximum mesh spacings $h_0 \leq h_1 \leq \cdots \leq h_J$, and let A_j denote the coefficient matrix for the problem at level j. Let subscripts on vectors denote the grid level as well. The linear system on the finest level is $A\mathbf{u} = \mathbf{f}$, where $A \equiv A_0$. The multigrid *V-cycle* consists of the following steps:

Given an initial guess $\mathbf{u}^{(0)}$, compute $\mathbf{r}^{(0)} \equiv \mathbf{r}_0^{(0)} = \mathbf{f} - A\mathbf{u}^{(0)}$.
For $k = 1, 2, \ldots,$

For $j = 1, \ldots, J-1$,
Project $\mathbf{r}_{j-1}^{(k-1)}$ onto grid level j; that is, set

$$\mathbf{f}_j = P_{j-1}^{j} \mathbf{r}_{j-1}^{(k-1)},$$

where P_{j-1}^{j} is the restriction matrix from grid level $j-1$ to grid level j.
Perform a Gauss–Seidel iteration (with $\mathbf{0}$ as initial guess) on grid level j; that is, solve

$$G_j \delta_j^{(k-1)} = \mathbf{f}_j,$$

where G_j is the lower triangle of A_j, and compute

$$\mathbf{r}_j^{(k-1)} = \mathbf{f}_j - A_j \delta_j^{(k-1)}.$$

Endfor

Project $\mathbf{r}_{J-1}^{(k-1)}$ onto grid level J by setting $\mathbf{f}_J = P_{J-1}^{J} \mathbf{r}_{J-1}^{(k-1)}$ and solve on the coarsest grid $A_J \mathbf{d}_J^{(k-1)} = \mathbf{f}_J$.

For $j = J-1, \ldots, 1$,
Interpolate $\mathbf{d}_{j+1}^{(k-1)}$ to grid level j and add to $\delta_j^{(k-1)}$; that is, replace

$$\delta_j^{(k-1)} \longleftarrow \delta_j^{(k-1)} + Q_{j+1}^{j} \mathbf{d}_{j+1}^{(k-1)},$$

where Q_{j+1}^{j} is the prolongation matrix from grid level $j+1$ to grid level j.
Perform a Gauss–Seidel iteration (with initial guess $\delta_j^{(k-1)}$) on grid level j; that is, set

$$\mathbf{d}_j^{(k-1)} = \delta_j^{(k-1)} + G_j^{-1}(\mathbf{f}_j - A_j \delta_j^{(k-1)}).$$

Endfor

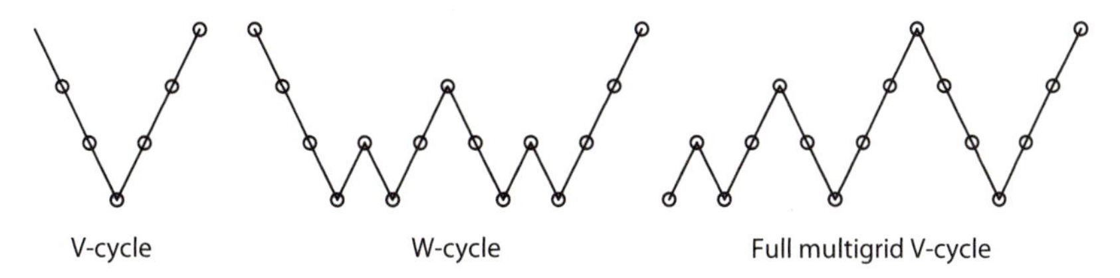

Figure 14.12. Multigrid cycling patterns.

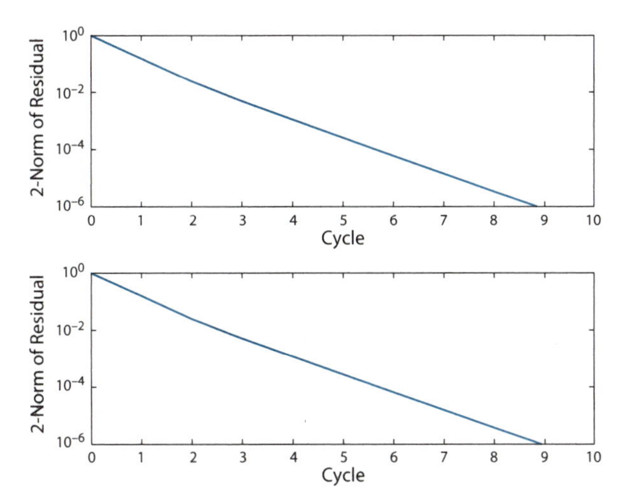

Figure 14.13. Convergence of a multigrid V-cycle with Gauss-Seidel relaxation $h = 1/64$ (*top*) and $h = 1/128$ (*bottom*).

Interpolate $\mathbf{d}_1^{(k-1)}$ to grid level 0 and replace $\mathbf{u}^{(k-1)} \longleftarrow \mathbf{u}^{(k-1)} + Q_1^0 \mathbf{d}_1^{(k-1)}$.

Perform a Gauss–Seidel iteration with initial guess $\mathbf{u}^{(k-1)}$ on grid level 0; that is, set $\mathbf{u}^{(k)} = \mathbf{u}^{(k-1)} + G^{-1}(\mathbf{f} - A\mathbf{u}^{(k-1)})$. Compute the new residual $\mathbf{r}^{(k)} \equiv \mathbf{r}_0^{(k)} = \mathbf{f} - A\mathbf{u}^{(k)}$.

This iteration is called a **V-cycle** because it consists of going down through the grids from fine to coarse performing a Gauss–Seidel iteration on each grid and then coming back up from coarse to fine again performing a Gauss–Seidel iteration at each level. Other patterns of visiting the grids are also possible, as pictured in figure 14.12.

Example 14.6.1. The plots in figure 14.13 show the convergence of the multigrid V-cycle, using a variant of Gauss–Seidel relaxation, for the linear system arising from Poisson's equation on a square with 63 and 127 interior grid points in each direction. The computations used five and six grid levels, respectively, and each solved directly on a grid with 3 interior grid points in each direction.

Note that the number of cycles required to reach a fixed level of accuracy is *independent of the mesh size h*! This is a remarkable feature of multigrid methods that is not shared by other iterative linear system solvers described in section 12.2. For the ICCG method (section 12.2.4), for example, the number of iterations is $O(h^{-1})$. The cost of a multigrid V-cycle is only about $\frac{5}{3}$ the cost of an iteration of the relaxation method on the fine grid, since one relaxation step is performed on the fine grid and two relaxation steps (one on the way down and one on the way up) are performed on each coarser grid. Since each coarser grid level has about $\frac{1}{4}$ as many points as the next finer level, the total work on coarse grids during a cycle is less than $2\sum_{j=1}^{\infty}\frac{1}{4^j} = \frac{2}{3}$ times the cost of a relaxation step on the fine grid. This does not account for the overhead of interpolating between grids, but even with this work included, the work for a multigrid V-cycle with Gauss–Seidel relaxation is less than $\frac{5}{3}$ that of, say, an ICCG iteration. A disadvantage of multigrid methods is that for more general problems of the form (14.1) the projection and interpolation operators, as well as the relaxation method used, may have to be specially tuned to the problem in order to achieve good performance. For problems on irregular grids, or problems that do not involve grids at all, *algebraic multigrid methods* attempt to achieve good convergence rates by working with smaller problems akin to the coarse grid problems in standard multigrid methods.

14.7 CHAPTER 14 EXERCISES

1. Let Δ denote the Laplacian operator: $\Delta u = u_{xx} + u_{yy}$. Show that a problem of the form

$$\Delta u = f \quad \text{in } \Omega, \quad u = g \quad \text{on } \partial\Omega,$$

can be solved as follows: First find any function v that satisfies $\Delta v = f$ in Ω (without regard to the boundary conditions). Then solve for w: $\Delta w = 0$ in Ω, $w = g - v$ on $\partial\Omega$. Then set $u = v + w$; i.e., show that $v + w$ satisfies the original boundary value problem.

2. Suppose you wish to solve Poisson's equation $\Delta u = f$ on the unit square with $u = g$ on the boundary. Using a 4 by 3 grid (that is, 3 interior mesh points in the x direction and 2 interior mesh points in the y direction), write in matrix form the system of linear equations that you would need to solve in order to obtain a second-order-accurate approximate solution.

3. On the book's web page is a code called `poisson.m` to solve Poisson's equation on the unit square with Dirichlet boundary conditions:

$$\Delta u = f \quad \text{in } \Omega, \quad u = g \quad \text{on } \partial\Omega,$$

where Ω is the unit square $(0, 1) \times (0, 1)$. Download this code and read through it to see if you understand what it is doing. Then try running it with some different values of h. (It uses the same number of subintervals in the x and y directions, so $h_x = h_y = h$.) Currently it is set up to solve a problem with $f(x, y) = x^2 + y^2$ and $g = 1$.

If you do not know an analytic solution to a problem, one way to check the code is to solve a problem on a fine grid and pretend that the result is the exact solution, then solve on coarser grids and compare your answers to the fine grid solution. However, you must be sure to compare solution values corresponding to the same grid points. Use this idea to check the accuracy of the code. Report the (approximate) L_2-norm or ∞-norm of the error for several different grid sizes. Based on your results, what would you say is the order of accuracy of the method?

4. Write a code to solve the heat equation in one spatial dimension:

$$u_t = u_{xx}, \quad 0 \le x \le 1, \quad t \ge 0,$$

$$u(x, 0) = u_0(x), \quad u(0, t) = u(1, t) = 0,$$

using forward differences in time and centered differences in space:

$$\frac{u_i^{(k+1)} - u_i^{(k)}}{\Delta t} = \frac{1}{h^2}\left(-2u_i^{(k)} + u_{i+1}^{(k)} + u_{i-1}^{(k)}\right).$$

Take $u_0(x) = x(1 - x)$ and go out to time $t = 1$. Plot the solution at time $t = 0$, at time $t = 1$, and at various times along the way. Experiment with different mesh sizes and different time steps, and report any observations you make about accuracy and stability.

5. Write a code to solve the wave equation:

$$u_{tt} = a^2 u_{xx}, \quad 0 \le x \le 1, \quad t \ge 0,$$

$$u(x, 0) = f(x), \quad u_t(x, 0) = g(x), \quad u(0, t) = u(1, t) = 0,$$

using the difference method:

$$\frac{u_j^{(k+1)} - 2u_j^{(k)} + u_j^{(k-1)}}{(\Delta t)^2} = \frac{a^2}{h^2}\left(-2u_j^{(k)} + u_{j+1}^{(k)} + u_{j-1}^{(k)}\right).$$

Take $f(x) = x(1 - x)$, $g(x) = 0$, and $a^2 = 2$, and go out to time $t = 2$. Plot the solution at each time step; you should see the solution behave like a plucked guitar string. Experiment with different values of h and Δt in order to verify numerically the stability condition

$$\Delta t \le \frac{h}{|a|}.$$

Turn in plots of your solutions for values of h and Δt that satisfy the stability conditions and also some plots showing what happens if the stability requirements are not met.

6. Consider the Lax–Friedrichs method

$$\frac{u_j^{(k+1)} - \frac{1}{2}\left(u_{j+1}^{(k)} + u_{j-1}^{(k)}\right)}{\Delta t} + a\frac{u_{j+1}^{(k)} - u_{j-1}^{(k)}}{2h} = 0$$

for the differential equation $u_t + au_x = 0$. Carry out a stability analysis to determine values of $\lambda = \Delta t / h$ for which the method is stable.

7. This is a *written* exercise. Let **f** be a vector of length 8 with components $f_0, f_1, \ldots, f_7$. Write down and clearly explain the sequence of operations you would use to compute the discrete Fourier transform of **f**, using the FFT technique. Your explanation should be sufficiently clear that someone unfamiliar with the FFT could follow your instructions and compute the FFT of any given vector of length 8.

APPENDIX A

REVIEW OF LINEAR ALGEBRA

A.1 VECTORS AND VECTOR SPACES

In geometry, a vector is an object comprising a magnitude and a direction. It is specified by giving the coordinates of its tip, assuming that it starts at the origin. For example, the vector $(1, 2)$ is a vector in the plane, $\mathbf{R}^2$, with magnitude $\sqrt{5}$ and pointing in the direction arctan(2) from the positive x-axis.

This concept can be generalized to $\mathbf{R}^n$, where a **vector** is given by an ordered n-tuple $(r_1, r_2, \ldots, r_n)$, where each r_i is a real number. Often it is written as a column of numbers instead of a row,

$$\begin{pmatrix} r_1 \\ \vdots \\ r_n \end{pmatrix},$$

which also can be denoted as $(r_1, \ldots, r_n)^T$, where the superscript ^{T} denotes the **transpose**.

The sum of two vectors, $\mathbf{u} = (u_1, \ldots, u_n)^T$ and $\mathbf{v} = (v_1, \ldots, v_n)^T$ in $\mathbf{R}^n$, is again a vector in $\mathbf{R}^n$, given by $\mathbf{u} + \mathbf{v} = (u_1 + v_1, \ldots, u_n + v_n)^T$. If α is a real number, then the product of α with the vector $\mathbf{u}$ is $\alpha\mathbf{u} = (\alpha u_1, \ldots, \alpha u_n)^T$.

A **vector space** V is any collection of objects for which addition and scalar multiplication are defined in such a way that the result of these operations is also in the collection and the operations satisfy some standard rules:

- A1 If $\mathbf{u}$ and $\mathbf{v}$ are in V, then $\mathbf{u} + \mathbf{v}$ is defined and is in V.
- A2 Commutativity: If $\mathbf{u}$ and $\mathbf{v}$ are in V, then $\mathbf{u} + \mathbf{v} = \mathbf{v} + \mathbf{u}$.
- A3 Associativity: If $\mathbf{u}$, $\mathbf{v}$, and $\mathbf{w}$ are in V, then $\mathbf{u} + (\mathbf{v} + \mathbf{w}) = (\mathbf{u} + \mathbf{v}) + \mathbf{w}$.
- A4 Additive identity: There is a vector $\mathbf{0} \in V$ such that $\mathbf{u} + \mathbf{0} = \mathbf{u}$ for all $\mathbf{u} \in V$.
- A5 Additive inverse: For every $\mathbf{u} \in V$, there is a vector $-\mathbf{u} \in V$ such that $\mathbf{u} + (-\mathbf{u}) = \mathbf{0}$.
- M1 If α is a scalar and $\mathbf{u}$ is in V, then $\alpha\mathbf{u}$ is defined and is in V.
- M2 Distributive and associative properties: If α and β are scalars and $\mathbf{u}$ and $\mathbf{v}$ are in V, then $\alpha(\mathbf{u} + \mathbf{v}) = \alpha\mathbf{u} + \alpha\mathbf{v}$, and $(\alpha + \beta)\mathbf{u} = \alpha\mathbf{u} + \beta\mathbf{u}$. Also $(\alpha\beta)\mathbf{u} = \alpha(\beta\mathbf{u})$.
- M3 $1\mathbf{u} = \mathbf{u}$, $0\mathbf{u} = \mathbf{0}$, and $(-1)\mathbf{u} = -\mathbf{u}$.

The scalar α must come from a specified *field*, usually the real or complex numbers. Unless otherwise stated, we will assume that the field is the set of real numbers $\mathbf{R}$. You should convince yourself that $\mathbf{R}^n$, with the rules given above for vector addition and multiplication by real scalars, is a vector space because it satisfies these rules. Other vector spaces include, for example, the space of polynomials of degree 2 or less: $a_0+a_1x+a_2x^2$. When we add two such

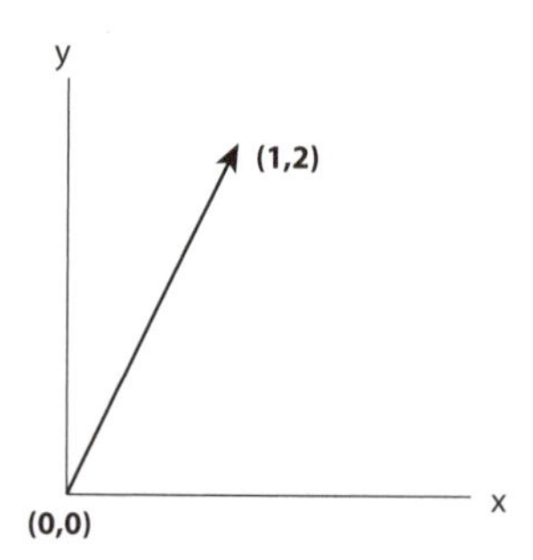

Figure A.1. A vector in the plane.

polynomials, $(a_0+a_1x+a_2x^2)+(b_0+b_1x+b_2x^2) = (a_0+b_0)+(a_1+b_1)x+(a_2+b_2)x^2$, we obtain a polynomial of degree 2 or less, and when we multiply such a polynomial by a scalar c, we obtain $c(a_0+a_1x+a_2x^2) = (ca_0)+(ca_1)x+(ca_2)x^2$, which is again a polynomial of degree 2 or less. It is easy to check that the other rules hold as well and so this is a vector space.

A **subset** S of a vector space V is a collection of some or all (or none if S is the empty set) of the elements of V; that is, S is a subset of V if every element of S is an element of V. If S is also a vector space (i.e., if it is closed under addition and scalar multiplication), then S is said to be a **subspace** of V. For example, suppose V is the vector space $\mathbf{R}^2$, which consists of all ordered pairs $(c_1, c_2)^T$. Then the subset of V consisting of all vectors of the form $(c, c)^T$, where the first and second coordinates are the same, is a *subspace* since the sum of two vectors $(c, c)^T$ and $(d, d)^T$ in this set is $(c+d, c+d)^T$, which is in the set, and the product of such a vector with a scalar α is $\alpha(c, c)^T = (\alpha c, \alpha c)^T$, which is in the set.

A.2 LINEAR INDEPENDENCE AND DEPENDENCE

A set of vectors $\mathbf{v}_1, \ldots, \mathbf{v}_m$ in a vector space V is said to be **linearly independent** if the only scalars $c_1, \ldots, c_m$ for which $c_1\mathbf{v}_1 + \ldots + c_m\mathbf{v}_m = \mathbf{0}$ are $c_1 = \ldots = c_m = 0$. If a set of vectors is not linearly independent; that is, if there exist scalars $c_1, \ldots, c_m$, *not* all zero, such that $c_1\mathbf{v}_1 + \ldots + c_m\mathbf{v}_m = \mathbf{0}$, then $\mathbf{v}_1, \ldots, \mathbf{v}_m$ are said to be **linearly dependent.** In $\mathbf{R}^2$, the vectors $(1, 0)^T$ and $(0, 1)^T$ are linearly independent because if $c_1(1, 0)^T + c_2(0, 1)^T \equiv (c_1, c_2)^T = (0, 0)^T$, then $c_1 = c_2 = 0$. On the other hand, the vectors $(1, 1)^T$ and $(2, 2)^T$ are linearly dependent because if, for example, $c_1 = -2$ and $c_2 = 1$, then $c_1(1, 1)^T + c_2(2, 2)^T = (0, 0)^T$.

A sum of scalar multiples of vectors, such as $c_1\mathbf{v}_1 + \ldots + c_m\mathbf{v}_m \equiv \sum_{j=1}^{m} c_j\mathbf{v}_j$, is called a **linear combination** of the vectors. If the scalars are not all 0, then it is called a *nontrivial* linear combination. The vectors $\mathbf{v}_1, \ldots, \mathbf{v}_m$ are linearly dependent if there is a *nontrivial* linear combination of them that is equal to the zero vector.

A.3 SPAN OF A SET OF VECTORS; BASES AND COORDINATES; DIMENSION OF A VECTOR SPACE

The **span** of a set of vectors $\mathbf{v}_1, \ldots, \mathbf{v}_m$ is the set of all linear combinations of $\mathbf{v}_1, \ldots, \mathbf{v}_m$. If $\mathbf{v}_1, \ldots, \mathbf{v}_m$ lie in a vector space V, then their span, denoted $\text{span}(\mathbf{v}_1, \ldots, \mathbf{v}_m)$, is a subspace of V since it is closed under addition and multiplication by scalars. The vectors $\mathbf{v}_1, \ldots, \mathbf{v}_m$ span all of V if every vector $\mathbf{v} \in V$ can be written as a linear combination of $\mathbf{v}_1, \ldots, \mathbf{v}_m$.

If the vectors $\mathbf{v}_1, \ldots, \mathbf{v}_m$ are *linearly independent* and *span* the vector space V, they are said to form a **basis** for V. For example, the vectors $(1, 0)^T$ and $(0, 1)^T$ form a basis for $\mathbf{R}^2$. The vectors $(1, 0)^T$, $(0, 1)^T$, and $(1, 1)^T$ span $\mathbf{R}^2$, but they are *not* a basis for $\mathbf{R}^2$ because they are linearly dependent: $1(1, 0)^T + 1(0, 1)^T - 1(1, 1)^T = (0, 0)^T$. The vectors $(1, 1)^T$ and $(2, 2)^T$ do not span $\mathbf{R}^2$, since all linear combinations of these vectors have the first and second coordinate equal and hence, for example, the vector $(1, 0)^T$ cannot be written as a linear combination of these vectors; thus they do *not* form a basis for $\mathbf{R}^2$.

While there are infinitely many choices for a basis of a vector space V, it can be shown that *all bases have the same number of elements*. This number is the **dimension** of the vector space V. Thus the vector space $\mathbf{R}^2$ has dimension 2, since the vectors $(1, 0)^T$ and $(0, 1)^T$ form a basis for $\mathbf{R}^2$. The space of polynomials of degree 2 or less has dimension 3, since the polynomials 1, x, and x^2 form a basis; they are linearly independent and any polynomial of degree 2 or less can be written as a linear combination $a_0 1 + a_1 x + a_2 x^2$ of these vectors. In $\mathbf{R}^n$, the set of basis vectors $\mathbf{e}_1, \ldots, \mathbf{e}_n$, where $\mathbf{e}_j$ has a 1 in position j and 0s everywhere else, is called the **standard basis** for $\mathbf{R}^n$. It can be shown, in addition, that in an n-dimensional vector space V, *any* set of n linearly independent vectors is a basis for V, and any set of n vectors that spans V is linearly independent and hence a basis for V.

Given a basis $\mathbf{v}_1, \ldots, \mathbf{v}_m$ for a vector space V, any vector $\mathbf{v} \in V$ can be written *uniquely* as a linear combination of the basis vectors: $\mathbf{v} = \sum_{j=1}^{m} c_j \mathbf{v}_j$. The scalars $c_1, \ldots, c_m$ are called the **coordinates** of $\mathbf{v}$ with respect to the basis $\mathbf{v}_1, \ldots, \mathbf{v}_m$. The coordinates of the vector $(1, 2, 3)^T$ with respect to the standard basis for $\mathbf{R}^3$, for example, are $1, 2, 3$ since $(1, 2, 3)^T = 1\mathbf{e}_1 + 2\mathbf{e}_2 + 3\mathbf{e}_3$. The coordinates of $(1, 2, 3)^T$ with respect to the basis $\{(1, 1, 1)^T, (1, 1, 0)^T, (1, 0, 0)^T\}$ are $3, -1, -1$, since $(1, 2, 3)^T = 3(1, 1, 1)^T - (1, 1, 0)^T - (1, 0, 0)^T$.

A.4 THE DOT PRODUCT; ORTHOGONAL AND ORTHONORMAL SETS; THE GRAM–SCHMIDT ALGORITHM

The **dot product** or **inner product** of two vectors $\mathbf{u} = (u_1, \ldots, u_n)^T$ and $\mathbf{v} = (v_1, \ldots, v_n)^T$ in $\mathbf{R}^n$ is

$$\mathbf{u} \cdot \mathbf{v} \equiv \langle \mathbf{u}, \mathbf{v} \rangle = \sum_{j=1}^{n} u_j v_j.$$

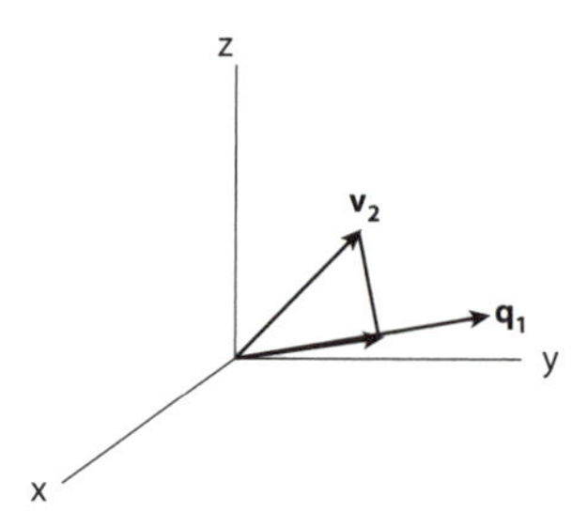

Figure A.2. The orthogonal projection of a vector onto a one-dimensional subspace.

In other vector spaces, inner products may be defined differently. For example, in the vector space of continuous functions on the interval $[0, 1]$—you should check that this *is* a vector space—the inner product of two functions f and g can be defined as $\int_0^1 f(x)g(x)\,dx$. An **inner product** is any rule that takes a pair of vectors $\mathbf{u}$ and $\mathbf{v}$ and maps them to a real number denoted $\langle \mathbf{u}, \mathbf{v} \rangle$ (or sometimes denoted $(\mathbf{u}, \mathbf{v})$ or $\mathbf{u} \cdot \mathbf{v}$), which has the following properties:

1. $\langle \mathbf{u}, \mathbf{v} \rangle = \langle \mathbf{v}, \mathbf{u} \rangle$.
2. $\langle \mathbf{u}, \mathbf{v} + \mathbf{w} \rangle = \langle \mathbf{u}, \mathbf{v} \rangle + \langle \mathbf{u}, \mathbf{w} \rangle$.
3. $\langle \mathbf{u}, \alpha\mathbf{v} \rangle = \alpha\langle \mathbf{u}, \mathbf{v} \rangle$, for any real number α.
4. $\langle \mathbf{u}, \mathbf{u} \rangle \geq 0$ with equality if and only if $\mathbf{u}$ is the zero vector.

An inner product gives rise to a **norm**. The norm of a vector $\mathbf{u}$ is defined as $\|\mathbf{u}\| = \langle \mathbf{u}, \mathbf{u} \rangle^{1/2}$. In $\mathbf{R}^n$, the norm arising from the inner product defined above is the usual Euclidean norm: $\|\mathbf{u}\| = \langle \mathbf{u}, \mathbf{u} \rangle^{1/2} = \sqrt{\sum_{j=1}^n u_j^2}$. In $\mathbf{R}^3$, for example, the norm of the vector $(1, 2, 3)^T$ is $\sqrt{1^2 + 2^2 + 3^2} = \sqrt{14}$. The inner product of the vector $(1, 2, 3)^T$ with the vector $(3, -2, 1)^T$ is $1 \cdot 3 - 2 \cdot 2 + 3 \cdot 1 = 2$.

Two vectors are said to be **orthogonal** if their inner product is 0. The vectors $(1, 2, 3)^T$ and $(3, -3, 1)^T$ are orthogonal since $1 \cdot 3 - 2 \cdot 3 + 3 \cdot 1 = 0$. The standard basis vectors for $\mathbf{R}^n$ are also orthogonal because the inner product of any pair $\mathbf{e}_i$ and $\mathbf{e}_j$, $i \neq j$, is 0. Orthogonal vectors are said to be **orthonormal** if they each have norm 1. The standard basis vectors for $\mathbf{R}^n$ are orthonormal since, in addition to satisfying the orthogonality conditions $\langle \mathbf{e}_i, \mathbf{e}_j \rangle = 0$ for $i \neq j$, they also satisfy $\|\mathbf{e}_i\| = 1$, $i = 1, \ldots, n$. The vectors $(1, 2, 3)^T$ and $(3, -3, 1)^T$ are orthogonal but *not* orthonormal since $\|(1, 2, 3)^T\| = \sqrt{14}$ and $\|(3, -3, 1)^T\| = \sqrt{19}$. Given a pair of orthogonal vectors, one can construct a pair of orthonormal vectors by dividing each vector by its norm and thus creating a vector of norm 1. The vectors $(1/\sqrt{14}, 2/\sqrt{14}, 3/\sqrt{14})^T$ and $(3/\sqrt{19}, -3/\sqrt{19}, 1/\sqrt{19})^T$ are orthonormal.

Given a set of linearly independent vectors $\mathbf{v}_1, \ldots, \mathbf{v}_m$, one can construct an orthonormal set $\mathbf{q}_1, \ldots, \mathbf{q}_m$ using the **Gram–Schmidt algorithm.** Start by setting $\mathbf{q}_1 = \mathbf{v}_1/\|\mathbf{v}_1\|$, so that $\mathbf{q}_1$ has norm 1 and points in the same direction as $\mathbf{v}_1$. Next, take $\mathbf{v}_2$ and subtract its orthogonal projection onto $\mathbf{q}_1$. The **orthogonal projection** of $\mathbf{v}_2$ onto $\mathbf{q}_1$ is the closest vector (in Euclidean norm) to $\mathbf{v}_2$ from $\text{span}(\mathbf{q}_1)$; geometrically, it is the scalar multiple of the vector $\mathbf{q}_1$ that one obtains by dropping a perpendicular from $\mathbf{v}_2$ to $\mathbf{q}_1$. The formula is $\langle \mathbf{v}_2, \mathbf{q}_1 \rangle \mathbf{q}_1$.

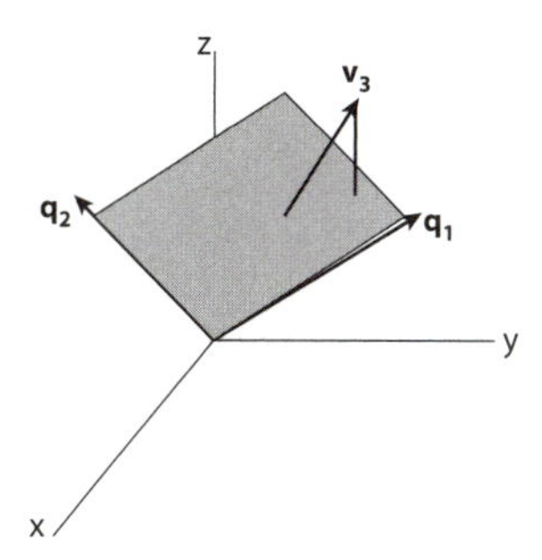

Figure A.3. The orthogonal projection of a vector onto a two-dimensional subspace.

Note that the vector $\tilde{\mathbf{q}}_2 = \mathbf{v}_2 - \langle \mathbf{v}_2, \mathbf{q}_1 \rangle \mathbf{q}_1$ is orthogonal to $\mathbf{q}_1$ since $\langle \tilde{\mathbf{q}}_2, \mathbf{q}_1 \rangle = \langle \mathbf{v}_2, \mathbf{q}_1 \rangle - \langle \mathbf{v}_2, \mathbf{q}_1 \rangle \langle \mathbf{q}_1, \mathbf{q}_1 \rangle = 0$. Now we must normalize $\tilde{\mathbf{q}}_2$ to obtain the next orthonormal vector $\mathbf{q}_2 = \tilde{\mathbf{q}}_2 / \|\tilde{\mathbf{q}}_2\|$. We have now constructed two orthonormal vectors $\mathbf{q}_1$ and $\mathbf{q}_2$ that span the same space as $\mathbf{v}_1$ and $\mathbf{v}_2$.

The next step is to take vector $\mathbf{v}_3$ and subtract its orthogonal projection onto the span of $\mathbf{q}_1$ and $\mathbf{q}_2$. This is the closest vector to $\mathbf{v}_3$ from $\text{span}(\mathbf{q}_1, \mathbf{q}_2)$; geometrically, it is the vector in the plane spanned by $\mathbf{q}_1$ and $\mathbf{q}_2$ that one obtains by dropping a perpendicular from $\mathbf{v}_3$ to this plane. The formula is $\langle \mathbf{v}_3, \mathbf{q}_1 \rangle \mathbf{q}_1 + \langle \mathbf{v}_3, \mathbf{q}_2 \rangle \mathbf{q}_2$.

Note that the vector $\tilde{\mathbf{q}}_3 = \mathbf{v}_3 - \langle \mathbf{v}_3, \mathbf{q}_1 \rangle \mathbf{q}_1 - \langle \mathbf{v}_3, \mathbf{q}_2 \rangle \mathbf{q}_2$ is orthogonal to both $\mathbf{q}_1$ and $\mathbf{q}_2$ since $\langle \tilde{\mathbf{q}}_3, \mathbf{q}_1 \rangle = \langle \mathbf{v}_3, \mathbf{q}_1 \rangle - \langle \mathbf{v}_3, \mathbf{q}_1 \rangle \langle \mathbf{q}_1, \mathbf{q}_1 \rangle - \langle \mathbf{v}_3, \mathbf{q}_2 \rangle \langle \mathbf{q}_2, \mathbf{q}_1 \rangle = 0$, and similarly $\langle \tilde{\mathbf{q}}_3, \mathbf{q}_2 \rangle = 0$. To obtain the next orthonormal vector we set $\mathbf{q}_3 = \tilde{\mathbf{q}}_3 / \|\tilde{\mathbf{q}}_3\|$.

This process is repeated. At each step j, we set:

$$\tilde{\mathbf{q}}_j = \mathbf{v}_j - \sum_{i=1}^{j-1} \langle \mathbf{v}_j, \mathbf{q}_i \rangle \mathbf{q}_i,$$

$$\mathbf{q}_j = \tilde{\mathbf{q}}_j / \|\tilde{\mathbf{q}}_j\|.$$

A.5 MATRICES AND LINEAR EQUATIONS

Consider the following system of 3 linear equations in the 3 unknowns x, y, and z:

$$\begin{aligned} x + y - 2z &= 1, \\ 3x - y + z &= 0, \\ -x + 3y - 2z &= 4. \end{aligned} \tag{A.1}$$

To solve this linear system, one eliminates variables by adding or subtracting appropriate multiples of one equation to another. For example, to eliminate the variable x from the second equation, one could subtract 3 times the first equation from the second to obtain

$$(3x - y + z) - 3(x + y - 2z) = 0 - 3, \quad \text{or} \quad -4y + 7z = -3.$$

To eliminate x from the third equation, one could add the first equation to the third to obtain

$$(-x+3y-2z)+(x+y-2z)=4+1, \quad \text{or} \quad 4y-4z=5.$$

Finally, one can eliminate y from this last equation by adding the previous one:

$$(4y-4z)+(-4y+7z)=5-3, \quad \text{or} \quad 3z=2.$$

From this it follows that $z=2/3$, and substituting this value into the previous equation: $4y-4(2/3)=5$, or, $y=23/12$. Finally, substituting these values into the first equation: $x+23/12-4/3=1$, or, $x=5/12$.

The process that we have carried out is called **Gaussian elimination** with **back substitution.** It can be written in an abbreviated form using matrix and vector notation. A **matrix** is just a rectangular array of numbers or other elements. In this case the matrix of coefficients is a 3 by 3 array, and the equations can be written as follows:

$$\begin{pmatrix} 1 & 1 & -2 \\ 3 & -1 & 1 \\ -1 & 3 & -2 \end{pmatrix} \begin{pmatrix} x \\ y \\ z \end{pmatrix} = \begin{pmatrix} 1 \\ 0 \\ 4 \end{pmatrix}. \tag{A.2}$$

The product of this 3 by 3 matrix with the vector $(x, y, z)^T$ of unknowns is defined to be the vector of length 3 whose entries are the expressions on the left-hand side of equations (A.2).

In general, an m by n matrix A has the form

$$A = \begin{pmatrix} a_{11} & a_{12} & \dots & a_{1n} \\ a_{21} & a_{22} & \dots & a_{2n} \\ \vdots & \vdots & & \vdots \\ a_{m1} & a_{m2} & \dots & a_{mn} \end{pmatrix},$$

where a_{ij} is referred to as the (i, j)-entry in A. Such a matrix is said to have m rows and n columns. The product of this matrix with an n-vector $(x_1, x_2, \dots, x_n)^T$ is the m-vector whose ith entry is $\sum_{j=1}^{n} a_{ij}x_j$, $i=1,\dots,m$. One also can define the product of A with an n by p matrix B whose (j, k)-entry is b_{jk}, $j=1,\dots n$, $k=1,\dots,p$. It is the m by p matrix whose columns are the products of A with each column of B. The (i, k)-entry in AB is $\sum_{j=1}^{n} a_{ij}b_{jk}$, $i=1,\dots,m$, $k=1,\dots,p$. In order for the product of two matrices to be defined, the number of columns in the first matrix must equal the number of rows in the second. In general, matrix–matrix multiplication is *not* commutative: $AB \neq BA$, even if the matrix dimensions are such that both products are defined.

Matrix addition is simpler to define. The sum of two m by n matrices A and B with entries a_{ij} and b_{ij}, respectively, is the m by n matrix whose (i, j)-entry is $a_{ij}+b_{ij}$. Two matrices must have the same dimensions in order to add them.

Returning to the 3 by 3 example (A.2), the process of Gaussian elimination can be carried out using shorthand matrix notation. First, append the right-hand side vector to the matrix in (A.2):

$$\left(\begin{array}{rrr|r} 1 & 1 & -2 & 1 \\ 3 & -1 & 1 & 0 \\ -1 & 3 & -2 & 4 \end{array}\right).$$

Subtracting 3 times the first equation from the second means replacing row two in this appended array by the difference between row two and 3 times row one. Adding the first equation to the third equation means replacing row three by the sum of row three and row one:

$$\left(\begin{array}{rrr|r} 1 & 1 & -2 & 1 \\ 0 & -4 & 7 & -3 \\ 0 & 4 & -4 & 5 \end{array}\right).$$

Note that this new appended matrix is just shorthand notation for the set of equations

$$\begin{aligned} x + y - 2z &= 1, \\ -4y + 7z &= -3, \\ 4y - 4z &= 5. \end{aligned}$$

Adding the second of these equations to the third means replacing row three in the appended matrix by the sum of rows two and three:

$$\left(\begin{array}{rrr|r} 1 & 1 & -2 & 1 \\ 0 & -4 & 7 & -3 \\ 0 & 0 & 3 & 2 \end{array}\right).$$

We now have a shorthand notation for the set of equations

$$\begin{aligned} x + y - 2z &= 1, \\ -4y + 7z &= -3, \\ 3z &= 2, \end{aligned}$$

which are solved as before by first solving the third equation for z, substituting that value into the second equation and solving for y, then substituting both values into the first equation and solving for x.

A.6 EXISTENCE AND UNIQUENESS OF SOLUTIONS; THE INVERSE; CONDITIONS FOR INVERTIBILITY

Under what conditions does a system of m linear equations in n unknowns have a solution, and under what conditions is the solution unique? Such a linear system can be written in the form $A\mathbf{x} = \mathbf{b}$, where A is an m by n matrix whose (i, j)-entry is a_{ij}, $\mathbf{b}$ is a given m-vector of the form $(b_1, \ldots, b_m)^T$, and $\mathbf{x}$ is the n-vector of unknowns $(x_1, \ldots, x_n)^T$.

First, suppose that $\mathbf{b}$ is the zero vector. This is then called a **homogeneous** linear system. Certainly $\mathbf{x} = \mathbf{0}$ is one solution. If there is another solution, say $\mathbf{y}$, whose entries are not all 0, then there are infinitely many solutions, since for any scalar α, $A(\alpha\mathbf{y}) = \alpha A\mathbf{y} = \alpha 0 = 0$. Now suppose that $\hat{\mathbf{b}}$ is a vector whose entries are not all 0, and consider the **inhomogeneous** linear system $A\hat{\mathbf{x}} = \hat{\mathbf{b}}$. By definition, this linear system has a solution if and only if $\hat{\mathbf{b}}$ lies in the **range** of A. If this linear system has one solution, say $\hat{\mathbf{y}}$, and if the homogeneous system $A\mathbf{x} = 0$ has a solution $\mathbf{y} \neq 0$, then the inhomogeneous system has infinitely many solutions since $A(\hat{\mathbf{y}} + \alpha\mathbf{y}) = A\hat{\mathbf{y}} + \alpha A\mathbf{y} = \hat{\mathbf{b}} + \alpha 0 = \hat{\mathbf{b}}$, for any scalar α. Conversely, if the homogeneous system has only the trivial solution $\mathbf{x} = 0$, then the inhomogeneous system cannot have more than one solution since if $\hat{\mathbf{y}}$ and $\hat{\mathbf{z}}$ were two different solutions then $\hat{\mathbf{y}} - \hat{\mathbf{z}}$ would be a nontrivial solution to the homogeneous system since $A(\hat{\mathbf{y}} - \hat{\mathbf{z}}) = A\hat{\mathbf{y}} - A\hat{\mathbf{z}} = \hat{\mathbf{b}} - \mathbf{hatb} = 0$. We summarize these statements in the following theorem.

Theorem A.6.1. Let A be an m by n matrix. The homogeneous system $A\mathbf{x} = 0$ either has only the trivial solution $\mathbf{x} = 0$ or it has infinitely many solutions. If $\hat{\mathbf{b}} \neq 0$ is an m-vector, then the inhomogeneous system $A\hat{\mathbf{x}} = \hat{\mathbf{b}}$ has:

1. No solutions if $\hat{\mathbf{b}}$ is not in the range of A.
2. Exactly one solution if $\hat{\mathbf{b}}$ is in the range of A and the homogeneous system has only the trivial solution.
3. Infinitely many solutions if $\hat{\mathbf{b}}$ is in the range of A and the homogeneous system has a nontrivial solution. In this case all solutions of $A\hat{\mathbf{x}} = \hat{\mathbf{b}}$ are of the form $\hat{\mathbf{y}} + \mathbf{y}$, where $\hat{\mathbf{y}}$ is a particular solution and $\mathbf{y}$ ranges over all solutions of the homogeneous system.

Consider now the case where $m = n$, so that we have n linear equations in n unknowns. The product of the matrix A with a vector $\mathbf{x}$ is a certain linear combination of the columns of A; if these columns are denoted $\mathbf{a}_1, \ldots, \mathbf{a}_n$, then $A\mathbf{x} = \sum_{j=1}^{n} x_j \mathbf{a}_j$. As noted in section A.2, the homogeneous system $A\mathbf{x} = 0$ has only the trivial solution $\mathbf{x} = 0$ if and only if the vectors $\mathbf{a}_1, \ldots, \mathbf{a}_n$ are linearly independent. In this case, these vectors span $\mathbf{R}^n$ since any set of n linearly independent vectors in an n-dimensional space span that space (and hence form a basis for the space), as noted in section A.3. Hence every n-vector $\hat{\mathbf{b}} \in \mathbf{R}^n$ lies in the range of A, and it follows from theorem A.6.1 that for every n-vector $\hat{\mathbf{b}}$ the linear system $A\hat{\mathbf{x}} = \hat{\mathbf{b}}$ has a unique solution.

The n by n **identity** matrix I has 1s on its main diagonal (i.e., the (j, j)-entries, $j = 1, \ldots n$, are 1s) and 0s everywhere else. This matrix is the identity for matrix multiplication because for any n by n matrix A, $AI = IA = A$. Also, for any n-vector $\hat{\mathbf{x}}$, $I\hat{\mathbf{x}} = \hat{\mathbf{x}}$.

The n by n matrix A is said to be **invertible** if there is an n by n matrix B such that $BA = I$; it can be shown that an equivalent condition is that the n by n matrix B satisfies $AB = I$. It also can be shown that the matrix B, which is called the **inverse** of A and denoted A^{-1}, is unique. If A is invertible, then the linear system $A\hat{\mathbf{x}} = \hat{\mathbf{b}}$ can be solved for $\hat{\mathbf{x}}$ by multiplying each side on the left by A^{-1}: $A^{-1}(A\hat{\mathbf{x}}) = (A^{-1}A)\hat{\mathbf{x}} = I\hat{\mathbf{x}} = \hat{\mathbf{x}} = A^{-1}\hat{\mathbf{b}}$. Thus, if A is invertible, then for any n-vector $\hat{\mathbf{b}}$, the linear system $A\hat{\mathbf{x}} = \hat{\mathbf{b}}$ has a unique solution. It follows from the previous arguments that the columns of A are therefore linearly independent. Conversely, if the columns of A are linearly independent, then each equation $A\mathbf{b}_j = \mathbf{e}_j$, $j = 1, \ldots, n$, where $\mathbf{e}_j$ is the jth standard basis vector (i.e., the jth column of I), has a unique solution $\mathbf{b}_j$. It follows that if B is the n by n matrix with columns $(\mathbf{b}_1, \ldots, \mathbf{b}_n)$, then $AB = I$; that is, A is invertible and $B = A^{-1}$.

The following theorem summarizes these results.

Theorem A.6.2. Let A be an n by n matrix. The following conditions are equivalent:

1. A is invertible.
2. The columns of A are linearly independent.
3. The equation $A\mathbf{x} = 0$ has only the trivial solution $\mathbf{x} = 0$.
4. For every n-vector $\hat{\mathbf{b}}$, the equation $A\hat{\mathbf{x}} = \hat{\mathbf{b}}$ has a unique solution.

An invertible matrix is also said to be **nonsingular**, while a **singular** matrix is an n by n matrix that is not invertible.

The **transpose** of A, denoted A^T, is the matrix whose (i, j)-entry is the (j, i)-entry in A; in other words, the columns of A become the rows of A^T. The transpose of a product AB is the product of the transposes in reverse order: $(AB)^T = B^T A^T$. To see this, note that the (i, j)-entry in $(AB)^T$ is the (j, i)-entry in AB, which is $\sum_k a_{jk} b_{ki}$, while the (i, j)-entry in $B^T A^T$ is $\sum_k (B^T)_{ik} (A^T)_{kj} = \sum_k b_{ki} a_{jk}$. It follows that if A is invertible and B is the inverse of A, then B^T is the inverse of A^T since $AB = I \Leftrightarrow (AB)^T = B^T A^T = I^T = I \Leftrightarrow B^T = (A^T)^{-1}$. Thus, an n by n matrix A is invertible if and only if A^T is invertible. From theorem A.6.2, we know that A^T is invertible if and only if its columns are linearly independent; that is, if and only if the rows of A are linearly independent. We thus have another equivalent condition for an n by n matrix A to be invertible:

5. The rows of A are linearly independent.

The **determinant** of a 2 by 2 matrix is defined by

$$\det \begin{pmatrix} a_{11} & a_{12} \\ a_{21} & a_{22} \end{pmatrix} = a_{11}a_{22} - a_{12}a_{21}.$$

For example,

$$\det \begin{pmatrix} 3 & 1 \\ 2 & -4 \end{pmatrix} = 3 \cdot (-4) - 2 \cdot 1 = -14.$$

The determinant of an n by n matrix is defined inductively in terms of determinants of $(n-1)$ by $(n-1)$ matrices. Let A be an n by n matrix and define A_{ij} to be the $(n-1)$ by $(n-1)$ matrix obtained from A by deleting row i and column j. Then

$$\det(A) = \sum_{j=1}^{n} (-1)^{i+j} a_{ij} \det(A_{ij}) = \sum_{i=1}^{n} (-1)^{i+j} a_{ij} \det(A_{ij}).$$

This is called the **Laplace expansion** for the determinant; the first summation is the expansion along row i, and the second is the expansion along column j. It can be shown that the expansion along any row i or any column j gives the same value, det(A). For example, if A is the 3 by 3 matrix

$$A = \begin{pmatrix} 3 & 1 & 2 \\ 2 & -2 & 3 \\ -1 & 0 & 1 \end{pmatrix},$$

then, expanding along the first row, we find

$$\begin{aligned} \det(A) &= 3 \cdot \det\begin{pmatrix} -2 & 3 \\ 0 & 1 \end{pmatrix} - 1 \cdot \det\begin{pmatrix} 2 & 3 \\ -1 & 1 \end{pmatrix} + 2 \cdot \det\begin{pmatrix} 2 & -2 \\ -1 & 0 \end{pmatrix} \\ &= 3 \cdot [(-2) \cdot 1 - 3 \cdot 0] - [2 \cdot 1 - 3 \cdot (-1)] + 2 \cdot [2 \cdot 0 - (-1) \cdot (-2)] \\ &= -15, \end{aligned}$$

while expanding along the second column gives

$$\begin{aligned} \det(A) = -1 \cdot \det\begin{pmatrix} 2 & 3 \\ -1 & 1 \end{pmatrix} + (-2) \cdot \det\begin{pmatrix} 3 & 2 \\ -1 & 1 \end{pmatrix} - 0 \cdot \det\begin{pmatrix} 3 & 2 \\ 2 & 3 \end{pmatrix} \\ = -[2 \cdot 1 - 3 \cdot (-1)] - 2[3 \cdot 1 - 2 \cdot (-1)] = -15. \end{aligned}$$

It can be shown that still another equivalent condition for an n by n matrix A to be invertible is

6. $\det(A) \neq 0$.

When the n by n matrix A is nonsingular so that $\det(A) \neq 0$, the linear system $A\mathbf{x} = \mathbf{b}$ can be solved for $\mathbf{x}$ using **Cramer's rule.** Let $A \overset{j}{\leftarrow} \mathbf{b}$ denote the matrix obtained from A by replacing its jth column by the vector $\mathbf{b}$. Then the jth entry in the solution vector $\mathbf{x}$ is

$$x_j = \frac{\det(A \overset{j}{\leftarrow} \mathbf{b})}{\det(A)}, \quad j = 1, \ldots, n.$$

For example, to solve the linear system

$$\begin{pmatrix} 3 & 1 & 2 \\ 2 & -2 & 3 \\ -1 & 0 & 1 \end{pmatrix} \begin{pmatrix} x_1 \\ x_2 \\ x_3 \end{pmatrix} = \begin{pmatrix} 1 \\ -1 \\ 0 \end{pmatrix},$$

since we already know from above that the determinant of the coefficient matrix A is -15, evaluating the other determinants we find

$$\det\begin{pmatrix} 1 & 1 & 2 \\ -1 & -2 & 3 \\ 0 & 0 & 1 \end{pmatrix} = -1, \quad \det\begin{pmatrix} 3 & 1 & 2 \\ 2 & -1 & 3 \\ -1 & 0 & 1 \end{pmatrix} = -10, \quad \det\begin{pmatrix} 3 & 1 & 1 \\ 2 & -2 & -1 \\ -1 & 0 & 0 \end{pmatrix} = -1,$$

and so

$$x_1 = \frac{1}{15}, \quad x_2 = \frac{10}{15} = \frac{2}{3}, \quad x_3 = \frac{1}{15}.$$

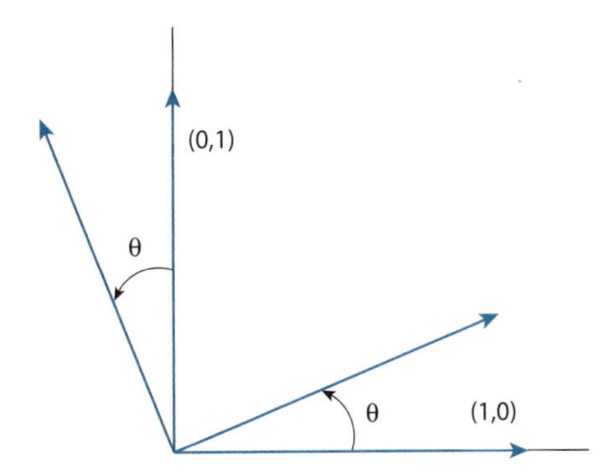

Figure A.4. Counterclockwise rotation of a vector.

A.7 LINEAR TRANSFORMATIONS; THE MATRIX OF A LINEAR TRANSFORMATION

A **linear transformation** is a mapping T from a vector space V to a vector space W that satisfies

$$T(\mathbf{u}+\mathbf{v}) = T(\mathbf{u}) + T(\mathbf{v}), \quad T(\alpha\mathbf{u}) = \alpha T(\mathbf{u}),$$

for all vectors $\mathbf{u}, \mathbf{v} \in V$ and for all scalars α.

Suppose the dimension of V is n and the dimension of W is m. Let $\mathcal{B}_1 = \{\mathbf{v}_1, \ldots, \mathbf{v}_n\}$ be a basis for V and let $\mathcal{B}_2 = \{\mathbf{w}_1, \ldots, \mathbf{w}_m\}$ be a basis for W. Then any vector $\mathbf{v} \in V$ can be written in the form $\mathbf{v} = \sum_{j=1}^{n} c_j \mathbf{v}_j$ for certain scalars $c_1, \ldots, c_n$, and likewise any vector $\mathbf{w} \in W$ can be written as a linear combination of the basis vectors $\mathbf{w}_1, \ldots, \mathbf{w}_m$. In particular, each vector $T(\mathbf{v}_j)$ can be written in the form $T(\mathbf{v}_j) = \sum_{i=1}^{m} a_{ij}\mathbf{w}_i$, for certain scalars $a_{1j}, \ldots, a_{mj}$. It follows from the linearity of T that

$$T(\mathbf{v}) = \sum_{j=1}^{n} c_j T(\mathbf{v}_j) = \sum_{j=1}^{n} c_j \left(\sum_{i=1}^{m} a_{ij}\mathbf{w}_i \right) = \sum_{i=1}^{m} \left(\sum_{j=1}^{n} a_{ij} c_j \right) \mathbf{w}_i .$$

Thus, the coordinates of $T(\mathbf{v})$ with respect to the basis $\mathcal{B}_2$ are the numbers $\sum_{j=1}^{n} a_{ij} c_j$, $i = 1, \ldots, m$. These numbers are the entries in $A\mathbf{c}$, where A is the m by n matrix whose (i, j)-entry is a_{ij} and $\mathbf{c}$ is the n-vector of coordinates of $\mathbf{v}$ with respect to the basis $\mathcal{B}_1$. This matrix A is called the *matrix of* T with respect to the bases $\mathcal{B}_1$ and $\mathcal{B}_2$, and it is sometimes denoted ${}_{\mathcal{B}_2}[T]_{\mathcal{B}_1}$.

When the vector spaces V and W are $\mathbf{R}^n$ and $\mathbf{R}^m$, respectively, and the standard basis vectors are used in both spaces, then if $\mathbf{v} = \sum_{j=1}^{n} c_j \mathbf{e}_j = (c_1, \ldots, c_n)^T = \mathbf{c}$, this just says that $T(\mathbf{v}) = \sum_{i=1}^{m} (A\mathbf{c})_i \mathbf{e}_i = A\mathbf{c}$.

As an example, consider the linear transformation T in the plane that rotates vectors counterclockwise through an angle θ. The image under T of the vector $(1, 0)^T$ is the vector $(\cos\theta, \sin\theta)^T$, and the image under T of the vector $(0, 1)^T$ is the vector $(\cos(\pi/2+\theta), \sin(\pi/2+\theta))^T = (-\sin\theta, \cos\theta)^T$, as illustrated in figure A.4.

Hence, for an arbitrary vector $\mathbf{v} = (c_1, c_2)^T$ in $\mathbf{R}^2$, we have

$$T\begin{pmatrix} c_1 \\ c_2 \end{pmatrix} = c_1 \begin{pmatrix} \cos\theta \\ \sin\theta \end{pmatrix} + c_2 \begin{pmatrix} -\sin\theta \\ \cos\theta \end{pmatrix} = \begin{pmatrix} \cos\theta & -\sin\theta \\ \sin\theta & \cos\theta \end{pmatrix} \begin{pmatrix} c_1 \\ c_2 \end{pmatrix},$$

and the matrix of T with respect to the standard basis is

$$\begin{pmatrix} \cos\theta & -\sin\theta \\ \sin\theta & \cos\theta \end{pmatrix}.$$

On the other hand, if, say $\mathcal{B}_1 = \{(1,1)^T, (1,-1)^T\}$ and $\mathcal{B}_2 = \{(-1,1)^T, (0,1)^T\}$, then the matrix of T with respect to these bases will be different. We have

$$T\begin{pmatrix} 1 \\ 1 \end{pmatrix} = \begin{pmatrix} \cos\theta - \sin\theta \\ \sin\theta + \cos\theta \end{pmatrix} = (\sin\theta - \cos\theta)\begin{pmatrix} -1 \\ 1 \end{pmatrix} + 2\cos\theta \begin{pmatrix} 0 \\ 1 \end{pmatrix},$$

$$T\begin{pmatrix} 1 \\ -1 \end{pmatrix} = \begin{pmatrix} \cos\theta + \sin\theta \\ \sin\theta - \cos\theta \end{pmatrix} = -(\cos\theta + \sin\theta)\begin{pmatrix} -1 \\ 1 \end{pmatrix} + 2\sin\theta \begin{pmatrix} 0 \\ 1 \end{pmatrix}.$$

Thus the matrix of T with respect to the bases $\mathcal{B}_1$ and $\mathcal{B}_2$ is

$${}_{\mathcal{B}_2}[T]_{\mathcal{B}_1} = \begin{pmatrix} \sin\theta - \cos\theta & -(\cos\theta + \sin\theta) \\ 2\cos\theta & 2\sin\theta \end{pmatrix}.$$

Note that if we form matrices S_1 and S_2 whose columns are the two basis vectors in $\mathcal{B}_1$ and $\mathcal{B}_2$, respectively,

$$S_1 = \begin{pmatrix} 1 & 1 \\ 1 & -1 \end{pmatrix}, \quad S_2 = \begin{pmatrix} -1 & 0 \\ 1 & 1 \end{pmatrix},$$

then if A is the matrix of T in the standard basis, then $AS_1 = S_2({}_{\mathcal{B}_2}[T]_{\mathcal{B}_1})$. In the next section, we consider the case where the two bases $\mathcal{B}_1$ and $\mathcal{B}_2$ are the same.

A.8 SIMILARITY TRANSFORMATIONS; EIGENVALUES AND EIGENVECTORS

Two n by n matrices that represent the same linear transformation T from $\mathbf{R}^n$ to $\mathbf{R}^n$ in different bases (but each using the *same* basis for the domain and range) are said to be **similar**. Suppose A is the matrix of T in the standard basis. Let $\mathcal{B}$ be a different basis for $\mathbf{R}^n$, and let S be the n by n matrix whose columns are the basis vectors in $\mathcal{B}$. As noted in the previous section, the matrices A and ${}_{\mathcal{B}}[T]_{\mathcal{B}}$ satisfy $AS = S({}_{\mathcal{B}}[T]_{\mathcal{B}})$, or $A = S({}_{\mathcal{B}}[T]_{\mathcal{B}})S^{-1}$, where we know that S is invertible since its columns are linearly independent. An equivalent definition of similarity is: Two n by n matrices A and C are **similar** if there is a nonsingular matrix S such that $A = SCS^{-1}$ (or, equivalently, $C = \mathcal{S}A\mathcal{S}^{-1}$, where $\mathcal{S} = S^{-1}$). If A and C are similar via this definition; that is, if $A = SCS^{-1}$, then if we think of A as being the matrix of a linear transformation T in the standard basis, then C is the matrix of T in the basis consisting of the columns of S; we know that these vectors are linearly independent and hence form a basis for $\mathbf{R}^n$ because S is invertible.

Let A be an n by n matrix. If there exists a nonzero n-vector $\mathbf{v}$ and a scalar λ (which can be complex) such that $A\mathbf{v} = \lambda\mathbf{v}$, then λ is said to be an **eigenvalue** of A and $\mathbf{v}$ a corresponding **eigenvector**. Similar matrices have the same eigenvalues but possibly different eigenvectors. To see this, suppose that A and C are similar so that there is a matrix S such that $A = SCS^{-1}$. Then if $A\mathbf{v} = \lambda\mathbf{v}$, for some nonzero vector $\mathbf{v}$, then $SCS^{-1}\mathbf{v} = \lambda\mathbf{v}$, or, $C(S^{-1}\mathbf{v}) = \lambda(S^{-1}\mathbf{v})$. Thus λ is an eigenvalue of C with eigenvector $S^{-1}\mathbf{v}$ (which is nonzero since $\mathbf{v}$ is nonzero and S^{-1} is nonsingular). Conversely, if $C\mathbf{u} = \lambda\mathbf{u}$ for some nonzero vector $\mathbf{u}$, then $S^{-1}AS\mathbf{u} = \lambda\mathbf{u}$, or $A(S\mathbf{u}) = \lambda(S\mathbf{u})$. Thus λ is an eigenvalue of A with corresponding eigenvector $S\mathbf{u}$ (which is nonzero since $\mathbf{u}$ is nonzero and S is nonsingular).

The equation $A\mathbf{v} = \lambda\mathbf{v}$ can be written in the form $(A - \lambda I)\mathbf{v} = \mathbf{0}$. This equation has a nonzero solution vector $\mathbf{v}$ if and only if $A - \lambda I$ is singular (see Section A.6). Thus λ is an eigenvalue of A if and only if $A - \lambda I$ is singular, and an equivalent condition is $\det(A - \lambda I) = 0$. The eigenvectors associated with an eigenvalue λ of A are the vectors in the **null space** of $A - \lambda I$; that is, all nonzero vectors $\mathbf{v}$ for which $(A - \lambda I)\mathbf{v} = \mathbf{0}$. Note that if $\mathbf{v}$ is an eigenvector associated with eigenvalue λ, then so is $\alpha\mathbf{v}$ for any nonzero scalar α.

The polynomial in λ defined by $\det(A - \lambda I)$ is called the **characteristic polynomial** of A. Its roots are the eigenvalues of A. For example, to find the eigenvalues of the 2 by 2 matrix

$$A = \begin{pmatrix} 2 & 1 \\ 1 & 2 \end{pmatrix},$$

we form the characteristic polynomial

$$\det\begin{pmatrix} 2-\lambda & 1 \\ 1 & 2-\lambda \end{pmatrix} = (2-\lambda)^2 - 1 = \lambda^2 - 4\lambda + 3,$$

and note that since this can be factored in the form $(\lambda - 3)(\lambda - 1)$, its roots are 3 and 1. These are the eigenvalues of A. To find the eigenvectors associated with the eigenvalue 3, for example, we look for all solutions of the homogeneous system $(A - 3I)\mathbf{v} = \mathbf{0}$:

$$\begin{pmatrix} -1 & 1 \\ 1 & -1 \end{pmatrix}\begin{pmatrix} x_1 \\ x_2 \end{pmatrix} = \begin{pmatrix} 0 \\ 0 \end{pmatrix}.$$

The solutions consist of all nonzero scalar multiples of the vector $(1, 1)^T$. To find the eigenvectors associated with the eigenvalue 1, we look for all solutions of the homogeneous system $(A - I)\mathbf{v} = \mathbf{0}$:

$$\begin{pmatrix} 1 & 1 \\ 1 & 1 \end{pmatrix}\begin{pmatrix} x_1 \\ x_2 \end{pmatrix} = \begin{pmatrix} 0 \\ 0 \end{pmatrix}.$$

The solutions consist of all nonzero scalar multiples of the vector $(1, -1)^T$.

Even real matrices may have complex eigenvalues. For example, the matrix

$$\begin{pmatrix} 2 & 1 \\ -1 & 2 \end{pmatrix}$$

has characteristic polynomial

$$\det \begin{pmatrix} 2-\lambda & 1 \\ -1 & 2-\lambda \end{pmatrix} = (2-\lambda)^2 + 1 = \lambda^2 - 4\lambda + 5.$$

The roots of this polynomial are $2 \pm i$, where $i = \sqrt{-1}$.

The eigenvalues of a *diagonal* matrix are just the diagonal entries and the corresponding eigenvectors are the standard basis vectors $\mathbf{e}_1, \ldots, \mathbf{e}_n$, since

$$\begin{pmatrix} \lambda_1 & & & & \\ & \ddots & & & \\ & & \lambda_j & & \\ & & & \ddots & \\ & & & & \lambda_n \end{pmatrix} \begin{pmatrix} 0 \\ \vdots \\ 1 \\ \vdots \\ 0 \end{pmatrix} = \lambda_j \begin{pmatrix} 0 \\ \vdots \\ 1 \\ \vdots \\ 0 \end{pmatrix}.$$

If an n by n matrix A has n linearly independent eigenvectors, then we say that A is **diagonalizable**, because it is similar to a diagonal matrix of eigenvalues. To see this, let $V = (\mathbf{v}_1, \ldots, \mathbf{v}_n)$ be an n by n matrix whose columns are eigenvectors of A. Then V is invertible since its columns are linearly independent. We can write $AV = V\Lambda$, where $\Lambda = \mathrm{diag}(\lambda_1, \ldots, \lambda_n)$, or $A = V\Lambda V^{-1}$.

Not all matrices are diagonalizable. For example, the matrix

$$A = \begin{pmatrix} 1 & 1 \\ 0 & 1 \end{pmatrix}$$

has characteristic polynomial

$$\det \begin{pmatrix} 1-\lambda & 1 \\ 0 & 1-\lambda \end{pmatrix} = (1-\lambda)^2,$$

which has a double root at $\lambda = 1$. We say that this matrix has eigenvalue 1 with (algebraic) **multiplicity** 2. The corresponding eigenvectors are nonzero solutions of

$$\begin{pmatrix} 0 & 1 \\ 0 & 0 \end{pmatrix} \begin{pmatrix} x_1 \\ x_2 \end{pmatrix} = \begin{pmatrix} 0 \\ 0 \end{pmatrix};$$

that is, all nonzero scalar multiples of $(1, 0)^T$. Since this matrix does not have two *linearly independent* eigenvectors, it is *not* similar to a diagonal matrix. Note that the presence of a repeated eigenvalue (i.e., an eigenvalue of multiplicity greater that one) does not necessarily mean that a matrix is not diagonalizable; the 2 by 2 identity matrix (which is already diagonal) has 1 as an eigenvalue of multiplicity 2, but it also has two linearly independent eigenvectors associated with this eigenvalue, for example, the two standard basis vectors $(1, 0)^T$ and $(0, 1)^T$.

The following theorem states some important facts about eigenvalues and eigenvectors of an n by n matrix A.

Theorem A.8.1. Let A be an n by n matrix.

1. If A is real *symmetric* (i.e., $A = A^T$), then the eigenvalues of A are real and A has n linearly independent eigenvectors that can be taken to be *orthonormal*. Hence if Q is the n by n matrix whose columns are these orthonormal eigenvectors and if Λ is the diagonal matrix of eigenvalues, then $A = Q\Lambda Q^T$ since $Q^T = Q^{-1}$.
2. If A has n *distinct* eigenvalues (i.e., the roots of the characteristic polynomial are all simple roots), then A is diagonalizable.
3. If A is a *triangular matrix* (one whose entries above the main diagonal or below the main diagonal are all 0) then the eigenvalues are the diagonal entries.
4. The *determinant* of A is the product of its eigenvalues (counted with multiplicities).
5. The *trace* of A, which is the sum of its diagonal entries $\sum_{j=1}^{n} a_{jj}$, is equal to the sum of its eigenvalues (counted with multiplicities).

APPENDIX B

TAYLOR'S THEOREM IN MULTIDIMENSIONS

Recall that Taylor's theorem with remainder in one dimension says that a function $f : \mathbf{R} \to \mathbf{R}$ that is $k+1$ times differentiable can be expanded about a point $a \in \mathbf{R}$ as follows:

$$f(a+h) = f(a) + hf'(a) + \frac{h^2}{2!} f''(a) + \ldots + \frac{h^k}{k!} f^{(k)}(a) + \frac{h^{k+1}}{(k+1)!} f^{(k+1)}(\xi),$$

where ξ is some point between a and $a+h$.

Suppose now that f: $\mathbf{R}^2 \to \mathbf{R}$ is a function of two independent variables x_1 and x_2, and we wish to expand $f(a_1+h_1, a_2+h_2)$ about the point (a_1, a_2). Consider the line segment joining the points (a_1, a_2) and (a_1+h_1, a_2+h_2). This line segment can be described as the set of points $\{(a_1+th_1, a_2+th_2) : 0 \leq t \leq 1\}$. Let us define a function g of the single real variable t by

$$g(t) = f(a_1+th_1, a_2+th_2).$$

Differentiating g using the chain rule, we find

$$g'(t) = \frac{\partial f}{\partial x_1} h_1 + \frac{\partial f}{\partial x_2} h_2,$$

$$g''(t) = \frac{\partial^2 f}{\partial x_1^2} h_1^2 + \frac{\partial^2 f}{\partial x_1 \partial x_2} h_1 h_2 + \frac{\partial^2 f}{\partial x_2 \partial x_1} h_2 h_1 + \frac{\partial^2 f}{\partial x_2^2} h_2^2$$

$$= \sum_{i_1, i_2 = 1}^{2} \frac{\partial^2 f}{\partial x_{i_1} \partial x_{i_2}} h_{i_1} h_{i_2},$$

$$g'''(t) = \sum_{i_1, i_2, i_3 = 1}^{2} \frac{\partial^3 f}{\partial x_{i_1} \partial x_{i_2} \partial x_{i_3}} h_{i_1} h_{i_2} h_{i_3},$$

$$\vdots$$

$$g^{(k)}(t) = \sum_{i_1, i_2, \ldots, i_k = 1}^{2} \frac{\partial^k f}{\partial x_{i_1} \partial x_{i_2} \cdots \partial x_{i_k}} h_{i_1} h_{i_2} \cdots h_{i_k},$$

$$g^{(k+1)}(t) = \sum_{i_1, i_2, \ldots, i_{k+1} = 1}^{2} \frac{\partial^{k+1} f}{\partial x_{i_1} \partial x_{i_2} \cdots \partial x_{i_{k+1}}} h_{i_1} h_{i_2} \cdots h_{i_{k+1}},$$

where all partial derivatives of f are evaluated at the point (a_1+th_1, a_2+th_2).

Now using a one-dimensional Taylor series expansion for $g(1) = f(a_1 + h_1, a_2 + h_2)$ about 0 (where $g(0) = f(a_1, a_2)$), namely,

$$g(1) = \sum_{\ell=0}^{k} \frac{1}{\ell!} g^{(\ell)}(0) + \frac{1}{(k+1)!} g^{(k+1)}(\tau), \quad 0 < \tau < 1, \tag{B.1}$$

and substituting the above expressions for the derivatives of g, we find

$$\begin{aligned} f(a_1 + h_1, a_2 + h_2) = {} & f(a_1, a_2) + \left(\frac{\partial f}{\partial x_1} h_1 + \frac{\partial f}{\partial x_2} h_2 \right) \\ & + \left(\frac{1}{2!} \sum_{i_1, i_2 = 1}^{2} \frac{\partial^2 f}{\partial x_{i_1} \partial x_{i_2}} h_{i_1} h_{i_2} \right) + \cdots \\ & + \left(\frac{1}{k!} \sum_{i_1, i_2, \ldots, i_k = 1}^{2} \frac{\partial^k f}{\partial x_{i_1} \partial x_{i_2} \cdots \partial x_{i_k}} h_{i_1} h_{i_2} \cdots h_{i_k} \right) \\ & + \left(\frac{1}{(k+1)!} \sum_{i_1, i_2, \ldots, i_{k+1} = 1}^{2} \right. \\ & \left. \frac{\partial^{k+1} f(a_1 + \tau h_1, a_2 + \tau h_2)}{\partial x_{i_1} \partial x_{i_2} \cdots \partial x_{i_{k+1}}} h_{i_1} h_{i_2} \cdots h_{i_{k+1}} \right). \end{aligned} \tag{B.2}$$

Here all partial derivatives of f of order up through k are evaluated at the point (a_1, a_2). This is Taylor's theorem with remainder for $\mathbf{R}^2$.

If we assume that the partial derivatives of f of order up through $k + 1$ are all continuous, then it can be shown that the order of differentiation does not matter; for example, $\frac{\partial^2 f}{\partial x_1 \partial x_2} = \frac{\partial^2 f}{\partial x_2 \partial x_1}$. In this case, the formula (B.2) can be written as

$$\begin{aligned} f(a_1 + h_1, a_2 + h_2) = {} & f(a_1, a_2) + \left(\frac{\partial f}{\partial x_1} h_1 + \frac{\partial f}{\partial x_2} h_2 \right) \\ & + \left(\frac{1}{2!} \sum_{i=0}^{2} \binom{2}{i} \frac{\partial^2 f}{\partial x_1^{2-i} \partial x_2^{i}} h_1^{2-i} h_2^{i} \right) + \cdots \\ & + \left(\frac{1}{k!} \sum_{i=0}^{k} \binom{k}{i} \frac{\partial^k f}{\partial x_1^{k-i} \partial x_2^{i}} h_1^{k-i} h_2^{i} \right) \\ & + \left(\frac{1}{(k+1)!} \sum_{i=0}^{k+1} \binom{k+1}{i} \right. \\ & \left. \times \frac{\partial^{k+1} f(a_1 + \tau h_1, a_2 + \tau h_2)}{\partial x_1^{k+1-i} \partial x_2^{i}} h_1^{k+1-i} h_2^{i} \right). \end{aligned}$$

Again, all partial derivatives of f of order up through k are evaluated at the point (a_1, a_2).

Formula (B.2) can be generalized to n dimensions using the same technique. Consider a line segment from the point $(a_1, \ldots, a_n)$, about which the expansion is to be based, to the point $(a_1 + h_1, \ldots, a_n + h_n)$. Such a line segment can be expressed as $\{(a_1 + th_1, \ldots, a_n + th_n) : 0 \leq t \leq 1\}$. Define a function $g(t)$ of the single real variable t by

$$g(t) = f(a_1 + th_1, \ldots, a_n + th_n),$$

and consider the one-dimensional Taylor expansion for $g(1) = f(a_1 + h_1, \ldots, a_n + h_n)$ about 0 (where $g(0) = f(a_1, \ldots, a_n)$), given in (B.1). Differentiate g using the chain rule and substitute the expressions involving partial derivatives of f for the derivatives of g. The result is

$$f(a_1 + h_1, \ldots, a_n + h_n) = \sum_{\ell=0}^{k} \frac{1}{\ell!} \sum_{i_1, \ldots, i_\ell = 1}^{n} \frac{\partial^\ell f}{\partial x_{i_1} \cdots \partial x_{i_\ell}} h_{i_1} \cdots h_{i_\ell} + \frac{1}{(k+1)!} \times \sum_{i_1, \ldots, i_{k+1} = 1}^{n} \frac{\partial^{k+1} f(a_1 + \tau h_1, \ldots, a_n + \tau h_n)}{\partial x_{i_1} \cdots \partial x_{i_{k+1}}} h_{i_1} \cdots h_{i_{k+1}}. \tag{B.3}$$

Suppose now that $\mathbf{f}: \mathbf{R}^n \to \mathbf{R}^m$ is a vector-valued function of n variables:

$$\mathbf{f}(\mathbf{x}) = \begin{pmatrix} f_1(x_1, \ldots, x_n) \\ \vdots \\ f_m(x_1, \ldots, x_n) \end{pmatrix}.$$

Expansion (B.3) can be carried out for each of the scalar-valued functions $f_1, \ldots, f_m$:

$$f_i(a_1 + h_1, \ldots, a_n + h_n) = f_i(a_1, \ldots, a_n) + \sum_{j=1}^{n} \frac{\partial f_i}{\partial x_j} h_j + O(h^2), \quad i = 1, \ldots, m, \tag{B.4}$$

where $h = \max\{h_1, \ldots, h_n\}$ and $O(h^2)$ denotes terms that involve products of two or more h_j's. Using matrix–vector notation, the system of equations (B.4) can be written compactly as

$$\mathbf{f}(\mathbf{a} + \mathbf{h}) = \mathbf{f}(\mathbf{a}) + J_f(\mathbf{a})\mathbf{h} + \mathbf{O}(\|\mathbf{h}\|)^2,$$

where $\mathbf{a} = (a_1, \ldots, a_n)^T$, $\mathbf{h} = (h_1, \ldots, h_n)^T$, and J_f is the m by n *Jacobian* matrix for f:

$$J_f(\mathbf{a}) = \begin{pmatrix} \partial f_1/\partial x_1 & \ldots & \partial f_1/\partial x_n \\ \vdots & & \vdots \\ \partial f_m/\partial x_1 & \ldots & \partial f_m/\partial x_n \end{pmatrix}.$$

The $\mathbf{O}(\|\mathbf{h}\|)^2$ notation is meant to represent vector terms whose norm is $O(\|\mathbf{h}\|^2)$.

REFERENCES

[1] Anderson, E., Z. Bai, C. Bischof, S. Blackford, J. Demmel, J. Dongarra, J. D. Croz, A. Greenbaum, S. Hammarling, A. McKenney, and D. Sorensen (1999). *LAPACK Users' Guide* (3rd ed.). Philadelphia, PA: SIAM.

[2] Anderson, M., R. Wilson, and V. Katz (Eds.) (2004). *Sherlock Holmes in Babylon and Other Tales of Mathematical History*. Mathematical Association of America.

[3] Anton, H., I. Bivens, and S. Davis (2009). *Calculus* (9th ed.). Wiley.

[4] Ariane (2004, October). Inquiry board traces Ariane 5 failure to overflow error. *SIAM News 29*(8). http://www.cs.clemson.edu/ steve/Spiro/arianesiam.htm, accessed September 2011.

[5] Arndt, J. and C. Haenel (2001). *Pi-Unleashed*. Springer.

[6] Barber, S. and T. Chartier (2007, July/August). Bending a soccer ball with CFD. *SIAM News 40*(6), 6.

[7] Barber, S., S. B. Chin, and M. J. Carré (2009). Sports ball aerodynamics: a numerical study of the erratic motion of soccer balls. *Computers and Fluids 38*(6), 1091–1100.

[8] Bashforth, F. (1883). *An Attempt to Test the Theories of Capillary Action by Comparing the Theoretical and Measured Forms of Fluid. With an Explanation of the Method of Integration Employed in Constructing the Tables Which Give the Theoretical Forms of Such Drops, by J. C. Adams*. Cambridge, England: Cambridge Univeristy Press.

[9] Battles, Z. and L. Trefethen (2004). An extension of MATLAB to continuous functions and operators. *SIAM J. Sci. Comput. 25*(5), 1743–1770.

[10] Berggren, L., J. Borwein, and P. Borwein (2000). *Pi: A Source Book* (2nd ed.). Springer.

[11] Berrut, J.-P. and L. N. Trefethen (2004). Barycentric Lagrange interpolation. *SIAM Review 46*(3), 501–517.

[12] Berry, M. W. and M. Browne (2005). *Understanding Search Engines: Mathematical Modeling and Text Retrieval (Software, Environments, Tools)* (Second ed.). Philadelphia, PA: SIAM.

[13] Björck, A., C. W. Gear, and G. Söderlind (2005, May). Obituary: Germund Dahlquist. *SIAM News*. http://www.siam.org/news/news.php?id=54.

[14] Blackford, L. S., J. Choi, A. Cleary, E. D'Azevedo, J. Demmel, I. Dhillon, J. Dongarra, S. Hammarling, G. Henry, A. Petitet, K. Stanley, D. Walker, and R. C. Whaley (1997). *ScaLAPACK Users' Guide*. Philadelphia, PA: SIAM.

[15] Blackford, L. S., J. Demmel, J. Dongarra, I. Duff, S. Hammarling, G. Henry, M. Heroux, L. Kaufman, A. Lumsdaine, A. Petitet, R. Pozo, K. Remington, and

R. C. Whaley (2002). An updated set of basic linear algebra subprograms (BLAS). *ACM Trans. Math. Soft. 28*, 135–151.

[16] Bradley, R. E. and C. E. Sandifer (Eds.) (2007). *Leonhard Euler: Life, Work and Legacy*. Elsevier Science.

[17] Brezinski, C. and L. Wuytack (2001). *Numerical Analysis: Historical Developments in the 20th Century*. Gulf Professional Publishing.

[18] Brin, S. and L. Page (1998). The anatomy of a large-scale hypertextual web search engine. *Computer Networks and ISDN Systems 30*, 107–117.

[19] Bultheel, A. and R. Cools (2010). *The Birth of Numerical Analysis*. World Scientific: Hackensack, NJ.

[20] Chabert, J. L. (Ed.) (1999). *A History of Algorithms: From the Pebble to the Microchip*. New York: Springer. (Translation of 1994 French edition).

[21] Chartier, T. (2006, April). Googling Markov. *The UMAP Journal 27*, 17–30.

[22] Chartier, T. and D. Goldman (2004, April). Mathematical movie magic. *Math Horizons*, 16–20.

[23] Chowdhury, I. and S. ~P. Dasgupta (2008). *Dynamics of Structure and Foundation: A Unified Approach*. Vol. 2. Boca Raton, FL: CRC Press.

[24] Citation Classic (1993, December). This week's citation classic. *Current Contents 33*. http://www.garfield.library.upenn.edu/classics1993/A1993MJ84500001.pdf.

[25] comScore Report (2010, January). comScore reports global search market growth of 46 percent in 2009. http://www.comscore.com/Press_Events/Press_Releases/2010/1/Global_Search_Market_Grows_46_Percent_in_2009.

[26] Cooley, J. W. (1997, March). James W. Cooley, an oral history conducted by Andrew Goldstein.

[27] Cooley, J. W. and J. W. Tukey (1965). An algorithm for the machine calculation of complex Fourier series. *Math. Comput. 19*, 297–301.

[28] Coughran, W. (2004, November). Google's index nearly doubles. http://googleblog.blogspot.com/2004/11/googles-index-nearly-doubles.html.

[29] Curtiss, C. F. and J. O. Hirschfelder (1952). Integration of stiff equations. *Proc. US Nat. Acad. Sci. 38*, 235–243.

[30] Danielson, G. C. and C. Lanczos (1942). Some improvements in practical Fourier analysis and their application to X-ray scattering from liquids. *J. Franklin Inst.*, 365–380 and 435–452.

[31] Davis, P. (1996). B-splines and geometric design. *SIAM News 29*(5).

[32] Demmel, J. W. (1997). *Applied Numerical Linear Algebra*. Philadelphia, PA: SIAM.

[33] Devaney, R. L. (2006, September). Unveiling the Mandelbrot set. *Plus magazine 40*. http://plus.maths.org/content/unveiling-mandelbrot-set, accessed May 2011.

[34] Devlin, K. Devlin's angle: Monty Hall. http://www.maa.org/devlin/devlin_07_03.html, accessed August 2010.

[35] Diefenderfer, C. L. and R. B. Nelsen (Eds.) (2010). *Calculus Collection: A Resource for AP and Beyond*. MAA.

[36] Dovidio, N. and T. Chartier (2008). Searching for text in vector space. *The UMAP Journal 29*(4), 417–430.

[37] Edelman, A. (1997). The mathematics of the Pentium division bug. *SIAM Review 39*, 54–67.

[38] Farin, G. and D. Hansford (2002, March). Reflection lines: Shape in automotive design. *Ties Magazine*.

[39] Floudas, C. A. and P. M. Pardalos (Eds.) (2001). *Encyclopedia of Optimization*. Kluwer.

[40] Forsythe, G. E. (1953). Solving linear algebraic equations can be interesting. *Bull. Amer. Math. Soc. 59*, 299–329.

[41] Frankel, S. P. (1950). Convergence rates of iterative treatments of partial differential equations. *Math. Tables and Other Aids to Comput. 4*(30), 65–75.

[42] Freund, R. W. and N. M. Nachtigal (1991). QMR: A quasi-minimal residual method for non-Hermitian linear systems. *Numer. Math. 60*, 315–339.

[43] Garwin, R. L. (2008, November). U.S. Science Policy at a Turning Point. *Plenary talk at the 2008 Quadrennial Congress of Sigma Pi Sigma*. http://www.fas.org/rlg/20.htm.

[44] Gear, C. W. (1971). *Numerical Initial Value Problems in Ordinary Differential Equations*. Englewood Cliffs, NJ: Prentice-Hall.

[45] Gleick, J. (1996, December). A bug and a crash: Sometimes a bug is more than a nuisance. *New York Times Magazine*. http://www.around.com/ariane.html, accessed September 2005.

[46] Goldstine, H. H. (1972). *The Computer from Pascal to von Neumann*. Princeton University Press.

[47] Grandine, T. A. (2005). The extensive use of splines at Boeing. *SIAM News 38*(4).

[48] Greenbaum, A. (1997). *Iterative Methods for Solving Linear Systems*. Philadelphia, PA: SIAM.

[49] Hall, A. (1873). On an experimental determination of π. *Messenger of Mathematics 2*, 113–114.

[50] Haveliwala, T. and S. Kamvar (2003). The second eigenvalue of the Google matrix.

[51] Hestenes, M. R. and E. Stiefel (1952). Methods of conjugate gradients for solving linear systems. *J. Res. Nat. Bur. Standards 49*, 409–436.

[52] Higham, D. J. and N. J. Higham (2005). *MATLAB Guide*. Philadelphia, PA: SIAM.

[53] Hotelling, H. (1943). Some new methods in matrix calculations. *Annals of Mathematical Statistics 14*, 1–34.

[54] Howle, V. E. and L. N. Trefethen (2001). Eigenvalues and musical instruments. *Journal of Computational and Applied Mathematics 135*, 23–40.

[55] Janeba, M. The Pentium problem. http://www.willamette.edu/~mjaneba/pentprob.html, accessed September 2005.

[56] Julia, G. (1919). Mémoire sur l'itération des fonctions rationelles. *Journal de Mathématiques Pures et Appliquées 8*, 47–245.

[57] Katz, V. (1998). *A History of Mathematics: An Introduction* (2nd ed.). Addison Wesley.

[58] Kleinberg, J. (1999). Authoritative sources in a hyperlinked environment. *Journal of the ACM 46*(5), 604–632.

[59] Kleinberg, J., R. Kumar, P. Raghavan, S. Rajagopalan, and A. Tomkins (1999). The Web as a graph: Measurements, models and methods. *Lecture Notes in Computer Science 1627*, 1–18.

[60] Kollerstrom, N. (1992). Thomas Simpson and 'Newton's method of approximation': An enduring myth. *The British Journal for the History of Science 25*, 347–354.

[61] Lanczos, C. (1950). An iteration method for the solution of the eigenvalue problem of linear differential and integral operators. *J. Res. Nat. Bur. Standards 45*, 255–282.

[62] Lanczos, C. (1952). Solutions of linear equations by minimized iterations. *J. Res. Nat. Bur. Standards 49*, 33–53.

[63] Langville, A. and C. Meyer (2004). Deeper inside PageRank. *Internet Mathematics 1*(3), 335–380.

[64] Langville, A. and C. Meyer (2006). *Google's PageRank and Beyond: The Science of Search Engine Rankings*. Princeton University Press.

[65] Lilley, W. (1983, November). Vancouver stock index has right number at last. *Toronto Star*, C11.

[66] MacTutor. MacTutor history of mathematics - indexes of biographies. http://www.gap-system.org/~history/BiogIndex.html.

[67] Massey, K. (1997). Statistical models applied to the rating of sports teams. *Bluefield College*.

[68] Math Whiz. Math whiz stamps profound imprint on computing world. http://www.bizjournals.com/albuquerque/stories/2009/02/02/story9.html, accessed May 2011.

[69] Mathews, J. H. and R. W. Howell (2006). *Complex Analysis for Mathematics and Engineering* (5th ed.). Jones and Bartlett Mathematics.

[70] McCullough, B. D. and H. D. Vinod (1999, June). The numerical reliability of econometric software. *Journal of Economic Literature 37*(2), 633–665.

[71] McDill, J. and B. Felsager (1994, November). The lighter side of differential equations. *The College Mathematics Journal 25*(5), 448–452.

[72] Meijering, E. (2002, March). A chronology of interpolation: From ancient astronomy to modern signal and image processing. *Proceedings of the IEEE 90*, 319–342.

[73] Meijerink, J. A. and H. A. van der Vorst (1977). An iterative solution method for linear systems of which the coefficient matrix is a symmetric M-matrix. *Math. Comp. 31*, 148–162.

[74] Meurant, G. (2006). *The Lanczos and Conjugate Gradient Algorithms: From Theory to Finite Precision Computations*. Philadelphia, PA: SIAM.

[75] Meyer, C. D. (2000). *Matrix Analysis and Applied Linear Algebra: Solutions Manual*, Volume 2. Philadelphia, PA: SIAM.

[76] Moler, C. B. (1995, January). A tale of two numbers. *SIAM News 28*(1), 16.

[77] Newton, I. (1664–1671). Methodus fluxionum et serierum infinitarum.

[78] Nicely, T. Dr. Thomas Nicely's Pentium email. http://www.emery.com/bizstuff/nicely.htm, accessed September 2005.

[79] Overton, M. L. (2001). *Numerical Computing with IEEE Floating Point Arithmetic*. Philadelphia, PA: SIAM.

[80] Overture. https://computation.llnl.gov/casc/Overture/, accessed August 2010.

[81] Page, L., S. Brin, R. Motwani, and T. Winograd (1998, November). The PageRank citation ranking: Bringing order to the web. Technical report, Stanford Digital Library Technologies Project, Stanford University, Stanford, CA, USA.

[82] Peterson, I. (1997, May). Pentium bug revisited. http://www.maa.org/mathland/mathland_5_12.html, accessed September 2005.

[83] Press, W. H., S. A. Teukolski, W. T. Vettering, and B. H. Flannery (1992). *Numerical Recipes in Fortran 77* (Second ed.). Cambridge, England: Cambridge Univeristy Press.

[84] Quinn, K. (1983, November). Ever had problems rounding off figures? This stock exchange has. *The Wall Street Journal*, 37.

[85] Roozen-Kroon, P. (1992). *Structural Optimization of Bells*. Ph. D. thesis, Technische Universiteit Eindhoven.

[86] Rosenthal, J. S. (2008, September). Monty Hall, Monty Fall, Monty Crawl. *Math Horizons*, 5–7.

[87] Saad, Y. and M. H. Schultz (1986). GMRES: A generalized minimal residual algorithm for solving nonsymmetric linear systems. *SIAM J. Sci. Statist. Comput. 7*, 856–869.

[88] Salzer, H. (1972). Lagrangian interpolation at the Chebyshev points $x_{n,\nu} = \cos(\nu\pi/n)$, $\nu = 0(1)n$; some unnoted advantages. *Computer J. 15*, 156–159.

[89] Selvin, S. A problem in probability (letter to the editor). *The American Statistician 29*(1), 67.

[90] Slaymaker, F. H. and W. F. Meeker (1954). Measurements of tonal characteristics of carillon bells. *J. Acoust. Soc. Amer. 26*, 515–522.

[91] Sonneveld, P. and M. ~B. van Gijzen (2008). A family of simple and fast algorithms for solving large nonsymmetric linear systems. *SIAM J. Sci. Comput.* 31, 1035–1062.

[92] Stearn, J. L. (1951). Iterative solutions of normal equations. *Bull. Geodesique*, 331–339.

[93] Stein, W. and D. Joyner (2005). SAGE: System for algebra and geometry experimentation. *ACM SIGSAM Bulletin 39*(2), 61–64.

[94] The Mathworks, Inc. MATLAB website. http://www.mathworks.com, accessed September 2011.

[95] Trefethen, L. N. (2008). Is Gauss quadrature better than Clenshaw–Curtis? *SIAM Review 50*(1), 67–87.

[96] Trefethen, L. N. and D. Bau (1997). *Numerical Linear Algebra*. Philadelphia, PA: SIAM.

[97] Trefethen, L. N., N. Hale, R. B. Platte, T. A. Driscoll, and R. Pachón (2009). Chebfun version 4. http://www.maths.ox.ac.uk/chebfun/. Oxford University.

[98] van der Vorst, H. A. (1992). Bi-CGSTAB: A fast and smoothly converging variant of Bi-CG for the solution of nonsymmetric linear systems. *SIAM J. Sci. Comput. 13*, 631–644.

[99] von Neumann, J. and H. H. Goldstine (1947). Numerical inverting of matrices of high order. *Bulletin of the American Mathematical Society 53*, 1021–1099.

[100] vos Savant, M. (1990a, September). Ask Marilyn. *Parade Magazine*.

[101] vos Savant, M. (1990b, December). Ask Marilyn. *Parade Magazine*.

[102] vos Savant, M. (1991, February). Ask Marilyn. *Parade Magazine*.

[103] Weisstein, E. W. Buffon–Laplace needle problem. http://www.mathworld.wolfram.com/Buffon-LaplaceNeedleProblem.html, accessed May 2011.

[104] Weisstein, E. W. Buffon's needle problem. http://www.mathworld.wolfram.com/BuffonNeedleProblem.html, accessed May 2011.

[105] Weisstein, E. W. Cauchy, Augustin (1789–1857). http://scienceworld.wolfram.com/biography/Cauchy.html, accessed May 2011.

[106] Weisstein, E. W. Simpson, Thomas (1710–1761). http://scienceworld.wolfram.com/biography/Simpson.html, accessed July 2006.

[107] Wilkinson, J. H. (1961). Error analysis of direct methods of matrix inversion. *Journal of the ACM 8*, 281–330.

[108] Wilkinson, J. H. (1965). *The Algebraic Eigenvalue Problem*. Oxford, UK: Oxford University Press.

[109] Young, D. M. (1950). *Iterative Methods for Solving Partial Differential Equations of Elliptical Type*. Ph. D. thesis, Harvard University.

[110] Young, D. M. (1990) A historical review of iterative methods. http://history.siam.org/pdf/dyoung.pdf Reprinted from *A History of Scientific Computing*. S. G. Nash (ed.), 180–194. Reading, MA: ACM Press, Addison–Wesley.

INDEX

Abel, 301
absolute error, 124, 126, 128
Adams, John Couch, 276
Adams–Bashforth methods, 275–276, 278, 279
Adams–Moulton methods, 276–279
Archimedes, 221, 224
Ariane 5 disaster, 109–110

backward differentiation formulas. *See* BDF methods, 289
backward Euler method, 277, 288, 290, 299
barycentric interpolation, 183–185, 196, 218
  weights, 183, 193
Bashforth, Francis, 276
basis, 185, 286, 367, 368, 375, 377, 386–387
  A-orthogonal, 337, 338
  orthogonal, 338, 375
  orthonormal, 168, 344
BCS (Bowl Championship Series), 179
BDF methods, 289–290, 296
Bessel's correction, 54
Bessel, Friedrich, 54
BiCGSTAB method, 344
bisection, 75–80, 102, 103
  rate of convergence, 78–79
BLAS (Basic Linear Algebra Subroutines), 148
  BLAS1—vector operations, 148
  BLAS2—matrix–vector operations, 148
  BLAS3—matrix–matrix operations, 149
Bolzano–Weierstrass theorem, 333
Brin, Sergey, 17, 64, 320
Bubnov, I. G., 367
Buffon needle problem, 56–58, 69
Cauchy, Augustin, 169, 255, 258
chebfun, 192–197, 208, 218–220, 225, 241, 250, 375, 378, 414
Chebyshev points, 192, 197, 208, 218, 219, 240, 241, 375, 376, 414
Chebyshev polynomials, 196, 208, 218–220, 225, 247, 249, 375, 414
Cholesky factorization, 144
Clenshaw–Curtis quadrature, 227, 240, 242, 247
companion matrix, 282
compound interest, 129–130
computer animation, 2–4, 274
  cloth movement, 2–4, 274
  key-frame animation, 2
  Yoda, 2–4, 37, 274
conditioning of linear systems, 154–164, 256, 336
  absolute condition number, 160, 162
  componentwise error bounds, 162
  condition number, 160–162, 164, 165, 167, 175–177, 336, 340, 360
  ill-conditioned matrix, 181, 208
  relative condition number, 160
conditioning of problems, 125–126, 220–221
  absolute condition number, 125–126, 129
  ill-conditioned problem, 220–221, 301
  relative condition number, 125–126, 129, 130
conjugate gradient method, 336–344, 347, 360, 384
  convergence analysis, 340
  preconditioned, 341–344, 384, 415, 418
constant slope method, 89–90
Cooley, James W., 409, 412
Cotes, Roger, 228, 229

Cramer's rule, 151–153, 164, 166, 177, 430
cubic spline interpolation, 201–206, 209
  complete spline, 203
  natural cubic spline, 203, 210
  not-a-knot spline, 203, 204, 210

Dahlquist equivalence theorem, 283, 290
Dahlquist, Germund, 283
Danielson, 412
Darwin, Charles, 251
diffusion equation. *See* heat equation, 379
Dirichlet boundary conditions, 351, 362, 363, 365, 370, 376, 377, 384, 386, 399, 401, 408–410, 418
discretization error. *See* truncation error, 213
divided differences, 187–191
dot product. *See* inner product, definition, 423
drunkard's walk, electrical networks, 15–16

ecological models, 8–11
  harvesting strategies, 10
  Leslie matrix, 9
Edelman, Alan, 109
Ehrenfests' urn, 16–17
eigenfunctions, definition, 359
eigenvalue problems, 300–326
  characteristic polynomial, 301
  deflation, 312–313, 319
  Gerschgorin's theorem, 308–310, 335, 346
  Hessenberg form, 317–318
  inverse iteration, 312–315, 318; rate of convergence, 314; shifted, 313–315, 318, 346
  power method, 310–313, 315, 317, 318, 320, 324–325, 345–347;
  rate of convergence, 311–312, 314, 325, 346 shifted, 312–313, 318, 346
  QR algorithm, 315–319, 346; rate of convergence, 317–319, 346; shifts, 318–319
  Rayleigh quotient iteration, 315–316, 318
  Schur's theorem, 307
  similarity transformation, 305–318, 377; Householder reflection, 318 orthogonal, 306, 317; unitary, 307
  tridiagonal form, 318
eigenvalue, definition, 300, 433
  algebraic multiplicity, 307, 434
  characteristic polynomial, 282, 301–303, 306, 307, 433–435
  geometric multiplicity, 302, 307
eigenvector, definition, 300, 433
  eigenspace, 301–303, 306
  left, 305, 322
  right, 305, 322, 323
Euler, 185
Euler's method, 257–263, 265, 270, 271, 274, 276, 284, 295–297
Euler, Leonhard, 244, 257, 261
Euler–Maclaurin formula, 243–247

fast Fourier transform. *See* FFT, 193
fast Poisson solver, 384, 409, 414
Fatou, Pierre, 98, 101
Fermat near-miss, 122
FFT, 193, 241, 409, 411–414, 419
  cosine transform, 414
  inverse cosine transform, 414
  inverse FFT, 414
Fibonacci numbers, 14–15, 280
finite difference methods, 212, 327, 348, 352–365, 370, 375, 376, 381–386, 404–408
  backward difference formula, 394, 405
  centered difference formula, 352, 360–362, 370, 376, 395, 405, 408, 419
  forward difference formula, 390, 404, 405, 419
  nodes (mesh points), 352, 356, 360
  one-sided difference formula, 362, 376
  upwind differencing, 362
finite element methods, 6, 365–376, 386–388
  accuracy, 372–374, 377
  collocation methods, 366, 377
  Galerkin approximation, 366, 373, 377, 386
  piecewise bilinear, 386
  piecewise linear, 367–368, 370, 374, 377, 387
  Ritz approximation, 372–373
  test space, 366, 386
  trial space, 366, 378, 386
finite precision. *See* floating-point arithmetic, 2

fixed point iteration, 93–98, 104, 330
complex plane, 98–101; Julia set, 98–101; Mandelbrot set, 101; orbit, 98
conditions for convergence, 94–98
fixed-point representation, 112
floating-point arithmetic, 2, 107–123, 340
absolute rounding error, 116
binary, 107, 110–112, 121
exceptions, 119–120; $\pm\infty$, 119; NaN (not a number), 119; overflow, 119, 311; underflow, 119, 311
floating-point representation, 112–114; double precision, 113; floating-point number, 112, 121, 159; hidden-bit representation, 112; single precision, 112; subnormal numbers, 114, 120; toy system, 113–114
IEEE standard, 114, 117, 121; correctly rounded arithmetic, 117–121; double precision, 114, 120, 121; exceptions, 115 extended precision, 114; single precision, 114, 129
machine precision, 113, 120, 142, 213, 215, 216, 223, 319, 378
relative rounding error, 116–117
rounding errors, 107, 121, 261–262
rounding modes, 113, 116–117; round down, 116; round to nearest, 116, 120; round towards 0, 116; round up, 116
Fourier series, 399, 400
cosine series, 247
sine series, 399
Fourier transform, 393, 411–412
discrete, 409, 411–414, 419; inverse, 411, 412
fractals, 34, 38, 101, 222
Julia set, 99
Koch's snowflake, 34
Frankel, H., 330

Galerkin, Boris, 367
Galileo's inclined plane experiment, 174–175, 178
Garwin, Richard, 409
Gauss quadrature, 227, 234–241, 247, 291
weighted, 239–240
Gauss, Carl Friedrich, 133, 235, 349, 409, 412
Gauss–Lobatto points, 192
Gauss–Seidel method, 330–331, 335–336, 346–349, 414, 416–418
Gaussian elimination, 133, 151, 181, 327, 328, 343, 353, 426–427
banded matrix, 145–148, 176, 383, 409
Cholesky factorization, 144, 175
high performance, 148–151; block algorithms, 149–151; LAPACK, 150; parallelism, 150–151; ScaLAPACK, 151
LU factorization, 135, 139–141, 175; multiple right-hand sides, 139, 141; solution of triangular systems, 135, 140
operation counts, 137–139, 176
pivoting, 141–144; full pivoting, 165; partial pivoting, 137, 142–144, 176, 177; PLU factorization, 144; stability, 164, 166
Gear, William, 289
generalized minimal residual method. *See* GMRES method, 344
Geri's Game, 206
Gerschgorin's theorem, 308–310, 335, 346, 362
GMRES method, 344–345, 348
restarted, 344, 345
Goldstine, H. H., 131
Google's PageRank, 11, 13–14, 17, 18, 320–326
Markov chain model, 321–326; transition matrix, 321–324
Monte Carlo simulation of web surfing, 64–67
Gram, Jorgen Pedersen, 169
Gram–Schmidt algorithm, 169–171, 236, 249, 317, 318, 424–425
modified, 344–345
Green's theorem, 386, 388
Gregory, James, 234

heat equation, 379, 380, 388, 392, 397, 409, 419
Hestenes, M., 337
Heun's method, 265–266, 270, 295
high performance computing, 148–151
LAPACK, 150
memory hierarchy, 148, 150

parallelism, 150–151; distributed memory, 150
ScaLAPACK, 151
Hooke's law, 252–253
Hooke, Robert, 253
Horner's rule, 183, 186, 207
Hotelling, H., 131
Householder reflection, 317, 318
Huygens, 221

implicit Runge–Kutta methods, 290–292
incomplete Cholesky factorization, 342–344
incomplete LU factorization, 343
information retrieval, 11–14
vector space model of relevance to query, 11–13; document matrix, 11, 12, 17
initial value problem. *See* ordinary differential equations—initial value problem, 251
inner product, definition, 423–424
Intel Pentium bug, 108–109
Intermediate Value Theorem, 76, 104
interpolation, 2
inverse iteration, 312–315, 318
rate of convergence, 314
shifted, 313–315, 318, 346
invertible matrix. *See* matrix, nonsingular, 428
IRK methods. *See* implicit Runge–Kutta methods, 290
iterative methods for linear systems, 5, 327–345, 383
BiCGSTAB method, 344
conjugate gradient method, 336–344, 347, 360, 384; convergence analysis, 340; preconditioned, 341–344, 384, 415, 417, 418
error, definition, 328
Gauss–Seidel method, 330–331, 335–336, 340, 342, 344, 346–349, 414, 416–418
GMRES method, 344–345, 348; restarted, 344, 345
Jacobi method, 329–331, 335, 340, 344, 346–348, 414
Krylov space, 337–339, 344
preconditioners, 328–330, 336, 337, 341, 345, 414, 415; Gauss–Seidel, 330, 342, 414; incomplete Cholesky, 342–344; incomplete LU, 343; Jacobi, 329, 342, 414; SOR, 330, 342; SSOR, 342
QMR method, 344
residual, definition, 329
simple iteration, 328–337, 342, 344, 347, 414, 415; convergence analysis, 332–336
SOR method, 330–331, 340, 342, 344
SSOR method, 342

Jacobi method, 329–331, 335, 346–348, 414
Jacobi, Carl Gustav, 348, 349
Jiuzhang Suanshu, 133, 134
Julia set, 98–101
Julia, Gaston, 98–99, 101

Kahan, William, 114–115
Kepler, 297
Kleinberg, Jon, 320
Krylov space, 337–339, 344

Lagrange, 185, 255, 380
Lagrange interpolation, 181, 185, 207, 275
barycentric form, 183–186, 196, 208, 218; weights, 183, 193, 208
Lanczos, Cornelius, 337, 412
LAPACK, 150
Laplace, 169, 255, 380
Laplace's equation, 379, 418
Lax–Richtmyer equivalence theorem, 406, 407
least squares problems, 166–175, 337, 344, 366
fitting polynomials to data, 171–175, 178
normal equations, 167–168, 171–173, 178, 336
QR decomposition, 168–172, 178
Leclerc, G-L (Comte de Buffon), 56
Legendre, 380
linear difference equations, 279–283, 357
characteristic polynomial, 280–283, 290, 357
root condition, 282, 283
linear transformation, 431
matrix of, 431–433
Lipschitz continuity, 255, 264, 270, 279, 292, 295
Lotka, Alfred, 253

Lotka–Volterra equations, 253, 258, 295, 296

Machin, John, 83, 106
Maclaurin series, 81, 82, 106
Maclaurin, Colin, 244
Malthus, Thomas, 251
Malthusian growth model, 251
Mandelbrot set, 101
Mandelbrot, Benoit, 99, 101, 222
Markov chain, 320–324
  transition matrix, 321–324
Massey, K., 179
Mathematical Contest in Modeling (MCM), 17
mathematical modeling, 1–18
MATLAB, 19–40
  `cond`, 161, 165, 175, 177, 178
  `eig`, 319
  `eps`, 113
  `ode23s`, 298
  `ode45`, 267, 269, 296, 298
  `polyfit`, 207, 208
  `polyval`, 207, 208
  `qr`, 346
  `rand`, 52, 69
  `randn`, 52, 347
  `spline`, 204
  clearing variables, 31
  comments, 25
  conditionals, 29, 30
  functions, 27, 28
  getting help, 22, 23
  getting started, 19, 20
  LAPACK, 150
  logging your session, 31
  loops, 21, 29, 30
  M-files, 24, 25
  matrices, 23, 24
  plotting, 25, 27
  printing, 28, 29
  vectors, 20, 22
matrix
  banded, 145, 176, 327, 342, 383
  block Toeplitz, 383
  block tridiagonal, 383
  block TST, 409, 410
  characteristic polynomial, 282, 296, 301–303, 306
  companion, 282
  dense, 376
  determinant, 296, 429–430, 435; Laplace expansion, 429
  diagonal, 177, 308, 342, 348, 434
  diagonalizable, 305–307, 310, 434, 435
  diagonally dominant, 144, 145, 335, 346, 362, 364
  Hermitian, 132
  Hermitian transpose, 132
  Hessenberg, 318
  Hilbert, 158, 161
  ill conditioned, 158, 181, 208
  inverse, 428
  irreducible, 325
  Jordan form, 306–307, 309
  M-matrix, 343
  negative definite, 384
  nonsingular, 133, 327, 335, 336, 343, 347, 362, 364, 428–429
  nonsymmetric, 361
  orthogonal, 177, 307, 317, 318, 410
  permutation, 144
  positive definite, 384
  primitive, 325
  rotation, 36, 37
  Schur complement, 149
  singular, 141, 364, 429, 433
  sparse, 328, 375
  symmetric, 132, 203, 312, 316, 318, 338, 342, 356, 361, 369, 370, 377, 384, 435
  symmetric positive definite, 144–145, 335–336, 339–342, 344, 359, 361, 415
  Toeplitz, 356, 383
  trace, 435
  translation, 37
  transpose, 132, 429
  triangular, 135, 170, 186, 187, 308, 317, 318, 333, 341–343, 348, 435
  tridiagonal, 145, 203, 318, 353, 356, 361, 369–371, 383, 410
  tridiagonal symmetric Toeplitz (TST), 356–357, 400, 409, 410
  unitary, 307, 333
  Vandermonde, 181, 183, 186
  well conditioned, 165
matrix multiplication, 132–133
matrix splitting. *See* iterative methods for linear systems, preconditioners, 328
matrix, definition, 426
Mean Value Theorem, 229

Meijerink, 343
midpoint method, 262–264, 270, 295
Moler, Cleve, 33
Monte Carlo methods, 4, 6, 41–70, 320
  importance sampling, 56
  integration, 56, 65
  Texas Holdem, 42–47, 55–56
  web surfing, 64, 67
Monty Hall problem, 48, 68
Moulton, Forest Ray, 276
Muller's method, 207
multigrid methods, 414–418
  algebraic, 418
  interpolation (prolongation), 415–418
  projection (restriction), 415, 416, 418
  V-cycle, 416–418

Neumann boundary conditions, 351, 363, 364, 388
Newton interpolation, 185, 190, 207
  divided differences, 187–190, 207
Newton's method, 83, 89, 102–105
  complex plane, 98, 101–102
  rate of convergence, 85–89
Newton, Isaac, 83, 185, 228, 296, 297
Newton–Cotes formulas, 227, 235, 240, 241, 248
Nicely, Thomas, 108
nonlinear equation in one unknown, 71–106, 264
  bisection, 75–80, 102, 103
  fixed point iteration, 93–98, 104
  Muller's method, 207
  Newton's method, 83, 89, 102–105
  quasi-Newton methods, 89, 93; constant slope method, 89–90; secant method, 90–93, 103, 207
  regula falsi, 79, 93
  Steffensen's method, 104
nonlinear systems, 291–296
  fixed point iteration, 292–294
  Newton's method, 293–296, 299
  quasi-Newton methods, 294
norms, 124, 126, 154–158, 356
  L_2-norm, 356, 366, 374, 418
  ∞-norm, 418
  energy norm, 372, 374
  matrix norm, 155–158, 332–334; ∞-norm, 156–158, 176, 334, 335; 1-norm, 156–158, 176, 334; 2-norm, 156–158, 359; induced by a vector norm, 156, 332–334 submultiplicative, 156; subordinate to (compatible with) a vector norm, 156 vector norm, 154–155, 332, 334, 347; *A*-norm, 337–342, 344; ∞-norm, 154, 176, 354, 365, 372; 1-norm, 154, 176; 2-norm (Euclidean norm), 154, 176, 337, 338, 340, 344, 347, 348, 356, 424; inner product norm, 155, 424; *p*-norm, 154; unit circles in $\mathbf{R}^2$, 155
numerical differentiation, 212–226
  centered difference formula, 214–216, 221–223
  forward difference formula, 215, 226
numerical integration, 227–248, 369, 371
  Clenshaw–Curtis quadrature, 227, 240, 242, 247
  composite midpoint rule, 249
  Euler–Maclaurin formula, 243–247
  Gauss quadrature, 227, 234–241, 247; weighted, 239–240
  method of undetermined coefficients, 229–230, 238, 240
  midpoint rule, 371
  Newton–Cotes formulas, 227, 235, 240, 241, 248
  Romberg integration, 242–243
  Simpson's rule, 227, 230–232, 234, 266, 372; composite, 233–235, 242, 250
  singularities, 247–248
  trapezoid rule, 198, 227–229, 264, 265; composite, 232–234, 242–247, 250

$O(\cdot)$ notation, 81, 128, 138
ODE. *See* ordinary differential equations—initial value problem, 251
optimization, 1
ordinary differential equations—initial value problem, 3, 4, 251–299, 350, 389, 403
  A-stable method, 287–289, 291, 293
  absolute stability region, 285–290, 293, 299
  backward Euler method, 277, 288, 290, 299
  BDF methods, 289–290, 296
  consistent method, 259, 270–272, 283

convergent method, 259, 272, 283, 290, 296
Euler's method, 257–263, 265, 270, 271, 274, 276, 284–286, 293, 295–297
existence and uniqueness of solutions, 253–257
explicit method, 264, 276, 293, 299
global error, 259–261, 271, 272, 279, 283, 290
Heun's method, 265–266, 270, 271, 295
implicit method, 264, 276, 289, 291, 293, 294, 298; nonlinear equations, 291–295
implicit Runge–Kutta methods, 290–292
local truncation error, 259, 262–266, 271–273, 276–279, 283, 285, 290, 295, 296
midpoint method, 262–264, 270, 271, 295
multistep methods, 257, 271, 275–283, 289–291, 296; Adams–Bashforth, 275–276, 278, 279; Adams–Moulton, 276–279; Dahlquist equivalence theorem, 283, 290; Dahlquist equivalence theorem, 283; linear difference equations, 279–283; root condition, 283; zero-stable method, 283, 290, 296
one-step methods, analysis, 270–272
Runge–Kutta methods, 265–267, 273, 288, 295, 297, 298
Runge–Kutta–Fehlberg methods, 267, 273
second-order Taylor method, 262
stable method, 270–272, 292, 295
stiff equations, 267, 284–285, 290, 291, 293, 294
systems, 274–275, 291, 297, 298, 300, 389
trapezoidal method, 264, 265, 277, 278, 288, 289, 291, 293, 296
unstable method, 279, 281
well-posed problem, 256, 258
orthogonal polynomials, 236–240, 291
weighted, 238–240, 249
orthogonal projection, 168, 424, 425
orthogonal vectors, definition, 424
orthonormal vectors, 435
orthonormal vectors, definition, 424
Page, Larry, 17, 64, 320
Painlevé, Paul, 255
parallelism. *See* high performance computing, parallelism, 150
partial differential equations, 1, 4, 327, 328, 352, 379–408
consistent method, 391, 406, 407
convergent method, 406, 407
Crank–Nicolson method, 395, 396
Dirichlet boundary conditions, 384, 386, 399, 401, 408–410, 418
discriminant, 379
elliptic equations, 379, 381, 409
explicit method, 394
finite difference methods, 381–386, 404–408; CFL condition, 407; five-point operator, 410; global error, 391; instability, 392; local truncation error, 385, 390, 391, 394, 396, 405, 406, 408; natural ordering, 327, 382, 410; stability, 391–394, 396, 401, 405–408, 419; von Neumann stability analysis, 393
finite element methods, 386–388; Galerkin approximation, 386; piecewise bilinear, 386; piecewise; linear, 387; test space, 386; trial space, 386
fundamental modes, 398–400
heat equation, 379, 380, 388, 392, 397, 409, 419
hyperbolic equations, 379–381, 402–408; characteristics, 402–404; shock wave, 402; systems, 403
implicit method, 394–396, 409
initial boundary value problem (IBVP), 389, 397, 399, 400, 404
initial value problem (IVP), 402
Laplace's equation, 379, 418
Lax–Friedrichs method, 405, 419
Lax–Richtmyer equivalence theorem, 406, 407
leapfrog method, 405
multistep method, 405, 408
Neumann boundary conditions, 388
one-step method, 405
parabolic equations, 379–380, 388, 405, 407, 409
periodic boundary conditions, 397, 398
Poisson's equation, 379, 381,

409–410, 414, 417, 418
semidiscretization (method of lines), 389
separation of variables, 396–402
stable method, 395, 396, 406, 407
unconditionally stable method, 394, 396
unstable method, 393, 408
wave equation, 380, 381, 401, 403, 419; one-way, 402, 404; two-way, 408
well-posed problem, 406
Perron's theorem, 323, 325
Perron–Frobenius theorem, 325
piecewise polynomial interpolation, 197–206, 374
bilinear, 414, 415
cubic Hermites, 200–201, 209,; 374
cubic splines, 201–206, 209, 374; complete cubic spline, 203; natural; cubic spline, 203, 210; not-a-knot spline, 203, 204, 210
piecewise linear functions, 197–199, 208, 209, 374; accuracy of, 198–199
piecewise quadratics, 199–200, 374; accuracy of, 199
Pixar, 206
Poisson's equation, 145, 327–328, 331, 340, 344, 379, 381, 409–410, 414, 417, 418
Poisson, Siméon-Denis, 255, 380
polynomial interpolation, 181–197, 218–220, 225
accuracy of, 190–197
Chebyshev points, 192, 197, 208, 218, 219
Lagrange form, 185, 186, 207, 275;
barycentric form, 181, 183–186, 196, 208, 218
Newton form, 185, 190, 207; divided differences, 187–190, 207
Vandermonde system, 181
positive definite operator, 372, 373
power method, 310–313, 318, 320, 324–325, 345–347
rate of convergence, 311–312, 325, 346
shifted, 312–313, 318, 346
preconditioners. *See* iterative methods for linear systems, preconditioners, 328
Principal Component Analysis (PCA), 180

QMR method, 344
QR algorithm, 316–319, 346
rate of convergence, 317–319, 346
shifts, 318–319; Wilkinson shift, 319
QR decomposition, 168–171, 177, 181, 317, 346
full, 170, 171
reduced, 170, 171
quadrature. *See* numerical integration, 227
quasi-minimal residual method. *See* QMR method, 344
quasi-Newton methods, 89, 93
constant slope method, 89–90
secant method, 90–93, 103, 207; rate of convergence, 91–93

radiation transport, 4–6
Boltzmann equation, 4–6
Raphson, Joseph, 84
Rayleigh quotient, 316
Rayleigh quotient iteration, 315–316, 318
regula falsi, 79, 93
relative error, 124, 126, 177
Richardson extrapolation, 221–225, 242, 273
Richardson, Lewis Fry, 221, 222
Ritz, Walter, 367
Robin boundary conditions, 351
Romberg integration, 242–243
rounding error, 142, 213–218, 222, 261–262, 279, 285, 313, 315
rounding error analysis, 126–127
backward error analysis, 127, 131
forward error analysis, 126–127
Runge function, 192–196, 207–208, 219–220, 240
Runge–Kutta methods, 265–267, 273, 288, 295, 297, 298
Runge–Kutta–Fehlberg methods, 267, 273

ScaLAPACK, 151
Schmidt, Erhard, 169
Schur's theorem, 307, 333
Schur, Issai, 307
secant method, 90–93, 103, 207

rate of convergence, 91–93
self-adjoint operator, 361, 372, 373
self-similarity, 99
shear mapping, 303–304, 306
similar matrices, 432–434
Simpson's rule, 227, 230–232, 234, 266, 372
composite, 233–235, 242, 250
Simpson, Thomas, 83, 84, 234
soccer simulation, 6–8
computational fluid dynamics (CFD), 6, 7
Soccer Sim, 7
SOR method, 330–331
Southwell, Richard, 329
spectral methods, 374–376
spectral collocation, 375–376
spectral radius, 333–336, 348, 415
splines, 1, 201–206, 209, 211
bicubic spline, 210
complete spline, 203
natural cubic spline, 203, 210
not-a-knot spline, 203, 204, 210
NURBS, 205
parametric spline, 211
quadratic spline, 209
SSOR method, 342
stability of algorithms, 126–129
backward stable algorithm, 127, 164–166, 177
unstable algorithm, 127, 128
statistics, 46–56
Bessel's correction, 54
Central Limit Theorem, 46, 53–56, 60, 68, 69
confidence intervals, 54, 60
continuous random variables, 51–53
covariance, 50
cumulative distribution function, 51–52
discrete random variables, 48–51, 67
expected value (mean), 48–51, 53–55, 60, 67–69
functions of a random variable, 49
independent identically distributed (iid) random variables, 53, 54, 60
independent random variables, 50, 51
Law of Large Numbers, 53
normal (Gaussian) distribution, 52–54, 68
probability density function (pdf), 51–52
sample average, 53–56, 60, 67
standard deviation, 49, 54–56, 60, 61, 68–70
sum of random variables, 49–50
uniform distribution, 51–53, 67–69
variance, 49–51, 53–56, 60, 67–69
Steffensen's method, 104
Stiefel, E., 337
Strutt, John William (Lord Rayleigh), 315
successive overrelaxation. *See* SOR method, 330
surface subdivision, 206
symmetric successive overrelaxation method. *See* SSOR method, 342

Taylor's theorem, 80, 83, 85, 90, 213, 214, 216, 226, 257
multivariate, 265, 436–438
series expansion, 81–82, 105, 106
Taylor polynomials, 82, 105, 106
with remainder, 80, 104, 436, 437
Taylor, Brook, 83, 106
trapezoid rule, 198, 227–229, 264, 265
composite, 232–234, 242–247, 250, 356
trapezoidal method, 264, 265, 277, 278, 288, 289, 291, 293, 296
triangular linear systems, 135, 140–141, 175, 176, 186, 187
truncation error, 213–218, 222, 223
Tukey, John W., 409, 412
two-point boundary value problems, 350–377
Dirichlet boundary conditions, 351, 362, 363, 365, 370, 376, 377
finite difference methods, 352–365, 376; global error, 354–360; local truncation error, 354, 355, 360; upwind differencing, 362
finite element methods, 365–376;
accuracy, 372–374, 377; collocation methods, 366, 377; Galerkin approximation, 366, 373, 377
piecewise linear, 367–368, 370, 374, 377; Ritz approximation, 372–373; test space, 366; trial space, 366, 378
Neumann boundary conditions, 351, 363, 364

Robin boundary conditions, 351
spectral methods, 374–376; spectral collocation, 375–376

Ulam, Stanislaw, 41

van der Vorst, 343
Vancouver stock exchange correction, 117
Vandermonde matrix, 181, 183, 186
Vandermonde, Alexandre-Théophile, 181
vector space, definition, 421–422, 424
basis, 423, 428, 431; coordinates, 423, 431
dimension, 423
linear combination of vectors, 422–423, 428
linearly dependent vectors, 422–423
linearly independent vectors, 422–423, 428, 429
span, 423, 428
subspace, 422–423
vector, definition, 421
Volterra, Vito, 253
von Neumann, J., 41, 131

Wallis, 185
Waring, 185
wave equation, 380, 381, 401, 403, 419
one-way, 402, 404
two-way, 408
weak form of differential equation, 366
Wien radiation law, 102
Wilkinson, J. H., 131, 165

Young, David M., 329–330, 348